Hydrogen Peroxide Metabolism in Health and Disease

OXIDATIVE STRESS AND DISEASE

Series Editors

Lester Packer, PhD
Enrique Cadenas, MD, PhD

UNIVERSITY OF SOUTHERN CALIFORNIA SCHOOL OF PHARMACY
LOS ANGELES, CALIFORNIA

1. Oxidative Stress in Cancer, AIDS, and Neurodegenerative Diseases, *edited by Luc Montagnier, René Olivier, and Catherine Pasquier*
2. Understanding the Process of Aging: The Roles of Mitochondria, Free Radicals, and Antioxidants, *edited by Enrique Cadenas and Lester Packer*
3. Redox Regulation of Cell Signaling and Its Clinical Application, *edited by Lester Packer and Junji Yodoi*
4. Antioxidants in Diabetes Management, *edited by Lester Packer, Peter Rösen, Hans J. Tritschler, George L. King, and Angelo Azzi*
5. Free Radicals in Brain Pathophysiology, *edited by Giuseppe Poli, Enrique Cadenas, and Lester Packer*
6. Nutraceuticals in Health and Disease Prevention, *edited by Klaus Krämer, Peter-Paul Hoppe, and Lester Packer*
7. Environmental Stressors in Health and Disease, *edited by Jürgen Fuchs and Lester Packer*
8. Handbook of Antioxidants: Second Edition, Revised and Expanded, *edited by Enrique Cadenas and Lester Packer*
9. Flavonoids in Health and Disease: Second Edition, Revised and Expanded, *edited by Catherine A. Rice-Evans and Lester Packer*
10. Redox–Genome Interactions in Health and Disease, *edited by Jürgen Fuchs, Maurizio Podda, and Lester Packer*
11. Thiamine: Catalytic Mechanisms in Normal and Disease States, *edited by Frank Jordan and Mulchand S. Patel*
12. Phytochemicals in Health and Disease, *edited by Yongping Bao and Roger Fenwick*
13. Carotenoids in Health and Disease, *edited by Norman I. Krinsky, Susan T. Mayne, and Helmut Sies*
14. Herbal and Traditional Medicine: Molecular Aspects of Health, *edited by Lester Packer, Choon Nam Ong, and Barry Halliwell*
15. Nutrients and Cell Signaling, *edited by Janos Zempleni and Krishnamurti Dakshinamurti*
16. Mitochondria in Health and Disease, *edited by Carolyn D. Berdanier*
17. Nutrigenomics, *edited by Gerald Rimbach, Jürgen Fuchs, and Lester Packer*
18. Oxidative Stress, Inflammation, and Health, *edited by Young-Joon Surh and Lester Packer*

Hydrogen Peroxide Metabolism in Health and Disease

Edited by
Margreet C.M. Vissers, Mark B. Hampton,
and Anthony J. Kettle

CRC Press
Taylor & Francis Group
Boca Raton London New York

CRC Press is an imprint of the
Taylor & Francis Group, an **informa** business

CRC Press
Taylor & Francis Group
6000 Broken Sound Parkway NW, Suite 300
Boca Raton, FL 33487-2742

First issued in paperback 2020

ISBN-13: 978-1-4987-7615-8 (hbk)
ISBN-13: 978-0-367-65758-1 (pbk)

Visit the Taylor & Francis Web site at
http://www.taylorandfrancis.com

and the CRC Press Web site at
http://www.crcpress.com

This book is dedicated to Christine Winterbourn, pioneer in the study of oxidative stress and longtime colleague and friend to many.

Contents

SECTION I Chemistry and Biochemistry of H_2O_2

SECTION II Biological Sources of H_2O_2

SECTION III   Myeloperoxidase and Derived Oxidants

SECTION IV H_2O_2 in Cellular Metabolism and Signaling

Series Preface

HYDROGEN PEROXIDE METABOLISM IN HEALTH AND DISEASE

Oxidative stress is an underlying factor in health and disease. In this series of books the importance of oxidative stress and disease associated with cell and organ systems of the body is highlighted by exploring the scientific evidence and the clinical applications of this knowledge. This series is intended for researchers in the biomedical sciences, clinicians, and all persons with interest in the health sciences. The potential of such knowledge for healthy development and aging and disease prevention warrants further understanding on how oxidants and antioxidants modulate cell and tissue function.

It is a privilege for the editors of the series Oxidative Stress and Disease to have this book dedicated to Christine Winterbourn. The editors, Margreet C.M. Vissers, Mark B. Hampton, and Tony J. Kettle, completed their PhD studies under Dr. Winterbourn's mentorship and have honored her with this book, *Hydrogen Peroxide Metabolism in Health and Disease*, that compiles important and state-of-the-art insights on the ever-growing field of redox biology. The book also contains a foreword (*A Perspective on Hydrogen Peroxide*) by Christine Winterbourn, a nicely documented path from free radicals to hydrogen peroxide biology and underscoring the significance of chemistry to understand the complexity of biological systems.

Lester Packer
Enrique Cadenas
Oxidative Stress and Disease Series Coeditors

Foreword

A PERSPECTIVE ON HYDROGEN PEROXIDE

Hydrogen peroxide impacts life in numerous ways, and I am delighted to see so much of its biological chemistry brought together in this impressive volume. I feel greatly honored that the book has been dedicated to me, and I would like to pay tribute to my long-term colleagues and friends, Margreet Vissers, Tony Kettle, and Mark Hampton, who, under Margreet's guidance, have done a sterling job in making it a reality. I am also grateful to Lester Packer for seeding the idea and to all my colleagues who agreed to contribute. Most authors I have known for years, some I have collaborated with, others have been competitors (in the best sense of the word), and together their chapters give us a wide-ranging overview of hydrogen peroxide and its relatives. Thank you all.

Hydrogen peroxide is no stranger to biology. As Koppenol and Bechara recount in more detail in subsequent chapters, it has been known as a chemical species for 200 years. Catalase was named more than 100 years ago and was one of the first enzymes to be crystallized, and anyone wanting to demonstrate its existence has only to add reagent peroxide to blood and observe the bubbles of oxygen that evolve. Hydrogen peroxide production by bacteria was demonstrated in the 1920s and by phagocytosing neutrophils in the 1960s, and a decade later Sies and Chance detected it inside living cells. However, investigation into its biology remained relatively low key until the seminal discovery of superoxide dismutase by Fridovich and McCord ignited interest in superoxide and the discipline of free radical biology and medicine. Hydrogen peroxide was seen as one of the players in the field, but with the early emphasis on free radicals and toxicity, it remained more in the background. However, as it has become appreciated that biological reactions of reactive oxidants are not necessarily detrimental and have an important cell regulatory role, hydrogen peroxide has come to receive more of the attention.

Hydrogen peroxide is produced by a multitude of biological mechanisms. Currently, the focus tends to be mainly on release from mitochondria, from activated NADPH oxidases, and from autoxidation or redox cycling reactions, with the context being cell signaling or toxicity. Yet it is important to recognize that hydrogen peroxide is also an end product of metabolic activities that go on in the background of this activity. These include peroxisomal oxidation of a range of substrates, as well as disulfide formation in the endoplasmic reticulum (ER), where one hydrogen peroxide is produced for each disulfide formed. Whereas aberrant peroxisomal or ER function can have redox consequences, it would appear that basal peroxide formation from these sources is well handled by cell metabolism without invoking signaling or toxicity mechanisms. Therefore, to understand how cells respond to hydrogen peroxide, it is a matter of knowing not only how much they are exposed to it but also where it is produced, how far it can travel, and what is in the vicinity for it to react with. This volume addresses these aspects of hydrogen peroxide biology, but there is still much to be learned about how the disparate activities involving hydrogen peroxide are distinguished.

It seems hard to believe, but my career spans the field of free radical biology and medicine right from the early days. It has covered free radicals, nonradical oxidants, cells that produce these species and enzymes that react with them, and the vast majority of the papers that I have published on these topics would mention hydrogen peroxide. It first captured my interest in my work with red blood cells, when we observed that superoxide radicals are produced during hemoglobin autoxidation and, following the discovery of superoxide dismutase, wanted to know how this radical is handled by the cells. It was already known from studying glucose-6-phosphate dehydrogenase deficiency that red cells need defenses against hydrogen peroxide, and mechanisms involving superoxide and other radicals could not be understood without considering the role of hydrogen peroxide. My interest in free radicals also brought me to study hydrogen peroxide in another context: oxidant production by neutrophils. Although there were earlier indications from Eyer and Quastel that these cells produce hydrogen peroxide, and Klebanoff had identified the myeloperoxidase antimicrobial system, it was the report from Babior's group that neutrophils release superoxide radicals that attracted most attention. We investigated the fate of this superoxide, leading into a large body of work on interactions of superoxide and hydrogen peroxide with myeloperoxidase and characterizing reactions of the hypochlorous acid generated in this system.

Although we initially expected the superoxide radical to be highly reactive and cytotoxic, it came to be realized that its biological activity is limited. As a consequence, there is now a tendency to consider superoxide as merely a precursor of hydrogen peroxide. To me, this is unwise as it ignores the more selective yet important reactions that superoxide undergoes, just as I think it is important not to consider the two species as a ROS (reactive oxygen species) that reacts as a single entity. Nevertheless, toxicity and redox regulatory mechanisms do commonly involve hydrogen peroxide, and much of my research has been directed at understanding the reactions involved. These include reactions with heme peroxidases and transition metal complexes, which are mostly fast and generate either one-electron (radical) oxidants or, in the case of peroxidases, hypohalous acids, which are more reactive than hydrogen peroxide. Direct, two-electron oxidation of biological substrates by hydrogen peroxide is restricted mainly to thiol compounds. This specificity is a good attribute for signaling, and we know some of the factors that determine reactivity. However, there are still many unanswered questions about how selectivity between different thiol proteins is achieved.

When explaining how hydrogen peroxide can be both a signaling molecule and a toxic agent, it is often proposed that free radicals and one-electron reactions are the bad guys, whereas signaling is good and involves nonradical, two-electron oxidation steps. While this concept has some merit, it may be too clear a distinction, particularly as many of the most favorable reactions of hydrogen peroxide are with metal centers and there is evidence that one-electron reactions occur even under conditions of low oxidant exposure. Also, considering how any outcome is likely to depend on the nature of the particular oxidant(s) and where they are generated, the adage that low concentrations of oxidants are good and high concentrations are bad seems simplistic. I prefer to think more of a continuum over which an oxidant can participate in metabolic, signaling, and destructive pathways. It is then more appropriate to change our concept of "antioxidants" from agents that remove oxidants to agents that regulate this redox metabolism.

Incredible advances in our understanding of the biological chemistry of hydrogen peroxide and other reactive oxidants have occurred during my time in the redox field. What may be less evident is the feeling I have that as the field expands with the excitement of the biology, some of the known chemistry gets left behind. We do need the chemistry in order to understand the more complex biological systems, and I am optimistic that this book will help bridge the gap.

Christine Winterbourn

Preface

The idea for this book was first proposed in 2015 by Lester Packer, longtime colleague and champion of the oxidative stress field. The invitation was for a volume in the CRC Press series on oxidative stress to be dedicated to Professor Christine Winterbourn, to acknowledge her substantial contribution to this research area over many decades. Professor Winterbourn's scientific career began in the 1960s when, as a graduate with a specific talent for physical chemistry, she undertook studies into the aging of red blood cells. It was the exploration of Heinz body formation in red cells as a result of oxidative damage mediated by mutant hemoglobins that triggered her interest in oxidative reactions in a biological setting. These findings coincided with the discovery of superoxide dismutase in 1969; the realization that this enzyme was both ubiquitous and necessary for survival in an oxygen-rich environment was the genesis of the free radical and oxidative stress research area. The 1970s, 1980s, and 1990s were exciting decades for discovery in redox biology. Professor Winterbourn's curiosity, energy, and drive led her to engage with many aspects of this research area—from the chemistry and enzymology of oxidant generation in a biological environment to the cellular and pathological impacts of oxidant generation by inflammatory cells and xenobiotics.

We three editors completed our PhD studies under Professor Winterbourn's mentorship, all in different aspects of oxidative stress. Over the decades, our collaboration resulted in a complementary and collaborative research program identified locally as the Centre for Free Radical Research, with interests and expertise in redox and peroxidase chemistry, white cell–derived oxidants and inflammation, and redox influences on cell biology. This diversity serves as an illustration of Professor Winterbourn's broad scientific interests and drive for new knowledge. The challenge of understanding, investigating, and validating biochemical reactions in complex biological environments requires a breadth of understanding that few in our field have managed. A particular strength of Professor Winterbourn's research has been the relentless rigor with which she approaches all aspects of her work—her insistence that a particular reaction be both biologically possible and chemically feasible before it can be considered a possibility is legendary, as is her tendency to scribble calculations in the margins of published papers to both check and extend the data analysis!

When considering the makeup of a prospective book dedicated to Professor Winterbourn, we wished to reflect the breadth of her interest in, and influence on, the oxidative stress field. The diversity of potential topics provided us with a significant challenge when identifying a unifying theme for the book. We eventually settled on hydrogen peroxide as the central protagonist. H_2O_2 can be considered the archetypical reactive oxygen species, generated in many different ways in biological systems, the substrate for a range of peroxidases, and a potent signaling molecule through the oxidation of specific thiol proteins. However, there are many unanswered questions, and we believe it is fitting and timely to provide a volume that refocuses attention on the central role that H_2O_2 plays in redox biology.

We have divided the book into subsections. The first topic is the biological chemistry of H_2O_2 and includes reviews of H_2O_2 itself (Chapter 1), its detection and chemistry in biological settings (Chapters 2 through 4), and its capacity to react with proteins, both generally (Chapter 5) and specifically (Chapter 6). The reader of this section should emerge with an appreciation of not only the historical research context, with the earliest reference in Chapter 1 dating back to 1818, but also the continued biological relevance of H_2O_2 and the more recent identification of biological targets such as the peroxiredoxins. Section II focuses on the sources of H_2O_2 in biology. As summarized in Chapter 7, mitochondria are recognized as a continual and important source of cellular reactive oxidants, and evidence is emerging that controlled leakage from the electron transport chain enables communication between mitochondria and the rest of the cell. The generation of H_2O_2 by quinones is the topic of a review in Chapter 8, and reading this should deflect the reader from any impression that cell-derived oxidants are the only source of oxidative stress. Undoubtedly, a major source of H_2O_2 in biology are the NOX enzymes, and Chapters 9 and 10 provide overviews of the current state of knowledge of these protein complexes.

Much of the influence of H_2O_2 is mediated by peroxidases that are abundant throughout biology. None is more important or abundant than myeloperoxidase, the green enzyme that predominates in inflammatory cells. It is present in sufficient concentrations to lend its color to the neutrophils that contain it—these "white" cells are actually pale green—and to the biological fluids of inflammation, for example, the green tinge of pus. Myeloperoxidase and other peroxidases, such as osinophil and salivary peroxidase, and lactoperoxidase, use H_2O_2 to generate a variety of halogenated oxidants and free radicals that are predominant sources of oxidative stress. In Section III, Chapters 11 through 13 provide an overview of the enzymology of myeloperoxidase and of the generation and reactivity of the poorly understood halogenated oxidants and their impact on inflammation and the resulting tissue injury.

Oxidant production was long thought to have deleterious consequences, but the identification of the NOX enzymes in many cellular settings led to the realization that it can be a signaling mechanism. This depends on the existence of redox sensors, and reactive cysteine residues in proteins are the most prominent mechanism. Section IV contains a smorgasbord of papers that are an introduction to the area of H_2O_2-mediated cellular signaling. These range from an up-to-date perspective of the aquaporins that assist with delivery and transport of H_2O_2 across membranes to descriptions on how H_2O_2 impacts thioredoxin and glutathione systems, and finishing with a discussion of the role of H_2O_2 in regulating vascular tone.

We are grateful to the authors of the contributed chapters for their generosity—each has taken time out from an extremely busy schedule to provide us with an overview of many important aspects of the oxidative stress field. The contributors to this book are all long-standing colleagues and friends of Professor Winterbourn, and each is an expert in her/his field, having made significant research contributions. As a consequence, we have amassed a volume that presents both a historical context and updated information on the most recent developments in redox biology. We believe the diversity of this book is a fitting tribute to the broad interests and expertise of Professor Winterbourn, and we hope that it will be of value to those new to the field and to experts alike.

Margreet Vissers
Mark Hampton
Tony Kettle

Contributors

Elias S.J. Arnér
Division of Biochemistry
Department of Medical Biochemistry
and Biophysics
Karolinska Institutet
Stockholm, Sweden

Ohara Augusto
Departamento de Bioquímica
Instituto de Química
Universidade de São Paulo
São Paulo, Brazil

Wilhelm J. Baader
Departamento de Química
Fundamental
Instituto de Química
Universidade de São Paulo
São Paulo, Brazil

Etelvino J.H. Bechara
Departamento de Química
Fundamental
Instituto de Química
Universidade de São Paulo
São Paulo, Brazil

Gerd Patrick Bienert
Metalloid Transport Group
Leibniz Institute of Plant Genetics and
Crop Plant Research
Gatersleben, Germany

Sebastián Carballal
Departamento de Bioquímica
Facultad de Medicina
and
Center for Free Radical and Biomedical
Research
Universidad de la República
Montevideo, Uruguay

Michael J. Davies
Department of Biomedical Sciences
Panum Institute
University of Copenhagen
Copenhagen, Denmark

Becky A. Diebold
Department of Pathology and
Laboratory Medicine
Emory University
Atlanta, Georgia

Louisa V. Forbes
Department of Pathology
Center for Free Radical Research
University of Otago, Christchurch
Christchurch, New Zealand

Henry Jay Forman
Professor of Research Gerontology
University of Southern California
Distinguished Professor Emeritus of
Biochemistry
University of California
Merced, California

Shivaprakash Gangappa
Influenza Division
Centers for Disease Control and
Prevention
Atlanta, Georgia

Mark B. Hampton
Department of Pathology
Center for Free Radical
Research
University of Otago, Christchurch
Christchurch, New Zealand

Clare L. Hawkins
Department of Biomedical Sciences
University of Copenhagen
Sydney, New South Wales, Australia

and

Department of Biomedical Sciences
Panum Institute
University of Copenhagen
Copenhagen, Denmark

Elizabeth C. Hinchy
MRC Mitochondrial Biology Unit
University of Cambridge
Cambridge, United Kingdom

Amelia R. Hofstetter
Ruminant Diseases and Immunology
    Research Unit
National Animal Disease Center
United States Department of Agriculture
Agricultural Research Service
Ames, Iowa

P. Andrew Karplus
Department of Biochemistry and
    Biophysics
Oregon State University
Corvallis, Oregon

Anthony J. Kettle
Department of Pathology
Center for Free Radical Research
University of Otago, Christchurch
Christchurch, New Zealand

Willem H. Koppenol
Department of Chemistry and Applied
    Biological Sciences
Swiss Federal Institute of Technology
Zürich, Switzerland

J. David Lambeth
Department of Pathology and
    Laboratory Medicine
Emory University
Atlanta, Georgia

Ghassan J. Maghzal
Vascular Biology Division
Victor Chang Cardiac Research
    Institute
and
St Vincent's Clinical School
University of New South Wales
Sydney, New South Wales, Australia

Matilde Maiorino
Department of Molecular Medicine
University of Padova
Padova, Italy

Iria Medraño-Fernandez
Division of Genetics and Cell Biology
Vita-Salute San Raffaele University
San Raffaele Hospital (IRCCS-OSR)
Milan, Italy

Hugo P. Monteiro
Department of Biochemistry
Center for Cellular and Molecular
    Therapy
Escola Paulista de Medicina
Universidade Federal de São Paulo
São Paulo, Brazil

Rex Munday (Deceased)
AgResearch Limited
Ruakura Research Centre
Hamilton, New Zealand

Michael P. Murphy
MRC Mitochondrial Biology Unit
University of Cambridge
Cambridge, United Kingdom

William M. Nauseef
Department of Medicine
Roy J. and Lucille A. Carver College
    of Medicine
University of Iowa and Veterans
    Administration Medical Center
Iowa City, Iowa

Fernando T. Ogata
Department of Biochemistry
Center for Cellular and Molecular
    Therapy
Escola Paulista de Medicina
Universidade Federal de São Paulo
São Paulo, Brazil

Alexander V. Peskin
Department of Pathology
Center for Free Radical Research
University of Otago, Christchurch
Christchurch, New Zealand

Leslie B. Poole
Department of Biochemistry
and
Centers for Redox Biology and
    Medicine, Molecular Signaling, and
    Structural Biology
Wake Forest School of Medicine
Winston-Salem, North Carolina

Juliet M. Pullar
Department of Pathology
Center for Free Radical Research
University of Otago, Christchurch
Christchurch, New Zealand

Rafael Radi
Departamento de Bioquímica
Facultad de Medicina
and
Center for Free Radical and Biomedical
    Research
Universidad de la República
Montevideo, Uruguay

Benjamin S. Rayner
The Heart Research Institute
and
Sydney Medical School
University of Sydney
Sydney, New South Wales, Australia

Roberto Sitia
Division of Genetics and Cell Biology
Vita-Salute San Raffaele University
San Raffaele Hospital (IRCCS-OSR)
Milan, Italy

Christopher P. Stanley
Vascular Biology Division
Victor Chang Cardiac Research
    Institute
Sydney, New South Wales, Australia

Roland Stocker
Vascular Biology Division
Victor Chang Cardiac Research
    Institute
and
St Vincent's Clinical School
University of New South Wales
Sydney, New South Wales,
    Australia

Arnold Stern
School of Medicine
New York University
New York, New York

Cassius V. Stevani
Departamento de Química
    Fundamental
Instituto de Química
Universidade de São Paulo
São Paulo, Brazil

Madia Trujillo
Departamento de Bioquímica
Facultad de Medicina
and
Center for Free Radical and Biomedical
    Research
Universidad de la República
Montevideo, Uruguay

Daniela Ramos Truzzi
Departamento de Bioquímica
Instituto de Química
Universidade de São Paulo
São Paulo, Brazil

Fulvio Ursini
Department of Molecular Medicine
University of Padova
Padova, Italy

Margreet C.M. Vissers
Department of Pathology
Center for Free Radical Research
University of Otago, Christchurch
Christchurch, New Zealand

Ari Zeida
Facultad de Ciencias Exactas y Naturales
Departamento de Química Inorgánica
Analítica y Química-Física and
    INQUIMAE CONICET
Universidad de Buenos Aires
Buenos Aires, Argentina

and

Departamento de Bioquímica
Facultad de Medicina and Center for Free
    Radical and Biomedical Research
Universidad de la República
Montevideo, Uruguay

Section I

Chemistry and Biochemistry of H_2O_2

1 Hydrogen Peroxide, a Molecule with a Janus Face

Its History, Chemistry, and Biology

Willem H. Koppenol

CONTENTS

INTRODUCTION

In biology, H_2O_2 plays an important role as a component of oxidative stress and as a signaling molecule. After a description of the discovery and properties of H_2O_2, where I stress that this molecule is not as reactive as often thought, I focus on the Fenton reaction, as for decades this reaction illustrated the bad side of H_2O_2. The more recently discovered role of H_2O_2 as a redox mediator and messenger represents its good side. These two aspects define H_2O_2, a molecule with a Janus face. I come to the conclusion that the way we now perceive H_2O_2 is quite different from that 40 years ago.

DISCOVERY AND PROPERTIES OF HYDROGEN PEROXIDE

"C'est en traitant le peroxide de barium par les acides que je suis parvenu à faire ces nouvelles combinaisons, qui, pour la plupart, sont très-remarquable et dignes de fixer l'attention des chimistes" ("While treating the peroxide of barium with acids, I have achieved to make these mixtures, which are mostly very remarkable and worthy to draw the attention of chemists"). The citation is the first sentence of Thenard's first article, published nearly 200 years ago, on what later turned out to be the discovery of H_2O_2 [1]. And yes, hydrogen peroxide has drawn our attention for nearly 200 years. When Thenard (1777–1857) heated barium in O_2 and subsequently dissolved the solid in acid, he obtained acidic solutions that contained extra oxygen. Initially, he thought that the acid he used became oxygenated, but by the end of 1818, he had already realized that the water was oxygenated [2] by removing all ions: he had dissolved BaO_2 in HCl, precipitated Ba^{2+} with dilute H_2SO_4, and removed Cl^- with Ag_2O. He concentrated the solution by exposing it to reduced pressure [2] and gave the density as 1.453 [3], which corresponds, in essence, to anhydrous H_2O_2 (see Table 1.1). He refined his synthesis, which made it into textbooks. In 1953, one could still find a lecture demonstration based on his research [4]: First, 100 mL of 20% H_2SO_4 in a 300 mL beaker is placed in an ice bath. To the cooled sulfuric acid some ice is added, and BaO_2 is slowly added until the solution is still weakly acidic. One lets the solution stand and frees it from $BaSO_4$ by filtration. Properties of H_2O_2, such as decomposition upon heating, catalytic decomposition by KI, by catalase, by silver colloids, oops, nanoparticles, and various oxidations were then demonstrated with the synthesized H_2O_2. Higher concentrations of H_2O_2 can be achieved by distillation under reduced pressure, as H_2O boils at a lower temperature than H_2O_2 (Table 1.1). In the 1960s, some undergraduate students at the University of Utrecht, the author being one of them, were tasked with preparing 100% H_2O_2 from a commercial solution of approximately 30% (w/w) [5]. It was considered safe to distill off water under reduced pressure to a concentration of 90% H_2O_2. Further purification was attempted by repeated crystallizations, as H_2O_2 freezes at a slightly lower temperature than water (Table 1.1). As reviewed by Schumb et al. [6], frozen or liquid H_2O_2 will not explode; only when the concentration in the gas phase is too high, H_2O_2 will decompose and possibly explode on a rough surface. Other properties of H_2O_2 are summarized in Table 1.1.

H_2O_2 is remarkably stable; Thenard already showed that.

Today H_2O_2 is no longer produced by burning Ba in O_2. Synthesis by electrochemistry was attempted, but is not very efficient. Instead, the anthraquinone process is used: the quinone is reduced with H_2 to the hydroquinone with the help of a metal catalyst; then the hydroquinone is oxidized by O_2 to yield the quinone and H_2O_2. This process was first commercially used by IG Farbenindustrie, Ludwigshafen, Germany, during World War II; the quinone used was 2-methyl-anthraquinone [6].

As oxygen is transferred from H_2O_2 during oxidation, one of the structures originally proposed was $H_2O=O$. In 1927, Raikow discussed this structure with many others and concluded that $H-O-O-H$ was in equilibrium with $H_2O=O$. He argued that $H-O-O-H$ was reducing and proposed the name hydrogen pseudoperoxide, while $H_2O=O$ was the real, oxidizing, hydrogen peroxide [7]. We now know that H_2O_2 can be both oxidizing and reducing (Table 1.1). Please note that at the time this structure was proposed, there were no generally accepted theories of chemical bonding. The valence shell electron pair repulsion theory predicts the correct structure: imagine a book open at 90°C, then one hydrogen is on the upper left page and the other on the lower right page. The O–O bond is 145 pm, the H–O bond is 99 pm, the H–O–O angle is 103° [8,9]. These parameters apply to frozen H_2O_2 with a purity greater than 99%.

The properties of H_2O_2 have been known for over 60 years.

TABLE 1.1
Physical Properties of H_2O_2

Physical Property	Value	Comments
Melting point	−0.43°C[a]	Sat. solutions tend to supercool
Boiling point	150.2°C[a]	Extrapolated, decomposes
Eutectic points	1st 45.2%,[b] −52.2°C[a]	Formation of $H_2O_2 \cdot 2H_2O$
	2nd 61.2%,[b] −56.1°C[a]	
Density, g mL^{-1}	1.4425[a]	At 25°C
pK_a	10.65[c]	At 25°C
$\Delta_f G°$, kJ mol^{-1}	−134.03[d]	
$\Delta_f H°$, kJ mol^{-1}	−191.17[d]	
$S°$, J/(mol K)	+143.9[d]	
$E°'$ (H_2O_2, H$^+$/HO$^\bullet$, H_2O)	+0.39 V[e]	At 25°C, pH 7
$E°'$ (O_2, 2H$^+$/H_2O_2)	+0.28 V[e]	At 25°C, pH 7, pO_2 = 100 kPa
$E°'$ (H_2O_2, 2H$^+$/2H_2O)	+1.35 V[e]	At 25°C, pH 7
O–O bond strength	232 kJ mol^{-1} ($\Delta H°$)[f]	
Dipole moment	2.26 Debye (2.26 × 10^{-18} esu)[a]	7.54 × 10^{-30} cm

[a] [6].
[b] By weight.
[c] [10].
[d] [11].
[e] [12].
[f] [13].

H_2O_2 is unstable with respect to disproportionation. The energy released at pH 7 is 2F[$E°'$(H_2O_2, 2H$^+$/2H_2O) − $E°'$(O_2, 2H$^+$/H_2O_2)], or 103 kJ mol^{-1} H_2O_2, in which F is the Faraday, $E°'$ is the electrode potential at pH 7 (Table 1.1 [12]), and the number 2 preceding F is the number of electrons in the redox equation. The high value of +1.35 V for the two-electron electrode potential of $E°'$(H_2O_2, 2H$^+$/2H_2O) does not necessarily indicate that H_2O_2 is a strong oxidant: kinetically, in dilute solution, electron transfers generally take place one electron at a time. For that reason, the one-electron electrode potential of the H_2O_2, H$^+$/HO$^\bullet$, H_2O couple, +0.39 V [12], is more relevant to one-electron reductants, for example, redox-active metal complexes.

H_2O_2 is a thermodynamically strong two-electron oxidant, but kinetically a weak one-electron oxidant.

REACTION OF HYDROGEN PEROXIDE WITH IRON(II)

During the 1970s, the Fenton reaction was proposed as the causative agent of oxidative damage because it produced the reactive hydroxyl radical. The following is a short summary of Fenton's findings [14]. Fenton (1854–1929) studied chemistry at the University of Oxford; King's College, London; and the University of Cambridge, where, in 1878, he obtained the post of "Additional Demonstrator in Chemistry." In 1899, he became a fellow of the Royal Society on the basis of his work with H_2O_2 and iron(II). He was appointed to the post of University Lecturer in Chemistry at the University of Cambridge in 1904.

The origin of the Fenton reaction was a case of serendipity. A fellow student, whose identity is unfortunately not known, showed a solution with a violet color to Fenton. The student obtained this violet solution by mixing reagents at random. Present in the solution were H_2O_2 and tartaric acid; upon the addition of a ferrous salt, immediately followed by a base, an intense violet color developed. In 1876, Fenton, still an undergraduate student, sent a brief note to *Chemical News* [15], which was published 10 days after submission! The same note also mentioned that the same color was obtained with HOCl instead of H_2O_2. A full description, delayed by his heavy teaching load, was published in 1894 [16]. A number of other organic compounds, such as citric, succinic, oxalic, and acetic acids, do not produce this color, although these compounds can be modified by iron(II) and H_2O_2. A ferric salt also failed to produce the violet color; iron(II) is essential. In 1893, he isolated the product, and he reported its empirical formula as $C_2H_2O_3$ [17]. He mentioned [16] that the amount of iron is not very important, because it acts catalytically: Fe(III) is reduced during the reaction. The product of the oxidation of tartaric acid turned out to be dihydroxymaleic acid, $C_4H_4O_6$. Fenton did not publish a reaction mechanism for this reaction.

Fenton did not know that hydroxyl radicals are formed when Fe(II) reduces H_2O_2.

Two years after Fenton's death in 1929, the two Nobel Prize winners Fritz Haber (1868–1934) and Richard M. Willstätter (1872–1942) wrote a hypothesis paper on radical chain reactions in organic chemistry and biochemistry [18]: they assumed that the first step in the reaction of catalase with H_2O_2 is Reaction 1.1, now known as the Fenton reaction, followed by the chain reactions (1.2) and (1.3). These two reactions—they further assumed—did not require an enzyme; thus, catalase was only necessary to initiate the chain reactions (1.2) and (1.3). Haber and his assistant Joseph Weiss (1905–1972) adopted this mechanism for the decay of H_2O_2 by iron salts at low pH and concluded that since more than $1H_2O_2$ per $2Fe^{2+}$ is consumed when H_2O_2 is present in excess, indeed the chain reactions (1.2) and (1.3) take place [19,20]. Chain termination takes place by Reaction 1.4, as reviewed [21].

$$Fe^{2+} + H_2O_2 + H^+ \rightarrow Fe^{3+} + H_2O + HO^\bullet \qquad (1.1)$$

$$HO^\bullet + H_2O_2 \rightarrow H_2O + HO_2^\bullet + H^+ \qquad (1.2)$$

$$HO_2^\bullet + H^+ + H_2O_2 \rightarrow O_2 + HO^\bullet + H_2O \qquad (1.3)$$

$$Fe^{2+} + HO^\bullet + H^+ \rightarrow Fe^{3+} + H_2O \qquad (1.4)$$

Because of being of Jewish descent, Haber decided to leave Germany in 1933 with his assistant J. Weiss. They worked for a few months in Cambridge, and while visiting Basel, Haber died of heart disease at the age of 65 [22]. Weiss stayed in England and later became professor of chemistry at the University of Newcastle upon Tyne [23]. Reactions 1.2 and 1.3 have become known as the Haber–Weiss cycle, although it would be more appropriate to name it the Haber–Willstätter cycle [21]. Neither Haber and Willstätter [18] nor Haber and Weiss [20] cited Fenton. Given that superoxide does not reduce hydrogen peroxide, the expression "Haber–Weis reaction" should not be used. Furthermore, given the definition of superoxide by Haber and Willstätter ("und die Eigenschaft beigelegt, Hydroperoxyd unter Entstehung von molekularem Sauerstoff, Wasser und ungeladenetn Hydroxyl zu zersetzen" or "has been attributed the property to destroy hydrogen peroxide under formation of molecular oxygen, water and hydroxyl"), the expression of "Fenton-catalyzed Haber–Weiss reaction" is, given this definition, equally unacceptable.

The use of a chain reaction was an innovation.

By the late 1940s, the mechanism proposed by Haber and Weiss had been criticized by Philip George (1920–2008), who showed that $O_2^{\cdot-}$ does not react with H_2O_2 [24]. He did so by dropping pieces of KO_2 in a concentrated solution of H_2O_2. The oxygen produced corresponded entirely to what would be expected from the dismutation of $O_2^{\cdot-}$, not to what would have been released if H_2O_2 underwent the chain reactions (1.2) and (1.3). An improved mechanism for the decomposition of H_2O_2 at low pH proposed by George and coworkers in 1949 [25] is shown in Table 1.2. Weiss and Humphrey admitted that the reaction of Fe^{3+} with $HO_2^{\cdot}$ "largely replaces reaction 1.3" [26]. More details were published in 1951 [27,28]. For the first time, reference is made to Fenton's full report of the oxidation of tartaric acid [16]. George moved from the United Kingdom to the United States and, in 1955, he became professor of biophysical chemistry at the University of Pennyvania (see http://academictree.org/chemistry/peopleinfo.php?pid=80406&expand=bio)

George and coworkers showed that $HO_2^{\cdot}$ does not reduce H_2O_2.

TABLE 1.2

Rate Constants for the Reactions in the Mechanism of Iron-Catalyzed Decomposition of H_2O_2 at Low pH

Reactions	Reaction No.	k (M^{-1} s^{-1})[a]
$Fe^{2+} + H_2O_2 + H^+ \rightarrow Fe^{3+} + H_2O + HO^{\cdot}$	1.1	41.5
$HO^{\cdot} + H_2O_2 \rightarrow H_2O + HO_2^{\cdot}$	1.2	$2.7 \cdot 10^7$
$Fe^{2+} + HO^{\cdot} + H^+ \rightarrow Fe^{3+} + H_2O$	1.4	$4.3 \cdot 10^8$
$Fe^{2+} + HO_2^{\cdot} + H^+ \rightarrow Fe^{3+} + H_2O_2$	1.5	$1.2 \cdot 10^6$
$Fe^{3+} + HO_2^{\cdot} \rightarrow Fe^{2+} + O_2 + H^+$	1.6	$2.0 \cdot 10^4$ (pH 1)

[a] Rate constants from [29–31].

FENTON REACTION IN BIOLOGY

The discovery in 1969 of an enzymatic function for hemocuprein [32], namely catalysis of the disproportionation of $O_2^{\cdot-}$ [33], started a new field of study—free radical biochemistry. Mn- and Fe-containing superoxide dismutases were discovered subsequently [34–36]. Given that aerobic organisms contain superoxide dismutase(s), that the reaction of $O_2^{\cdot-}$ with—at least Cu/Zn—superoxide dismutases is extremely efficient [37–39] and fast [40–44], that, in the cytosol, the concentration of CuZn superoxide dismutase is in the low micromolar range, that the uncatalyzed dismutation of $O_2^{\cdot-}$ is already rapid: k = ca. 10^6 M^{-1} s^{-1} at neutral pH [45], the conclusion is warranted that $O_2^{\cdot-}$ is reactive and, thus, dangerous. However, at that time, $O_2^{\cdot-}$ was known to react with, aside from the superoxide dismutases, iron complexes [46], copper [47], quinones and similar molecules [48–50], and the electron transfer protein cytochrome c [51]. Later, proteins with Fe_4S_4 clusters [52] and $NO^{\cdot}$ [53] were added to this list. However, at the end of the 1970s, there was no evidence from radiation–chemical studies that $O_2^{\cdot-}$ damages proteins, lipids, or DNA. Thus, $O_2^{\cdot-}$ begets a more harmful molecule, and Reaction 1.3, which yields the very reactive hydroxyl radical, is resurrected. History repeated itself: It was pointed out that Reaction 1.3 is too slow to be significant, as reviewed [21]. When $O_2^{\cdot-}$ is not allowed to disproportionate, as in an organic solvent [54], or in the gas phase [55], then Reaction 1.3 may occur. To save the Haber–Weiss cycle, iron complexes were invoked to act as catalysts, as in the mechanism proposed by George et al. [25], and Reactions 1.1 and 1.7 became known as the "Fenton-catalyzed Haber–Weiss reaction," although $O_2^{\cdot-}$ was defined by Haber and Willstätter [18] as capable of reducing H_2O_2.

$$Fe^{3+}X + O_2^{\cdot-} \rightarrow Fe^{2+}X + O_2 \tag{1.7}$$

In Reaction 1.7, at pH 7, X stands for an uncharacterized ligand, or ligands, of iron. At low pH, Fe^{3+}(aq) does oxidize $HO_2^{\cdot}$ (see Reaction 1.6 in Table 1.2). However, there is no evidence for a fast reduction of a physiological iron(III) complex by superoxide. Later, it was realized and accepted that other reductants, for example, monohydrogen ascorbate, are more likely candidates to reduce Fe^{3+} complexes in vivo, yet the identity of intracellular iron complexes and the kinetics of their reactions with $O_2^{\cdot-}$, ascorbate, and H_2O_2 were never established. This is not to say that excess iron is not harmful. In iron-overload diseases, the iron is present in serum as non-transferrin-bound iron, consisting of iron, citrate, and albumin at micromolar concentration. Furthermore, in a cellular compartment with a low pH and devoid of enzymes that scavenge H_2O_2, such as the lysosome, the Fenton reaction may be relevant [56]. We recently investigated whether iron citrate complexes—mostly corresponding to Fe(III)(citrate)$_2$—could participate in the Fenton reaction and be re-reduced by ascorbate. Such iron complexes may be present in blood. We found an electrode potential of -0.03 V $< E^{\circ\prime} > +0.01$ V, that is, ca. 0 V, for the (Fe^{3+}cit/Fe^{2+}cit) couple potential at pH 7. Redox cycling does occur, with re-reduction by ascorbate being rate limiting, $k \approx 3$ M^{-1} s^{-1}. Most of the ascorbyl radical formed is sequestered by complexation with iron and remains EPR silent. We concluded that when H_2O_2 is available, iron-citrate complexes may contribute to pathophysiological manifestations of iron-overload diseases [57]. In general, not much is known about the coordination of intracellular iron. Often it is referred to as "free" or "labile" iron, whereby the use of the term "labile" runs counter to its definition in inorganic chemistry, where it indicates a metal ion that rapidly exchanges ligands, not necessarily a metal complex that is chelatable or redox-active.

The role of "free" iron in oxidative stress may have been overestimated.

Given that superoxide dismutase alone could be protective during oxidative challenges, Beckman and coworkers proposed in 1990 that $ONOO^-$ was formed from $NO^•$ and $O_2^{•-}$ [58], both of which are enzymatically produced. In addition, $O_2^{•-}$ is formed by leakage of the electron transport chain. The rate of the reaction of $O_2^{•-}$ with $NO^•$ is diffusion controlled [53]. ONOOH does not react very fast with biomolecules and is therefore more selective in its reactions than $HO^•$. This molecule oxidizes, hydroxylates, and nitrates biomolecules; an example is tyrosine [59]. Evidence for the formation of $ONOO^-$ in vivo followed from the observation of nitrated tyrosines [60,61]. Given that the nitration of tyrosine is first order in ONOOH and zero order in tyrosine, as well as in many other molecules that are modified by ONOOH, it was believed that ONOOH undergoes homolysis to form $HO^•$ and $NO_2^•$ to an extent of 30%. Although an extensive, if not complete, review of the literature [62] found no evidence for such a percentage—homolysis may occur to an extent of at most 5%—the 30% homolysis hypothesis is still being perpetuated in the literature [63,64]. More important is that the anion, $ONOO^-$, reacts rapidly with the CO_2 in the tissues, which may lead to the strongly oxidizing radicals $CO_3^{•-}$ and $NO_2^•$ [65]. Although peroxynitrite is formed in vivo from $O_2^{•-}$ and $NO^•$, for in vitro experiments peroxynitrite has often been synthesized by mixing acidified H_2O_2 with NO_2^-, followed by rapid quenching with HO^- [66]. It should not been forgotten that this preparations is not pure, and contains nitrite, nitrate and, unless it has been removed with MnO_2, unreacted hydrogen peroxide.

Formation of $ONOO^-$ is kinetically far more feasible than that of $HO^•$ via the Fenton reaction. ONOOH is more reactive than H_2O_2.

HYDROGEN PEROXIDE IN BIOLOGY

Thenard tested whether H_2O_2 reacted with many compounds—inorganic and organic. He observed that the decomposition of H_2O_2 was initiated by "fibrin blanche" that he had recently obtained from blood [67]. It is safe to assume that the white fibrin still contained catalase. At the time, the action of fibrin could not be explained. The discovery of catalase in 1900 [68] implied that H_2O_2 is formed in vivo, but for decades to come nobody could detect H_2O_2 in tissues. As reviewed [69], Wieland postulated in 1913 that O_2 directly abstracts two hydrogen atoms from an organic substrate [70], a view that was ridiculed by Warburg [71]. Possibly because of this feud, Oppenheimer wrote [72], cited by Raikow [7]: H_2O_2 ist sehr wahrscheinlich eine biologisch höchst wichtige Substanz, aber ihre Rolle ist um so weniger klar, als es trotz aller Bemühungen bisher niemals in lebenden Geweben nachgewiesen worden ist ("H_2O_2 is quite probably a biologically very important compound, but its role is much less clear, because in spite of all efforts its presence in living tissues has not been demonstrated"). Although it was shown that H_2O_2 was produced by a bacterium [73], a fungus [74], and spermatozoa [75], it was not shown to be present in higher organisms. Indeed, Wieland failed to find H_2O_2 in a freshly sacrificed dog [76]. Britton Chance (1913–2010) invented the stopped-flow spectrophotometer [77] and used it in 1947 to mix a catalase preparation from horse with H_2O_2. He obtained the spectrum of an intermediate named Compound I, which has a maximum at 405 nm. The rate constant for the formation of Compound I was 3×10^7 M^{-1} s^{-1}.

He found that the intermediate reacted with H_2O_2, as expected, but also with other compounds such as ascorbate and alcohols [78]. Five years later, he found the 405 nm band of Compound I in respiring *Micrococcus lysodeikticus* [79]. The Soret band of Compound 1 is not suitable to detect catalase in mammalian cells, because UV light does not penetrate tissue very far, and because other heme proteins absorb in that region of the spectrum. Sies had the idea to use the near-infrared band of Compound I (660–640 nm), a methodology that avoids these complications. The approach was successful and resulted in 1970 in a publication, with Chance as coauthor, in which formation of H_2O_2 in perfused rat liver was demonstrated [80] (see also [81]). This was a considerable achievement as the detection of H_2O_2 in vivo was considered not feasible [82]. In the meantime, another enzyme had been discovered that disposed of H_2O_2 and hydroperoxides at the expense of glutathione, and was thus named glutathione peroxidase [83]. At present, 8 different glutathione peroxidases are known. H_2O_2 oxidizes a selenocysteine [84] followed by re-reduction with 2 glutathione in a "ter uni ping pong" mechanism [85]. In 1994, another family of enzymes that removes H_2O_2, the peroxiredoxins, was discovered [86]. These enzymes have a cysteine that is oxidized to a sulfenic acid. The fast reaction with H_2O_2—$k = 1 \times 10^5$ to 1×10^7 M^{-1} s^{-1}—is ascribed to the unusual low pK_a of the cysteine and hydrogen bonding interactions between H_2O_2 and the enzyme (see [87] for review).

Scavenging of H_2O_2 by catalase and glutathione peroxidase is fast. Nature uses a trick to accelerate the reaction with the peroxidative cysteine in peroxiredoxin.

Where these enzymes are present, H_2O_2 will have a limited lifetime. As a messenger molecule, it needs to be terminated, the termination being the message. As a component of the Fenton reaction, it needs to be eliminated before it causes damage. Under different conditions, H_2O_2 is necessary. In search for the mechanism of neutrophils, phagocytic white blood cells, Klebanoff reported at the end of the 1960s that a solution of halides (but not of F$^-$), myeloperoxidase, and H_2O_2 was microbicidal [88,89]. As part of the phagocytotic process, extra O_2 is consumed—the respiratory burst—and undergoes one electron to $O_2^{\bullet-}$ [90], which is expelled into the phagosomal space. Catalyzed disproportionation by superoxide dismutase yields O_2 and H_2O_2. H_2O_2 converts myeloperoxidase to its Compound I state, which transfers an O to Cl$^-$. HOCl, which has a pK_a of 7.4, reacts slowly and is thus more selective. It was Winterbourn who demonstrated that the oxidant was HOCl [91]. Responsible for the one-electron reduction of O_2 is a NADPH oxidase. Research in this area yielded a family of NADPH oxidases, present in phagocytic cells, but also in other cells [92], with the one in neutrophils, NOX2, being the most active [93]. Macrophages also contain a NADPH oxidase that produces $O_2^{\bullet-}$ [94], but in those cells disproportionation is undesirable: $O_2^{\bullet-}$ is to react with NO$^\bullet$ to form ONOO$^-$. Remarkably, when it comes to microbicidal agents, Nature turns to inorganic oxidants [95].

HYDROGEN PEROXIDE: FRIEND OR FOE?

Our thinking about H_2O_2 has changed over the years. First, we thought that it is formed by rogue electrons that decided to leave the mitochondrial electron transport chain before cytochrome *c* oxidase [96], clearly a harmful event that leads to oxidative stress, defined as a condition in which defense mechanisms cannot cope with oxidant processes [97]. At that time, H_2O_2 was depicted as a strong oxidant and responsible for the generation of $HO^•$ via the Fenton reaction. We understood why there were enzymes that disposed of H_2O_2. But it was also realized that H_2O_2 was also necessary to kill invading organisms. More recently, it was discovered that H_2O_2 is a signaling molecule that links the redox couples $NAD(P)^+/NAD(P)$H with RSSR/2RSH [98] and modulates transcription factors [99]. The local concentration of H_2O_2, and thus the response, may be manipulated by reversibly inhibiting peroxiredoxins [87]. Its movement from origin to target may be helped by aquaporins [100]. More details about H_2O_2 as a signaling molecule can be found in a contemporary review [101].

H_2O_2 can play the role of a signaling molecule because it is not very reactive, a fact already known to its discoverer, Louis Jacques Thenard.

ACKNOWLEDGMENT

I thank Helmut Sies for helpful comments on the first draft of this manuscript.

REFERENCES

1. Thenard L-. J. 1818. Observations sur des combinaisons nouvelles entre l'oxygène et divers acides. *Annales de Chimie et de Physique* 8: 306–313.
2. Thenard L-. J. 1818. Observations sur l'influence de l'eau dans la formation des acides oxigénés. *Annales de Chimie et de Physique* 9: 314–317.
3. Thenard L-. J. 1819. Nouveaux résultats sur la combination de l'oxigène avec l'eau. *Annales de Chimie et de Physique* 10: 335–336.
4. Rheinboldt H. 1953. *Chemische Unterichtsversuche. Ausgewählte Beispiele für den Gebrauch an Hochschulen und höheren Lehranstalten.* Berlin, Germany: Springer Verlag.
5. Brauer G. 1960. *Handbuch der Präparativen Anorganischen Chemie.* Stuttgart, Germany: Ferdinand Enke Verlag.
6. Schumb W. C., Satterfield C. N., and Wentworth R. L. 1955. *Hydrogen Peroxide.* New York: Reinhold.
7. Raikow P. N. 1927. Eine neue Theorie über die Struktur des Wasserstoffdioxyds und über den Mechanismus der Reaktionen bei seinen chemischen Umwandlungen. *Zeitschrift für anorganische und allgemeine Chemie* 168: 297–304.
8. Abrahams S. C., Collin R. L., and Lipscomb, W. N. 1951. The crystal structure of hydrogen peroxide. *Acta Crystallographica* 4: 15–20.
9. Busing W. R. and Levy H. A. 1965. Crystal and molecular structure of hydrogen peroxide: A neutron-diffraction study. *The Journal of Chemical Physics* 42: 3054–3059.
10. Evans M. G. and Uri N. 1949. The dissociation constant of hydrogen peroxide and the electron affinity of the HO_2 radical. *Transactions of the Faraday Society* 45: 224–230.
11. Wagman D. D., Evans W. H., Parker V. B., Schumm R. H., Halow I., Bailey S. M., Churney K. L., and Nuttal R. L. 1982. Selected values for inorganic and C_1 and C_2 organic substances in SI units. *Journal of Physical Chemical Reference Data* 11 (Suppl. 2): 37–38.
12. Koppenol W. H., Stanbury D. M., and Bounds P. L. 2010. Electrode potentials of partially reduced oxygen species, from dioxygen to water. *Free Radical Biology and Medicine* 49(3): 317–322.

13. Evans M. G., Baxendale J. H., and Uri N. 1949. The heat of the reaction between ferrous ions and hydrogen peroxide in aqueous solution. *Transactions of the Faraday Society* 45: 236–239.
14. Koppenol W. H. 1993. The centennial of the Fenton reaction. *Free Radical Biology and Medicine* 15(6): 645–651.
15. Fenton H. J. H. 1876. On a new reaction of tartaric acid. *Chemical News* 33: 190.
16. Fenton H. J. H. 1894. Oxidation of tartaric acid in the presence of iron. *Journal of the Chemical Society, Transactions* 65: 899–910.
17. Fenton H. J. H. 1893. The oxidation of tartaric acid in presence of iron. *Journal of Chemical Society, Proceedings* 9: 113.
18. Haber F. and Willstätter R. 1931. Unpaarigheit und Radikalketten im Reaktionsmechanismus organischer und enzymatischer Vorgänge. *Chemische Berichte* 64: 2844–2856.
19. Haber F. and Weiss J. 1932. Über die Katalyse des Hydroperoxydes. *Naturwissenschaften* 51: 948–950.
20. Haber F. and Weiss J. 1934. The catalytic decomposition of hydrogen peroxide by iron salts. *Proceedings of the Royal Society of London* 147: 332–351.
21. Koppenol W. H. 2001. The Haber-Weiss cycle—70 years later. *Redox Report* 6(4): 229–234.
22. Stoltzenberg D. 1994. *Fritz Haber, Chemiker, Nobeltreisträger, Deutscher, Jude.* Weinheim, Germany: Wiley VCH.
23. Scholes G. 1972. Professor J. J. Weiss, 1905–1972. *International Journal of Radiation Biology and Related Studies in Physics, Chemistry and Medicine* 22: 311–312.
24. George P. 1947. Some experiments on the reactions of potassium superoxide in aqueous solutions. *Discussions of the Faraday Society* 2: 196–205.
25. Barb W. G., Baxendale J. H., George P., and Hargrave K. R. 1949. Reactions of ferrous and ferric ions with hydrogen peroxide. *Nature* 163: 692–694.
26. Weiss J. and Humphrey C. W. 1949. Reaction between hydrogen peroxide and iron salts. *Nature* 163: 691.
27. Barb W. G., Baxendale J. H., George P., and Hargrave K. R. 1951. Reactions of ferrous and ferric ions with hydrogen peroxide II. The ferric ion reaction. *Transactions of the Faraday Society* 47: 591–616.
28. Barb W. G., Baxendale J. H., George P., and Hargrave K. R. 1951. Reactions of ferrous and ferric ions with hydrogen peroxide. Part I. The ferrous ion reaction. *Transactions of the Faraday Society* 47: 462–500.
29. Hardwick T. J. 1957. The rate constant of the reaction between ferrous ions and hydrogen peroxide in acid solution. *Canadian Journal of Chemistry* 35: 428–436.
30. Ross A. B., Bielski B. H. J., Buxton G. V., Cabelli D. E., Greenstock C. L., Helman W. P., Huie R. E., Grodkowski J., and Neta P. 1994. NDRL-NIST Solution Kinetics Database: Ver. 2 (http://kinetics.nist.gov/solution/). Gaithersburg, MD: National Institute of Standards and Technology.
31. Rush J. D. and Bielski B. H. J. 1985. Pulse radiolytic studies of the reactions of hydrodioxyl/superoxide with Fe(II)/Fe(III) ions. The reactivity of hydrodioxyl/superoxide with ferric ions and its implication on the occurrence of the Haber-Weiss reaction. *The Journal of Physical Chemistry* 89: 5062–5066.
32. Mann T. and Keilin D. 1938. Haemocuprein, a copper-protein compound of red blood corpuscles. *Nature* 142: 148.
33. McCord J. M. and Fridovich I. 1988. Superoxide dismutase: The first twenty years (1968–1988). *Free Radical Biology and Medicine* 5(5–6): 363–369.
34. Keele B. B., McCord J. M., and Fridovich I. 1970. Superoxide dismutase from *Escherichia coli* B. *Journal of Biological Chemistry* 245(22): 6176–6181.
35. Marklund S. 1978. Purification and characterization of a manganese containing superoxide dismutase from bovine heart mitochondria. *International Journal of Biochemistry* 9(5): 299–306.

36. Yost F. J. and Fridovich I. 1973. An iron-containing superoxide dismutase from *Escherichia coli*. *Journal of Biological Chemistry* 248(14): 4905–4908.

37. Benovic J., Tillman T., Cudd A., and Fridovich I. 1983. Electrostatic facilitation of the reaction catalyzed by the manganese-containing and the iron-containing superoxide dismutases. *Archives of Biochemistry and Biophysics* 221(2): 329–332.

38. Getzoff E. D., Tainer J. A., Weinter P. K., Kollman P. A., Richardson J. S., and Richardson C. 1983. Electrostatic recognition between superoxide and copper, zinc superoxide dismutase. *Nature* 306(5940): 287–290.

39. Koppenol W. H. 1981. The physiological role of the charge distribution on superoxide dismutase. In *Oxygen and Oxyradicals in Chemistry and Biology*, eds. M. A. J. Rodgers and E. L. Powers, pp. 671–674. New York: Academic Press.

40. Fielden E. M., Roberts P. B., Bray R. C., Lowe D. J., Mautner G. N., Rotilio G., and Calabrese L. 1974. The mechanism of action of superoxide dismutase from pulse radiolysis and electron paramagnetic resonance. *Biochemical Journal* 139(1): 49–60.

41. Klug-Roth D., Fridovich I., and Rabani J. 1973. Pulse radiolysis investigations of superoxide catalyzed disproportionation. Mechanism for bovine superoxide dismutase. *Journal of the American Chemical Society* 95: 2786–2791.

42. Lavelle F., McAdam M. E., Fielden E. M., Roberts P. B., Puget K., and Michelson A. M. 1977. A pulse-radiolysis study of the catalytic mechanism of the iron-containing superoxide dismutase from *Photobacterium leiognathi*. *Biochemical Journal* 161(1): 3–11.

43. McAdam M. E., Lavelle F., Fox R. A., and Fielden E. M. 1977. A pulse-radiolysis study of the manganese-containing superoxide dismutase from *Bacillus stearothermophilus*. Further studies on the properties of the enzyme. *Biochemical Journal* 165(1): 81–97.

44. Pick M., Rabani J., Yost F., and Fridovich I. 1974. The catalytic mechanism of the manganese containing SOD of *Escherichia coli* studied by pulse radiolysis. *Journal of the American Chemical Society* 96(23): 7329–7332.

45. Bielski B. H. J., Cabelli D. E., Arudi R. L., and Ross A. B. 1985. Reactivity of HO_2/O_2^- radicals in aqueous solution. *Journal of Physical Chemical Reference Data* 14: 1041–1100.

46. Ilan Y. A. and Czapski G. 1977. The reaction of superoxide radical with iron complexes of EDTA studied by pulse radiolysis. *Biochimica et Biophysica Acta* 498(1): 386–394.

47. Rabani J., Klug-Roth D., and Lilie J. 1973. Pulse radiolytic investigations of the catalyzed disproportionation of peroxy radicals. Aqueous cupric ions. *The Journal of Physical Chemistry* 77: 1169–1175.

48. Ilan Y. A., Meisel D., and Czapski G. 1974. The redox potential of the $O_2–O_2$ system in aqueous media. *Israel Journal of Chemistry* 12: 891–895.

49. Meisel D. and Czapski G. 1975. One-electron transfer equilibria and redox potentials of radicals studied by pulse radiolysis. *The Journal of Physical Chemistry* 79: 1503–1509.

50. Misra H. P. and Fridovich I. 1972. The univalent reduction of oxygen by reduced flavins and quinones. *Journal of Biological Chemistry* 247(1): 188–192.

51. Koppenol W. H., van Buuren K. J. H., Butler J., and Braams R. 1976. The kinetics of the reduction of cytochrome *c* by the superoxide anion radical. *Biochimica et Biophysica Acta* 449(2): 157–168.

52. Liochev S. I. 1996. The role of iron-sulfur clusters in *in vivo* hydroxyl radical production. *Free Radical Research* 25(5): 369–384.

53. Nauser T. and Koppenol W. H. 2002. The rate constant of the reaction of superoxide with nitrogen monoxide: Approaching the diffusion limit. *The Journal of Physical Chemistry A* 106: 4084–4086.

54. MacManus-Spencer L. A. and McNeill K. 2005. Quantification of singlet oxygen production in the reaction of superoxide with hydrogen peroxide using a selective chemiluminescent probe. *Journal of the American Chemical Society* 127: 8954–8955.

55. Blanksby S. J., Bierbaum V. M., Ellison G. B., and Kato S. 2007. Superoxide does react with peroxides: Direct observation of the Haber-Weiss reaction in the gas phase. *Angewandte Chemie International Edition* 46(26): 4948–4950.

56. Yu Z. Q., Persson H. L., Eaton J. W., and Brunk U. T. 2003. Intra-lysosomal iron: A major determinant of oxidant-induced cell death. *Free Radical Biology and Medicine* 34(10): 1243–1252.

57. Adam F. I., Bounds P. L., Kissner R., and Koppenol W. H. 2015. Redox properties and activity of iron-citrate complexes: Evidence for redox cycling. *Chemical Research in Toxicology* 28(4): 604–614.

58. Beckman J. S., Beckman T. W., Chen J., Marshall P. A., and Freeman B. A. 1990. Apparent hydroxyl radical production by peroxynitrite: Implications for endothelial injury from nitric oxide and superoxide. *Proceedings of the National Academy of Sciences of the United States of America* 87(4): 1620–1624.

59. Ramezanian M. S., Padmaja S., and Koppenol W. H. 1996. Hydroxylation and nitration of phenolic compounds by peroxynitrite. *Chemical Research in Toxicology* 9(1): 232–240.

60. Beckman J. S., Ye Y. Z., Anderson P. G., Chen J., Accavitti M. A., Tarpey M. M., and White C. R. 1994. Extensive nitration of protein tyrosines in human atherosclerosis detected by immunohistochemistry. *Biological Chemistry Hoppe-Seyler* 375(2): 81–88.

61. Radi R. 2013. Protein tyrosine nitration: Biochemical mechanisms and structural basis of functional effects. *Accounts of Chemical Research* 46(2): 550–559.

62. Koppenol W. H., Bounds P. L., Nauser T., Kissner R., and Rüegger H. 2012. Peroxynitrous acid: Controversy and consensus surrounding an enigmatic oxidant. *Dalton Transactions* 41(45): 13779–13787.

63. Carballal S., Bartesaghi S., and Radi R. 2014. Kinetic and mechanistic considerations to assess the biological fate of peroxynitrite. *Biochimica et Biophysica Acta* 1840(2): 768–780.

64. Radi R. 2013. Peroxynitrite, a stealthy biological oxidant. *Journal of Biological Chemistry* 288(37): 26464–26472.

65. Lymar S. V. and Hurst J. K. 1995. Rapid reaction between peroxonitrite ion and carbon dioxide: Implications for biological activity. *Journal of the American Chemical Society* 117: 8867–8868.

66. Reed J. W., Ho H. H., and Jolly W. L. 1974. Chemical synthesis with a quenched flow reactor. Hydroxytrihydroborate and peroxynitrite. *Journal of the American Chemical Society* 96: 1248–1249.

67. Thenard L-. J. 1819. Nouvelles observations sur l'eau oxigénée. *Annales de Chimie et de Physique* 11: 85–87.

68. Loew O. 1900. A new enzyme of general occurrence in organisms. A preliminary note. *Science* 11: 702–703.

69. Koppenol W. H. 2016. Hydrogen peroxide, from Wieland to Sies. *Archives of Biochemistry and Biophysics* 595: 9–12.

70. Wieland H. 1913. Über den Mechanismus der Oxydationsvorgänge. *Berichte der deutschen chemischen Gesellschaft* 46: 3327–3342.

71. Warburg O. 1923. Über die Grundlagen der Wielandschen Atmungstheorie. *Biochemische Zeitschrift* 142: 518–523.

72. Oppenheimer C. N. 1923. *Kurzes Lehrbuch der Chemie in Natur und Wissenschaft.* Leipzig, Germany: Georg Thieme.

73. McLeod J. W. and Gordon J. 1922. Production of hydrogen peroxide by bacteria. *Biochemical Journal* 16(4): 499.

74. Pearce A. A. 1940. On the so-called "iodide oxidase". Mechanism of iodide oxidation by *Aspergillus*. *Biochemical Journal* 34(10–11): 1493–1500.

75. Tosic J. and Walton A. 1946. Formation of hydrogen peroxide by spermatozoa and its inhibitory effect on respiration. *Nature* 158: 485.

76. Wieland H. 1925. Über den Mechanismus der Oxydationsvorgänge IX. *Justus Liebigs Annalen der Chemie* 445: 181–201.

77. Chance B. 2004. The stopped-flow method and chemical intermediates in enzyme reactions—A personal essay. *Photosynthesis Research* 80(1–3): 387–400.

78. Chance B. 1947. An intermediate compound in the catalase-hydrogen peroxide reaction. *Acta Chemica Scandinavica* 1: 236–267.

79. Chance B. 1952. The state of catalase in the respiring bacterial cell. *Science* 116(3008): 202.

80. Sies H. and Chance B. 1970. The steady state level of catalase compound I in isolated hemoglobin-free perfused rat liver. *FEBS Letters* 11(3): 172–176.

81. Chance B., Sies H., and Boveris A. 1979. Hydroperoxide metabolism in mammalian organs. *Physiological Reviews* 59(3): 527–605.

82. Sies H. 2014. Role of metabolic H$_2$O$_2$ generation: Redox signalling and oxidative stress. *Journal of Biological Chemistry* 289(13): 8735–8741.

83. Mills G. C. 1957. Hemoglobin catabolism. I. Glutathione peroxidase, an erythrocyte enzyme which protects hemoglobin from oxidase breakdown. *Journal of Biological Chemistry* 229(1): 189–197.

84. Forstrom J. W., Zakowski J. J., and Tappel A. L. 1978. Identification of the catalytic site of rat liver glutathione peroxidase as selenocysteine. *Biochemistry* 17(13): 2639–2644.

85. Flohé L., Loschen G., Eichele E., and Gunzler W. A. 1972. Glutathione peroxidase. V. The kinetic mechanism. *Hoppe-Seyler's Zeitschrift für Physiologische Chemie* 353(6): 987–999.

86. Chae H. Z., Chung S. J., and Rhee S. G. 1994. Thioredoxin-dependent peroxide reductase from yeast. *Journal of Biological Chemistry* 269(44): 27670–27678.

87. Rhee S. G., Woo H. A., Kil I. S., and Bae S. H. 2012. Peroxiredoxin functions as a peroxidase and a regulator and sensor of local peroxides. *Journal of Biological Chemistry* 287(7): 4403–4410.

88. Klebanoff S. J. 1967. A peroxidase-mediated antimicrobial system in leucocytes. *Journal of Clinical Investigation* 46: 1078.

89. Klebanoff S. J. 1968. Myeloperoxidase-halide-hydrogen peroxide antibacterial system. *Journal of Bacteriology* 95(6): 2131–2138.

90. Babior B. M., Kipnes R. S., and Curnutte J. T. 1974. Biological defense mechanisms. The production by leukocytes of superoxide, a potential bactericidal agent. *Journal of Clinical Investigation* 52(3): 741–744.

91. Winterbourn C. C. 1985. Comparative reactivities of various biological compounds with myeloperoxidase-hydrogen peroxide-chloride, and similarity of oxidant to hypochlorite. *Biochimica et Biophysica Acta* 840(2): 204–210.

92. Lambeth J. D. 2004. NOX enzymes and the biology of reactive oxygen. *Nature Reviews Immunology* 4(3): 181–189.

93. Winterbourn C. C., Kettle A. J., and Hampton M. B. 2016. Reactive oxygen species and neutrophil function. *Annual Review of Biochemistry* 85: 765–792.

94. Hoffman M. and Autor A. P. 1980. Production of superoxide anion by an NADPH-oxidase from rat pulmonary macrophages. *FEBS Letters* 121(2): 352–354.

95. Hurst J. K. and Lymar S. V. 1999. Cellularly generated inorganic oxidants as natural microbicidal agents. *Accounts of Chemical Research* 32: 520–528.

96. Loschen G., Azzi A., Richter C., and Flohé L. 1974. Superoxide radicals as precursors of mitochondrial hydrogen peroxide. *FEBS Letters* 42(1): 68–72.

97. Sies H. 1985. Oxidative stress: Introductory remarks. In *Oxidative Stress*, pp. 1–8. London, U.K.: Academic Press.

98. Jones D. P. and Sies H. 2015. The redox code. *Antioxidants & Redox Signaling* 23(9): 734–746.

99. Marinho H. S., Real C., Cyrne L., Soares H., and Antunes F. 2014. Hydrogen peroxide sensing, signaling and regulation of transcription factors. *Redox Biology* 2: 535–562.
100. Henzler T. and Steudle E. 2000. Transport and metabolic degradation of hydrogen peroxide in *Chara corallina*: Model calculations and measurements with the pressure probe suggest transport of H_2O_2 across water channels. *Journal of Experimental Botany* 51: 2053–2066.
101. Sies H. 1985–2017. Hydrogen peroxide as a central redox signaling molecule in physiological oxidative stress: Oxidative eustress. *Redox Biology* 11: 613–617.

2 Hydrogen Peroxide and Other Peroxides in Chemiluminescence, Bioluminescence, and Photo(bio)chemistry in the Dark

Etelvino J.H. Bechara, Wilhelm J. Baader, and Cassius V. Stevani

CONTENTS

INTRODUCTION

Hydrogen peroxide and other peroxides are directly or indirectly involved in the electronic excitation of oxidation products of many chemiluminescent and bioluminescent processes. The intense peroxyoxalate chemiluminescence ($\Phi_{CL} \sim 10\%$) and firefly bioluminescence ($\Phi_{BL} \sim 45\%$), both resulting from the generation of excited singlet products, are iconic and radiant examples. In this chapter, we emphasize the *in vitro* and *in vivo* intermediacy of 1,2-dioxetane and 1,2-dioxetanone cyclic peroxides—sources of excited carbonyls by thermolysis—and consequent light emission or photochemistry. Of particular interest for their radical-like reactivity and long lifetimes are triplet carbonyls, which allow energy transfer to biomolecules or photochemistry in the absence of light and eventually trigger biological responses. Expounded upon in this chapter are mechanisms of electronic chemiexcitation in chemi- and bioluminescent processes including biomolecules catalyzed by peroxidases and accompanied by cleavage products, energy transfer to dioxygen yielding singlet oxygen, plant colchicine isomerization in the dark, phosphate-catalyzed permeation of mitochondria, initiation of lipid peroxidation of polyunsaturated fatty acids, and DNA pyrimidine dimerization in the dark. A large field of investigation of the roles of triplet carbonyls in living organisms and in organic synthesis remains to be deeply explored.

HISTORIC ROOTS OF EXCITED STATES AND FREE RADICALS

The chemistry of the luminescence of organic compounds emerged largely in parallel to that of free radicals and bioluminescence in the second half of the nineteenth century. Molecular oxygen, peroxides, and light emission interweave throughout the history of chemi- and bioluminescence, of which the occurrence from decaying fishes (bacteria) and rotten wood (fungi) had early been registered by Aristotle (384–322 BC), according to the Herculean book by E.N. Harvey *A History of Luminescence. From the Earliest Times Until 1900* [1]. Robert Boyle, in 1667, unveiled the first clues of the chemistry of bioluminescence when he observed that shining firefly larvae became dim "in a vacuum" and turned bright under aeration. In 1892, Raphael Dubois established the luciferin/luciferase basis of bioluminescence from a clever experiment with extracts prepared from the lanterns of a *Pyrophorus* click beetle (Elateridae). Intense light emission occurred when mixing a previously boiled substrate-containing extract with a "cold" extract rich in a thermolabile component, the luciferase. In fact, the "hot" and the "cold" extract protocol continues to be used to investigate the mechanism involved in bioluminescent organisms and to follow the isolation of their components. A relevant and recent example is the now-clarified controversy regarding the chemiluminescent nature of fungal bioluminescence versus a classical enzymatic process in favor of the latter mechanism [2]. The participation of a luciferin and two enzymes, namely, a NAD(P)H-dependent hydroxylase (formerly postulated as reductase) and a luciferase, was definitely demonstrated. A comprehensive account of the chemical and biological diversity of terrestrial and marine bioluminescence is offered by the masterpiece book *Bioluminescence. Living Lights, Lights for Living* by Wilson and Hastings

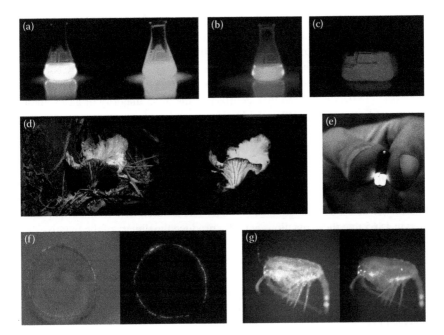

FIGURE 2.1 Chemiluminescence of (a) the peroxyoxalate system using perylene (blue-greenish) or rubrene (orange) as activator; (b) the induced decomposition of a phenoxyl-substituted 1,2-dioxetane; (c) and horseradish peroxidase–catalyzed oxidation of an isobutanal (IBAL) silyl enol ether in the presence of 9,10-dibromoanthracene-2-sulfonate anion (DBAS); bioluminescence of (d) the Brazilian fungus *Neonothopanus gardneri*, which grows on the base of palms in Northern coconut forests; (e) firefly *Macrolampis omissa* found in the Brazilian savannas; (f) jellyfish *Aequorea macrodactyla*; and (g) the copepod *Metridia pacifica*. ([e] and [f]: Photos courtesy of Yuichi Oba, Nagoya University, Nagoya, Japan.)

[3] (Figure 2.1). The firefly *Photinus pyralis* and the jellyfish *Aequorea victoria* are the most explored luminous beings pertaining to reaction mechanisms, analytical applications, and bioengineering.

Still quoted by Harvey, Radziszewski, in the late 1880s, first described chemiluminescent reactions from organic compounds, including lophine (2,4,5-triphenylimidazole), by shaking their alkaline alcohol solutions with air, and subsequently characterized luminogenic compounds by their emission spectrum upon the addition of H_2O_2: blue from luminol, yellow from lucigenin, and red from metalloporphyrins. At present, the chemiluminescence mechanisms of several natural and synthetic compounds when exposed to molecular oxygen, H_2O_2, or hypochlorite continue to attract the attention of investigators for their challenging chemical and technological interest. Among them, the highest quantum yields of chemiluminescence were measured for luminol, acridine derivatives, and peroxyoxalates [4–7] (Figure 2.1a through c).

The nineteenth century also witnessed the first attempt to isolate highly reactive molecular fragments, the putative "free" radicals. A compelling, comprehensive essay about the history of free radicals was written by Ihde [8] and ref. therein for the International Union of Pure and Applied Chemistry (IUPAC). He initiates the article by recalling Gay-Lussac's (1815) identification of the gas obtained from heating $Hg(CN)_2$ as *cyanogène ou radical de l'acide prussique*, that is, "CN" radical, a kind of "compound element," later recognized as its dimer $(CN)_2$. Years before, Gay-Lussac (1810) discovered that, contrary to the great majority of gases, molecular oxygen (dioxygen) was attracted by an external magnet—it was paramagnetic. Indeed, we now know that dioxygen is a stable diradical, with two solitary electrons in two degenerate antibonding orbitals—oxygen is a triplet species. Also of historical interest are the attempts by Wöhler and Liebig in 1832 and Wurtz in 1855 to isolate alkyl radicals from the treatment of alkyl halides with sodium ($RX + Na^o \rightarrow R^{\bullet} + NaX$) and Kolbe's efforts to obtain "free" radicals from the anodic oxidation of carboxylic acid salts ($RCO_2^{\bullet-} - e^- \rightarrow CO_2 + R^{\bullet}$). Later, Schorlemmer [9] demonstrated in 1865 that these reactions actually produce the dimers of the putative "free" radicals, that is, the alkanes R–R. Interestingly, Fenton found out in late 1890s that Thénard's *l'eau oxigénée* (H_2O_2), described in 1810s, is highly oxidant when in the presence of iron salts [10,11]. Today, we know that the hydroxylating property of the "Fenton reagent" over a cornucopia of chemicals and biomolecules, attributing to the $HO^{\bullet}$ radical, is implicated in a high number of physiological and pathogenic processes [12].

The hypothesis of radicals with independent existence turned to be a reality, thanks to Gomberg in 1900 when he identified a free, resonance-stabilized triphenylmethyl radical ($Ph_3C^{\bullet}$) as the product of sodium-promoted displacement of bromide anion from Ph_3CBr [13]. It should be mentioned, however, that Casimir Wurster (1854–1913) crystalized the one-electron oxidation product of *N,N,N',N'*-tetramethyl-*p*-phenylenediamine (TMPD) as labile blue needles three decades before, without identifying this compound as a "free" radical [14]. The currently named "Wurster Blue" radical was decades later properly identified as the highly stable, resonant $TMPD^{\bullet+}$ cation radical, which is commercially available, for example, as its perchlorate salt. TMPD in aerated phosphate buffer rapidly turns blue from one-electron transfer to the dissolved oxygen yielding $TMPD^{\bullet+}$ plus superoxide anion radical. In 1972, TMPD was reported to promote the oxidation of an *N-n*-propyl NADH model to the respective pyridinium cation (the NAD^+ model), coupled to ATP synthesis in pyridine solvent when in the presence of inorganic phosphate and ADP [15]. This reaction stood as a model for ATP synthesis at site I of the respiratory chain, among several others published between 1960 and 1970 (Slater's chemical hypothesis of ATP synthesis).

This discovery launched intense efforts of biochemists to evidence the occurrence of radicals in living organisms, from bacteria up to mammals in the evolutionary tree. The findings of Gerschman et al. [16] that indicated oxygen- and water-derived radicals might be responsible for the similarly observed biochemical and physiological toxic effects of pure oxygen and x-rays in rats and mice were followed by the groundbreaking recognition by McCord and Fridovich of ubiquitous cupreins as superoxide dismutases (SOD) [17]. These enzymes convert the superoxide radical anion ($O_2^{\bullet-}$) into molecular oxygen and H_2O_2, a key agent of redox imbalance, cell regulation, and signal transduction. In this regard, rarely recollected is the fact that oxidative damage to biomolecules related to dioxygen was preceded by findings on H_2O_2 by Thénard [10], a friend of Gay-Lussac, 150 years before McCord and Fridovich's report on superoxide dismutases. Thénard discovered that bacteria, fungi, plants, and animal extracts contain information antagonistic to highly oxidant H_2O_2, once they trigger evolution of dioxygen bubbles by decomposing H_2O_2 into dioxygen and water. This antioxidant factor, an enzyme, was much later named catalase by Loew [18] and crystallized by Sumner and Dounce [19]. Despite being discovered almost two centuries ago and having eluded its presence in living organisms, the real-time concentration of H_2O_2 in cells, tissues, and fluids remains unsettled, controversial, and very often probably meaningless once it is constantly produced and used up at rates depending on the biological compartment and metabolic status [20].

About 20 years after the discovery of SOD, Louis Ignarro [21] and Salvador Moncada [22] independently identified the endothelium relaxing factor as being another radical—nitrogen monoxide or nitric oxide (NO$^{\bullet}$)—a discovery that opened new doors in biology and medicine. Aside from controlling vasodilation and cardiovascular function, NO$^{\bullet}$ radical reportedly acts in neurotransmission, platelet adhesion, inflammatory processes, and insulin liberation, among other biological functions and dysfunctions. Superoxide and NO$^{\bullet}$ radicals were later shown to react with each other at a rate controlled by diffusion ($>10^9$ M^{-1} s^{-1}) to form the peroxynitrite anion (ONOO$^-$). Peroxynitrite rapidly decomposes to nitrate (~70%) and is protonated to peroxynitrous acid (ONOOH), whose pK_a ≥6.8, depending on the ionic strength. In cells, the main targets of peroxynitrite are ferroproteins (e.g., hemoglobin), which are oxidized to the corresponding ferriproteins, and the CO_2/HCO_3^- pair yielding two radicals: the highly oxidant carbonate radical, $CO_3^{\bullet-}$ ($E_0' = 1.71$ V), and $NO_2^{\bullet}$ radical, capable of promoting the nitrosation of protein tyrosine residues [23,24]. Conversely, HOONO can oxidize various thiols to sulfenic acids while being reduced to nitrite, and suffers homolysis to $NO_2^{\bullet}$ and HO$^{\bullet}$ ($E_0' = 2.31$ V) radicals, by which it acts as a one-electron oxidant of several biomolecules. The NO$^{\bullet}$ and $O_2^{\bullet-}$ radicals can thus generate a diverse family of highly reactive species—peroxides and radicals—capable of modifying biomolecules in distinct ways [25].

The discovery of the identity and plural functions of the $O_2^{\bullet-}$ and NO$^{\bullet}$ radicals constitutes a turning point in biology, medicine, and correlated sciences. The molecular bases underlying normal and adverse metabolic conditions associated with redox imbalance in aging, obesity, inflammation, diabetes, cardiopathy, neurodegeneration, and cancer have benefited from the sophistication of technology, including nanoHPLC, NMR, EPR, MS, as well as numerous fluorescent probes for imaging and analytical methods and synthetic routes for biological markers. Terms like oxidative stress, reactive oxygen species, and antioxidants have been continuously clarified and redefined [26].

This chapter will revisit and update the mechanisms of electronic chemiexcitation in chemi- and bioluminescent processes; the involvement of H_2O_2, 1,2-dioxetanes, and other peroxides in these phenomena; and potential biological roles of triplet carbonyls generated by peroxide-mediated enzymatic systems in "photochemistry in the absence of light."

1,2-DIOXETANES AND 1,2-DIOXETANONES: SOURCES OF EXCITED PRODUCTS IN CHEMISTRY AND BIOLOGY

In the search for a common mechanism of electronic excitation and light emission in chemi- and bioluminescence, the intermediacy of unstable peroxides was a promising hypothesis. Once molecular oxygen or hydrogen peroxide is consumed, the O–O bond is much weaker than the C–C bond, and a very stable carbonyl product is formed. It was soon realized that indeed an energy-rich intermediate must be involved in the chemiexcitation step of such reactions [3]. In both chemi- and bioluminescent (luciferase-catalyzed) processes, production of an energy-rich intermediate by the oxidation of a luminogenic reagent (luciferin, in bioluminescence) by dioxygen or H_2O_2 is followed by its cleavage to the electronic excited product, which decays to the ground state by visible, cold light [6].

With no exception, all bioluminescent systems consume molecular oxygen, whereas classical chemiluminescent compounds (e.g., acridines, luminol, oxalates, aldehydes) require either O_2 or H_2O_2 to initiate the reactions [3,27]. The marine acorn worm *Balanoglossus biminiensis* studied by Dure and Cormier and in the early sixties is the only bioluminescent system quoted to be dependent on H_2O_2 (likely to be originated from molecular oxygen); therefore, its luciferase has a peroxidase activity [28,29]. Nevertheless, the mechanism involved in bioluminescence of acorn worms still remains a mystery [30].

The "energy-rich" intermediates of luminescent systems were then proposed to be linear and cyclic peroxides. Strong candidates, however, were the 1,2-dioxetanes, highly unstable tetratomic cyclic peroxides (Figure 2.2a). This hypothesis was endorsed by thermodynamic calculations of their thermal cleavage to two carbonyl products, showing that one of them can be formed in its excited singlet or triplet state, whose decay to the ground state would be accompanied by light emission. The first 1,2-dioxetane to be synthesized, crystallized, and shown to emanate perceptible blue light ($\lambda_{max} \sim 450$ nm) under heating above room temperature was 3,3,4-trimethyl-1,2-dioxetane [31]. Three years later, Adam and Liu [32] succeeded to synthesize 3,3-dimethyl-1,2-dioxetanone, which is much less stable than 1,2-dioxetanes, and therefore a more acceptable model for bioluminescence.

1,2-Dioxetanes with bulky substitution or rotation restriction are more stable. These are the cases of tetraethyl-1,2-dioxetane and bis(adamantyl)-1,2-dioxetane, which are stable at room temperature in solution and show appreciable decomposition only at temperatures above 100°C [33]. The unimolecular decomposition of these cyclic peroxides (Figure 2.2a) was shown to produce preferentially triplet excited carbonyl products (up to 50%) and much lower yields of singlet-excited states (less than 1%) [34]. The thermolysis of 1,2-dioxetanes is thought to proceed via a mixed concerted and diradical mechanism involving concerted, but not simultaneous, O–O and C–C bond cleavage leading to two carbonyl products, one of them possibly in an electronically excited state, most preferably the triplet state [35]. An important advance in the design and synthesis of 1,2-dioxetane models for chemi- and bioluminescence arose when Koo and Schuster [35] discovered that certain cyclic and linear organic peroxides undergo catalyzed decomposition in the presence of fluorescence compound with low oxidation potentials (called activators, ACT),

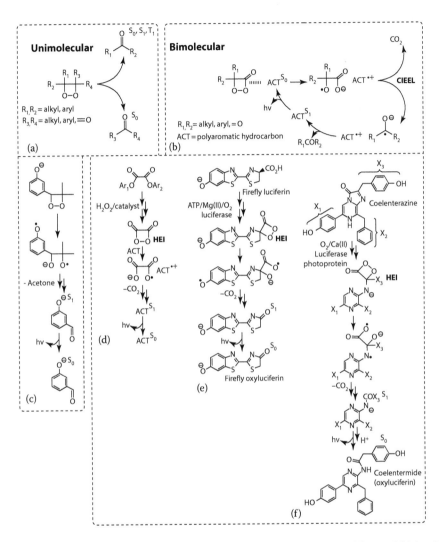

FIGURE 2.2 Chemiexcitation mechanisms in 1,2-dioxetane decomposition and bioluminescence transformations. (a) Unimolecular 1,2-dioxetane thermolysis. (b) Activator (ACT)-
catalyzed 1,2-dioxetanone decomposition via CIEEL (Chemically Initiated Electron Exchange
Luminescence) mechanism. (c) Intramolecular electron transfer-induced decomposition of a
phenoxyl-substituted 1,2-dioxetane. (d) Base-catalyzed reaction of an activated oxalate ester
with hydrogen peroxide in the presence of an activator. (e) Chemical transformation in the
excitation process of the luciferin/luciferase firefly system. (f) Chemiexcitation in the coelenterazine reaction. HEI, high energy intermediate; CIEEL, chemically initiated electron
exchange luminescence.

leading to the formation of the ACT's singlet excited state during peroxide decomposition (Figure 2.2b). Though 1,2-dioxetanes do not show this kind of ACT-catalyzed decomposition, 1,2-dioxetanones (containing a carbonyl function in the four-membered peroxide ring) show this behavior and are considered the model compounds for the so-called Chemically Initiated Electron Exchange Luminescence (CIEEL) mechanism, which was shown to operate in firefly lanterns [36]. The originally proposed reaction for CIEEL mechanism implied the intermediacy of a 1,2-dioxetanodione (Figure 2.2b), that is, a hypothetical CO_2 dimer, arrested in a solvent cage together with a highly fluorescent electron donor—the "activator" (ACT). Electron transfer from ACT to the hypothetical $(CO_2)_2$ dimer would form CO_2 and $ACT^{\bullet}+/CO_2^{\bullet-}$ cation–anion radical pair, which decays by back electron transfer from the high-energy singly occupied molecular orbital (SOMO) of $CO_2^{\bullet-}$ to the low-energy SOMO orbital of $ACT^{\bullet+}$, thus rendering singlet ACT in the first electronically excited singlet state (S_1) [37].

Although more recent studies have shown that the initially determined singlet quantum yields of around 10% for ACT-catalyzed 1,2-dioxetanone decomposition have been overestimated in several orders of magnitude [38], the involvement of electron-transfer processes in these reactions has been confirmed [39]. Contrarily, 1,2-dioxetanes containing easily oxidizable substituents can undergo an intramolecular version of the CIEEL mechanism leading to the efficient generation of singlet-excited carbonyl compounds, as reported for model phenoxyl-substituted 1,2-dioxetane derivative in the 1980s [40,41], which are actually utilized as chemiluminescence detection systems in many immunoassays [42,43] (Figure 2.2c). Figure 2.2d depicts the reaction mechanism reported to operate in the chemiluminescent peroxyoxalate system and Figures 2.2e and f illustrate the reaction mechanisms of firefly and coelenterate bioluminescence, respectively, long reported to occur via a 1,2-dioxetanone intermediate [3].

OTHER POTENTIAL BIOLOGICAL SOURCES OF TRIPLET CARBONYLS: LIPIDS, PROTEINS, AND NUCLEIC ACIDS

In this context, worthy to note is that triplet carbonyls can also be generated from the annihilation of alkoxyl and alkylperoxyl radicals, the latter bearing a geminal hydrogen atom, although at lower but biologically significant yields: <0, 1%–8% and <0, 1%, respectively (Figure 2.3) [44] and ref. therein. The annihilation of alkylperoxyl radicals, named the Russell reaction, is long known to concomitantly produce up to 14% of red-emitting singlet dioxygen [45–47]. Scarce evidence exists for the generation of a triplet carbonyl product by a retro-Paternó–Büchi reaction [48].

Recently, the formerly reported generation of alkoxyl and alkylperoxyl radicals by peroxidation of lipids known to be accompanied by ultraweak chemiluminescence [49] and singlet oxygen formation has also been observed during peroxidation of proteins and DNA [50,51]. Biological peroxidations are known to take place not only enzymatically but can also be initiated by chemically generated radicals ($HO^{\bullet}$, $NO^{\bullet}$) and several nonradical oxidants (metal ions, peroxynitrite, hypochlorite, cytochrome c), as well as by triplet carbonyls [52]. In addition, the cyclization

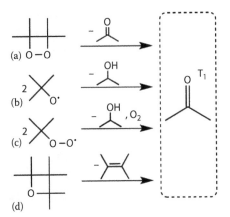

FIGURE 2.3 Possible biological sources of triplet carbonyls by (a) thermolysis of 1,2-dioxetane intermediate; (b) alkyloxyl radical dismutation; (c) alkylperoxyl radical annihilation (Russell reaction) during peroxidation of lipids, proteins and DNA; (d) retro-Paternó–Büchi reaction of oxetanes.

of the $(CH_3)_2C(OO^\bullet)C(CH_3)CH_2$ allylperoxyl radical, produced by the oxidation of the corresponding "ene" hydroperoxide by peroxynitrite or HRP, to a hypothetical 1,2-dioxetane radical followed by its cleavage was also shown to be weakly chemiluminescent [53].

H₂O₂-DEPENDENT CHEMILUMINESCENCE

Undoubtedly, the most mechanistically defying, brightest, and commercially explored ("light sticks") chemiluminescent system is that of peroxyoxalate esters treated with H_2O_2 in alkaline viscous solvents, in the presence of an activator (ACT) with a low oxidation potential [54], which follows the CIEEL pathway (Figure 2.2d).

The peroxyoxalate systems consist in the base-catalyzed reaction of activated oxalic aryl esters with hydrogen peroxide, in the presence of chemiluminescence activators mentioned before. Various mechanistic aspects of this complex transformation have been clarified. Kinetic studies of this system in anhydrous solvents revealed the mechanism of the interaction of hydrogen peroxide with the oxalic esters, leading to several reaction intermediates and rate constants that could be determined for several of the rate-limiting reaction steps [55,56]. The study of synthesized and characterized reactive intermediates contributed to the definition of the exact reaction mechanism, including the identification of the probable high-energy intermediate (HEI) of the transformation [55,57]. Additional studies were performed on the mechanism of the chemiexcitation step, where the HEI interacts with the ACT, leading to excited state formation, showing definitively that this process involves electron transfer steps according to the CIEEL mechanism [5,54,56] and an explanation of the extremely high efficiency of this system (up to 50% of excited state formation) was finally available.

The peroxyoxalate reaction is one of the most efficient chemiluminescent reactions known and has found widespread use in practical ("light-sticks") and analytical applications. It can be utilized for H_2O_2 detection in a wide variety of analytical and bioanalytical systems and is also useful for peroxide quantification in systems that produce or destroy H_2O_2 [54]. Additionally, several oxalic esters and chemiluminescent activators, which emit in different colors (e.g., rubrene, yellow; perylene, green; 9,10-diphenylanthracene, blue), are commercialized in the form of light sticks and utilized as attractors for fishing, in emergency kits and recreation objects. Although their content is highly cyto- and genotoxic (essentially all activators), they are labeled as safe and no indication for discharge is given. Millions of light sticks utilized for pelagic fishery can be found as beach-discard in northeastern Brazil. Under-educated, naive local population is inadvertently using this highly toxic material in the hope of alleviating joint pains and other finalities [38]. Regarding the believed analgesic properties of the light stick content, it is worth to mention the chlorosalicylate nature of its oxalic ester substrate.

The chemiluminescent oxidation of luminol (5-amino-2,3-dihydrophthalazine-1,4-dione) in aqueous medium is one of the oldest chemiluminescent described and occurs generally in an aqueous medium with hydrogen peroxide as the oxidant; it is catalyzed by hemin and several transition metal cations. The emission is due to 3-aminophthalate dianion fluorescence [5]. The efficiency of the luminol reaction with a quantum yield of around 1% can be considered as moderate, but it is low compared to the peroxyoxalate reaction, yet much higher than many biological chemiluminescent systems. As the reaction can be catalyzed by several metal ions but also by peroxidases like horseradish peroxidase, innumerous analytical and bioanalytical applications have been found, being the first chemiluminescent transformation utilized for the development of commercial, analytical assay kits [58–62]. The reaction in aqueous medium appears to depend on hydrogen peroxide to be efficient [5], and therefore this oxidant can be selectively quantified by luminol chemiluminescence in diverse chemical and biochemical systems [63].

Fluorescence probes for spatial and temporal detection of hydrogen peroxide are commonly related to redox processes of an appropriately reduced probe and hydrogen peroxide. Upon oxidation by the peroxide, the prefluorescence probe is transformed into its fluorescent analogue. Several novel fluorescent probes have been described with the ability to detect hydrogen peroxide with high selectivity, and some of them can be utilized to detect intracellular hydrogen peroxide. The most generally utilized probes for selective hydrogen peroxide detection are aryl boronates, which suffer deprotection/oxidation in the presence of hydrogen peroxide and are thereby transformed into a fluorescent phenol derivative [64,65]. In some cases, it was possible to use chemiluminescent or bioluminescent probes for the *in vivo* detection of hydrogen peroxide, enabling inclusively whole animal studies [66–71].

DIOXETANONE-MEDIATED BIOLUMINESCENCE

Dubois' discovery of the luciferin–luciferase involvement in light emission by click beetles (Coleoptera) had laid dormant for more than two centuries until its chemical mechanism was revealed. In parallel to intense research by biologists to characterize

the anatomy, physiology, habitats, and roles of bioluminescence in mating, preda-
tion, camouflage, and other inter- and intraspecific communication (Figure 2.1),
many chemists have attempted to explicate the molecular mechanisms underneath
the luminescence. At this point, we will focus on luminescent beetles and coelenterazine-dependent systems, two emblematic examples of bioluminescence where the
high-energy intermediate is most likely a 1,2-dioxetanone.

The isolation and structural elucidation of firefly *Photinus pyralis* luciferin, as
well as its chemical synthesis, were reported by White et al. [72]. Soon after, it was
found that the bioluminescence spectrum closely matches the oxyluciferin fluorescence spectrum, the reaction depends on MgATP with the release of CO_2 and AMP,
and the color is modulated by the pH and divalent cations such as Ca^{2+}. The reaction
products are those expected from a 1,2-dioxetanone intermediate, thus suggesting
that it was transiently formed by the addition of molecular oxygen to the α-CH-
CO_2AMP bond in the luciferin thiazole moiety (Figure 2.2e). Firefly luciferase
exhibits two consecutive enzymatic activities. First, it acts as an ATP-dependent synthetase (a ligase) by activating the carboxylic group to an adenylate derivative plus
pyrophosphate, similar to the activation of fatty acids and amino acids biosynthesis
in triglyceride and protein biosynthesis, and second, as a dioxygenase by catalyzing
the insertion of molecular oxygen into the C-4 luciferin. Cyclization of the resulting
α-hydroperoxide by displacement of AMP, a good leaving group, would produce a
1,2-dioxetanone, whose cleavage yields CO_2 plus excited oxyluciferin (Figure 2.2e).

Unequivocal demonstration of a 1,2-dioxetanone intermediate in firefly bioluminescence was reported by Shimomura and Johnson [73] using double-labeling
of CO_2 and oxyluciferin products with $^{18}O_2$ and $H_2^{17}O$, to correct fast exchange of
oxygen atoms between water, CO_2, and oxyluciferin. The operation of a CIEEL
mechanism involving electron transfer from the luciferin phenolate moiety to the
thiazole-dioxetanone ring had been previously proposed by Koo et al. [35].

The modulation of the emission color—from green in click beetles (Elateridae) to
red in railroad worms (Phengodidae)—by the partition of the chemiexcitation energy
between the ketone and enolate forms of oxyluciferin has been shown to be controlled by the presence of Arg and Glu residues and the content of water in the active
site [74]. Luciferin is the same in all luminous beetles and putatively biosynthesized
from *p*-benzoquinone and cysteine (Figure 2.2e) [75–77]. Design and isolation of
numerous luciferase mutants able to elicit diverse colors have allowed their use as
gene reporters in bioengineering and to engender powerful analytical tools in biology, medicine, and environmental sciences. Important features of firefly bioluminescence are its high quantum yield of light emission (~45%) and ATP dependence that
can be used to quantify ATP in food as well as biological, environmental, and clinical
samples below femtomolar concentrations [78].

Interestingly, hyperoxia imposed on larval elaterid *Pyrearinus termitilluminans*
induces CuZn- and Mn-superoxide dismutase, luciferase, and catalase activities in
its prothoracic lantern, but not as much in the dim segments of the larva, which suggests cooperative participation of these enzymes in the protection of the larva against
redox imbalance [79]. *In vivo* EPR experiments carried out with paramagnetic lith-
ium phthalocyanine implanted in the prothorax of *P. termitilluminans* larvae revealed
about 20% consumption of the local dioxygen for the larva to shine when lured with a

termite prey [53]. In this regard, larvae of phylogenetically close nonluminescent bee-tles such as the mealworm *Tenebrio molitor* (Coleoptera: Tenebrionidae) were shown to possess a ligase activity that mimics luciferase by eliciting red light emission upon injection of firefly luciferin [80]. This supports the hypothesis that luciferase evolved from AMP-CoA ligases by acquiring an oxygenase with luminogenic property [81].

Nitric oxide was reported to control the light flashing by fireflies through the induction of a NO synthase activity in the terminal tracheoles surrounding the lan-terns [82]. Nitric oxide diffusion to the lantern mitochondria toward and binding to cytochrome oxidase would cause its temporary inhibition. The sudden local increase of oxygen concentration could then trigger the millisecond flash of light. This mecha-nism, however, is not compatible with the continuous light emission by firefly larvae during prey attraction, nor does it explain the second lasting light emission by flying elaterids for mating. In addition, with the respiratory chain blocked by NO·-inhibited cytochrome oxidase, there is no production of the ATP essential for the biolumines-cent reaction [83].

Many marine bioluminescent organisms use coelenterazine as a substrate to pro-duce light (Figures 2.1g and 2.2f). Among these living beings, only the luciferase of the sea pansy *Renilla reniformis* and the copepods *Gaussia princeps* and *Metridia longa* have been frequently applied for research on cancer and neural circuits, viral and bacterial diseases, drug screening, and immunoassay development, among others [78]. Unlike the jellyfish *Aequorea victoria*, whose bioluminescence requires a photo-protein as the energy acceptor from initially excited oxyluciferin, these three marine creatures emit light upon a classic luciferin–luciferase reaction. It is noteworthy that photoproteins are formed by the reaction of the apophotoprotein with luciferin in the presence of molecular oxygen yielding a stable luciferin hydroperoxide, which is attached to the protein. The light emission from a photoprotein is triggered by a metal cation (e.g., calcium ions in the case of coelenterazine-dependent systems), and the protein has a very low turnover. On the other hand, most luciferases react directly with luciferins in the presence of molecular oxygen according to classical principles of enzymology and present a much faster turnover. Whether a 1,2-dioxetanone intermediate is involved in a photoprotein or luciferase-mediated reaction still has to be unequivocally demonstrated. The first step of high-energy intermediate forma-tion probably involves the reaction of molecular oxygen with coelenterazine, yield-ing a coelenterazine hydroperoxide, whose cyclization promoted by calcium ions putatively leads to a 1,2-dioxetanone. Firefly and coelenterazine-based systems are two more examples where the CIEEL sequence is postulated to be involved. The decomposition of a 1,2-dioxetanone from either the firefly luciferin or coelenterazine can generate the respective oxyluciferin in its electronic excited state and release CO_2. As proposed by the intramolecular CIEEL pathway, the O–O bond of luciferin–dioxetanone moiety cleaves upon an electron transfer from the electron-rich part of the molecule yielding a radical anion, whose subsequent C–C cleavage leads to the formation of $CO_2^{·-}$ and the radical cation of oxyluciferin. Carbon dioxide radical anion has a free energy change of 235 kJ mol^{-1}, estimated by its half-wave reduction potential [57]; this energy is likely to be high enough to transfer its electron back to the low-energy SOMO orbital of oxyluciferin radical, thus rendering singlet oxylu-ciferin in the fluorescent, first electronically excited singlet state.

TRIPLET CARBONYLS, A NEGLECTED OXYGEN-DERIVED REACTIVE SPECIES

Independent of one another, Cilento [84–86], Lamola [87], and White et al. [88] raised the hypothesis of the generation of typically photochemical products in tissues of living organisms not exposed to light, which was coined "photochemistry in the dark" or "photochemistry without light." Moreover, the biosynthesis of some of the putative "photo" products reported in the literature disobeyed the symmetry orbital rules of Woodward–Hoffman, unless in the electronically excited state, such as [2 + 2]-cycloadditions yielding cyclobutanes. According to these researchers, energy transfer from chemically and enzymatically generated excited carbonyls to several biological acceptors could drive photochemical processes in the dark. Triplet ketones and aldehydes were of particular interest due to their long lifetimes (>µs), radical-like reactivity, and efficient and easy production from 1,2-dioxetanes and other per-oxides. Nevertheless, triplet carbonyls are low-emissive species that are difficult to detect for their low quantum yield of phosphorescence and high rate constant of quenching by dioxygen ($k_q \sim 10^8$ M^{-1} s^{-1}) when generated in aerated solvents.

Triplet carbonyls can be considered as diradicals, presumed from the distorted structure of this sort of electronic excitation, formally presenting a solitary electron in each atom of the carbonyl bond [89]. Indeed, both radicals and triplets undergo oxygen suppression, isomerization, (cyclo)addition, polymerization, α- and β-cleavage, and hydrogen abstraction reactions (Figure 2.4). For example, the *tert*-butoxy radical suf-fers β-cleavage to acetone and methyl radical; can abstract hydrogen from alcohols and unsaturated fatty acids, thereby initiating their peroxidation; and adds to olefins, thus

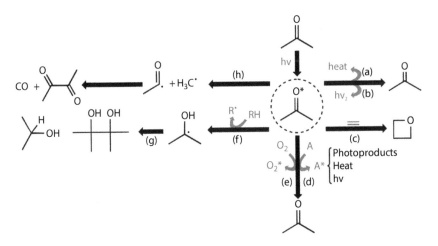

FIGURE 2.4 Photophysical and photochemical fates of triplet acetone: (a) heat dissipation; (b) light emission (hν_2); (c) [2 + 2]-cycloaddition to olefins; (d) energy transfer to adequate acceptors such as DBAS; (e) energy transfer to ground state dioxygen yielding singlet oxygen; (f) hydrogen abstraction from several biomolecules, include polyunsaturated fatty acids; (g) generation of isopropanol and pinacol; (h) Norrish type I (α-cleavage) yielding methyl and acetyl radicals followed by diacetyl generation and decarbonylation [141] and ref. therein.

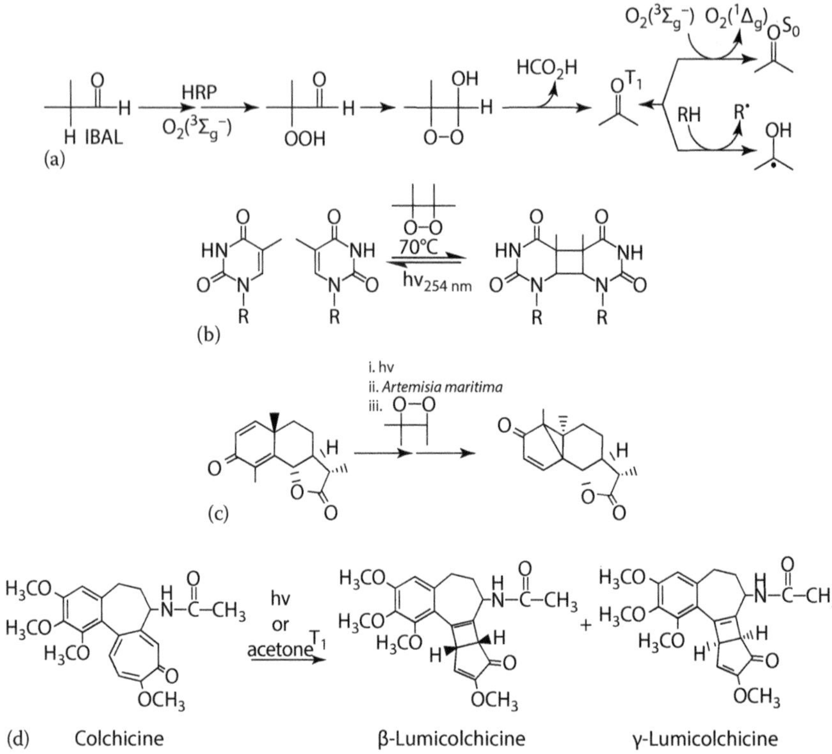

FIGURE 2.5 Literature examples of "photochemistry in the dark" [141] and ref therein. (a) Generation of triplet acetone and singlet oxygen by HRP-catalyzed aerobic oxidation of isobutanal (IBAL). (b) Thymine dimerization to cyclobutane derivatives achieved by tetramethyl-1,2-dioxetane thermolysis. (c) Isomerization of santonin to lumisantonin induced by light or 3,3,4-trimethyl-1,2-dioxetane decomposition, as well as in the plant *Artemisia*. (d) Conversion of alkaloid colchicine to lumicolchicines under either irradiation, exposure to triplet acetone generated by 3,3,4-trimethyl-1,2-dioxetane thermolysis or in the dark within *Colchicum autumnale* corms infused with radiolabeled ^{14}C-colchicine.

forming radical adducts that can be trapped by organic nitroxyls [90]. Triplet species overwhelm chemically free radicals as a result of their ability to transfer energy to adequate acceptors (e.g., polycondensed aromatics, xanthene dyes, chlorophyll, flavins) by either the long-range Foster mechanism or collisional quenching leading to their excited states from which light emission or diverse photochemistry can arise [91–93].

Further discussed at this point are the emblematic historical studies employed to validate the hypothesis of "photochemistry in the dark" driven by triplet acetone-generated dioxetanes (Figure 2.5). The classical photochemical rearrangement of santonin into lumisantonin was accomplished by White and Wei [94] also in the absence of light by coupling the "photo"chemical reaction to the thermolysis of trimethyl-1,2-dioxetane (Figure 2.5c). Hypothetically, the santonin isomerization occurred by a triplet–triplet energy transfer process from excited acetone, similar to the UV-induced rearrangement of 4,4-diphenylcyclohexadienone. Noteworthy is the fact that both compounds, santonin and its photoproduct, are found in leaves and flowers of absinthe (*Artemisia* spp.), although whether this is really a "dark" process in the plant has not been reported to the best of our knowledge.

A very simple, clever, and biologically relevant experiment conducted by Lamola [87] was the attempt to dimerize thymine to its cyclobutane dimer by heating a mixture of trimetyl-1,2-dioxetane with ^{14}C-2-thymine labeled *Escherichia coli* DNA in phosphate buffer pH 7.0 at 70°C (Figure 2.5b). This would imply that the dioxetane-generated triplet acetone collides with DNA thymine residues leading to its dimer. Accordingly, the radiochromatogram of the spent reaction mixture presented a peak with retention time assignable to thymine dimer, which disappeared upon UVB irradiation at 254 nm due to a retro-[2 + 2]-cycloaddition reaction as expected, based on the literature. Lamola then hypothesized that "dark" cyclobutane pyrimidine dimerization (CPD), known to be mutagenic and eventually carcinogenic, may take place in cells not exposed to UV light and accentuated that, if real, "this would require the presence of DNA repair systems which are not exposed to light."

Additional support for the dark photochemistry hypothesis was provided by Turro and Lechtken [94], who reported *trans*-1,2-dicyanoethylene (DCE) isomerization to its *cis*-configuration and [2 + 2]-cycloaddition to excited acetone, yielding an oxetane when heated together with tetramethyl-1,2-dioxetane. The DCE isomerization would result from energy transfer from triplet acetone and the Paternó–Büchi cycloaddition with singlet acetone. In addition, White et al. [88] suggested that the cyclobutane dimerization of plant cinnamates to truxillates, both found in *Erythroxylum coca*, might proceed via a similar mechanism.

In the late 1970s, Cilento and coworkers reported the first dark "photo"chemical experiments using enzyme-catalyzed reactions as efficient sources of triplet ketones. They demonstrated that micromolar horseradish peroxidase (HRP) together with traces of H_2O_2 can drive weakly luminescent reactions of millimolar α-alkyl- and α-aryl-substituted propanal, as well as ethanoic and 3-oxobutanoic acids in normally aerated phosphate-buffered solutions (Figure 2.5a). Some of these reactions could be monitored only in a liquid scintillation counter or upon addition of a fluorescent acceptor [27].

The most reactive substrates were found to bear a carbonyl-activated hydrogen atom and an α-alkyl or α-aryl substituent, hence prone to hydrogen abstraction by HRP-I, followed by dioxygen attack and, consecutively, leading to α-hydroperoxide, 1,2-dioxetane, and cleavage products. In most cases, (1) the reaction products are spectrophotometrically similar to those expected from a 1,2-dioxetane intermediate, (2) the light emission is intensified by triplet counters to values detectable in a spectrofluorimeter, and (3) chemical modification of the enzyme is inferred from the loss of catalytic activity. Interestingly, myoglobin was shown to catalyze the aerobic oxidation of acetoacetate leading to methylglyoxal probably in the excited state as the observed protein damage was similar to that caused by methylglyoxal irradiation at 270 nm [95]. This reaction was later revisited using EPR spin trapping techniques and confirmed to be initiated by α-hydrogen abstraction from acetoacetate anion by ferrimyoglobin/H_2O_2, yielding the secondary radical and propagated by oxygen insertion [96]. Ultraweak chemiluminescence occurred, and loss of secondary and tertiary myoglobin structure was revealed by its characteristic CD spectra.

Also noteworthy is the Zinner et al. [97] report that HRP/Mn^{2+}, in aerated medium, catalyzes the oxidation of the tuberculostatic drug isoniazid to pyridine-4-carboxal-dehyde and N$_2$, presumably through a diazene intermediate. Direct light emission by the product was not detected at room temperature as expected; however, the reaction sensitized the fluorescence of added eosin, a xanthine dye. The two bromine atoms in the eosin molecule allow triplet energy transfer due to the so-called heavy atom effect, although one cannot exclude hydrogen abstraction by radical intermediates yielding eosin semiquinone, whose dismutation has long been reported to be luminescent in alkaline eosin solutions in the presence of H$_2$O$_2$ [98].

1980s ADVANCES IN THE GENERATION OF TRIPLET CARBONYLS IN THE DARK AND BEYOND

The first evidence for the *in vivo* operation of a "dark" photochemical process came from the study of the intramolecular 1,4-cycloaddition of the tropolone ring of the alkaloid colchicine into α- and β-lumicolchicines in underground corms of meadow saffron (*Colchicum autumnale*) [99] (Figure 2.5d). *Colchicum* is a short-day plant whose extracts have been used to treat gout since ancient times. In the winter, ^{14}C-colchicine was infused through the plant underground corms, and their extracts were prepared two days thereafter in a dark room illuminated with a red lamp. Using HPLC, radiolabeled β- and γ-lumicolchicines, respectively, the colchicine *trans*- and *cis*-cyclobutene isomers, reportedly formed by exposure of colchicine to daylight, were detected in the corm extracts by HPLC. This reaction is expected to proceed via the disrotatory electrocyclic ring closure of the tropolone moiety, and forbidden to take place in the ground state. Previous studies by TLC with colchicine and tetramethyl-1,2-dioxetane in the dark suggested that colchicine "photo"isomerization is induced by dioxetane-generated triplet acetone. Instead, flash photolysis of colchicine revealed that it requires excitation to the singlet state to be isomerized [100]. Whether a *Colchicum* dioxetane metabolite intermediate acts as a source of excited carbonyls to "photo"activate colchicine is a challenging question that deserves further investigation.

The IBAL/HRP/H$_2$O$_2$ reaction is undoubtedly the most explored model source of triplet carbonyls, namely, triplet acetone, from mechanistic, chemiluminescence, and reactivity points of view (Figure 2.5a) [92]. Triplet acetone was identified by (1) its chemiluminescence spectrum ($\lambda_{max} \sim 430$ nm), which matches the phosphorescence spectrum of acetone; (2) energy transfer to the DBAS attested by the enhancement of the fluorescence emission; (3) quenching by *trans,trans*-sorbate, a water-soluble conjugated diene; (4) formate and acetone analysis as the main products, thereby supporting the intermediacy of a 1,2-dioxetane intermediate; (5) detection of iso-propanol and pinacol as IBAL by-products possibly formed by hydrogen abstraction from the sugar portion (18%) of HRP; and (6) phosphate acid–base catalysis of IBAL enolization to a more easily oxidized substrate [101,102]. Accordingly, the stable IBAL silyl enol ether derivative reacts much faster with dioxygen than IBAL itself in the presence of HRP/H$_2$O$_2$ [103].

Also noteworthy were the attempts to generate acetophenone and benzophe-none at the triplet state by means of HRP/H$_2$O$_2$-catalyzed aerobic oxidation of

2-phenylpropanal (PPA) and diphenylacetaldehyde (DPPA), respectively [104]. The observed PPA-elicited ultraweak chemiluminescence in the blue region was amplified by DBAS and that from DPPA emitted red light ($\lambda_{max} \sim 630$ nm). The latter chemiluminescence was then attributed to energy transfer to dioxygen yielding singlet oxygen, whose visible bimolecular emission reportedly peaks at 634 and 703 nm. In parallel, the enzyme-generated triplet benzophenone was shown to damage DNA, phospholipids, and proteins of isolated rat liver mitochondria [105].

3-Methylacetoacetone (MAA) treated with HRP was shown to emit orange light due to the production of triplet diacetyl ($\lambda_{max} \sim 520$ nm, shoulder ~ 550 nm), ten times longer lived than triplet acetone ($\tau^0 \sim 1$ μs in deaerated water) [106]. In turn, the catabolites acetoacetate and 2-methylacetoacetate, known to be overproduced in diabetes and isoleucinemia, respectively, underwent myoglobin-catalyzed aerobic oxidation to either excited methylglyoxal or diacetyl [96]. A secondary carbon-centered radical originated by ^{13}C-labeled acetoacetate was detected by EPR hyperfine spin trapping with methylnitrosopropane, MNP ($a_N = 1.46$ mT), thereby confirming the operation of a mechanism initiated by α-hydrogen abstraction from the substrate. Using 2-methylacetoacetate, as expected, the EPR spectrum displayed two MNP spin adducts assignable to a tertiary carbon-centered radical and to the acetyl radical ($a_N = 1.54$ and 0.83 mT, respectively), the latter formed from α-cleavage of the excited diacetyl product.

Importantly, the HRP-catalyzed aerobic oxidation of n-pentanal lead to acetone and ethylene, a plant hormone, both putatively originated by β-cleavage of the triplet n-butanal produced by a dioxetane intermediate followed by γ-hydrogen abstraction (Norrish type II reaction) [107].

Moreover, phosphate-induced permeabilization of isolated rat liver mitochondria was suggested to be mediated and amplified by triplet carbonyls formed during membrane lipid peroxidation [108]. Accordingly, mitochondrial swelling, H_2O_2 production, and associated lipid peroxidation were only partially prevented in the presence of butylhydroxytoluene and cyclosporine, but strongly inhibited upon addition of the sodium sorbate anion, a collisional quencher of triplet carbonyls. For comparison, sorbate addition was also capable of blocking concentration-dependent phosphate-catalyzed peroxidation of phosphatidylcholine/dicetyl-phosphate liposomes.

Indirect strategies are helpful for suggesting the nature of reaction intermediates or final productions of non- or low-emissive triplet carbonyls in chemical and biological systems. Among them, we mention: (1) depending upon the intensity of light emission, to use either a liquid scintillation counter, a photon-counting, a simple spectrofluorimeter, or a luminometer; (2) when possible, to match the spectral distribution of chemiluminescence with that of photoexcited product; (3) enhancement of ultraweak chemiluminescence by water-soluble highly fluorescent acceptors such as sodium DBAS for triplets (k_{ET}, electron transfer rate constant $\sim 10^9$ M^{-1} s^{-1}) and sodium 9,10-diphenylanthracene-2-sulfonate (DPAS) for singlets, with (4) concomitant determination of $k_{ET} \times \tau$ values from double reciprocal plots of intensity versus acceptor concentration, where τ is the lifetime of the excited donor [109]; (5) to verify whether the products are those expected from a 1,2-dioxetane intermediate (Figure 2.4a); (6) to look for adducts of radical intermediates of the studied reaction using EPR nitroxide or nitroso spin traps [110]; (7) to identify typical photoproducts in the final reaction mixture expected from photocleavage, isomerization, dimerization, or hydrogen abstraction from adequate donors (Figure 2.5a); and (8) to conduct Stern–Volmer quenching studies of triplet donors by sorbate (*trans, trans*-2,4-hexadienoate) salts or esters ($k_q \sim 10^9\,M^{-1}\,s^{-1}$) and search for their *trans, cis-, cis, cis-, and cis, trans*-isomers in the spent reaction mixtures.

CLASSES OF CARBONYL SUBSTRATES PRONE TO GENERATE EITHER EXCITED SPECIES OR REACTIVE SPECIES

The oxidation routes undertaken by carbonyl metabolites or their models in aerated media have been shown to bifurcate to a final production of either excited products or reactive oxygen species [111] and ref. therein (Figure 2.6).

On one hand, carbonyl compounds (aldehydes, ketones, esters) in their phosphate-catalyzed enol form, exposed to strong oxidants (e.g., HRP/H_2O_2, peroxynitrite), tend to undergo α-electron abstraction to form a carbon-centered radical (Figure 2.6, Class I). This could then be followed by dioxygen insertion to the radical, cyclization to a hypothetical 1,2-dioxetane, and finally cleavage to two products, one of them in the triplet state (Figure 2.6). This is the case of metabolites like IBAL, *n*-pentanal, 2-phenylpropanal, acetoacetic acid, indoleacetic acid, and phenylacetic acid [92]. Some of these compounds or their analogs are known to occur in biological systems, such as indoleacetic acid (a plant hormone) and acetoacetic acid (a ketonic body) accumulated in ketoacidoses.

Otherwise, the presence of a hydroxyl or amino group at the α-carbon will lead to an enediol, whose hydrogen abstraction yields a resonant enolyl radical analogous to an *ortho*-semiquinone. Further electron transfer to dioxygen is expected to ultimately yield $O_2^{\cdot-}$ radical, H_2O_2, HO$^{\cdot}$ radical, and an α-dicarbonyl product, which is a biologically reactive electrophile toward proteins and DNA (Figure 2.6, Class II) [112]. This mechanism has been proposed to operate in the metal- or xanthine/xanthine oxidase-catalyzed aerobic oxidation of dihydroxyacetone phosphate leading to methylglyoxal [113].

Both classes where the carbonyl or the α-hydroxy atom is replaced by nitrogen, in imines (Schiff bases) or α-amino carbonyls, respectively, exhibit similar chemical behavior. This was reported for the conjugate of either albumin, lysozyme, or protamine with glyoxaldehyde exposed to HRP in aerated buffer, accompanied by ultraweak chemiluminescence [114] (Class I), and for hemoglobin-treated 5-aminolevulinic acid, a heme precursor accumulated in porphyric disorders (Class II) [110]. Accordingly, significantly elevated plasmatic levels of SOD and glutathione peroxidase were measured in both latent and symptomatic carriers of intermittent acute porphyria [115].

CHEMICAL AND ENZYMATIC "DARK" GENERATION OF SINGLET OXYGEN

Singlet oxygen was discovered by Kautsky and De Bruijn [116] as a "reactive, metastable state of the oxygen molecule" capable of promoting the photooxidation of dyes sensitized by fluorescent compounds like trypaflavine, hematoporphyrin,

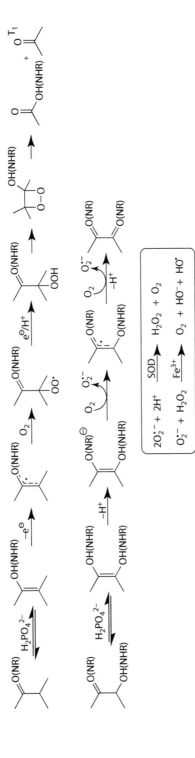

FIGURE 2.6 Potential sources of excited states and radicals from carbonyl and imine metabolites.

and chlorophyll. Using a creative experiment with a solid aerated mixture of trypafla-vine and green leucomalachite adsorbed in silica beads, they noted that the mixture turned blue after 10 min of illumination. Under a nitrogen atmosphere, however, the experiment was unsuccessful. Decades later, singlet oxygen ($O_2\,^1\Delta_g$) was identified as the red-emitting product ($\lambda_{max} \sim 634$ nm) from the reaction of H_2O_2 with hypo-chlorite. Soon, the 634 nm band and another one at 703 nm were assigned to the bimolecular decay of singlet oxygen, and a near-infrared (NIR) emission at 1268 nm was identified as its monomolecular emission [117]. Henceforth, singlet oxygen has been synthesized by a plethora of chemical and photosensitized processes and implicated in beneficial (e.g., phagocytosis, photodynamic therapy, cell signaling) and pathogenic responses (e.g., drug photosensitivity, inborn protoporphyria IX, skin burns, and mutagenesis) [118,119]. The highly electrophilic properties and long lifetime of singlet oxygen (microseconds in water to over 100 µs in organic solvents) imbues it with the ability to promptly react with unsaturated compounds yielding either 1,2-dioxetanes by 1,2-cycloaddition to olefins deprived of α-hydrogen atoms such as tetramethoxyethylene and bis-adamantene; or "ene" hydroperoxides from 1,3-addition to alkyl-substituted olefins; or 1,4-endoperoxides from conjugated dienes (Diels–Alder type reaction). Histidine, nucleotides, unsaturated fatty acids, furanes, tocopherols, anthracenes, tertiary amines, organic sulfides, and many other synthetic and endogenous compounds are reportedly targets of singlet oxygen.

The chemical generation and biological roles of singlet oxygen *in vivo* are very often polemic matters due mainly to its short lifetime and low biological concentra-tions, despite numerous reports on its involvement in lipid, amino acid, and nucleo-tide peroxidation; lipoxygenase-, myeloperoxidase-, and cyclooxygenase-catalyzed reactions; and hemeprotein-catalyzed reactions with excess H_2O_2 [120]. Quenching by carotenoids, azides, and histidine were long suggested as tools to test singlet oxygen generation in biological systems [121]. Detection in D_2O medium was also shown to increase 20 times its NIR phosphorescence emission at ~1270 nm. The most efficient carotenoid to quench singlet oxygen is lycopene, which is found in tomatoes and other vegetables and is reported to prevent prostate cancer [122,123].

Numerous chemical sources of singlet oxygen have been described, and among them the reaction of $ONOO^-$ with H_2O_2 [124]. The observed chemiluminescence peaking above 610 nm (dimol emission) and at 1270 nm (monomol emission) enhanced in D_2O and quenching by sodium azide attested to the formation of excited dioxygen. This reaction was then suggested to contribute to the cytotoxicity mediated by H_2O_2 and $ONOO^-$. Later, the reaction at near-neutral pH was shown to proceed via peroxynitrate (O_2NOO^-) as intermediate, whose decomposition ultimately pro-duces singlet oxygen and nitrite [125]. Independently, Miyamoto et al. [51] reported that the decomposition of $ONOO^-$ or O_2NOOH at neutral to alkaline pH generates $O_2(^1\Delta_g)$ in a yield of ca. 1% and 2%–10%, respectively.

$$ONOOH + ONOO^- \rightarrow NO_2^- + O_2NOO^- + H^+ \rightarrow 2\,NO_2^- + O_2\,[O_2(^1\Delta_g)]$$

In recent years, production of singlet oxygen via hydroperoxides of proteins, lipids, and nucleotides by the Russell mechanism when treated with metal ions, peroxynitrite, HOCl, and cytochrome *c* has been well documented [50]. These

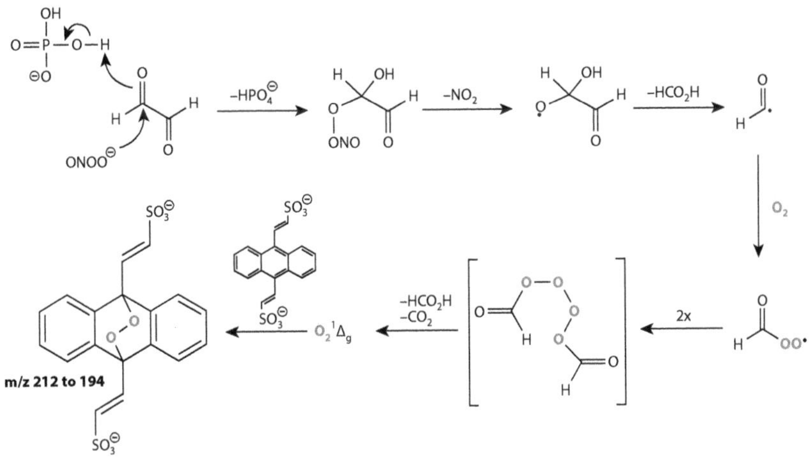

FIGURE 2.7 Chemical quenching of singlet oxygen generated from the glyoxal/peroxynitrite with AVS, in aerated D_2O phosphate buffer enriched with $^{18}O_2$ (marked in red), followed by MS identification of the product.

findings have mobilized efforts to disclose the involvement of such processes *in vivo*, both normal and pathogenic.

Taking for granted that (1) alkyl radicals bearing a geminal hydrogen atom undergo dimolecular annihilation by the Russell reaction yielding up to 3%–14% of singlet oxygen [46,47], and (2) peroxynitrite adds to α-dicarbonyl compounds (e.g., diacetyl, methylglyoxal) ultimately rendering acyl radicals and their carboxylic acid products [126,127], the reaction of glyoxal with peroxynitrite was investigated as a candidate metabolic source of singlet oxygen [128] and ref. therein. Indeed, nucleophilic addition of peroxynitrite to glyoxal was shown to generate formate as the main product and singlet oxygen as by-product identified by its characteristic NIR emission at 1270 nm, quenching by azide and histidine, and chemical trapping of $^{18}O_2(^1\Delta_g)$ with anthracene-9,10-divinylsulfonate (AVS), using HPLC/MS/MS for detection of the corresponding 9,10-endoperoxide derivative (Figure 2.7).

Other glyoxals of biological interest are diacetyl and methylglyoxal, both related to "carbonyl stress" or "electrophilic stress" and putatively linked to aging-associated disorders (obesity, diabetes, cardiopathy, neurodegeneration) and environmental toxicants [112,129,130]. These dicarbonyls originate from the oxidative degradation of carbohydrates and lipids and undergo nucleophilic addition of basic protein amino acid residues, leading to protein cross-linking and "advanced glycation end products" (AGES) and advanced lipid end products (ALES) by the Maillard reactions. They also add to nucleobases, forming ethane and propane adducts related to mutagenesis and carcinogenesis [131–134]. Recently, both diacetyl and methylglyoxal were reported to undergo nucleophilic addition of peroxynitrite (($k_2 \approx 1.0 \times 10^4$ and 1.0×10^5 M^{-1} s^{-1}, respectively), thus yielding the acetyl radical from the homolysis of peroxynitrosocarbonyl adduct intermediates, and acetate or acetate plus formate ions, respectively [126,127]. In the presence of amino acids (His, Lys), and Lys-containing tetrapeptides, the dicarbonyl reaction could be coupled to the acetylation of both α- and ε-amino acid groups [135]. Whether acetyl radical-promoted posttranslational modification of peptides and proteins occurs is a matter of speculation that deserves further investigation.

Singlet oxygen ($^1\Delta_g$) was also unequivocally demonstrated to be generated by triplet–triplet energy transfer from triplet acetone produced by either tetramethyl-1,2-dioxetane thermolysis or the IBAL/HRP system to ground state dioxygen ($^3\Sigma_g$). Singlet oxygen was detected in the NIR region at 1270 nm and trapped by water-soluble anthracene-9,10-diyldiethane-2,1-diyl-disulfate (EAS). HPLC-ESI-MS/MS studies with $^{18}O_2$ corroborated the chemical trapping of dioxygen yielding the corresponding double-[^{18}O]-labeled 9,10-endoperoxide (EAS$^{18}O_2$).

TRIPLET CARBONYLS AND SINGLET OXYGEN ARE INVOLVED IN MELANOGENESIS?

Oxidative damage of epithelial cells was recently demonstrated to occur under irradiation with visible, and not only UV, light. The skin pigments—eumelanin in dark individuals, and pheomelanin in those with blond or red hair—are long known to absorb sunlight and protect the skin against sunburns, DNA oxidations, base dimerization, and eventually skin cancer. Thermal relaxation of electronically excited melanin to the ground state provides skin photoprotection. Paradoxically, Chiarelli Neto et al. [136] found that a fraction of pheomelanin in the triplet state, more efficiently than eumelanin, transfers energy to ground state ($O_2 \, ^3\Sigma_g$) yielding singlet oxygen ($O_2 \, ^1\Delta_g$), which diffuses throughout the cells causing extensive damage. 2'-Deoxyguanosine oxidation by excited dioxygen and DNA single-strand breaks were verified denoting concomitant operation of triplet melanin-initiated type I radical reactions (Foote type I photosensitization). The authors reported decreased melanocyte cell viability, and augmented membrane permeability, DNA oxidation, and necro-apoptotic cell death. Premutagenic DNA lesions that may trigger photoaging and skin cancer were also observed.

Highly intriguing effects of UVAB light on melanocytes have recently been revealed by [137]. Specifically, DNA pyrimidine dimerization induced by UVA light persisted for 3–4 hours "in the dark," after irradiation. That melanin was implicated in the formation of cyclobutane pyrimidine dimers (CPDs) after irradiation was indicated by the lack of "dark" pyrimidine dimerization in fibroblasts and albino melanocytes and diminished dimerization in the presence of kojic acid known to inhibit melanin synthesis. In addition, decreased "dark" CPDs in melanocytes were found by inhibiting NADPH oxidase and nitric oxide synthase, sources of $O_2^{\cdot-}$ and NO$^\cdot$ radicals. Nitrotyrosine-containing proteins were imaged in the cell nuclei of dark, but not in albino, melanocytes. These data confirmed peroxynitrite as a mediator of "dark" pyrimidine dimerization. Particularly noteworthy in addition to these findings was that the dimer ratio (TC + CT)/TT was found to be 10%–30% under direct UVA and 3–4 times higher in the delayed CPDs. Accordingly, augmented CPDs resulted from silencing the *Xpa* or *Xpc* excision repair systems. Also important is that cytosine-containing CPDs are reportedly the UV signature of carcinogenic C $\rightarrow$ T mutations putatively involved in melanoma development (Figure 2.8).

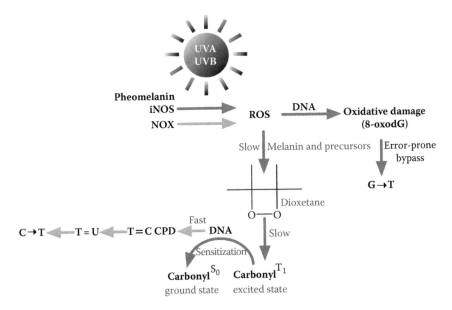

FIGURE 2.8 "Dark" generation of cyclobutane pyrimidine dimers (CPDs) for several hours after melanocyte exposure to UVA and UVB light. The dimers ratio (TC+CT)/TT is 10%–30% under direct UVA, but increases fourfold after irradiation and lasts for 3–4 hours. DNA C → T mutations are putative UV-signature DNA for melanoma.

With respect to the reaction mechanism of "dark" generation of CPDs, a proposal came out from the "photochemistry without light" hypothesis. First, the observed ultraweak chemiluminescence detected from the melanocyte cultures was enhanced by DBAS, a fluorescent triplet carbonyl acceptor, with concomitant abatement of CPD formation. Pyrimidine dimerization was also partially blocked upon addition of sorbate, a collisional quencher of triplet carbonyls. Melanin granules surrounding the nucleus of melanocytes not exposed to UVA were found to cross the nucleus membrane. Accordingly, the HRP- or peroxynitrite-triggered aerobic oxidation of the melanin precursor 5,6-dihydroxyindole-2-carboxylic acid and caused the cleavage of the pyrrole ring to a product analogous to the oxidation of tryptophan to kynurenine.

In summary, the mechanism of melanin-induced pyrimidine dimerization in the dark was envisaged as resulting from peroxynitrite-promoted oxidation of melanin polymer, whose fragments diffused toward the nuclear volume. Hydrogen abstraction from the dihydroxyindole moiety by nuclear peroxynitrite to a resonance-stabilized melanin-derived radical could be followed by molecular oxygen addition to the catechol or pyrrole moiety of the indole ring and cyclization of the peroxy-radical formed to a hypothetical 1,2-dioxetane. Spontaneous thermolysis of the 1,2-dioxetane intermediate is expected to produce a triplet carbonyl product, which can excite DNA thymine or cytosine residues and ultimately form CPDs. Accordingly, melanin exposed to H_2O_2 has been previously reported to undergo degradation via a 1,2-dioxetane intermediate and triplet carbonyl products [138,139].

Altogether, the data discussed here reinforces the notion that melanin is protective against both visible and UV sunlight, but nevertheless, concomitantly triggers adverse responses that continue to cause mutagenic and carcinogenic lesions after sun exposure.

More recently, "photochemistry in the dark" was also implicated by Kamal and Komatsu [140] in the ultraweak chemiluminescence and other stress protein responses observed from the roots of soybean plants submitted to continuous flood conditions, either under light or dark exposure. Root light emission measured with a photocounter significantly increased in the second day of flooding exposure while decreasing in the following few days. Combined protein damage and light emission were suggested to be originated by reactive oxygen species (H_2O_2, and $O_2^{\cdot-}$ and $HO^{\cdot}$ radicals) and chemically excited species produced in the absence of light such as singlet oxygen and carbonyls. The modified proteins associated with biophoton emission were related to ascorbate peroxidase and H_2O_2 changes and identified by proteomics analysis. Decreased malate, lactate, and succinate dehydrogenase activities were measured, whereas the lysine-ketoglutarate reductase increased in both light and dark conditions, leading to the conclusion that both chemical or enzymatic electronic "photo"excitation of products may take place in plant flooding stress.

FINAL REMARKS

Dioxygen, hydrogen peroxide, and organic peroxides are the oxidants of many luminescent reactions. Those involving excitation of the product to the singlet, fluorescent state can often be appreciated visually, as is the case of bioluminescent processes and luminol, acridine, and oxalate chemiluminescence. On the other hand, excited triplet products are poor emitters and promptly quenched by dioxygen, thus requiring addition of adequate light enhancers like DBAS to make possible their detection.

Although triplet aldehydes and ketones are as reactive as free radicals, long-lived (microseconds), and endowed with the potential to participate in biological processes, they have been disregarded in the studies of oxidative and electrophilic stresses. This is also the case of singlet oxygen that despite being investigated for several decades, the knowledge accumulated about its biological sources, targets, and responses is still insufficient. The development of commercial ultrasensitive photomultipliers to register their *in vivo* ultraweak chemiluminescence spectra, both in the visible (for triplet carbonyls) and near-infrared regions (singlet oxygen), has been accomplished only in more recent years. However, if these equipment are not available, several analytical tools can be employed to visualize excited carbonyl triplets in biological samples, including the amplification of light emission by adequate highly fluorescent triplet energy acceptors such as water-soluble DBAS and emission quenching by conjugated dienes like sorbates. In the former case, double reciprocal plots of reaction rates and temperature allow the estimation of rate constants and lifetimes of the triplet donors [27] and references therein. On the other hand, in the presence of, for example, *trans,trans*-sorbic acid, the detection of the *cis,trans*, *trans,cis*, and *cis,cis*-sorbic acids attests to their chemiexcitation origin. Additionally, the detection of typical photoproducts via the Norrish cleavage from excited carbonyls, produced chemically or enzymatically, can also be argued as evidence for their existence. The radical-like and energy transfer properties of triplet carbonyls produced by peroxidase-catalyzed reactions could well be harnessed to design organic synthetic routes involving carbon-carbonyl cleavage.

In this article, several model reactions support the proposal of triplet carbonyls involvement in biological processes—for example, the HRP-catalyzed production of isopropanol and pinacol from isobutanal, the phosphate-induced permeabilization of mitochondria, ethylene hormone generation from *n*-pentanal, methylglyoxal excitation by acetoacetate oxidation. However, only a few cases of *in vivo* "photochemistry in the dark" were reported, among them, the conversion of colchicine to lumicolchicines in *Colchicum* corms, pyrimidine cyclobutane dimerization in melanocytes, and ultraweak light emission by soybean roots under flooding stress. More literature inspection, insight, and experimental efforts are required to explain the presence and possible roles of "dark" photoproducts in living organisms.

ACKNOWLEDGMENTS

We here pay a tribute to Dr. Christine Winterbourn for her seminal contributions to the understanding of the molecular bases of the cell redox balance, long-lasting collaboration with Brazilian research groups, and constant attention and friendship. We thank the Fundação de Amparo à Pesquisa do Estado de São Paulo FAPESP (EJHB 2006/56530-4, WJB 2014/22136-4, CVS 2013/16885-1), the Conselho Nacional de Desenvolvimento Científico e Tecnológico CNPq (EJHB 302326/2011-1), and the FAPESP/INCT Redoxoma (EJHB 573530/2008-4) for financial support. We also thank Mr. Timothy Palmer for reading this manuscript.

REFERENCES

1. Harvey, E. N. 1957. *A History of Luminescence—From the Earliest Times Until 1900*. Philadelphia, PA: The American Philosophical Society.
2. Oliveira, A. G., Stevani, C. V., Waldenmaier, H. E., Viviani, V., Emerson, J. M., Loros, J. J., and Dunlap, J. C. 2015. Circadian control sheds light on fungal bioluminescence. *Curr. Biol.* 25:964–968.
3. Wilson, T. and J. W. Hastings 2013. *Bioluminescence—Living Lights, Lights for Living*. Cambridge, MA: Harvard University Press.
4. Campbell, A. K. 1988. *Chemiluminescence—Principles and Applications in Biology and Medicine*. Chichester, U.K.: Ellis Horwood.
5. Baader, W. J., Stevani, C. V., and Bastos, E. L. 2006. Chemiluminescence of organic peroxides. In: *The Chemistry of Peroxides*, Vol. 2, ed. Z. Rappoport, pp. 1211–1278. Chichester, U.K.: Wiley & Sons.
6. Roda, A. 2011. *Chemiluminescence and Bioluminescence: Past, Present and Future*. Cambridge, U.K.: RSC Publishing.
7. Augusto, F. A., Souza, G. A., Souza, J. S. P., Khalid, M., and Baader, W. J. 2013. Efficiency of electron transfer initiated chemiluminescence. *Photochem. Photobiol.* 89:1299–1317.
8. Ihde, A. 1967. The history of free radicals and Moses Gomberg's contributions. *Pure Appl. Chem.* 15:1–14.
9. Schorlemmer, C. 1865. Researches on the hydrocarbons of the series C_nH_{2n+2}. *Proc. R. Soc. Lond.* 14:164–176.

10. Thénard, L. 1818–1819 Nouvelles recherches sur l'eau oxigénée. *Ann. Chim.* 9:441–443.
11. Fenton, H. J. H. 1894. Oxidation of tartaric acid in presence of iron. *J. Chem. Soc. Trans.* 65:899–911.
12. Halliwell, B. and Gutteridge, J. M. C. 2015. *Free Radicals in Biology and Medicine*, 5th edn. New York: Oxford University Press.
13. Gomberg, M. 1900. An instance of trivalent carbon: Triphenylmethyl. *J. Am. Chem. Soc.* 22:757–771.
14. Wurster, C. and Schobig, E. 1879. Ueber die Einwirkung oxydirender Agentien auf Tetramethylparaphenylendiamin. *Ber. Dtsch. Chem. Ges.* 12:1807–1813.
15. Bechara, E. J. H. and Cilento, G. 1972. Formation of adenosine triphosphate in the oxidation of a model for the reduced pyridine nucleotides. *Biochemistry* 11:2606–2610.
16. Gerschman, R., Gilbert, D. L., Nye, S. W., Dwyer, P., and Fenn, W. O. 1954. Oxygen poisoning and X-irradiation: A mechanism in common. *Science* 119:623–626.
17. McCord, J. and Fridovich, I. 1969 Superoxide dismutase: An enzymic function for erythrocuprein (hemocuprein). *J. Biol. Chem.* 244:6049–6055.
18. Loew, O. 1900. A new enzyme of general occurrence in organisms. *Science* 11:701–702. Sauerstoffmoleküle durch Sensibilisierung. *Naturwissenschaften* 19:1043.
19. Sumner, J. B. and Dounce, A. L. 1937. Crystalline catalase. *Science* 85:366–367.
20. Forman, H. J., Bernardo, A., and Davis, K. J. A. 2016. What is the concentration of hydrogen peroxide in blood and plasma. *Arch. Biochem. Biophys.* 603:48–53.
21. Ignarro, L. J., Buga, G. M., Wood, K. S., Byrns, R. E., and Chaudhuri, G. 1986. Endothelium-derived relaxing factor produced and released from artery and vein is nitric oxide. *Proc. Natl. Acad. Sci. USA* 84:9265–9269.
22. Palmer, R. M. J., Ferrige, A. G., and Moncada, S. 1986. Nitric oxide release accounts for the biological activity of endothelium-derived relaxing factor. *Nature* 327:524–526.
23. Toledo, J. C. and Augusto, O. 2012. Connecting the chemical and biological properties of nitric oxide. *Chem. Res. Toxicol.* 25:975–999.
24. Radi, R. 2013. Peroxynitrite, a stealthy biological oxidant. *J. Biol. Chem.* 288:26464–26467.
25. Medinas, D. B., Cerchiaro, G., Trindade, D. F., and Augusto, O. 2007. The carbonate radical and related oxidants derived from bicarbonate buffer. *IUBMB Life* 59:255–262.
26. Winterbourn, C. C. 2008. Reconciling the chemistry and biology of reactive oxygen species. *Nat. Chem. Biol.* 4:278–286.
27. Adam, W. and Cilento, G. 1982. *Chemical and Biological Generation of Excited States*. New York: Academic Press.
28. Dure, L. S. and Cormier, M. J. 1964. Studies on the bioluminescence of *Balanoglossus biminknsis* extracts III. Kinetic comparison of luminescent and nonluminescent peroxidation reactions and a proposed mechanism for peroxidase action. *J. Biol. Chem.* 239:2351–2359.
29. Shimomura, O. 2012. *Bioluminescence. Chemical Principles and Methods*. Singapore: World Scientific.
30. Oba, Y., Stevani, C. V., Oliveira, A. G., Tsarkova, A. S., Chepurnykh, T. V., and Yampolsky, I. V. 2017. Selected least studied but not forgotten bioluminescent systems. *Photochem. Photobiol.* 93:405–415.
31. Kopecky, K. R. and Mumford, C. 1969. Luminescence in thermal decomposition of 3,3,4-trimethyl-1,2-dioxetane. *Can. J. Chem.* 47:709–711.
32. Adam, W. and Liu, J. C. 1972. Alpha-peroxy lactones. Synthesis and chemiluminescence. *J. Am. Chem. Soc.* 94:2894–2895.
33. Bechara, E. J. H. and Wilson, T. 1980. Alkyl substituent effects on dioxetane properties—Tetraethyl-, dicyclohexylidene-, and 3,4-dimethyl-3,4-di-*n*-butyldioxetanes—A discussion of decomposition mechanisms. *J. Org. Chem.* 45:5261–5268.

34. Adam, W. and Baader, W. J. 1985. Effects of methylation on the thermal stability and chemiluminescence properties of 1,2-dioxetanes. *J. Am. Chem. Soc.* 107:410–416.

35. Koo, J.-Y., Schmidt, S. P., and Schuster, G. B. 1978. Bioluminescence of the firefly: Key steps in the formation of the electronically excited state for model systems. *Proc. Natl. Acad. Sci. USA* 75:30–33.

36. Koo, J.-Y. and Schuster, B. B. 1978. Chemiluminenescence of diphenoyl peroxide. Chemically initiated electron exchange luminescence. A new general mechanism for chemical production of electronically excited states. *J. Am. Chem. Soc.* 100:4496–4503.

37. Rauhut, M. M., Bollyky, L. J., Roberts, B. G., Loy, M., Whitman, R. H., Iannotta, A. V., Semsel, A. M., and R.A. Clarke. 1967. Chemiluminescence from reactions of electronegatively substituted aryl oxalates with hydrogen peroxide and fluorescent compounds. *J. Am. Chem. Soc.* 89:6515–6522.

38. Oliveira, T. F., Silva, A. L. M., Moura, R. A., Bagattini, R. F., Medeiros, M. H. G., Di Mascio, P., Arruda Campos, I. P., Barretto, F. P., Bechara, E. J. H., and Melo Loureiro, A. P. 2014. Luminescent threat: Toxicity of light stick attractors used in pelagic fishery. *Sci. Rep.* 4:5359.

39. Bartoloni, F. H., Oliveira, M. A., Ciscato, L. F. M. L., Augusto, F. A., Bastos, E. L., and Baader, W. J. 2015. Chemiluminescence efficiency of catalyzed 1,2-dioxetanone decomposition determined by steric effects. *J. Org. Chem.* 80:3745–3751.

40. Schaap, A. P. and Gagnon, S. D. (1982) Chemiluminescence from a phenoxide-substituted 1,2-dioxetane: A model for firefly bioluminescence. *J. Am. Chem. Soc.* 104:3504–3506.

41. Nery, A. L. P., Weiss, D., Catalani, L. H., and Baader, W. J. 2000. Studies on the intramolecular electron transfer catalyzed thermolysis of 1,2-dioxetanes. *Tetrahedron* 56:5317–5327.

42. Adam, W., Reinhardt, D., and Saha-Möller, C. R. 1996. From the firefly bioluminescence to the dioxetane-based (AMPPD) chemiluminescence immunoassay: A retroanalysis. *Analyst* 121:1527–1531.

43. Matsumoto, M. 2004. Advanced chemistry of dioxetane-based chemiluminescent substrates originating from bioluminescence. *J. Photochem. Photobiol., C* 5:27–53.

44. Velosa, A. C., Baader, W. J., Stevani, C. V., Mano, C. M., and Bechara, E. J. H. 2013. 1,3-Diene probes for detection of triplet carbonyls in biological systems. *Chem. Res. Toxicol.* 20:1162–1169.

45. Russell, G. A. J. 1957. Deuterium-isotope effects in the autoxidation of aralkyl hydrocarbons—Mechanism of the interaction of peroxy radicals. *J. Am. Chem. Soc.* 79:3871–3877.

46. Howard, J. A. and Ingold, K. U. 1968. Self-reaction of sec-butylperoxy radicals. Confirmation of Russell mechanism. *J. Am. Chem. Soc.* 90:1056–1058.

47. Kanofsky, J. R. 1986. Singlet oxygen production from the reactions of alkylperoxy radicals—Evidence from 1268-nm chemiluminescence. *J. Org. Chem.* 51:3386–3388.

48. Farneth, W. E. and Johnson, D. G. 1984. Chemiluminescence in the infrared photochemistry of oxetanes: The formal reverse of ketone photocycloaddition. *J. Am. Chem. Soc.* 106:1875–1876.

49. Di Mascio, P., Catalani, L. H., and Bechara, E. J. H. 1992. Are dioxetanes chemiluminescent intermediates in lipoperoxidation? *Free Radic. Biol. Med.* 12:471–478.

50. Miyamoto, S., Martinez, G. R., Medeiros, M. H. G., and Di Mascio, P. 2014. Singlet molecular oxygen generated by biological hydroperoxides. *J. Photochem. Photobiol. B-Biol.* 139:24–33.

51. Miyamoto, S., Ronsein, G. E., Correa, T. C., Martinez, G. R., Medeiros, M. H. G., and Di Mascio, P. 2009. Direct evidence of singlet molecular oxygen generation from peroxynitrate, a decomposition product of peroxynitrite. *Dalton Trans.* 29:5720–5729.

52. Indig, G., Campa, A., Bechara, E. J. H., and Cilento, G. 1988. Conjugate diene formation promoted by triplet acetone acting upon arachidonic acid. *Photochem. Photobiol.* 48:719–723.

53. Timmins, G. S., Penatti, C. A. A., Bechara, E. J. H., and Swartz, H. M. 1999. Measurement of oxygen partial pressure, its control during hypoxia and hyperoxia, and its effect upon light emission in a bioluminescent elaterid larva. *J. Exp. Biol.* 202:2631–2638.

54. Ciscato, L. F. M. L., Augusto, F. A., Weiss, D., Bartoloni, F. H., Bastos, E. L., Albrecht, S., Brandl, H., Zimmermann, T., and Baader, W. J. 2012. The chemiluminescent peroxyoxalate system: State of the art almost 50 years from its discovery. *Arkivoc*, 2012 (iii):391–430.

55. Stevani, C. V., Lima, D. F., Toscano, V. G., and Baader, W. J. 1996. A kinetic study on the peroxyoxalate reaction: Imidazole as nucleofilic catalyst. *J. Chem. Soc. Perkin Trans.* 2:989–995.

56. Silva, S. M., Wagner, K., Weiss, D., Beckert, R., and Baader, W. J. 2002. Studies on the chemiexcitation step in peroxyoxalate chemiluminescence using steroid-substituted activators. *Luminescence* 17:362–369.

57. Stevani, C. V. and Baader, W. J. 2002. Preparation and characterization of 2,2,2-triphenyl-$2\lambda^5$-1,3,2-dioxastilbolane-4,5-dione as standard for an attempt to trap 1,2-dioxetanedione, a possible high-energy intermediate in peroxyoxalate chemiluminescence. *J. Chem. Res.* 2002:430–432.

58. Seitz, W. R. and Hercules, D. M. 1972. Determination of trace amounts of iron(II) using chemiluminescence analysis. *Anal. Chem.* 44:2143–2149.

59. Dodeigne, C., Thunus, L., and Lejeune R. 2000. Chemiluminescence as diagnostic tool: A review. *Talanta* 51:415–439.

60. Ma, M., Diao, F., Zheng, X., and Guo, Z. 2012. Selective light- triggered chemiluminescence between fluorescent dyes and luminol, and its analytical application. *Anal. Bioanal. Chem.* 404:585–592.

61. Giokas, D. L., Christodouleas, D. C., Vlachou, I., Vlessidis A. G., and Calokerinos, A. C. 2013. Development of a generic assay for the determination of total trihydroxy-benzoate derivatives based on gold-luminol chemiluminescence. *Anal. Chim. Acta* 764:70–77.

62. Zhang, X., He, S., Chen, Z., and Huang, Y. 2013. Nanoparticles as oxidase mimic-mediated chemiluminescence of aqueous luminol for sulfite in white wines. *J. Agric. Food Chem.* 61: 840–847.

63. Marquette, C. A. and Blum, L. J. 2006. Applications of the luminol chemiluminescent reaction in analytical chemistry. *Anal. Bioanal. Chem.* 385:546–554.

64. Guo, H., Aleyasin, H., Dickinson, B. C., Haskew-Layton, R. E., and Ratan, R. R. 2014. Recent advances in hydrogen peroxide imaging for biological applications. *Cell Biosci.* 4:64.

65. Lippert, A. R., Van de Bittner, G. C., and Chang, C. J. 2011. Boronate oxidation as a bioorthogonal reaction approach for studying the chemistry of hydrogen peroxide in living systems. *Acc. Chem. Res.* 44:793–804.

66. Van de Bittner, G. C., Bertozzi, C. R., and Chang, C. J. 2013. Strategy for dual-analyte luciferin imaging: In vivo bioluminescence detection of hydrogen peroxide and caspase activity in a murine model of acute inflammation. *J. Am. Chem. Soc.*, 135:1783–1795.

67. Karton-Lifshin, N., Segal, E., Omer, L., Portnoy, M., Satchi-Fainaro, R., and Shabat, D. 2011. A unique paradigm for a turn-on near-infrared cyanine-based probe: Noninvasive intravital optical imaging of hydrogen peroxide. *J. Am. Chem. Soc.*, 133:10960–10965.

68. Lee, D., Khaja, S., Velasquez-Castano, J. C., Dasari, M., Sun, C., Petros, J., Taylor, W. R., and Murthy, N. 2007. In vivo imaging of hydrogen peroxide with chemiluminescent nanoparticles. *Nat. Mater.* 6:765–769.

69. Lee, D., Venkata, R., Erigala, V. R., Dasari, M., Yu, J., Dickson, R. M., and Murthy, N. 2008. Detection of hydrogen peroxide with chemiluminescent micelles. *Int. J. Nanomedicine* 3:471–476.

70. Dasari, M., Lee, D., Erigala, V. R., and Murthy, N. 2009. Chemiluminescent PEG-PCL micelles for imaging hydrogen peroxide. *J. Biomed. Mater. Res.* 89A:561–566.
71. Chen, R., Zhang, L. Z., Gao, J., Wu, W., Hu, Y., and Jiang, X. Q. 2011. Chemiluminescent nanomicelles for imaging hydrogen peroxide and self-therapy in photodynamic therapy. *J. Biomed. Biotechnol.* 2011:679492.
72. White, E. H., McCapra, F., Field, G. F., and McElroy, W. D. 1961. The structure and synthesis of firefly luciferin. *J. Am. Chem. Soc.* 83:2402–2403.
73. Shimomura, O. and Johnson, F. H. 1979. Elimination of the effect of contaminating CO_2 in the O^{18}-labeling of the CO_2 produced in bioluminescent reactions. *Photochem. Photobiol.* 30:89–91.
74. Viviani, V. R. and Amaral, D. T. 2016. First report of *Pyrearinus larvae* (Coleoptera: Elateridae) in clayish canga caves and luminous termite mounds in the Amazon Forest with a preliminary molecular-based phylogenetic analysis of the *P. pumilus* group. *Ann. Entomol. Soc. Am.* 109:534–541.
75. Colepicolo Neto, P., Pagni, D., and Bechara, E. J. H. 1988 Luciferin biosynthesis in larval *Pyrearinus termitilluminans* (Coleoptera: Elateridae). *Comp. Biochem. Physiol.* 91B:143–147.
76. Viviani, V. R. and Bechara, E. J. H. 1993. Biophysical and biochemical aspects of phengodid (railroad worm) bioluminescence. *Photochem. Photobiol.* 58:615–621.
77. Kanie, S., Nishikawa, T., Ojika, M., and Oba, Y. 2016. One-pot non-enzymatic formation of firefly luciferin in a buffer from *p*-benzoquinone and cysteine. *Sci. Rep.* 6:24794.
78. Kaskova, Z. M., Tsarkova, A. S., and Yampolsky, I. V. 2016. 1001 lights: Luciferins, luciferases, their mechanisms of action and applications in chemical analysis, biology and medicine. *Chem. Soc. Rev.* 45:6048–6077.
79. Barros, M. P. and Bechara, E. J. H. 1998. Bioluminescence as a possible auxiliary oxygen detoxifying mechanism in elaterid larvae. *Free Radic. Biol. Med.* 24:767–777.
80. Viviani, V. R. and Bechara, E. J. H. 1996. Larval *Tenebrio molitor* (Coleoptera: Tenebrionidae) fat body extracts catalyze firefly D-luciferin and ATP dependent chemiluminescence: A luciferase-like enzyme. *Photochem. Photobiol.* 63:713–718.
81. Prado, R. A., Barbosa, J. A., Ohmiya, Y., and Viviani, V. R. 2011. Structural evolution of luciferase activity in Zophobas mealworm AMP/CoA-ligase (protoluciferase) through site-directed mutagenesis of the luciferin binding site. *Photochem. Photobiol. Sci.* 10:1226–1232.
82. Trimmer, B. A., Aprille, J. R., Dudzinski, D. M., Lagace, C. J., Lewis, S. M., Michel, T., Qazi, S., and Zayas, R. M. 2001. Nitric oxide and the control of firefly flash. *Science* 292:2486–2488.
83. Bechara, E. J. H., Costa, C., Colepicolo, P., Viviani, V., Barros, M. P., Timmins, G. S., Lall, A. B. et al. 2007. Chemical, biological and evolutionary aspects of beetle bioluminescence. *Arch. Org. Chem.* viii:311–323.
84. Cilento, G. 1965. On the possibility of generation and transfer of electronic energy in biochemical systems. *Photochem. Photobiol.* 4:1243–1247.
85. Cilento, G. 1973. Excited electronic states in dark biological processes. *Q. Rev. Biophys.* 6:485–501.
86. Cilento, G. 1975. Dioxetanes as intermediates in biological processes. *J. Theor. Biol.* 55:471–479.
87. Lamola, A. 1971. Production of pyrimidine dimers in DNA in the dark. *Biochem. Biophys. Res. Commun.* 43:893–898.
88. White, E. H., Miano, J. D., Watkins, C. J., and Breaux, E. J. 1974. Chemically produced excited states. *Angew. Chem. Int. Ed.* 13:229–243.
89. Turro, N. J., Ramamurthy, V., and Scaiano, J. C. 2010. *Modern Molecular Photochemistry of Organic Molecules.* Sausalito, CA: University Science Books.

90. Cuthbertson, M. J., Rizzardo, E., and Solomon, D. H. 1985. The addition of t-butoxy radicals to haloolefins. *Aust. J. Chem.* 38:315–324.

91. Cilento, G. 1984. Generation of electronically excited triplet species in biochemical systems. *Pure Appl. Chem.* 56:1179–1190.

92. Cilento, G. and Adam, W. 1995. From free radicals to electronically excited species. *Free Radic. Biol. Med.* 19:103–114.

93. Turro, N. J. and Lechtken, P. 1972. Molecular photochemistry. LII. Thermal decomposition of tetramethyl-1,2-dioxetane. Selective and efficient chemelectronic generation of triplet acetone. *J. Am. Chem. Soc.* 94:2886–2888.

94. White, E. H. and Wei, C. C. 1970. A possible role for chemically-produced excited states in biology. *Biochem. Biophys. Res. Commun.* 39:1219–1223.

95. Vidigal, C. C. and Cilento, G. 1975. Evidence for the generation of excited methylglyoxal in the myoglobin catalyzed oxidation of acetoacetate. *Biochem. Biophys. Res. Commun.* 62:184–190.

96. Ganini, D., Christoff, M., Ehrenshaft, M., Kadiiska, M., Mason, R. P., and Bechara, E. J. H. 2011. Myoglobin-H_2O_2 catalyzes the oxidation of beta-ketoacids to alpha-dicarbonyls: Mechanism and implications in ketosis. *Free Radic. Biol. Med.* 51:733–743.

97. Zinner, K., Vidigal, C. C., Durán, N., and G. Cilento. 1977. Oxidation of isonicotinic acid hydrazide by the peroxidase system. The formation of an excited product. *Arch. Biochem. Biophys.* 180:452–458.

98. Kamiya, I. and Iwaki, R. 1966. Studies of the chemiluminescence of several xanthene dyes. I. Kinetic studies of uranine and eosine chemiluminescence. *Bull. Chem. Soc. Jpn.* 39:257–263.

99. Brunetti, I. L., Bechara, E. J. H., Cilento, G., and White, E. H. 1982. Possible in vivo formation of lumicolchicines from colchicine by endogeneously generated triplet species. *Photochem. Photobiol.* 36:245–249.

100. Nery, A. L. P., Quina, F. H., Moreira, P. F., Medeiros, C. E. R., Baader, W. J., Shimizu, K., Catalani, L. H., and Bechara, E. J. H. 2001. Does the photochemical conversion of colchicine into lumicolchicines involve triplet transients? A solvent dependence study. *Photochem. Photobiol.* 73:213–218.

101. Bechara, E. J. H., Oliveira, O. M. M. F., Duran, N., Baptista, R. C., and Cilento, G. 1979. Peroxidase catalyzed generation of triplet acetone. *Photochem. Photobiol.* 30:101–110.

102. Baader, W. J., Bohne, C., Cilento, G., and Dunford, H. B. 1985. Peroxidase-catalyzed formation of triplet acetone and chemiluminescence from isobutyraldehyde and molecular oxygen. *J. Biol. Chem.* 260:10217–10225.

103. Adam, W., Baader, W. J., and Cilento, G. 1986. Enols of aldehydes in the peroxidase/oxidase—Promoted generation of excited triplet species. *Biochim. Biophys. Acta* 881:330–336.

104. Nantes, I. L., Bechara, E. J. H., and Cilento, G. 1996. Horseradish peroxidase-catalyzed generation of acetophenone and benzophenone in the triplet state. *Photochem. Photobiol.* 63:702–708.

105. Almeida, A. M., Vercesi, A. E., Bechara, E. J. H., and Nantes, I. L. 1999. Diphenylacetaldehyde-generated excited states promote damage to isolated rat liver mitochondrial DNA, phospholipids and proteins. *Free Radic. Biol. Med.* 27:744–751.

106. Soares, C. H. and Bechara, E. J. H. 1982. Enzymatic generation of triplet biacetyl. *Photochem. Photobiol.* 36:117–119.

107. Knudsen, F. S., Campa, A., Stefani, H. A., and Cilento, G. 1994. Plant hormone ethylene is a Norrish type II product from enzymically generated triplet n-butanal. *Proc. Natl. Acad. Sci. USA* 91:410–412.

108. Kowaltowski, A. L., Castilho, R. F., Grijalba, M. T., Bechara, E. J. H., and Vercesi, A. E. 1995. Effect of inorganic phosphate concentration on the nature of inner mitochondrial membrane alterations mediated by Ca^{2+} ions. A proposed model for phosphate-stimulated lipid peroxidation. *J. Biol. Chem.* 271:2929–2934.

109. Catalani, L. H., Wilson, T., and Bechara, E. J. H. 1987. Two water soluble fluorescence probes for chemiexcitation studies: Sodium 9,10-dibromo- and 9,10-diphenylanthracene-2-sulfonate. Synthesis, properties and application to triplet acetone and tetramethyldioxetane. *Photochem. Photobiol.* 45:273–281.

110. Monteiro, H. P., Abdalla, D. S. P., Augusto, O., and Bechara, E. J. H. 1989. Free radical generation during δ-aminolevulinic acid autoxidation: Induction by hemoglobin and connections with porphyrinpathies. *Arch. Biochem. Biophys.* 271: 206–216.

111. Bechara, E. J. H., Dutra, F., Cardoso, V. E. S., Sartori, A., Olympio, K. P. K., Penatti, C. A. A., Adhikari, A., and Assunção, N. A. 2006. The dual face of endogenous α-aminoketones: Pro-oxidizing metabolic weapons. *Comp. Biochem. Physiol. Part C* 146:88–110.

112. Thornalley, P. J. 2008. Protein and nucleotide damage by glyoxal and methylglyoxal in physiological systems: Role in ageing and disease. *Drug Metabol. Drug Interact.* 25:125–150.

113. Mashino, T. and Fridovich, I. 1987. Superoxide radical initiates the autoxidation of dihydroxyacetone. *Arch. Biochem. Biophys.* 254:547–551.

114. Medeiros, M. H. G. and Bechara, E. J. H. 1986. Chemiluminescent oxidation of protein adducts with glycolaldehyde. *Arch. Biochem. Biophys.* 248:435–439.

115. Medeiros, M. H. G., Marchiori, P. E., and Bechara, E. J. H. 1982. Superoxide dismutase, glutathione peroxidase and catalase activities in the erythrocytes of patients with intermittent acute porphyria. *Clin. Chem.* 28:242.

116. Kautsky, H. and de Bruijn, H. 1931. Die Aufklärung der Photoluminescenztilgung fluorescierender Systeme durch Sauerstoff; die Bildung aktiver, diffusionfähiger.

117. Khan, A. U. and Kasha, M. 1970. Chemiluminescence arising from simultaneous transitions in pairs of singlet oxygen molecules. *J. Am. Chem. Soc.* 92:3293–3300.

118. Foote, C. S. and Clennan, E. L. 1995. Properties and reactions of singlet oxygen. In *Active Oxygen in Chemistry*, eds. C. S. Foote, G. S. Valentine, A. Greenberg, and J. F. Liebman, pp. 105–140. London, U.K.: Chapman & Hall.

119. Greer, A. 2006. Christopher Foote's discovery of the role of singlet oxygen [$^1O_2(^1\Delta_g)$] in photosensitized oxidation reactions. *Acc. Chem. Res.* 39:797–804.

120. Packer, L. and Sies, H. eds. 2000. *Singlet Oxygen, UV-A, and Ozone*, Vol. 319: Methods in Enzymology. New York: Academic Press.

121. Krinsky, N. I. 1977. Singlet oxygen in biological systems. *Trends Biochem. Sci.* 2:35–38.

122. Di Mascio, P., Kaiser, S., and Sies, H. 1989. Lycopene as the most efficient biological carotenoid singlet oxygen quencher. *Arch. Biochem. Biophys.* 274:532–538.

123. Rao, A. V. and Rao, L. G. 2007. Carotenoids and human health. *Pharmacol. Res.* 55:207–216.

124. Di Mascio, P., Bribiva, K., Bechara, E. J. H., and Sies, S. 1996. The reaction of peroxynitrite and hydrogen peroxide produces singlet molecular oxygen. *Methods Enzymol.* 269: 395–400.

125. Gupta, D., Harrish, B., Kissner, R., and Koppenol, W. H. 2009. Peroxynitrate is formed rapidly during decomposition of peroxynitrate at neutral pH. *Dalton Trans.* 29:5730–5736.

126. Massari, J., Fujiy, D. E., Dutra, F., Vaz, S. M., Ferreira, A. M. C., Micke, G. A., Tavares, M. F. M., Tokikawa, R., Assunção, N. A., and Bechara, E. J. H. 2008. Radical acetylation of 2′-deoxyguanosine and L-histidine coupled to the reaction of diacetyl with peroxynitrite in aerated medium. *Chem. Res. Toxicol.* 21:879–887.

127. Massari, J., Tokikawa, R., Zanolli, L., Tavares, M. F. M., Assunção, N. A., and Bechara, E. J. H. 2010. Acetyl radical production by the methylglyoxal–peroxynitrite system: A possible route for L-lysine acetylation. *Chem. Res. Toxicol.* 23:1762–1770.

128. Massari, J., Tokikawa, R., Medinas, D. B., Angeli, J. P. F., Di Mascio, P., Assunção, N. A., and Bechara, E. J. H. 2011. Generation of singlet oxygen by the glyoxal/peroxynitrite system. *J. Am. Chem. Soc.* 133:20761–20768.

129. Zimniak, P. 2011. Relationship of electrophilic stress to aging. *Free Radic. Biol. Med.* 51:1087–1105.
130. Ott, C., Jacobs, K., Haucke, E., Santos, A. N., Grune, T., and Simm, A. 2014. Role of advanced glycation end-products in cellular signaling. *Redox Biol.* 2:411–429.
131. Medeiros, M. H. G. 2009. Exocyclic DNA adducts as biomarkers of lipid oxidation and predictors of disease. Challenges in developing sensitive and specific methods for clinical studies. *Chem. Res. Toxicol.* 22:419–425.
132. Cadet, J., Douki, T., and Ravanat, J. -L. 2010. Oxidatively generated base damage to cellular DNA. *Free Radic. Biol. Med.* 49:9–21.
133. Rabbani, N. and Thornalley, P. J. 2015. Dicarbonyl stress in cell and tissue dysfunction contributing to ageing and disease. *Biochem. Biophys. Res. Commun.* 458:221–226.
134. Matafome, P., Rodrigues, T., Serra, C., and Seiça, R. 2017. Methylglyoxal in metabolic disorders: Facts, myths and promises. *Med. Res. Rev.* 37:368–403. doi: 10.1002/med.21410.
135. Tokikawa, R., Loffredo, C., Uemi, M., Machini, M. T., and Bechara, E. J. H. 2013. Radical acylation of L-lysine derivatives and L-lysine-containing peptides by peroxynitrite-treated diacetyl and methylglyoxal. *Free Radic. Res.* 48:357–370.
136. Chiarelli-Neto, O., Ferreira, A. L., Martins, W. K., Pavani, C., Severino, D., Faião-Flores, F., Maria-Engler, S. S. et al. 2014. Melanin photosensitization and the effect of visible light on epithelial cells. *PLoS One* 9:e113266.
137. Premi, S., Wallisch, S., Mano, C. M., Weiner, A. B., Bacchiocchi, A., Wakamatsu, K., Bechara, E. J. H., Halaban, R., Douki, T., and Brash, D. E. 2015. Chemiexcitation of melanin derivatives induces DNA photoproducts long after UV exposure. *Science* 47:842–847.
138. Slawinska, D. and Slawinski, J. 1982. Electronically excited molecules in the formation and degradation of melanins. *Physiol. Chem. Phys.* 14:363–374.
139. Wakamatsu, K., Nakanishi, Y., Miyazaki, N., Kolbe, L., and Ito, S. 2012. UVA-induced oxidative degradation of melanins: Fission of indole moiety in eumelanin and conversion to benzothiazole moiety in pheomelanin. *Pigment Cell Melanoma Res.* 25:434–445.
140. Kamal, A. H. M. and Komatsu, S. 2016. Proteins involved in biophoton emission and flooding stress responses in soybean under light and dark conditions. *Mol. Biol. Resp.* 43:73–89.
141. Baader, W. J., Stevani, C. V., and Bechara, E. J. H. 2015. "Photo" chemistry without light? *J. Braz. Chem. Soc.* 26:2430–2447.

3 Comparative Analysis of Hydrogen Peroxide and Peroxynitrite Reactivity with Thiols

Madia Trujillo, Sebastián Carballal,
Ari Zeida, and Rafael Radi

CONTENTS

INTRODUCTION

HYDROGEN PEROXIDE AND PEROXYNITRITE AS BIOLOGICALLY RELEVANT PEROXIDES

Peroxides are chemical compounds in which two oxygen atoms are linked together by a single covalent bond. Hydrogen peroxide (H_2O_2), the simplest peroxide, is formed through different routes in biological systems: (1) superoxide radical ($O_2^{\cdot-}$) dismutation, which can be spontaneous, or, depending on the organism and cellular compartment, catalyzed by different superoxide dismutases (SODs) [1–3]; (2) one-electron $O_2^{\cdot-}$ reduction, such as during aconitase oxidation [4,5]; or (3) direct two-electron reduction of oxygen, which can be catalyzed by the divalent

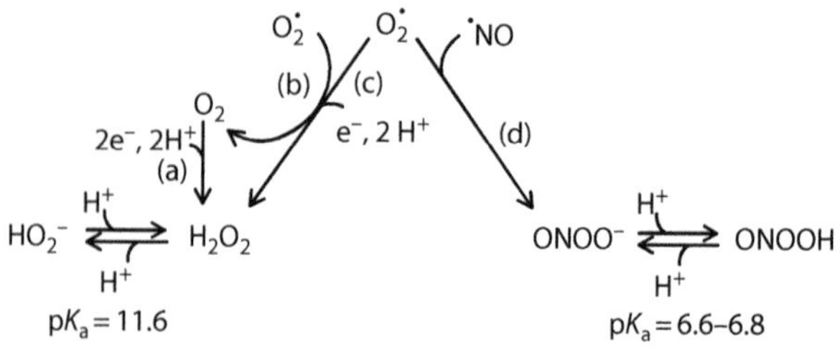

FIGURE 3.1 Routes of formation and acidity constants of H_2O_2 and ONOOH. In biological systems, H_2O_2 can be formed through the two-electron reduction of oxygen catalyzed by different oxidases (a), $O_2^{\cdot-}$ dismutation to H_2O_2 and O_2 catalyzed by SODs (b), or through the one-electron reduction of $O_2^{\cdot-}$ (c). Due to its high pK_a values (lower $pK_a = 11.6$ [83]), H_2O_2 is more than 99.99% protonated at physiological pH. The main route of ONOOH formation is the rapid recombination reaction between $O_2^{\cdot-}$ and $\cdot$NO that produces ONOO$^-$ (d), in equilibrium with its conjugated acid ONOOH. The pK_a value of this peroxyacid indicates that it will be mostly (~80%) deprotonated at physiological pH [11–13,16].

activity of several oxidases, including xanthine oxidase, Ero1, aldehyde oxidase, and monoamine oxidase [6–10] (Figure 3.1). In turn, peroxynitrous acid (ONOOH)* is a peroxy acid (or peracid), that is, a compound that contains an acidic –OOH group. ONOOH is the conjugated acid of peroxynitrite anion (ONOO$^-$) ($pK_a = 6.6$–6.8 [11–13]), whose main biological source is the rapid recombination reaction between $O_2^{\cdot-}$ and nitric oxide ($\cdot$NO) radicals [14–16][†] (Figure 3.1). $\cdot$NO is a small and lipophilic radical that can diffuse through membranes [20,21]. On the contrary, the charged nature of $O_2^{\cdot-}$ (the pK_a of the conjugated acid hydroperoxyl radical (HO$_2^{\cdot}$) is 4.8 [22]) limits its diffusion through membranes to those expressing anion channels, or to those delimiting compartments with acidic pH that allow $O_2^{\cdot-}$ protonation at a significant proportion [23]. Moreover, the diffusion distance for $O_2^{\cdot-}$ is estimated to be very short (~ 0.5 μm) [24]. Thus, ONOOH as well as H_2O_2 arising from $O_2^{\cdot-}$ are expected to be formed primary at the main sites of $O_2^{\cdot-}$ generation. Depending on the cell type, those sites could be mitochondria, phagosomes of inflammatory cells, as well as the extracellular space. Additionally, $O_2^{\cdot-}$ can be formed through redox-cycling of xenobiotics at different cell compartments [25].

PHYSICOCHEMICAL CHARACTERISTICS OF H_2O_2 AND PEROXYNITRITE

Both H_2O_2 and ONOOH are strong two-electron oxidants, with standard reduction potentials $E^{\circ\prime}(H_2O_2, H_2O) = 1.349$ V and $E^{\circ\prime}(ONOOH, H^+/NO_2^-) = 1.4$ V [26,27]. However, two-electron oxidations by H_2O_2 usually have higher activation energy than those by ONOOH and are therefore slower. As one-electron oxidants, H_2O_2

---

* IUPAC-recommended names for peroxynitrous acid (ONOOH) and its conjugated base, peroxynitrite anion (ONOO$^-$), are hydrogen oxoperoxonitrate and oxoperoxonitrate (1-), respectively. The term peroxynitrite is used to refer to the sum of ONOO$^-$ and ONOOH.

† An alternative possible route for peroxynitrite formation under biological conditions is the reaction of nitrosyl hydride (HNO), the one-electron reduction product of $\cdot$NO, with molecular oxygen, for which reported rate constants are in the range of 0.3–1.8 × 10^4 M^{-1}s^{-1} [17–19].

is weaker than ONOOH ($E^{\circ\prime}$(H_2O_2,$^{\bullet}$OH, H_2O = 0.32 V) vs $E^{\circ\prime}$(ONOOH, $H^+/^{\bullet}NO_2$, H_2O = 1.6 V), respectively [28,29]). In general, the bond dissociation energy (BDE) of a generic O–O bond is quite low (34 kcal/mol), relative to C–O bonds (~84 kcal/mol), that is, peroxides can suffer homolysis more easily [30]. Particularly, the BDE of H_2O_2 was determined as 50.5 kcal/mol at 25°C [30]. Therefore, H_2O_2 is susceptible to thermal, photolytic, and radiolytic homolysis as well as reduction by metals through Fenton chemistry [31]. Reported activation parameters of H_2O_2 homolysis indicate that the uncatalyzed reaction rate is very slow [32]. In turn, BDE of the O–O bond in ONOOH is even lower, 21 kcal/mol [27]. Therefore, ONOOH easily undergoes homolysis to produce two free radicals, hydroxyl radical ($^{\bullet}$OH) and nitrogen dioxide radical ($^{\bullet}NO_2$).* At 37°C and pH 7.4, ONOOH decay through homolysis occurs with an apparent, pH-dependent rate constant (k′) of 0.9 s^{-1} and a pH-independent rate constant (k) of 4.5 s^{-1} [27]. In most cellular and extracellular compartments, this route of ONOOH decay is of limited relevance due to the low rate of homolysis when compared with the rates of direct reactions of peroxynitrite with its main targets, namely, fast reacting thiols/selenols in cysteine- or selenocysteine-based peroxidases metals and CO_2 (see below). The reduction potential of $^{\bullet}NO_2$ is not as high as that of $^{\bullet}$OH, $E^{\circ\prime}$($^{\bullet}NO_2/NO_2^-$) = + 0.99 V vs $E^{\circ\prime}$ ($^{\bullet}$OH, H^+/H_2O) = 2.31 V [39,40]. Moreover, reactions involving $^{\bullet}$OH usually have no activation barrier and are diffusion-controlled, which results in nonspecific reactions with almost every moiety it encounters. $^{\bullet}NO_2$ is more selective and can diffuse longer distances before reacting with its targets [41].

The O–O bond is easily polarized, and thus H_2O_2 can act as a nucleophile (perhydroxyl anion (HO_2^-) is more nucleophilic than neutral H_2O_2 but its high pK_a indicates that it will be almost all protonated in biological media) and also as an electrophile [31]. In turn, reactions of peroxynitrite usually involve ONOOH acting as an electrophile or its conjugated basis ONOO$^-$ as a nucleophilic species. Compounds with thiol functional groups (R–SH, R being an alkyl group) are the main biological targets for peroxides. In the following sections, we will introduce the main thiol-containing compounds found in biological systems and discuss their oxidations by H_2O_2 and peroxynitrite that can occur either directly or indirectly, that is, through the generation of secondary oxidants. Diffusion properties and the expected half-lives of both oxidants in the presence of their main biological targets in the cytosol and mitochondrial matrix will be analyzed.

BIOLOGICALLY RELEVANT THIOL-CONTAINING COMPOUNDS

In biological systems, thiol functional groups can be found as part of low-molecular-weight (LMW) compounds and in protein Cys residues. Most living organisms contain millimolar concentrations of LMW thiols that play key biological functions, such as to keep an intracellular reducing environment, to provide electrons for redox enzymes, to react with electrophilic compounds, and to detoxify xenobiotics. In most eukaryotic

---

* The mechanism of the decomposition of ONOOH has been controversial [11,33]. Homolysis is presently accepted by most authors as radical formation in ~30% yields was confirmed by different methodologies [34–38].

cells and Gram-negative bacteria, these actions are performed by the tripeptide glutathione (GSH, gamma-glutamyl L-cysteinylglycine) [42], which is kept reduced by the flavoenzyme glutathione reductase at the expense of nicotinamide adenine dinucleotide phosphate (NADPH) [43]. Some microorganisms have other LMW thiols, such as trypanothione (TSH$_2$, bis-glutathionylspermidine) in Kinetoplastids; mycothiol (MSH, acetyl cysteine-glucosamine-*myo*-inositol) in Actinomycetes; bacillithiol (BSH, cysteine-glucosamine-malate) in Firmicutes [44–47]; coenzyme A in some spirochetes [48], for which specific disulfide reductases have been reported. Moreover, coenzyme A plays important metabolic roles in living systems [49]. In turn, free dihydrolipoic acid (DHLA) or its oxidized form, lipoic acid (LA), are almost undetectable in plasma and tissue of animals (unless they are orally supplemented) and are found as a lipoamide cofactor bound to dehydrogenase complexes such as mitochondrial pyruvate dehydrogenase and alpha-ketoglutarate complexes [50,51]. Ovothiols A-C (particularly abundant in marine invertebrate eggs and some trypanosomatids) and ergothioneine (produced by fungi and actinomycetes) are mercaptoimidazol-containing compounds [52–54]. Although mammals cannot synthesize ergothioneine, it is present in many dietary sources. A highly specific transporter in mammalian tissues allows it to be widely distributed [55,56]. These thiols are considered to play antioxidant functions, but enzymatic routes relying on them, including recycling systems, have not been reported so far [57,58]. A recent report indicated that at least some of the beneficial actions of ergothioneine are mediated through the upregulation of specific protein deacetylases [59]. Cysteine is another LMW thiol. The intracellular concentration of free Cys is quite low (reduced form: 125 µM and oxidized form: 31 µM in human HT29cells) and it is much more abundant as a component of proteins and glutathione [60]. It is, however, the more abundant LMW thiol in plasma where glutathione concentration is extremely low ([reduced Cys] = 8–10 µM and [GSH] = 4.7 µM) [61]. Moreover, many LMW thiol-containing compounds are used as drugs, examples being N-acetyl cysteine, antiarthritic drugs such as D-penicillamine, some inhibitors of metallo-β-lactamases, and inhibitors of angiotensin-converting enzyme such as captopril [62–65]. The reactions of many of these LMW thiols with oxidants formed *in vivo*, such as H$_2$O$_2$ and peroxynitrite, explain some of their antioxidant roles (Table 3.1). Finally, hydrogen sulfide (H$_2$S/HS$^-$), which has a pK_a of 7 [66], is produced in different organisms, including mammals, where it participates in signaling functions [67,68]. Although hydrogen sulfide is not properly a thiol, it shares some of their chemical properties, and its reactions with different peroxides including H$_2$O$_2$ and peroxynitrite have also been investigated. The reported rate constants have also been listed in Table 3.1 for comparative purposes [69,70]. Of note, the intrinsic reactivity of HS$^-$ is one order of magnitude lower than those of thiolates [70].

Cysteine, the single thiol-containing amino acid, is underrepresented in proteins of all organisms, and its abundance appears to correlate positively with the complexity of the life form [71]. The functions of Cys residues are diverse: they can (1) play structural roles, (2) coordinate metals, (3) be sites of posttranslational modifications with potential impact on protein structure/function; and (4) be critical for enzymatic activity, either changing or not the redox status of the thiol group during the catalytic cycle [72]. The reactions of protein Cys with peroxides are the bases of redox signaling and regulation of cellular functions, an

TABLE 3.1

Kinetics of Oxidation of Selected LMW Thiols H_2O_2 and Peroxynitrite

LMW Thiol	$k'_{H_2O_2}$ (M^{-1} s^{-1})	References	k'_{ONOOH} (M^{-1} s^{-1})	References
Cysteine ethyl ester	1.40[a]	[85]	6830	[80]
Trypanothione	5.37[b]	[203]	3600	[80]
Penicillamine	4.5	[75]	6420	[80]
Cysteine	2.9	[75]	4500	[16]
Mycothiol	0.6[c]	[117]	1670[d]	[87]
Glutathione	0.87	[75]	1360	[27]
Homocysteine	0.18[a]	[85]	700	[80]
N-Acetylcysteine	0.16	[75]	415	[80]
Dihydrolipoic acid	0.27	[204]	250	[80]
Hydrogen sulfide	0.73	[69]	6650	[205]

Shown values were reported at pH 7.4 and 37°C unless otherwise indicated. In the case of the dithiols trypanothione and dihydrolipoic acid, indicated values are per thiol group, assuming a similar reactivity for both thiols. ND = not determined. Rate constants for hydrogen sulfide are listed for comparative purposes.

[a] At pH 7.06 and 25°C.

[b] At pH 7.2 and 27°C.

[c] At pH 7.4 and 25°C.

[d] Estimated from Brønsted correlations.

expanding field of research, since disruptions in redox homeostasis can lead to oxidative stress-related diseases [73,74].

HYDROGEN PEROXIDE- AND PEROXYNITRITE-MEDIATED THIOL OXIDATIONS

LMW thiols and the thiol groups of protein cysteine residues are oxidized by different hydroperoxides including H_2O_2 and peroxynitrite [16,75]. The reactions can be direct (i.e., the reactive species are the peroxides themselves) or indirect (where the actual reactive species are secondary oxidants derived from the peroxides) [76–78]. The reaction mechanisms and biological consequences of these direct and indirect reactions are described below.

DIRECT REACTIONS OF H_2O_2 AND PEROXYNITRITE WITH THIOLS

Mechanisms. Hydroperoxides oxidize thiols by a two-electron oxidation mechanism leading to the formation of sulfenic acids (RSOH) [16,76,79]. The reactions occur by nucleophilic displacement mechanisms (S_N2), in which thiolates and protonated peroxides are the reactive species [75,80]. In the case of H_2O_2-mediated thiol oxidations, QM/MM molecular dynamic simulations indicated that the classical S_N2 reaction mechanism is modified by the transfer of proton from one peroxidic oxygen to the other, so that water and sulfenate are the direct products of the reaction as indicated in Equation 3.1. However, since this transfer occurs after the transition state in the reaction coordinate, it lacks kinetic consequences [81].

$$RS^- + H_2O_2 \rightarrow RSO^- + H_2O \tag{3.1}$$

In the case of ONOOH, the reaction proceeds directly to sulfenic acid and nitrite as indicated in Equation 3.2 [82] (Figure 3.2).

$$RS^- + ONOOH \rightarrow RSOH + NO_2^- \tag{3.2}$$

Since protonated peroxides and thiolates are the reagents, the rate constants of the reactions are affected by pH, as indicated in Equation 3.3

$$k' = k \cdot \frac{K_{aSH}}{K_{aSH} + \left[H^+\right]} \cdot \frac{\left[H^+\right]}{\left[H^+\right] + K_{aROOH}} \tag{3.3}$$

where
  k' is the apparent, pH-dependent rate constant of the reaction
  k is the real, pH-independent rate constant
  K_{aSH} is the acidity constant of the thiol
  K_{aROOH} is the acidity constant of the peroxide [80]

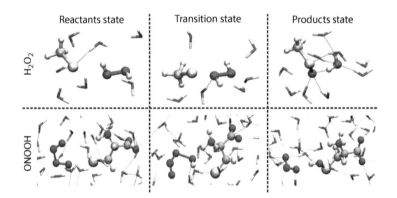

FIGURE 3.2 QM-MM molecular dynamic simulations of model thiolate oxidation by H_2O_2 and ONOOH. Representative structures of different states (reactant state, transition state, and product state) during the oxidation of model thiolates in aqueous microenvironment obtained through QM-MM molecular dynamic simulations. Top: reaction of methyl thiolate and H_2O_2 (reactant state) to form methyl sulfenate and H_2O (product state). Bottom: reaction of deprotonated cysteine and peroxynitrous acid in the more stable *cis*-conformation (reactant state) to form cysteine sulfenic acid and nitrite (product state). The following color code was used for representing atoms: S, yellow; O, red; C, cyan; H, white; N, blue. This figure was generated using data previously published in [81] and [82].

Thus, the terms $K_{aSH}/(K_{aSH} + [H^+])$ and $[H^+]/([H^+] + K_{aROOH})$ indicate the availability of thiolate and of protonated peroxide, at a given pH. Due to the high pK_a value of H_2O_2 (11.6 for the first deprotonation) [83], more than 99.9% is protonated at the cytosol, extracellular media, and even in the mitochondrial matrix where pH is ~7.6–8 [84]. Thus, H_2O_2 availability is practically 1. However, in the case of ONOOH, with a pK_a value close to physiological pH, the fraction of protonated peroxide is importantly affected by pH. Additionally, since thiol pK_a values are usually relatively close to physiological pH, the fraction of thiol that is deprotonated at a given pH also affects apparent rate constants. Figure 3.3 illustrates on the effect of pH on the apparent rate constants for the reactions of glutathione with H_2O_2 and peroxynitrite in a pH range from 5.5 to 9.5. Due to the minimal ionization of the former oxidant at those pHs, a bell-shaped profile is only evident for peroxynitrite-mediated thiol oxidation. Of note, while at pH = 7.4, 11% of free Cys is ionized, this can be as high as ~100% for thiols with $pK_a < 5.4$ or as low as <1% for thiol with $pK_a > 9.4$. Thus, for those LMW thiols with higher pK_a values, lower thiolate availability exists and therefore a lower pH-dependent rate constant is expected at physiological pH. The pH-independent rate constants, or in other words, the intrinsic reactivity of thiolates with a given peroxide also depends on thiol pK_a. The dependence is given by a Brønsted equation of the form

$$\log k_{RS^-} = \beta_{Nuc} \, pK_a + A \qquad (3.4)$$

where β_{Nuc} is the nucleophilic constant of the reactions.

For those LMW thiols that have been investigated, pH-independent rate constants for the reactions with H_2O_2 and ONOOH are in the order of 10^1–10^2 and 10^4–10^6 $M^{-1}\,s^{-1}$ in aqueous solutions at 25°C, respectively [75,80,85], with β_{Nuc} values of 0.27 and

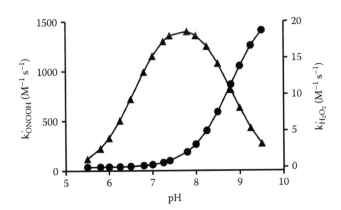

FIGURE 3.3 pH profile for the rate constants of GSH oxidation by H_2O_2 and peroxynitrite. Apparent pH-dependent rate constants (k′) for the reaction of glutathione with H_2O_2 (circles) and peroxynitrite (triangles) were calculated according to Equation 3.3 at each pH. The parameters used were pK_a $H_2O_2 = 11.6$, pK_a ONOOH = 6.6, pK_a glutathione = 8.8, and the pH-independent rate constant (k) values 22.7 and 2.6×10^5 $M^{-1}\,s^{-1}$ for H_2O_2 and peroxynitrite, respectively, at pH 7.4 and 37°C [75,80].

0.4, respectively [85,86]. β_{Nuc} values are considered indicators of the degree of charge transfer from the nucleophile to the electrophile at the transition state and are consistent with the charge redistribution at the transition state reported to occur both in H_2O_2- and ONOOH-mediated thiolate oxidations studied by QM-MM molecular dynamic simulations [81,82] (Figure 3.2). β_{Nuc} values were used to roughly estimate the rate constants of reactions of peroxides with other LMW thiols that are currently unavailable in the literature, provided their thiol pK_a is known [87].

Fast Reacting Thiols

For many protein Cys residues, pH-independent rate constants of oxidation by H_2O_2 and ONOOH follow the same Brønsted correlation as LMW thiols. However, there are some protein Cys residues that react with hydroperoxides much faster than expected according to their pK_a, even when considering pH-independent reactivities (i.e., disregarding thiolate availability).* Those have been named proteins with fast reacting thiols [89]. Most of these fast reacting thiols belong to nucleophilic Cys in thiol-dependent peroxidases, where they are collectively named peroxidatic Cys (Cys_P). These peroxidases include peroxiredoxins (Prxs), thiol-dependent glutathione peroxidases (GPx), and the bacterial organic hydroperoxide resistance protein (Ohr). They catalyze the reduction of different peroxides with particular specificities for oxidizing and reducing substrates [90–94]. They are bisubstrate enzymes with ping-pong kinetic mechanisms. Table 3.2 compiles available kinetic data on the reaction of protein Cys residues with H_2O_2 and peroxynitrite. Peroxiredoxins are a family of thiol-dependent peroxidases that are present in almost all organisms through life kingdoms and are distributed in different cell compartments. They are usually abundant, with concentrations that can reach hundreds of μM [92]. In the first part of the catalytic cycle, the oxidative part, they catalyze the two-electron reduction of H_2O_2 to H_2O, organic hydroperoxides to their corresponding alcohols, and ONOOH to nitrite, forming a sulfenate derivative at Cys_P. The mechanisms of reduction of this sulfenate back to thiol vary in the different Prxs. In most of them (2-Cys Prxs), a second protein Cys, the resolving Cys (Cys_R), forms a disulfide with Cys_P that is then usually reduced by thioredoxin or a related protein [95]. In 1-Cys Prxs, the sulfenic acid is reduced by alternative mechanisms [96–99]. The catalytic power of Prxs during H_2O_2 reduction relies on transition state stabilization by a hydrogen bond network involving Cys_P, the substrate, and conserved Thr and Arg side chains that are critical for activity [85,100–104]. However, different studies showed some differences in the interactions, which might reflect differences in the model Prx system utilized [103]. Furthermore, different Prxs have variable susceptibility to oxidative inactivation through the two-electron oxidation of sulfenate to form a sulfinic acid derivative of Cys_P, that, in some 2-Cys Prxs, can be reduced back to sulfenate by specific sulfiredoxins [105,106]. Both H_2O_2 and ONOOH can oxidize the sulfenate in Cys_P to sulfinic acid [107,108]. At least in the case of H_2O_2, for which several rate constants

---

* When considering apparent k at pH 7.4, reactivity of protein thiols with low pK_a compared with free Cys can increase by a maximum of 10-fold, considering the fractions of thiol as thiolate indicated above [88].

TABLE 3.2

Kinetics of the Reactions of Selected Thiol-Containing Proteins with H_2O_2 and Peroxynitrite

Protein Family	Protein	$k'_{H_2O_2}$ (M^{-1} s^{-1})	k'_{ONOOH} (M^{-1} s^{-1})	References
Peroxiredoxin	Prx2 (human RBC)	1.3×10^7– 1×10^8	1.4×10^7	[108,206]
Subfamily				
Prx1-AhpC	Prx 3(human)	2×10^7	8×10^{6d}	[201]
	T. cruzi (mit)TXNPx	6×10^6	1.8×10^7	[207]
	T. cruzi (cyt)TXNPx	3×10^7	1×10^6	[207]
	S. cerevisiae Tsa1	2.2×10^7	7.4×10^5	[208]
	S. cerevisiae Tsa2	1.3×10^7	5.1×10^5	[208]
	S. typhimurium AhpC	3.7×10^7	1.5×10^{6e}	[209,210]
Prx5	Human Prx5	3×10^5	7×10^{7d}– 1.2×10^8	[86,211]
AhpE	*M. tuberculosis* AhpE	8×10^4	1.9×10^7	[109,212]
Prx6	*A. marina* Prx6	1.1×10^7	2×10^6	[213]
	Human Prx6	3.4×10^7	3.7×10^5	[214]
Bcp/PrxQ	*X. fastidiosa* PrxQ	4.5×10^7	1×10^6	[215]
	E. coli Bcp/PrxQ	1.3×10^4		[216]
	M. tuberculosis PrxQB	6×10^3	1.4×10^6	[217]
Tpx	*M. tuberculosis* TPx	$>1 \times 10^5$	1.5×10^7	[218]
	E. coli TPx	4.4×10^4		[219]
Glutathione	*P. trichocarpa* GPx5	1×10^5	1.4×10^6	[110]
Peroxidase[a]	*T. brucei* GPx3	8.7×10^4		[220]
	C. glutamicum MPx	3.7×10^5		[221]
	Human GPx7	2.6×10^3		[222]
Ohr/OsmC	*X. fastidiosa* Ohr	3×10^3	2×10^7	[127]
Thiol-disulfide	*E. coli* thioredoxin	1.05		[204]
oxido-reductases	*M. tuberculosis* Thioredoxin C		1×10^4	[223]
	T. brucei tryparedoxin	3.6^b	3.5×10^3	[207,224]
OxyR	*E. coli* OxyR	2×10^5	ND	[128]
DksA	*P. aeruginosa* DksA2	65.1	1.4×10^3	[225]
	P. aeruginosa DksA	1.33	2.4×10^3	
	S. typhimurium	0.64	ND	
GADPH	Rabbit GADPH	$\sim500^c$	2.5×10^5	[226–228]
Serum albumin	Human SA	2.3	3.8×10^3	[229]
Tyrosine phosphatase	Human PTP1B	24^c	2.2×10^{7f}	[156,230]
	Rat PTP1	9.1^c		[231]
	Human Cdc25B	164^c	2.3×10^{7f}	[232]
	Human LAR	14^c		[230,231]

(Continued)

TABLE 3.2 (Continued)

Kinetics of the Reactions of Selected Thiol-Containing Proteins with H_2O_2 and Peroxynitrite

Protein Family	Protein	$k'_{H_2O_2}$ (M^{-1} s^{-1})	k'_{ONOOH} (M^{-1} s^{-1})	References
Creatine kinase	Rabbit muscle CK		8.85×10^{5g}	[233]
	Human brain CK	0.28^b		[234]
Arylamine N-acetyltransferase	Human NAT1	7^b	5×10^{4b}	[235,236]
DJ-1	Human DJ-1	0.56^c	2.7×10^5	[237]

Apparent rate constants (k') have been reported at physiological pH unless otherwise indicated. Since thiolates and protonated H_2O_2 or ONOOH are the reactive species [75,80], pH-independent rate constants can be calculated using Equations 3.2 and 3.3, respectively. Data regarding peroxidatic cysteine residues of thiol-based peroxidases are highlighted in gray. ND = not determined. Unless indicated, reactivities of multimeric enzymes are given per monomer, except for GADPH and CK for which are given per tetramer or dimer, respectively.

Abbreviations: AhpC, alkyl hydroperoxide reductase C; Prx, peroxiredoxin; RBC, red blood cell; *T. cruzi* (mit)TXNPx and (cyt)TXNPx, mitochondrial and cytosolic tryparedoxin peroxidase from *Trypanosoma cruzi*, respectively; *T. brucei* Px 3, *Trypanosoma brucei* peroxidase 3; *C. glutamicum* MPx, *Corynebacterium glutamicum* mycothiol peroxidase; *S. cerevisiae* Tsa1 and Tsa2, thiol-specific antioxidant proteins 1 and 2 from *Saccharomyces cerevisiae*, respectively; *S. typhimurium*, *Salmonella typhimurium*; *H. sapiens*, *Homo sapiens*; *M. tuberculosis*, *Mycobacterium tuberculosis*; AhpE, alkyl hydroperoxide reductase E; *A. marina*, *Arenicola marina*; *X. fastidiosa*, *Xylella fastidiosa*; *E. coli*, *Escherichia coli*; Bcp; bacterioferritin comigratory protein; TPx, thioredoxin peroxidase; *P. trichocarpa*, *Populus trichocarpa*; SA, serum albumin; PTP1B, protein tyrosine phosphatase 1B; Cdc25B, cell division cycle 25 B; LAR, leukocyte antigen-related; Trx, thioredoxin; GADPH, glyceraldehyde 3-phosphate dehydrogenase.

[a] Cysteine-dependent glutathione peroxidases. Rate constants of the reactions of selenocysteine-dependent glutathione peroxidases such as human GPxs1 and 3 with H_2O_2 are 4.1×10^7 and 4.0×10^7 M^{-1} s^{-1}, respectively (revised in [118]), and peroxynitrite oxidizes GPx1 with a rate constant of 8×10^6 M^{-1} s^{-1} per tetramer at pH 7.4 and 25°C [238].

[b] pH =7.5. Data on human brain CK was calculated from data shown in [234].

[c] pH = 7.

[d] pH = 7.8.

[e] pH = 6.8.

[f] Values determined by competition kinetics that need to be confirmed.

[g] pH = 6.9.

are available, rates constants for Cys_P overoxidation seem to parallel (i.e., follow the same trend) those of Cys_P oxidation [109]. No kinetic data on ONOOH-mediated Cys_P overoxidation has been reported so far.

Thiol-dependent glutathione peroxidases catalyze the 2-electron reduction of different peroxides, including H_2O_2 and peroxynitrite [110]. In spite of their name, in many cases, they usually prefer thioredoxin as the reducing substrate [111–113]. Moreover, GPx7 and GPx8 are thiol-dependent GPx expressed in the ER where they participate in oxidative protein folding using protein disulfide isomerase (PDI) as electron donor

[114] (PDI is also a reductant of Prx4). Thiol-dependent GPx are present in different life kingdoms [115–117], although not so universally distributed as Prxs. In many cases, they are structurally similar to the mammalian SeCys enzyme phospholipid hydroperoxide glutathione peroxidase, which has the unique ability of catalyzing the reduction of phospholipid hydroperoxides in oxidized membranes [115]. Cys-based GPxs are usually not so rapid in the reduction of H_2O_2 as those based in SeCys, with reactivities in the range of 10^4–10^5 M^{-1} s^{-1} for the former compared with 10^7 M^{-1} s^{-1} for the latter [110,117–119], and the molecular basis for this fact is unclear, despite intense investigation [120,121]. Indeed, reported rate constant values for the oxidation of LMW thiols by H_2O_2 are similar to the one obtained for free selenocysteine by quantum mechanics calculations [122]. More recently, the rate constants for the reactions of peroxynitrite with various LMW selenols were reported to be 250- to 830-fold higher than for the corresponding thiols at physiological pH [123]. Thus, the faster reactivity cannot be totally explained by the ~10-fold higher availability of selenolate at physiological pH (pK_a of the selenol group in selenocysteine = 5.2 [124] and pK_a of the thiol group in cysteine = 8.3) [125,126] and was ascribed to a higher nucleophilicity of selenolate compared to the corresponding thiolate [123]. In turn, the single report on the reactivity of peroxynitrite with a Cys-based GPx, poplar GPx5, yielded a rate constant of 1.4×10^6 M^{-1} s^{-1} at pH 7.4 and 25°C, which together with a pK_a value of 5.2 results in a pH-independent rate constant of 7×10^6 M^{-1} s^{-1} [110].

Bacterial Ohrs play key roles in the responses toward organic hydroperoxide-mediated oxidative stress [126]. They are particularly efficient in the reduction of organic hydroperoxides, whereas their catalytic efficiency toward H_2O_2 is much more limited (in the case of *Xylella fastidiosa* Ohr, $k_{cat}/K_{m\,t\text{-}buOOH} = 2 \times 10^6$ M^{-1} s^{-1} *versus* $k_{cal}/K_{m\,H_2O_2} = 2.4 \times 10^2$ M^{-1} s^{-1} using free lipoamide as the substrate). Ohr is also able to rapidly reduce ONOOH, and in the human pathogen *Pseudomona aeruginosa*, it seems to be important for the detoxification of this peroxide [127].

Besides Cys_P in thiol-dependent peroxidases, there are some other Cys residues in proteins with thiolates that react faster with hydroperoxides than expected according to their pK_a. For example, OxyR is a bacterial transcription factor that reacts with H_2O_2 with a rate constant of 1×10^5 M^{-1} s^{-1} [128]. Taking advantage of this fast reaction, the protein has been coupled with the redox-sensing green fluorescent protein, targeted to a different cell compartments and used as tool for real-time monitoring of cellular redox status [129]. However, the possible reactions of OxyR with peroxynitrite or other hydroperoxides have not been addressed so far, and, therefore, the precise knowledge of the nature of the oxidizing species being responsible for OxyR oxidation in cells requires alternative experimentation and controls. Other examples of Cys proteins showing high reactivity toward the oxidants are listed in Table 3.2.

Thiol Oxidation by Peroxides and Redox Signaling

It should be noted that although the oxidation of protein cysteine residues in transcription factors/signaling proteins is on the basis for H_2O_2-mediated redox signaling, in fact, most of these proteins react with H_2O_2 (and usually also peroxynitrite) at relatively slow rates (for comprehensive revisions on redox signaling, see

References 130–132). In some cases, Cys oxidation in response to increased peroxide levels is indirect, with Prxs or GPxs acting as primary sensors, that then oxidize
less reactive thiols in target signaling proteins [133]. This mechanism was initially
described in yeast but is nowadays considered to be more general, with examples
also described in mammalian cells [134–136]. Other possible explanations for less
reactive Cys residue oxidation in signaling proteins involve the inactivation of the
peroxidases through overoxidation of the sulfenic acid form of the enzyme to inactive sulfinic acid forms (or through protein phosphorylation) that then allows H_2O_2 to
react with less reactive Cys [137–139]. This mechanism would allow cells to respond
more slowly to changes in H_2O_2 levels [140]. Although peroxynitrite has also been
proposed to perform signaling actions [141,142], they are less well characterized,
and its actions have been mainly associated with oxidative damage.

Stoichiometry

As indicated above, the direct reaction of thiols with peroxides yields sulfenic acids
(RSOH). Unless stabilized by the protein microenvironment, RSOHs are usually
unstable species. They react with accessible thiol groups, which are usually in excess
to form of disulfides (RSSR). In that case, the net stoichiometry is two thiols oxidized per each peroxide molecule. In other cases, two RSOHs condense to yield a
thiosulfinate (RS(O)SR) [143]. Depending on protein topology, RSOH in proteins
can also undergo reaction with neighboring amide nitrogen yielding a cyclic sulfenylamide (RSNHR) [144–146]. The net stoichiometry of the latter reactions processes
one thiol oxidation per peroxide molecule. Finally, thiols can consume more than
one peroxide molecule, when they are exposed to oxidants in excess, producing sulfinic (RSO_2H) or even sulfonic acid (RSO_3H) derivatives [147,148]. In the case of
H_2O_2-mediated oxidation, and in the absence of transition metal ions, stoichiometry
is fairly simple. However, since ONOOH is unstable, the indicated stoichiometry
only applies when thiol concentration is high, so as to outcompete other routes of
peroxynitrite decay such as homolysis or reaction with CO_2.

Reactions with Carbon Dioxide

Biological media and systems contain high concentrations of CO_2 that upon either
uncatalyzed or carbonic anhydrase-catalyzed reaction with water produces H_2CO_3 in
equilibrium with HCO_3^- ($[CO_2] = 1.3$ mM and $[HCO_3^-] = 25$ mM in arterial plasma).
Indeed, due to its high concentration in biological compartments, the HCO_3^-/CO_2
pair constitutes a main buffer system both in intracellular and extracellular media.
H_2O_2 and peroxynitrite react with CO_2 to form secondary oxidizing species that are
able to oxidize thiols. However, the physiological importance and consequences of
the reactions are distinct as discussed here:

    a. *Reaction of hydrogen peroxide with CO_2: Peroxymonocarbonate formation*—
      Peroxymonocarbonate (HCO_4^-) results from the equilibrium perhydration reac-
      tion of CO_2 with both H_2O_2 and HO_2^-, with forward rate constants of 2×10^{-2}
      and 280 M^{-1} s^{-1}, respectively, such that both routes make a similar contribution
      at pH 8 [149]. HCO_4^- formation and decay as well as Keq values are affected
      by proteins and lipids (in favor of formation) [150]. Moreover, Zn^{+2}-containing

complexes and carbonic anhydrase also lead to increased HCO_4^- concentrations [150]. Formation of HCO_4^- is also increased by CuZnSOD, where it has been proposed as an intermediate of CuZnSOD-peroxidase activity leading to $CO_3^{•-}$ formation, although this mechanism has been a matter of debate [151–154]. Even though the reaction of CO_2 with H_2O_2 is slow, it seems that cellular components could facilitate it so that it might be of biological importance. Further work is required to get a better understanding of these effects.

Once formed, HCO_4^- oxidizes thiols by a two-electron mechanism to sulfenic acids [150]. The kinetics of the reaction between HCO_4^- with GSH as well as the single thiol group of bovine serum albumin indicated that, as expected for a nucleophilic displacement SN2 reaction, rate constants are higher than those for H_2O_2 but slower than those for peroxynitrite, following a Brønsted equation that correlates with the leaving group pK_a value [150] (3.5).

$$\log k_{RS^-} = \beta_{LG}\, pK_a + A \qquad (3.5)$$

where β_{LG} is the Brønsted coefficient applicable to the leaving group (the conjugated acid of the product formed upon peroxide reduction).

Of note, peroxidatic thiols in thiol-dependent peroxidases are not only more reactive toward hydroperoxides but also have oxidizing substrate specificities that do not follow the tendency expected from Brønsted correlations, and that vary depending on the peroxidase of interest [86,109,118,155]. In the case of the peroxiredoxin alkyl hydroperoxide reductase E from *Mycobacterium tuberculosis*, rate constants have been reported as 8.2×10^4, 1.1×10^7, and 1.9×10^7 M^{-1} s^{-1} for its reactions with H_2O_2, HCO_4^-, and peroxynitrite at pH 7.4, respectively. The rate constant for the inactivation of human protein tyrosine phosphatase 1B by H_2O_2 was also found to be higher in the presence of CO_2, and a mechanism involving critical Cys oxidation by HCO_4^- was proposed [156].

b. *Reaction of peroxynitrite with CO_2: Carbonate and nitrogen dioxide radical formation*—In biological systems, CO_2 is present at mM concentrations and constitutes an important target for peroxynitrite. The nucleophilic addition of $ONOO^-$ to CO_2 occurs with a second-order rate constant of 5.8×10^4 M^{-1} s^{-1} at 37°C [78,157] and yields a transient nitroso-peroxocarbonate ($ONOOCO_2^-$), which rapidly undergoes homolysis to produce $CO_3^{•-}$ and $^•NO_2$ in ~35% yields (Equation 3.6) and the remaining 65% yielding CO_2 and nitrate (NO_3^-) [158–160].

$$ONOO^- + CO_2 \rightarrow ONOOCO_2^- \rightarrow 0.35(^•NO_2 + CO_3^{•-}) + 0.65(CO_2 + NO_3^-) \qquad (3.6)$$

As mentioned previously, with reduction potentials of $E^{°\prime}(^•NO_2/NO_2^-) = +0.99$ V and $E^{°\prime}(CO_3^{•-}, H^+/HCO_3^-) = +1.78$ V, $^•NO_2$ is a moderate oxidant and also a nitrating agent, whereas $CO_3^{•-}$ is a relatively strong one-electron oxidant [27,161,162]. Thus, CO_2 should not be considered as a scavenger of peroxynitrite, since it promotes the formation of two short-lived one-electron oxidants. $CO_3^{\cdot-}$ and $\cdot NO_2$ exhibit different chemistries. They target mainly protein thiolates, methionine, and aromatic residues. Protein tyrosine nitration is an oxidative posttranslational modification frequently used as a biomarker for peroxynitrite formation in cells and tissues, although other pathways leading to tyrosine nitration exist [163–165]. Peroxynitrite-mediated tyrosine nitration does not involve a direct reaction but occurs through a radical process: $CO_3^{\cdot-}$, $\cdot NO_2$ as well as oxo-metal complexes formed from the reaction of peroxynitrite with metals perform the one-electron oxidation of tyrosine residues to tyrosyl radicals that then recombine with $\cdot NO_2$ yielding 3-nitrotyrosine (revised in [166]).

INDIRECT REACTIONS OF H_2O_2 AND PEROXYNITRITE WITH THIOLS

In the absence of catalysis by transition metals (Me), H_2O_2 on its own is stable. On the contrary, reduced transition metal complexes (mainly Fe^{2+} or Cu^{1+}) induce the formation of hydroxyl radical ($\cdot OH$) and higher oxidation states of the metal through Fenton chemistry [167] (Equations 3.7 and 3.8).

$$H_2O_2 + Me^{n+} \rightarrow OH^- + OH^\cdot + Me^{(n+1)+} \tag{3.7}$$

$$H_2O_2 + Me^{n+} \rightarrow MeO^{n+} + H_2O \tag{3.8}$$

However, ONOOH decays through homolysis of its peroxo bond to $\cdot NO_2$ and $\cdot OH$, or in the presence of CO_2 yields $CO_3^{\cdot-}$ and $\cdot NO_2$ (see above). These are one-electron oxidants that can oxidize thiols to thiyl radicals ($RS^\cdot$) (Equation 3.9)

$$RS^- + \begin{cases} \cdot OH \\ \cdot NO_2 \\ CO_3^{\cdot-} + H^+ \end{cases} \rightarrow RS^\cdot + \begin{cases} OH^- \\ NO_2^- \\ HCO_3^- \end{cases} \tag{3.9}$$

Since homolysis occurs with a rate constant of $0.9\ s^{-1}$ at pH 7.4 and 37°C, most peroxynitrite is expected to be involved in direct reactions with its main cellular targets such as metal-containing proteins, thiols/selenols in cysteine or selenocysteine-based peroxidases, and CO_2 before decaying. Therefore, little, if any, peroxynitrite would suffer homolysis in biological systems. Nevertheless, the reaction must be taken into account when analyzing peroxynitrite reactivity *in vitro*, since, due to the high reactivity and lack of specificity of $\cdot OH$, almost every molecule is prone to be oxidized by this radical. Interpretation of the results obtained using peroxynitrite addition to biomolecules *in vitro* should be done with care, since target oxidation can occur in the absence of a direct reaction with peroxynitrite. Discrimination between the two pathways requires the measurement of peroxynitrite decay rates: Direct targets accelerate peroxynitrite decomposition, whereas indirect targets do not [168]. Such measurements require the use of rapid-mixing equipment that allows peroxynitrite decomposition to be accurately followed (peroxynitrite anion absorbs UV light, $\varepsilon_{302\ nm} = 1670$ M^{-1} cm^{-1}, [16]). Additionally, specific scavengers that react with peroxynitrite-derived radicals but not with peroxynitrite itself, such as desferrioxamine and trolox, are also helpful to establish mechanisms of oxidation mediated by peroxynitrite [169–172]. Furthermore, one-electron oxidations of thiols usually lead to oxygen consumption in two phases, the second of which has been ascribed to be the result of long-lasting chain reactions mediated by secondary free radicals [170].

Among indirect reactions, those mediated by peroxynitrite-derived radicals formed in the presence of CO_2 are the most biologically relevant. As a charged species, $CO_3^{\bullet-}$ cannot penetrate lipid phases. Thus, it is unlike to be an efficient initiator of lipid peroxidation. In contrast, $CO_3^{\bullet-}$ is an important oxidant in aqueous environment. For instance, $CO_3^{\bullet-}$ and $^{\bullet}NO_2$ oxidize cysteine to cysteinyl radical with rate constants of 4.6×10^7 M^{-1} s^{-1} [173] and 5.0×10^7 M^{-1} s^{-1} [41] at pH 7.0, respectively. Second-order rate constants for the reactions of $CO_3^{\bullet-}$ and $^{\bullet}NO_2$ with several compounds in the range of $\sim 10^6$–10^9 M^{-1} s^{-1} have been reported (for a review, see Reference 40). These fast rate constant values indicate that $CO_3^{\bullet-}$ and $^{\bullet}NO_2$ can be significantly damaging agents in *vivo*. Of note, GSH, present at millimolar concentrations inside cells, is an important scavenger of $^{\bullet}NO_2$ ($k = 2 \times 10^7$ M^{-1} s^{-1}), and this is probably the most relevant reaction *in vivo* for the inhibition of tyrosine nitration [41]. In addition, $CO_3^{\bullet-}$ reacts with GSH and cysteine with rate constants of 5.3×10^6 and 4.6×10^7 M^{-1} s^{-1}, respectively, at pH 7.0 [174]. Also, the disulfide LA and its reduced form, DHLA (a dithiol), react rapidly with $CO_3^{\bullet-}$ ($k = 1.6 \times 10^9$ and 1.7×10^8 M^{-1} s^{-1} at pH 7.4, respectively) and $^{\bullet}NO_2$ ($k = 1.3 \times 10^6$ and 2.9×10^7 M^{-1} s^{-1} at pH 7.4, respectively) [175]. However, the antioxidant function of LA and DHLA would be limited due to their low concentrations compared to other cellular antioxidants [175].

Thiyl radicals are usually unstable and decay through different mechanisms depending on conditions ($^{\bullet}NO$, O_2 and thiol/ascorbate or other cellular reductant concentrations). In the case of protein thiyl radicals, local microenvironment and accessibility are also important (for a review on the formation and fates of thiyl radicals, see [176]).

DIFFUSION CAPABILITIES ACROSS BIOLOGICAL MEMBRANES

The diffusion coefficients of H_2O_2 in water are 1.43×10^3 and 1.83×10^3 $\mu m^2/s$ at 25°C and 37°C, respectively [177]. However, in highly crowded compartments, diffusion of small molecules can be reduced ($\sim 60\%$) [178]. Being small and neutral, H_2O_2 has been considered to freely diffuse through biological membranes. Results obtained using different cellular systems indicated that plasma membranes do impose a restriction on intracellular consumption of exogenous H_2O_2 [179–184]. Indeed, concentration gradients of H_2O_2 are formed across membranes and can reach $[H_2O_2]_{ext}/[H_2O_2]_{cyt}$ values as high as ~ 20 in *Saccharomyces cerevisiae* (stationary phase) or ~ 7 in mammalian (Jurkat) cells [183]. Even higher gradients (~ 650) have been recently measured in HeLa cells exposed to H_2O_2 using a genetically encoded sensor [185]. Permeability of plasma membrane to H_2O_2 varies with cell type and growth phase, as well as lipidic composition and particularly sterol membrane content [181,186]. For example, the permeability coefficient of H_2O_2 through the plasma membrane of Jurkat T cells was determined as 2×10^{-4} cm/s [179]. H_2O_2 is more polar than water, and as in the case of water, there are specific aquaporins that decrease the activation energy and facilitate H_2O_2 diffusion through membranes [187]. In consequence, aquaporins provide a route to control membrane permeability and to adjust H_2O_2 levels in different cellular compartments and in the extracellular space [187–190]. Of note, H_2O_2 is able to modulate its own transport by changing the expression profile of these protein channels [190].

Peroxynitrite is mostly deprotonated at physiological pH, and simple diffusion of ONOO$^-$ through membranes is hampered by its negative charge. On the contrary, ONOOH rapidly crosses phospholipid membranes, with a calculated permeability coefficient of $0.8–1 \times 10^{-3}$ cm/s [191,192]. In red blood cells, the 4,49-diisothio-cyanatostilbene-2,29-disulfonic acid–sensitive Cl$^-$/HCO$_3^-$ exchanger anion channel band 3 allows the diffusion of ONOO$^-$ across membranes [193,194]. The possible role of other band 3-related proteins found in various cell types including cardio-myocytes in the diffusion of ONOO$^-$ through membranes remains to be investigated [195,196].

PREFERENTIAL TARGETS FOR H_2O_2 AND PEROXYNITRITE IN DIFFERENT CELL COMPARTMENTS

The fates of H_2O_2 and peroxynitrite are influenced by cell compartmentalization. The fraction of the oxidant that is expected to be consumed by a given target A (F) depends not only on the rate constant of the reaction with A (k_A) compared to other targets (k_N), but also on the concentration of the target.

$$F = k_A [A] / \Sigma k_N [Target_n] \qquad (3.10)$$

In the cytosol, potential targets for both oxidants include* GSH (~5 mM), Prxs such as Prx2 (20 μM) and Prx5 (5 μM), and SeCys-dependent GPxs such as GPx1 and GPx4 (~2 μM) (concentration values compiled in Reference 197). In turn, in mitochondrial matrix, potential targets of both oxidants include GSH (~5 mM), Prx3 (60 μM) and Prx5 (20 μM), GPx1 (2 μM) and GPx4 (2 μM), PTP1B (0.1 μM) [197], and in the case of peroxynitrite, aconitase and MnSOD (metal-containing proteins that react with peroxynitrite with rate constants of 1.1×10^5 and 1×10^5 M^{-1} s^{-1}, respectively [198,199]); CO_2 (1.3 mM) should also be considered (values compiled in [200]).

Given the rate constants shown in Tables 3.1 and 3.2, as well as target concentrations indicated above, it has been calculated that Prxs and to a lesser extent GPxs are expected to trap almost all H_2O_2 both in the cytosol and in the mitochondrial

---

* To illustrate the point, we have indicated reported concentration and reactivities of selected targets for the oxidants, but many other potential target exist. Additionally, given concentrations may vary with cell types and even in the same cell depending on metabolic conditions.

TABLE 3.3
Half-Lives and Diffusion Distances of H_2O_2 and Peroxynitrite in Different Cellular Compartments

Cellular Compartment	Peroxide	$t_{1/2}$ (ms)	Travel Distance in $t_{1/2}$ (µm)
Cytosol	H_2O_2	0.3–1.6	0.5–1.2
	ONOOH	0.7	0.66
Mitochondrial matrix	H_2O_2	0.7	0.72
	ONOOH	0.3	0.5

Half-lives ($t_{1/2}$) of peroxynitrite and H_2O_2 were calculated assuming low steady-state concentrations of oxidants compared to cellular targets, and physiological conditions at which targets are mainly at reduced state, using the equation $t_{1/2} = \ln 2/\sum k[S]$. The travel distance before the oxidant concentration decay to half of the initial concentration was calculated from the relationship $\ln C/C_0 = 1 \sqrt{\sum k[S]/D}$ using rate constants (k) shown in Tables 3.1 and 3.2 and target concentrations ([S]) indicated in the text, as previously reported [76,239]. Diffusion coefficients of the oxidants were corrected (by a factor of ×0.6) to consider the effect of molecular crowding.

matrix [76,197]. Similarly, most peroxynitrite is expected to react directly with Prxs and GPx both in the cytosol and in the mitochondrial matrix (in this latter compartment, rate constants at the pH of the mitochondrial matrix (~ pH 7.8) were used for the calculations when available). A smaller but not negligible fraction of peroxynitrite would react with CO_2, which would be responsible for most of the peroxynitrite-dependent one-electron oxidation/nitration reactions [201].

The half-life of the oxidants in different biological compartments can be estimated from their reactivities as well as concentrations of different local targets (Table 3.3). Using Fick's second law and the above mentioned diffusion coefficient of H_2O_2 or a diffusion coefficient of nitrate (1500 µm²/s) for ONOOH [202], after correction for molecular crowding, the distances that the oxidants can travel before decaying to half of their initial concentration could been calculated for the different cellular compartments (Table 3.3). The half-lives of both oxidants are expected to be in the ms or less range, while calculated travel distances were less than 1 µm, that is, lower than most cell diameters. The situation may be different under conditions where peroxidase recycling by thioredoxin/thioredoxin reductase (TR)/NADPH or GSH/glutathione reductase (GR)/NADPH becomes rate-limiting, which would lead to the accumulation of oxidized forms of the enzymes, or under conditions where peroxidases get inactivated because of peroxidatic thiol overoxidation during severe oxidative stress.

CONCLUSIONS

Hydrogen peroxide and peroxynitrite are two biologically-relevant peroxides. Their reactivities, biological half-lives, and diffusion capabilities differ due to their dissimilar physicochemical properties. Hydrogen peroxide tends to be a more stable molecule in biological systems and to be part of signaling processes. Peroxynitrite is typically unstable, more reactive, and mostly related to oxidative damage.

Notably, while the reaction rates of peroxynitrite with LMW thiols are usually about three orders of magnitude higher than those of H_2O_2, in the case of many peroxidatic thiols of peroxidases, active site microenvironments make reactions rates to equalize. The presence of ·NO in sites of $O_2^{·-}$ formation switches the generation of H_2O_2 toward peroxynitrite. Both H_2O_2 and peroxynitrite promote two-electron oxidation of thiols to the corresponding sulfenic acids, but also can cause their one-electron oxidation to thiyl radical via secondary radicals. For H_2O_2, ·OH radicals usually arise from transition-metal catalyzed reactions and in the case of peroxynitrite, CO_2 is a key reactant to yield the radicals $CO_3^{·-}$ and ·NO_2. Thus, the redox mechanisms of thiol oxidation and the biological actions of H_2O_2 and peroxynitrite do not necessarily overlap.

ACKNOWLEDGMENTS

This work was supported by grants from Universidad de la República (CSIC Grupos 767 and CSIC i+d 367) and Espacio Interdisciplinario. Additional support was provided by the Programa de Desarrollo de Ciencias Básicas (PEDECIBA, Uruguay). A postdoctoral fellowship to Ari Zeida provided by Agencia Nacional de Investigación e Innovación (ANII, Uruguay) is gratefully acknowledged.

REFERENCES

1. Fridovich, I. 1978. The biology of oxygen radicals. *Science* 201 (4359):875–880.
2. Fridovich, I. 1978. Superoxide dismutases: Defence against endogenous superoxide radical. *Ciba Found Symp* 65:77–93.
3. McCord, J. M., B. B. Keele, Jr., and I. Fridovich. 1971. An enzyme-based theory of obligate anaerobiosis: The physiological function of superoxide dismutase. *Proc Natl Acad Sci U S A* 68 (5):1024–1027.
4. Gardner, P. R., I. Raineri, L. B. Epstein, and C. W. White. 1995. Superoxide radical and iron modulate aconitase activity in mammalian cells. *J Biol Chem* 270 (22):13399–13405.
5. Vasquez-Vivar, J., B. Kalyanaraman, and M. C. Kennedy. 2000. Mitochondrial aconitase is a source of hydroxyl radical. An electron spin resonance investigation. *J Biol Chem* 275 (19):14064–14069.
6. Sandri, G., E. Panfili, and L. Ernster. 1990. Hydrogen peroxide production by monoamine oxidase in isolated rat-brain mitochondria: Its effect on glutathione levels and Ca^{2+} efflux. *Biochim Biophys Acta* 1035 (3):300–305.
7. Simonson, S. G., J. Zhang, A. T. Canada, Jr., Y. F. Su, H. Benveniste, and C. A. Piantadosi. 1993. Hydrogen peroxide production by monoamine oxidase during ischemia-reperfusion in the rat brain. *J Cereb Blood Flow Metab* 13 (1):125–134.
8. Kellogg, E. W., 3rd and I. Fridovich. 1975. Superoxide, hydrogen peroxide, and singlet oxygen in lipid peroxidation by a xanthine oxidase system. *J Biol Chem* 250 (22):8812–8817.
9. Tu, B. P. and J. S. Weissman. 2002. The FAD- and O(2)-dependent reaction cycle of Ero1-mediated oxidative protein folding in the endoplasmic reticulum. *Mol Cell* 10 (5):983–994.
10. Badwey, J. A., J. M. Robinson, M. J. Karnovsky, and M. L. Karnovsky. 1981. Superoxide production by an unusual aldehyde oxidase in guinea pig granulocytes. Characterization and cytochemical localization. *J Biol Chem* 256 (7):3479–3486.

11. Kissner, R., T. Nauser, P. Bugnon, P. G. Lye, and W. H. Koppenol. 1997. Formation and properties of peroxynitrite as studied by laser flash photolysis, high-pressure stopped-flow technique, and pulse radiolysis. *Chem Res Toxicol* 10 (11):1285–1292.

12. Goldstein, S. and G. Czapski. 1995. The reaction of NO. with O_2^{-} and HO_2: A pulse radiolysis study. *Free Radic Biol Med* 19 (4):505–510.

13. Pryor, W. A. and G. L. Squadrito. 1995. The chemistry of peroxynitrite: A product from the reaction of nitric oxide with superoxide. *Am J Physiol* 268 (5 Pt 1): L699–L722.

14. Beckman, J. S., T. W. Beckman, J. Chen, P. A. Marshall, and B. A. Freeman. 1990. Apparent hydroxyl radical production by peroxynitrite: Implications for endothelial injury from nitric oxide and superoxide. *Proc Natl Acad Sci U S A* 87 (4):1620–1624.

15. Radi, R., J. S. Beckman, K. M. Bush, and B. A. Freeman. 1991. Peroxynitrite-induced membrane lipid peroxidation: The cytotoxic potential of superoxide and nitric oxide. *Arch Biochem Biophys* 288 (2):481–487.

16. Radi, R., J. S. Beckman, K. M. Bush, and B. A. Freeman. 1991. Peroxynitrite oxidation of sulfhydryls. The cytotoxic potential of superoxide and nitric oxide. *J Biol Chem* 266 (7):4244–4250.

17. Smulik, R., D. Debski, J. Zielonka, B. Michalowski, J. Adamus, A. Marcinek, B. Kalyanaraman, and A. Sikora. 2014. Nitroxyl (HNO) reacts with molecular oxygen and forms peroxynitrite at physiological pH. Biological implications. *J Biol Chem* 289 (51):35570–35581.

18. Miranda, K. M., N. Paolocci, T. Katori, D. D. Thomas, E. Ford, M. D. Bartberger, M. G. Espey et al. 2003. A biochemical rationale for the discrete behavior of nitroxyl and nitric oxide in the cardiovascular system. *Proc Natl Acad Sci U S A* 100 (16):9196–9201.

19. Liochev, S. I. and I. Fridovich. 2003. The mode of decomposition of Angeli's salt ($Na_2N_2O_3$) and the effects thereon of oxygen, nitrite, superoxide dismutase, and glutathione. *Free Radic Biol Med* 34 (11):1399–1404.

20. Vanderkooi, J. M., W. W. Wright, and M. Erecinska. 1994. Nitric oxide diffusion coefficients in solutions, proteins and membranes determined by phosphorescence. *Biochim Biophys Acta* 1207 (2):249–254.

21. Denicola, A., J. M. Souza, R. Radi, and E. Lissi. 1996. Nitric oxide diffusion in membranes determined by fluorescence quenching. *Arch Biochem Biophys* 328 (1):208–212.

22. Bielski, B. H. J., D. E. Cabelli, R. L. Arudi, and A. B. Ross. 1985. Reactivity of HO_2/O_2^- radicals in aqueous solution. *J Phys Chem Ref Data* 14:1041–1100.

23. Lynch, R. E. and I. Fridovich. 1978. Permeation of the erythrocyte stroma by superoxide radical. *J Biol Chem* 253 (13):4697–4699.

24. Mikkelsen, R. B. and P. Wardman. 2003. Biological chemistry of reactive oxygen and nitrogen and radiation-induced signal transduction mechanisms. *Oncogene* 22 (37):5734–5754.

25. Kehrer, J. P. and L. O. Klotz. 2015. Free radicals and related reactive species as mediators of tissue injury and disease: Implications for Health. *Crit Rev Toxicol* 45 (9):765–798.

26. Wood, P. M. 1988. The potential diagram for oxygen at pH 7. *Biochem J* 253 (1):287–289.

27. Koppenol, W. H., J. J. Moreno, W. A. Pryor, H. Ischiropoulos, and J. S. Beckman. 1992. Peroxynitrite, a cloaked oxidant formed by nitric oxide and superoxide. *Chem Res Toxicol* 5 (6):834–842.

28. Koppenol, W. H. and R. Kissner. 1998. Can O=NOOH undergo homolysis? *Chem Res Toxicol* 11 (2):87–90.

29. Kissner, R., T. Nauser, C. Kurz, and W. H. Koppenol. 2003. Peroxynitrous acid—Where is the hydroxyl radical? *IUBMB Life* 55 (10–11):567–572.

30. Cremer, D. 1983. General and theoretical aspects of the peroxide group. In *Peroxides*, S. Patai, ed., pp. 1–92. John Wiley & Sons, Chichester, UK.

31. Jones, C. W. 1999. Activation of hydrogen peroxide using inorganic and organic species. In *Applications of Hydrogen Peroxide and Derivatives*, C. W. Jones, ed., pp. 37–78. Cambridge, U.K.: RSC Clean Technologies Monographs.

32. Bach, R. D., P. Y. Ayala, and H. B. Schlegel. 1996. A reassessment of the bond dissociation energies of peroxides. *J Am Chem Soc* 118:12758–12765.

33. Gupta, D., B. Harish, R. Kissner, and W. H. Koppenol. 2009. Peroxynitrate is formed rapidly during decomposition of peroxynitrite at neutral pH. *Dalton Trans* 29:5730–5736.

34. Goldstein, S. and G. Merenyi. 2008. The chemistry of peroxynitrite: Implications for biological activity. *Methods Enzymol* 436:49–61.

35. Gerasimov, O. V. and S. V. Lymar. 1999. The yield of hydroxyl radical from the decomposition of peroxynitrous acid. *Inorg Chem* 38 (19):4317–4321.

36. Hodges, G. R. and K. U. Ingold. 1999. Cage-escape of geminate radical pairs can produce peroxynitrate from peroxynitrite under a wide variety of experimental conditions. *J Am Chem Soc* 121 (46):10695–10701.

37. Augusto, O., R. M. Gatti, and R. Radi. 1994. Spin-trapping studies of peroxynitrite decomposition and of 3-morpholinosydnonimine N-ethylcarbamide autooxidation: Direct evidence for metal-independent formation of free radical intermediates. *Arch Biochem Biophys* 310 (1):118–125.

38. Gatti, R. M., B. Alvarez, J. Vasquez-Vivar, R. Radi, and O. Augusto. 1998. Formation of spin trap adducts during the decomposition of peroxynitrite. *Arch Biochem Biophys* 349 (1):36–46.

39. Buettner, G. R. 1993. The pecking order of free radicals and antioxidants: Lipid peroxidation, alpha-tocopherol, and ascorbate. *Arch Biochem Biophys* 300 (2):535–543.

40. Augusto, O., M. G. Bonini, A. M. Amanso, E. Linares, C. C. Santos, and S. L. De Menezes. 2002. Nitrogen dioxide and carbonate radical anion: Two emerging radicals in biology. *Free Radic Biol Med* 32 (9):841–859.

41. Ford, E., M. N. Hughes, and P. Wardman. 2002. Kinetics of the reactions of nitrogen dioxide with glutathione, cysteine, and uric acid at physiological pH. *Free Radic Biol Med* 32 (12):1314–1323.

42. Sies, H. 1999. Glutathione and its role in cellular functions. *Free Radic Biol Med* 27 (9–10):916–921.

43. Carlberg, I. and B. Mannervik. 1975. Purification and characterization of the flavoenzyme glutathione reductase from rat liver. *J Biol Chem* 250 (14):5475–5480.

44. Newton, G. L., K. Arnold, M. S. Price, C. Sherrill, S. B. Delcardayre, Y. Aharonowitz, G. Cohen, J. Davies, R. C. Fahey, and C. Davis. 1996. Distribution of thiols in microorganisms: Mycothiol is a major thiol in most actinomycetes. *J Bacteriol* 178 (7):1990–1995.

45. Fairlamb, A. H., P. Blackburn, P. Ulrich, B. T. Chait, and A. Cerami. 1985. Trypanothione: A novel bis(glutathionyl)spermidine cofactor for glutathione reductase in trypanosomatids. *Science* 227 (4693):1485–1487.

46. Newton, G. L., M. Rawat, J. J. La Clair, V. K. Jothivasan, T. Budiarto, C. J. Hamilton, A. Claiborne, J. D. Helmann, and R. C. Fahey. 2009. Bacillithiol is an antioxidant thiol produced in *Bacilli. Nat Chem Biol* 5 (9):625–627.

47. Van Laer, K., C. J. Hamilton, and J. Messens. 2013. Low-molecular-weight thiols in thiol-disulfide exchange. *Antioxid Redox Signal* 18 (13):1642–1653.

48. Boylan, J. A., C. S. Hummel, S. Benoit, J. Garcia-Lara, J. Treglown-Downey, E. J. Crane, 3rd, and F. C. Gherardini. 2006. *Borrelia burgdorferi* bb0728 encodes a coenzyme A disulfide reductase whose function suggests a role in intracellular redox and the oxidative stress response. *Mol Microbiol* 59 (2):475–486.

49. Lipmann, F. 1953. On chemistry and function of coenzyme A. *Bacteriol Rev* 17 (1):1–16.

50. Packer, L., K. Kraemer, and G. Rimbach. 2001. Molecular aspects of lipoic acid in the prevention of diabetes complications. *Nutrition* 17 (10):888–895.
51. Smith, A. R., S. V. Shenvi, M. Widlansky, J. H. Suh, and T. M. Hagen. 2004. Lipoic acid as a potential therapy for chronic diseases associated with oxidative stress. *Curr Med Chem* 11 (9):1135–1146.
52. Genghof, D. S. 1970. Biosynthesis of ergothioneine and hercynine by fungi and Actinomycetales. *J Bacteriol* 103 (2):475–478.
53. Turner, E., R. Klevit, L. J. Hager, and B. M. Shapiro. 1987. Ovothiols, a family of redox-active mercaptohistidine compounds from marine invertebrate eggs. *Biochemistry* 26 (13):4028–4036.
54. Jang, J. H., O. I. Aruoma, L. S. Jen, H. Y. Chung, and Y. J. Surh. 2004. Ergothioneine rescues PC12 cells from beta-amyloid-induced apoptotic death. *Free Radic Biol Med* 36 (3):288–299.
55. Paul, B. D. and S. H. Snyder. 2010. The unusual amino acid L-ergothioneine is a physiologic cytoprotectant. *Cell Death Differ* 17 (7):1134–1140.
56. Grundemann, D., S. Harlfinger, S. Golz, A. Geerts, A. Lazar, R. Berkels, N. Jung, A. Rubbert, and E. Schomig. 2005. Discovery of the ergothioneine transporter. *Proc Natl Acad Sci U S A* 102 (14):5256–5261.
57. Marjanovic, B., M. G. Simic, and S. V. Jovanovic. 1995. Heterocyclic thiols as antioxidants: Why ovothiol C is a better antioxidant than ergothioneine. *Free Radic Biol Med* 18 (4):679–685.
58. Servillo, L., D. Castaldo, R. Casale, N. D'Onofrio, A. Giovane, D. Cautela, and M. L. Balestrieri. 2015. An uncommon redox behavior sheds light on the cellular antioxidant properties of ergothioneine. *Free Radic Biol Med* 79:228–236.
59. D'Onofrio, N., L. Servillo, A. Giovane, R. Casale, M. Vitiello, R. Marfella, G. Paolisso, and M. L. Balestrieri. 2016. Ergothioneine oxidation in the protection against high-glucose induced endothelial senescence: Involvement of SIRT1 and SIRT6. *Free Radic Biol Med* 96:211–222.
60. Jones, D. P., Y. M. Go, C. L. Anderson, T. R. Ziegler, J. M. Kinkade, Jr., and W. G. Kirlin. 2004. Cysteine/cystine couple is a newly recognized node in the circuitry for biologic redox signaling and control. *FASEB J* 18 (11):1246–1248.
61. Mansoor, M. A., A. M. Svardal, and P. M. Ueland. 1992. Determination of the in vivo redox status of cysteine, cysteinylglycine, homocysteine, and glutathione in human plasma. *Anal Biochem* 200 (2):218–229.
62. Cuperus, R. A., A. O. Muijsers, and R. Wever. 1985. Antiarthritic drugs containing thiol groups scavenge hypochlorite and inhibit its formation by myeloperoxidase from human leukocytes. A therapeutic mechanism of these drugs in rheumatoid arthritis? *Arthritis Rheum* 28 (11):1228–1233.
63. Roederer, M., F. J. Staal, P. A. Raju, S. W. Ela, and L. A. Herzenberg. 1990. Cytokine-stimulated human immunodeficiency virus replication is inhibited by N-acetyl-L-cysteine. *Proc Natl Acad Sci U S A* 87 (12):4884–4888.
64. Badarau, A., A. Llinas, A. P. Laws, C. Damblon, and M. I. Page. 2005. Inhibitors of metallo-beta-lactamase generated from beta-lactam antibiotics. *Biochemistry* 44 (24):8578–8589.
65. Pawlak, R., E. Chabielska, T. Matys, I. Kucharewicz, and W. Buczko. 2000. The role of the thiol group in the antithrombotic action of captopril. *Thromb Haemost* 84 (5):919–920.
66. Cotton, F. and G. Wilkinson. 1988. *Advanced Inorganic Chemistry*. John Wiley & Sons Inc, New York.
67. Lefer, D. J. 2007. A new gaseous signaling molecule emerges: Cardioprotective role of hydrogen sulfide. *Proc Natl Acad Sci U S A* 104 (46):17907–17908.

68. Li, L., P. Rose, and P. K. Moore. 2011. Hydrogen sulfide and cell signaling. *Annu Rev Pharmacol Toxicol* 51:169–187.
69. Carballal, S., M. Trujillo, E. Cuevasanta, S. Bartesaghi, M. N. Moller, L. K. Folkes, M. A. Garcia-Bereguiain et al. 2011. Reactivity of hydrogen sulfide with peroxynitrite and other oxidants of biological interest. *Free Radic Biol Med* 50 (1):196–205.
70. Cuevasanta, E., M. Lange, J. Bonanata, E. L. Coitino, G. Ferrer-Sueta, M. R. Filipovic, and B. Alvarez. 2015. Reaction of hydrogen sulfide with disulfide and sulfenic acid to form the strongly nucleophilic persulfide. *J Biol Chem* 290 (45):26866–26880.
71. Miseta, A. and P. Csutora. 2000. Relationship between the occurrence of cysteine in proteins and the complexity of organisms. *Mol Biol Evol* 17 (8):1232–1239.
72. Fomenko, D. E., S. M. Marino, and V. N. Gladyshev. 2008. Functional diversity of cysteine residues in proteins and unique features of catalytic redox-active cysteines in thiol oxidoreductases. *Mol Cells* 26 (3):228–235.
73. Jones, D. P. 2006. Redefining oxidative stress. *Antioxid Redox Signal* 8 (9–10): 1865–1879.
74. Sies, H. 2017. Hydrogen peroxide as a central redox signaling molecule in physiological oxidative stress: Oxidative eustress. *Redox Biol* 11:613–619.
75. Winterbourn, C. C. and D. Metodiewa. 1999. Reactivity of biologically important thiol compounds with superoxide and hydrogen peroxide. *Free Radic Biol Med* 27 (3–4):322–328.
76. Winterbourn, C. C. 2013. The biological chemistry of hydrogen peroxide. *Methods Enzymol* 528:3–25.
77. Trujillo, M., B. Alvarez, and R. Radi. 2016. One- and two-electron oxidation of thiols: Mechanisms, kinetics and biological fates. *Free Radic Res* 50 (2):150–171.
78. Denicola, A., B. A. Freeman, M. Trujillo, and R. Radi. 1996. Peroxynitrite reaction with carbon dioxide/bicarbonate: Kinetics and influence on peroxynitrite-mediated oxidations. *Arch Biochem Biophys* 333 (1):49–58.
79. Claiborne, A., H. Miller, D. Parsonage, and R. P. Ross. 1993. Protein-sulfenic acid stabilization and function in enzyme catalysis and gene regulation. *FASEB J* 7 (15): 1483–1490.
80. Trujillo, M. and R. Radi. 2002. Peroxynitrite reaction with the reduced and the oxidized forms of lipoic acid: New insights into the reaction of peroxynitrite with thiols. *Arch Biochem Biophys* 397 (1):91–98.
81. Zeida, A., R. Babbush, M. C. Lebrero, M. Trujillo, R. Radi, and D. A. Estrin. 2012. Molecular basis of the mechanism of thiol oxidation by hydrogen peroxide in aqueous solution: Challenging the SN2 paradigm. *Chem Res Toxicol* 25 (3):741–746.
82. Zeida, A., M. C. Gonzalez Lebrero, R. Radi, M. Trujillo, and D. A. Estrin. 2013. Mechanism of cysteine oxidation by peroxynitrite: An integrated experimental and theoretical study. *Arch Biochem Biophys* 539 (1):81–86.
83. Perrin, D. D. 1982. *Ionization Constants of Inorganic Acids and Bases in Aqueous Solution*, 2nd edn. Oxford, U.K.: Pergamon.
84. Santo-Domingo, J. and N. Demaurex. 2012. Perspectives on: SGP symposium on mitochondrial physiology and medicine: The renaissance of mitochondrial pH. *J Gen Physiol* 139 (6):415–423.
85. Portillo-Ledesma, S., F. Sardi, B. Manta, M. V. Tourn, A. Clippe, B. Knoops, B. Alvarez, E. L. Coitino, and G. Ferrer-Sueta. 2014. Deconstructing the catalytic efficiency of peroxiredoxin-5 peroxidatic cysteine. *Biochemistry* 53 (38):6113–6125.
86. Trujillo, M., A. Clippe, B. Manta, G. Ferrer-Sueta, A. Smeets, J. P. Declercq, B. Knoops, and R. Radi. 2007. Pre-steady state kinetic characterization of human peroxiredoxin 5: Taking advantage of Trp84 fluorescence increase upon oxidation. *Arch Biochem Biophys* 467 (1):95–106.

87. Reyes, A. M., B. Pedre, M. I. De Armas, M. A. Tossounian, R. Radi, J. Messens, and M. Trujillo. In Press. Chemistry and redox biology of mycothiol. *Antioxid Redox Signal.* (Epub ahead of print. doi: 10.1089/ars.2017.7074).
88. Ferrer-Sueta, G., B. Manta, H. Botti, R. Radi, M. Trujillo, and A. Denicola. 2011. Factors affecting protein thiol reactivity and specificity in peroxide reduction. *Chem Res Toxicol* 24 (4):434–450.
89. Trujillo, M., B. Alvarez, J. M. Souza, N. Romero, L. Castro, L. Thomson, and R. Radi. 2010. Mechanisms and biological cosnequences of peroxynitrite-dependent protein oxidation and nitration. In *Nitric Oxide. Biology and Pathobiology*, L. J. Ignarro, ed., pp. 61–102. San Diego, CA: Elsevier.
90. Cussiol, J. R., T. G. Alegria, L. I. Szweda, and L. E. Netto. 2010. Ohr (organic hydroperoxide resistance protein) possesses a previously undescribed activity, lipoyl-dependent peroxidase. *J Biol Chem* 285 (29):21943–21950.
91. Bryk, R., C. D. Lima, H. Erdjument-Bromage, P. Tempst, and C. Nathan. 2002. Metabolic enzymes of mycobacteria linked to antioxidant defense by a thioredoxin-like protein. *Science* 295 (5557):1073–1077.
92. Flohe, L., H. Budde, and B. Hofmann. 2003. Peroxiredoxins in antioxidant defense and redox regulation. *Biofactors* 19 (1–2):3–10.
93. Tripathi, B. N., I. Bhatt, and K. J. Dietz. 2009. Peroxiredoxins: A less studied component of hydrogen peroxide detoxification in photosynthetic organisms. *Protoplasma* 235 (1–4):3–15.
94. Sztajer, H., B. Gamain, K. D. Aumann, C. Slomianny, K. Becker, R. Brigelius-Flohe, and L. Flohe. 2001. The putative glutathione peroxidase gene of *Plasmodium falciparum* codes for a thioredoxin peroxidase. *J Biol Chem* 276 (10):7397–7403.
95. Hofmann, B., H. J. Hecht, and L. Flohe. 2002. Peroxiredoxins. *Biol Chem* 383 (3–4):347–364.
96. Djuika, C. F., S. Fiedler, M. Schnolzer, C. Sanchez, M. Lanzer, and M. Deponte. 2013. *Plasmodium falciparum* antioxidant protein as a model enzyme for a special class of glutaredoxin/glutathione-dependent peroxiredoxins. *Biochim Biophys Acta* 1830 (8):4073–4090.
97. Rouhier, N., E. Gelhaye, and J. P. Jacquot. 2002. Glutaredoxin-dependent peroxiredoxin from poplar: Protein-protein interaction and catalytic mechanism. *J Biol Chem* 277 (16):13609–13614.
98. Monteiro, G., B. B. Horta, D. C. Pimenta, O. Augusto, and L. E. Netto. 2007. Reduction of 1-Cys peroxiredoxins by ascorbate changes the thiol-specific antioxidant paradigm, revealing another function of vitamin C. *Proc Natl Acad Sci U S A* 104 (12):4886–4891.
99. Manevich, Y., S. I. Feinstein, and A. B. Fisher. 2004. Activation of the antioxidant enzyme 1-CYS peroxiredoxin requires glutathionylation mediated by heterodimerization with pi GST. *Proc Natl Acad Sci U S A* 101 (11):3780–3785.
100. Hall, A., K. Nelson, L. B. Poole, and P. A. Karplus. 2011. Structure-based insights into the catalytic power and conformational dexterity of peroxiredoxins. *Antioxid Redox Signal* 15 (3):795–815.
101. Zeida, A., A. M. Reyes, M. C. Lebrero, R. Radi, M. Trujillo, and D. A. Estrin. 2014. The extraordinary catalytic ability of peroxiredoxins: A combined experimental and QM/MM study on the fast thiol oxidation step. *Chem Commun (Camb)* 50 (70):10070–10073.
102. Nagy, P., A. Karton, A. Betz, A. V. Peskin, P. Pace, R. J. O'Reilly, M. B. Hampton, L. Radom, and C. C. Winterbourn. 2011. Model for the exceptional reactivity of peroxiredoxins 2 and 3 with hydrogen peroxide: A kinetic and computational study. *J Biol Chem* 286 (20):18048–18055.

103. Perkins, A., D. Parsonage, K. J. Nelson, O. M. Ogba, P. H. Cheong, L. B. Poole, and P. A. Karplus. 2016. Peroxiredoxin catalysis at atomic resolution. *Structure* 24 (10):1668–1678.
104. Flohe, L., H. Budde, K. Bruns, H. Castro, J. Clos, B. Hofmann, S. Kansal-Kalavar et al. 2002. Tryparedoxin peroxidase of *Leishmania donovani*: Molecular cloning, heterologous expression, specificity, and catalytic mechanism. *Arch Biochem Biophys* 397 (2):324–335.
105. Rhee, S. G. 2016. Overview on peroxiredoxin. *Mol Cells* 39 (1):1–5.
106. Woo, H. A., W. Jeong, T. S. Chang, K. J. Park, S. J. Park, J. S. Yang, and S. G. Rhee. 2005. Reduction of cysteine sulfinic acid by sulfiredoxin is specific to 2-cys peroxiredoxins. *J Biol Chem* 280 (5):3125–3128.
107. Peshenko, I. V. and H. Shichi. 2001. Oxidation of active center cysteine of bovine 1-Cys peroxiredoxin to the cysteine sulfenic acid form by peroxide and peroxynitrite. *Free Radic Biol Med* 31 (3):292–303.
108. Manta, B., M. Hugo, C. Ortiz, G. Ferrer-Sueta, M. Trujillo, and A. Denicola. 2009. The peroxidase and peroxynitrite reductase activity of human erythrocyte peroxiredoxin 2. *Arch Biochem Biophys* 484 (2):146–154.
109. Reyes, A. M., M. Hugo, A. Trostchansky, L. Capece, R. Radi, and M. Trujillo. 2011. Oxidizing substrate specificity of *Mycobacterium tuberculosis* alkyl hydroperoxide reductase E: Kinetics and mechanisms of oxidation and overoxidation. *Free Radic Biol Med* 51 (2):464–473.
110. Selles, B., M. Hugo, M. Trujillo, V. Srivastava, G. Wingsle, J. P. Jacquot, R. Radi, and N. Rouhier. 2012. Hydroperoxide and peroxynitrite reductase activity of poplar thioredoxin-dependent glutathione peroxidase 5: Kinetics, catalytic mechanism and oxidative inactivation. *Biochem J* 442 (2):369–380.
111. Koh, C. S., C. Didierjean, N. Navrot, S. Panjikar, G. Mulliert, N. Rouhier, J. P. Jacquot, A. Aubry, O. Shawkataly, and C. Corbier. 2007. Crystal structures of a poplar thioredoxin peroxidase that exhibits the structure of glutathione peroxidases: Insights into redox-driven conformational changes. *J Mol Biol* 370 (3):512–529.
112. Herbette, S., C. Lenne, N. Leblanc, J. L. Julien, J. R. Drevet, and P. Roeckel-Drevet. 2002. Two GPX-like proteins from *Lycopersicon esculentum* and *Helianthus annuus* are antioxidant enzymes with phospholipid hydroperoxide glutathione peroxidase and thioredoxin peroxidase activities. *Eur J Biochem* 269 (9):2414–2420.
113. Jung, B. G., K. O. Lee, S. S. Lee, Y. H. Chi, H. H. Jang, S. S. Kang, K. Lee et al. 2002. A Chinese cabbage cDNA with high sequence identity to phospholipid hydroperoxide glutathione peroxidases encodes a novel isoform of thioredoxin-dependent peroxidase. *J Biol Chem* 277 (15):12572–12578.
114. Nguyen, V. D., M. J. Saaranen, A. R. Karala, A. K. Lappi, L. Wang, I. B. Raykhel, H. I. Alanen, K. E. Salo, C. C. Wang, and L. W. Ruddock. 2011. Two endoplasmic reticulum PDI peroxidases increase the efficiency of the use of peroxide during disulfide bond formation. *J Mol Biol* 406 (3):503–515.
115. Bae, Y. A., G. B. Cai, S. H. Kim, Y. G. Zo, and Y. Kong. 2009. Modular evolution of glutathione peroxidase genes in association with different biochemical properties of their encoded proteins in invertebrate animals. *BMC Evol Biol* 9:72.
116. Arenas, F. A., W. A. Diaz, C. A. Leal, J. M. Perez-Donoso, J. A. Imlay, and C. C. Vasquez. 2010. The *Escherichia coli* btuE gene, encodes a glutathione peroxidase that is induced under oxidative stress conditions. *Biochem Biophys Res Commun* 398 (4):690–694.
117. Pedre, B., I. Van Molle, A. F. Villadangos, K. Wahni, D. Vertommen, L. Turell, H. Erdogan, L. M. Mateos, and J. Messens. 2015. The *Corynebacterium glutamicum* mycothiol peroxidase is a reactive oxygen species-scavenging enzyme that shows promiscuity in thiol redox control. *Mol Microbiol* 96 (6):1176–1191.

118. Toppo, S., L. Flohe, F. Ursini, S. Vanin, and M. Maiorino. 2009. Catalytic mechanisms and specificities of glutathione peroxidases: Variations of a basic scheme. *Biochim Biophys Acta* 1790 (11):1486–1500.
119. Schlecker, T., A. Schmidt, N. Dirdjaja, F. Voncken, C. Clayton, and R. L. Krauth-Siegel. 2005. Substrate specificity, localization, and essential role of the glutathione peroxidase-type tryparedoxin peroxidases in *Trypanosoma brucei*. *J Biol Chem* 280 (15):14385–14394.
120. Cardey, B., and M. Enescu. 2007. Selenocysteine versus cysteine reactivity: A theoretical study of their oxidation by hydrogen peroxide. *J Phys Chem A* 111 (4):673–678.
121. Hondal, R. J., S. M. Marino, and V. N. Gladyshev. 2013. Selenocysteine in thiol/disulfide-like exchange reactions. *Antioxid Redox Signal* 18 (13):1675–1689.
122. Cardey, B. and M. Enescu. 2005. A computational study of thiolate and selenolate oxidation by hydrogen peroxide. *Chemphyschem* 6 (6):1175–1180.
123. Storkey, C., D. I. Pattison, M. T. Ignasiak, C. H. Schiesser, and M. J. Davies. 2015. Kinetics of reaction of peroxynitrite with selenium- and sulfur-containing compounds: Absolute rate constants and assessment of biological significance. *Free Radic Biol Med* 89:1049–1056.
124. Stadtman, T. C. 1996. Selenocysteine. *Annu Rev Biochem* 65:83–100.
125. Wessjohann, L. A., A. Schneider, M. Abbas, and W. Brandt. 2007. Selenium in chemistry and biochemistry in comparison to sulfur. *Biol Chem* 388 (10):997–1006.
126. Atichartpongkul, S., S. Loprasert, P. Vattanaviboon, W. Whangsuk, J. D. Helmann, and S. Mongkolsuk. 2001. Bacterial Ohr and OsmC paralogues define two protein families with distinct functions and patterns of expression. *Microbiology* 147 (Pt 7):1775–1782.
127. Alegria, T. G., D. A. Meireles, J. R. Cussiol, M. Hugo, M. Trujillo, M. A. de Oliveira, S. Miyamoto et al. 2017. Ohr plays a central role in bacterial responses against fatty acid hydroperoxides and peroxynitrite. *Proc Natl Acad Sci U S A* 114 (2):E132–E141.
128. Aslund, F., M. Zheng, J. Beckwith, and G. Storz. 1999. Regulation of the OxyR transcription factor by hydrogen peroxide and the cellular thiol-disulfide status. *Proc Natl Acad Sci U S A* 96 (11):6161–6165.
129. Malinouski, M., Y. Zhou, V. V. Belousov, D. L. Hatfield, and V. N. Gladyshev. 2011. Hydrogen peroxide probes directed to different cellular compartments. *PLoS One* 6 (1):e14564.
130. Flohe, L. 2010. Changing paradigms in thiology from antioxidant defense toward redox regulation. *Methods Enzymol* 473:1–39.
131. Forman, H. J., F. Ursini, and M. Maiorino. 2014. An overview of mechanisms of redox signaling. *J Mol Cell Cardiol* 73:2–9.
132. Sies, H. 2014. Role of metabolic H_2O_2 generation: Redox signaling and oxidative stress. *J Biol Chem* 289 (13):8735–8741.
133. Flohe, L. 2015. The impact of thiol peroxidases on redox regulation. *Free Radic Res.* 50 (2):126–142.
134. Delaunay, A., D. Pflieger, M. B. Barrault, J. Vinh, and M. B. Toledano. 2002. A thiol peroxidase is an H_2O_2 receptor and redox-transducer in gene activation. *Cell* 111 (4):471–481.
135. Sobotta, M. C., W. Liou, S. Stocker, D. Talwar, M. Oehler, T. Ruppert, A. N. Scharf, and T. P. Dick. 2015. Peroxiredoxin-2 and STAT3 form a redox relay for H_2O_2 signaling. *Nat Chem Biol* 11 (1):64–70.
136. Wei, P. C., Y. H. Hsieh, M. I. Su, X. Jiang, P. H. Hsu, W. T. Lo, J. Y. Weng et al. 2012. Loss of the oxidative stress sensor NPGPx compromises GRP78 chaperone activity and induces systemic disease. *Mol Cell* 48 (5):747–759.
137. Day, A. M., J. D. Brown, S. R. Taylor, J. D. Rand, B. A. Morgan, and E. A. Veal. 2012. Inactivation of a peroxiredoxin by hydrogen peroxide is critical for thioredoxin-mediated repair of oxidized proteins and cell survival. *Mol Cell* 45 (3):398–408.

138. Wood, Z. A., L. B. Poole, and P. A. Karplus. 2003. Peroxiredoxin evolution and the regulation of hydrogen peroxide signaling. *Science* 300 (5619):650–653.

139. Woo, H. A., S. H. Yim, D. H. Shin, D. Kang, D. Y. Yu, and S. G. Rhee. 2010. Inactivation of peroxiredoxin I by phosphorylation allows localized H(2)O(2) accumulation for cell signaling. *Cell* 140 (4):517–528.

140. Marinho, H. S., C. Real, L. Cyrne, H. Soares, and F. Antunes. 2014. Hydrogen peroxide sensing, signaling and regulation of transcription factors. *Redox Biol* 2:535–562.

141. Klotz, L. O., P. Schroeder, and H. Sies. 2002. Peroxynitrite signaling: Receptor tyrosine kinases and activation of stress-responsive pathways. *Free Radic Biol Med* 33 (6):737–743.

142. Speckmann, B., H. Steinbrenner, T. Grune, and L. O. Klotz. 2016. Peroxynitrite: From interception to signaling. *Arch Biochem Biophys* 595:153–160.

143. Gupta, V. and K. S. Carroll. 2014. Sulfenic acid chemistry, detection and cellular lifetime. *Biochim Biophys Acta* 1840 (2):847–875.

144. Salmeen, A., J. N. Andersen, M. P. Myers, T. C. Meng, J. A. Hinks, N. K. Tonks, and D. Barford. 2003. Redox regulation of protein tyrosine phosphatase 1B involves a sulphenyl-amide intermediate. *Nature* 423 (6941):769–773.

145. van Montfort, R. L., M. Congreve, D. Tisi, R. Carr, and H. Jhoti. 2003. Oxidation state of the active-site cysteine in protein tyrosine phosphatase 1B. *Nature* 423 (6941):773–777.

146. Defelipe, L. A., E. Lanzarotti, D. Gauto, M. A. Marti, and A. G. Turjanski. 2015. Protein topology determines cysteine oxidation fate: The case of sulfenyl amide formation among protein families. *PLoS Comput Biol* 11 (3):e1004051.

147. Jacob, C., I. Knight, and P. G. Winyard. 2006. Aspects of the biological redox chemistry of cysteine: From simple redox responses to sophisticated signalling pathways. *Biol Chem* 387 (10–11):1385–1397.

148. Poole, L. B., P. A. Karplus, and A. Claiborne. 2004. Protein sulfenic acids in redox signaling. *Annu Rev Pharmacol Toxicol* 44:325–347.

149. Bakhmutova-Albert, E. V., H. Yao, D. E. Denevan, and D. E. Richardson. 2010. Kinetics and mechanism of peroxymonocarbonate formation. *Inorg Chem* 49 (24):11287–11296.

150. Trindade, D. F., G. Cerchiaro, and O. Augusto. 2006. A role for peroxymonocarbonate in the stimulation of biothiol peroxidation by the bicarbonate/carbon dioxide pair. *Chem Res Toxicol* 19 (11):1475–1482.

151. Bonini, M. G., S. A. Gabel, K. Ranguelova, K. Stadler, E. F. Derose, R. E. London, and R. P. Mason. 2009. Direct magnetic resonance evidence for peroxymonocarbonate involvement in the Cu,Zn-superoxide dismutase peroxidase catalytic cycle. *J Biol Chem* 284 (21):14618–14627.

152. Zhang, H., J. Joseph, M. Gurney, D. Becker, and B. Kalyanaraman. 2002. Bicarbonate enhances peroxidase activity of Cu,Zn-superoxide dismutase. Role of carbonate anion radical and scavenging of carbonate anion radical by metalloporphyrin antioxidant enzyme mimetics. *J Biol Chem* 277 (2):1013–1020.

153. Medinas, D. B., J. C. Toledo, Jr., G. Cerchiaro, A. T. do-Amaral, L. de-Rezende, A. Malvezzi, and O. Augusto. 2009. Peroxymonocarbonate and carbonate radical displace the hydroxyl-like oxidant in the Sod1 peroxidase activity under physiological conditions. *Chem Res Toxicol* 22 (4):639–648.

154. Liochev, S. I. and I. Fridovich. 2010. Mechanism of the peroxidase activity of Cu, Zn superoxide dismutase. *Free Radic Biol Med* 48 (12):1565–1569.

155. Zeida, A., A. M. Reyes, P. Lichtig, M. Hugo, D. S. Vazquez, J. Santos, F. L. Gonzalez Flecha, R. Radi, D. A. Estrin, and M. Trujillo. 2015. Molecular basis of hydroperoxide specificity in peroxiredoxins: The case of AhpE from *Mycobacterium tuberculosis*. *Biochemistry* 54 (49):7237–7247.

156. Zhou, H., H. Singh, Z. D. Parsons, S. M. Lewis, S. Bhattacharya, D. R. Seiner, J. N. LaButti, T. J. Reilly, J. J. Tanner, and K. S. Gates. 2011. The biological buffer bicarbonate/CO_2 potentiates H_2O_2-mediated inactivation of protein tyrosine phosphatases. *J Am Chem Soc* 133 (40):15803–15805.

157. Lymar, S. V. and J. K. Hurst. 1995. Rapid reaction between peroxonitrite ion and carbon dioxide: Implications for biological activity. *J Am Chem Soc* 117 (34):8867–8868.

158. Bonini, M. G., R. Radi, G. Ferrer-Sueta, A. M. Ferreira, and O. Augusto. 1999. Direct EPR detection of the carbonate radical anion produced from peroxynitrite and carbon dioxide. *J Biol Chem* 274 (16):10802–10806.

159. Goldstein, S. and G. Czapski. 1999. Viscosity effects on the reaction of peroxynitrite with CO_2: Evidence for radical formation in a solvent cage. *J Am Chem Soc* 121 (11):2444–2447.

160. Lymar, S. V. and J. K. Hurst. 1998. CO_2-catalyzed one-electron oxidations by peroxynitrite: Properties of the reactive intermediate. *Inorg Chem* 37 (2):294–301.

161. Bonini, M. G. and O. Augusto. 2001. Carbon dioxide stimulates the production of thiyl, sulfinyl, and disulfide radical anion from thiol oxidation by peroxynitrite. *J Biol Chem* 276 (13):9749–9754.

162. Lymar, S. V., H. A. Schwarz, and G. Czapski. 2000. Medium effects on reactions of the carbonate radical with thiocyanate, iodide, and ferrocyanide ions. *Radiat Phys Chem* 59 (4):387–392.

163. van der Vliet, A., J. P. Eiserich, B. Halliwell, and C. E. Cross. 1997. Formation of reactive nitrogen species during peroxidase-catalyzed oxidation of nitrite. A potential additional mechanism of nitric oxide-dependent toxicity. *J Biol Chem* 272 (12):7617–7625.

164. Eiserich, J. P., M. Hristova, C. E. Cross, A. D. Jones, B. A. Freeman, B. Halliwell, and A. van der Vliet. 1998. Formation of nitric oxide-derived inflammatory oxidants by myeloperoxidase in neutrophils. *Nature* 391 (6665):393–397.

165. Gunther, M. R., L. C. Hsi, J. F. Curtis, J. K. Gierse, L. J. Marnett, T. E. Eling, and R. P. Mason. 1997. Nitric oxide trapping of the tyrosyl radical of prostaglandin H synthase-2 leads to tyrosine iminoxyl radical and nitrotyrosine formation. *J Biol Chem* 272 (27):17086–17090.

166. Radi, R. 2013. Protein tyrosine nitration: Biochemical mechanisms and structural basis of functional effects. *Acc Chem Res* 46 (2):550–559.

167. Yamazaki, I. and L. H. Piette. 1990. ESR spin-trapping studies on the reaction of Fe^{2+} ions with H_2O_2-reactive species in oxygen toxicity in biology. *J Biol Chem* 265 (23):13589–13594.

168. Carballal, S., S. Bartesaghi, and R. Radi. 2014. Kinetic and mechanistic considerations to assess the biological fate of peroxynitrite. *Biochim Biophys Acta* 1840 (2):768–780.

169. Denicola, A., H. Rubbo, D. Rodriguez, and R. Radi. 1993. Peroxynitrite-mediated cytotoxicity to *Trypanosoma cruzi*. *Arch Biochem Biophys* 304 (1):279–286.

170. Quijano, C., B. Alvarez, R. M. Gatti, O. Augusto, and R. Radi. 1997. Pathways of peroxynitrite oxidation of thiol groups. *Biochem J* 322 (Pt 1):167–173.

171. Bartesaghi, S., M. Trujillo, A. Denicola, L. Folkes, P. Wardman, and R. Radi. 2004. Reactions of desferrioxamine with peroxynitrite-derived carbonate and nitrogen dioxide radicals. *Free Radic Biol Med* 36 (4):471–483.

172. Botti, H., M. Trujillo, C. Batthyany, H. Rubbo, G. Ferrer-Sueta, and R. Radi. 2004. Homolytic pathways drive peroxynitrite-dependent Trolox C oxidation. *Chem Res Toxicol* 17 (10):1377–1384.

173. Ross, A. B., W. G. Mallard, W. P. Helman, G. V. Buxton, R.T. Huie, and P. Neta. 1998. NDRL-NIST solution kinetics database, version 3. Notre Dame Radiation Laboratory, Notre Dame, IN and NIST Standard Reference Data, Gaithersburg, MD.

174. Chen, S. N. and M. Z. Hoffman. 1973. Rate constants for the reaction of the carbonate radical with compounds of biochemical interest in neutral aqueous solution. *Radiat Res* 56 (1):40–47.
175. Trujillo, M., L. Folkes, S. Bartesaghi, B. Kalyanaraman, P. Wardman, and R. Radi. 2005. Peroxynitrite-derived carbonate and nitrogen dioxide radicals readily react with lipoic and dihydrolipoic acid. *Free Radic Biol Med* 39 (2):279–288.
176. Winterbourn, C. 2013. Radical scavenging by thiols and the fate of thiyl radicals. In *Oxidative Stress and Redox Regulation*, U. Jacob and D. Reichmann, eds., pp. 43–58. Dordrecht, the Netherlands: Springer.
177. van Stroe-Blezen, S. A. M., F. M. Everaerts, F. J. J. Janssen, and R. A. Tacken. 1993. Diffusion coefficients of oxygen, hydrogen peroxide and glucose in a hydrogel. *Anal Chim Acta* 273:553–560.
178. Straube, R. and D. Ridgway. 2009. Investigating the effects of molecular crowding on Ca^{2+} diffusion using a particle-based simulation model. *Chaos* 19 (3):037110.
179. Antunes, F. and E. Cadenas. 2000. Estimation of H_2O_2 gradients across biomembranes. *FEBS Lett* 475 (2):121–126.
180. Seaver, L. C. and J. A. Imlay. 2001. Hydrogen peroxide fluxes and compartmentalization inside growing *Escherichia coli*. *J Bacteriol* 183 (24):7182–7189.
181. Sousa-Lopes, A., F. Antunes, L. Cyrne, and H. S. Marinho. 2004. Decreased cellular permeability to H_2O_2 protects *Saccharomyces cerevisiae* cells in stationary phase against oxidative stress. *FEBS Lett* 578 (1–2):152–156.
182. Branco, M. R., H. S. Marinho, L. Cyrne, and F. Antunes. 2004. Decrease of H_2O_2 plasma membrane permeability during adaptation to H_2O_2 in *Saccharomyces cerevisiae*. *J Biol Chem* 279 (8):6501–6506.
183. Marinho, H. S., L. Cyrne, E. Cadenas, and F. Antunes. 2013. The cellular steady-state of H_2O_2: Latency concepts and gradients. *Methods Enzymol* 527:3–19.
184. Makino, N., K. Sasaki, K. Hashida, and Y. Sakakura. 2004. A metabolic model describing the H_2O_2 elimination by mammalian cells including H_2O_2 permeation through cytoplasmic and peroxisomal membranes: Comparison with experimental data. *Biochim Biophys Acta* 1673 (3):149–159.
185. Huang, B. K. and H. D. Sikes. 2014. Quantifying intracellular hydrogen peroxide perturbations in terms of concentration. *Redox Biol* 2:955–962.
186. Pedroso, N., A. C. Matias, L. Cyrne, F. Antunes, C. Borges, R. Malho, R. F. de Almeida, E. Herrero, and H. S. Marinho. 2009. Modulation of plasma membrane lipid profile and microdomains by H_2O_2 in *Saccharomyces cerevisiae*. *Free Radic Biol Med* 46 (2):289–298.
187. Bienert, G. P. and F. Chaumont. 2014. Aquaporin-facilitated transmembrane diffusion of hydrogen peroxide. *Biochim Biophys Acta* 1840 (5):1596–1604.
188. Appenzeller-Herzog, C., G. Banhegyi, I. Bogeski, K. J. Davies, A. Delaunay-Moisan, H. J. Forman, A. Gorlach et al. 2016. Transit of H_2O_2 across the endoplasmic reticulum membrane is not sluggish. *Free Radic Biol Med* 94:157–160.
189. Medrano-Fernandez, I., S. Bestetti, M. Bertolotti, G. P. Bienert, C. Bottino, U. Laforenza, A. Rubartelli, and R. Sitia. 2016. Stress regulates aquaporin-8 permeability to impact cell growth and survival. *Antioxid Redox Signal* 24 (18):1031–1044.
190. Hooijmaijers, C., J. Y. Rhee, K. J. Kwak, G. C. Chung, T. Horie, M. Katsuhara, and H. Kang. 2012. Hydrogen peroxide permeability of plasma membrane aquaporins of *Arabidopsis thaliana*. *J Plant Res* 125 (1):147–153.
191. Marla, S. S., J. Lee, and J. T. Groves. 1997. Peroxynitrite rapidly permeates phospholipid membranes. *Proc Natl Acad Sci U S A* 94 (26):14243–14248.
192. Khairutdinov, R. F., J. W. Coddington, and J. K. Hurst. 2000. Permeation of phospholipid membranes by peroxynitrite. *Biochemistry* 39 (46):14238–14249.

193. Denicola, A., J. M. Souza, and R. Radi. 1998. Diffusion of peroxynitrite across erythrocyte membranes. *Proc Natl Acad Sci U S A* 95 (7):3566–3571.

194. Macfadyen, A. J., C. Reiter, Y. Zhuang, and J. S. Beckman. 1999. A novel superoxide dismutase-based trap for peroxynitrite used to detect entry of peroxynitrite into erythrocyte ghosts. *Chem Res Toxicol* 12 (3):223–229.

195. Drenckhahn, D., K. Zinke, U. Schauer, K. C. Appell, and P. S. Low. 1984. Identification of immunoreactive forms of human erythrocyte band 3 in nonerythroid cells. *Eur J Cell Biol* 34 (1):144–150.

196. Puceat, M., I. Korichneva, R. Cassoly, and G. Vassort. 1995. Identification of band 3-like proteins and Cl^-/HCO_3^- exchange in isolated cardiomyocytes. *J Biol Chem* 270 (3):1315–1322.

197. Cox, A. G., C. C. Winterbourn, and M. B. Hampton. 2009. Mitochondrial peroxiredoxin involvement in antioxidant defence and redox signalling. *Biochem J* 425 (2):313–325.

198. Tortora, V., C. Quijano, B. Freeman, R. Radi, and L. Castro. 2007. Mitochondrial aconitase reaction with nitric oxide, S-nitrosoglutathione, and peroxynitrite: Mechanisms and relative contributions to aconitase inactivation. *Free Radic Biol Med* 42 (7):1075–1088.

199. Demicheli, V., D. M. Moreno, G. E. Jara, A. Lima, S. Carballal, N. Rios, C. Batthyany et al. 2016. Mechanism of the reaction of human manganese superoxide dismutase with peroxynitrite: Nitration of critical tyrosine 34. *Biochemistry* 55 (24):3403–3417.

200. Ferrer-Sueta, G. and R. Radi. 2009. Chemical biology of peroxynitrite: Kinetics, diffusion, and radicals. *ACS Chem Biol* 4 (3):161–177.

201. Esteves, R.. 2015. Reducción de peroxinitrito por peroxirredoxina 3 mitocondrial humana. Graduate student thesis, Departamento de Bioquímica, Universidad de la República, Montevideo, Uruguay.

202. Romero, N., A. Denicola, J. M. Souza, and R. Radi. 1999. Diffusion of peroxynitrite in the presence of carbon dioxide. *Arch Biochem Biophys* 368 (1):23–30.

203. Ariyanayagam, M. R. and A. H. Fairlamb. 2001. Ovothiol and trypanothione as antioxidants in trypanosomatids. *Mol Biochem Parasitol* 115 (2):189–198.

204. Goldman, R., D. A. Stoyanovsky, B. W. Day, and V. E. Kagan. 1995. Reduction of phenoxyl radicals by thioredoxin results in selective oxidation of its SH-groups to disulfides. An antioxidant function of thioredoxin. *Biochemistry* 34 (14):4765–4772.

205. Cuevasanta, E., A. Zeida, S. Carballal, R. Wedmann, U. N. Morzan, M. Trujillo, R. Radi, D. A. Estrin, M. R. Filipovic, and B. Alvarez. 2015. Insights into the mechanism of the reaction between hydrogen sulfide and peroxynitrite. *Free Radic Biol Med* 80:93–100.

206. Peskin, A. V., F. M. Low, L. N. Paton, G. J. Maghzal, M. B. Hampton, and C. C. Winterbourn. 2007. The high reactivity of peroxiredoxin 2 with H(2)O(2) is not reflected in its reaction with other oxidants and thiol reagents. *J Biol Chem* 282 (16):11885–11892.

207. Pineyro, M. D., T. Arcari, C. Robello, R. Radi, and M. Trujillo. 2011. Tryparedoxin peroxidases from *Trypanosoma cruzi*: High efficiency in the catalytic elimination of hydrogen peroxide and peroxynitrite. *Arch Biochem Biophys* 507 (2):287–295.

208. Ogusucu, R., D. Rettori, D. C. Munhoz, L. E. Netto, and O. Augusto. 2007. Reactions of yeast thioredoxin peroxidases I and II with hydrogen peroxide and peroxynitrite: Rate constants by competitive kinetics. *Free Radic Biol Med* 42 (3):326–334.

209. Bryk, R., P. Griffin, and C. Nathan. 2000. Peroxynitrite reductase activity of bacterial peroxiredoxins. *Nature* 407 (6801):211–215.

210. Parsonage, D., D. C. Desrosiers, K. R. Hazlett, Y. Sun, K. J. Nelson, D. L. Cox, J. D. Radolf, and L. B. Poole. 2010. Broad specificity AhpC-like peroxiredoxin and its thioredoxin reductant in the sparse antioxidant defense system of *Treponema pallidum*. *Proc Natl Acad Sci U S A* 107 (14):6240–6245.

211. Dubuisson, M., D. Vander Stricht, A. Clippe, F. Etienne, T. Nauser, R. Kissner, W. H. Koppenol, J. F. Rees, and B. Knoops. 2004. Human peroxiredoxin 5 is a peroxynitrite reductase. *FEBS Lett* 571 (1–3):161–165.

212. Hugo, M., L. Turell, B. Manta, H. Botti, G. Monteiro, L. E. Netto, B. Alvarez, R. Radi, and M. Trujillo. 2009. Thiol and sulfenic acid oxidation of AhpE, the one-cysteine peroxiredoxin from *Mycobacterium tuberculosis*: Kinetics, acidity constants, and conformational dynamics. *Biochemistry* 48 (40):9416–9426.

213. Loumaye, E., G. Ferrer-Sueta, B. Alvarez, J. F. Rees, A. Clippe, B. Knoops, R. Radi, and M. Trujillo. 2011. Kinetic studies of peroxiredoxin 6 from *Arenicola marina*: Rapid oxidation by hydrogen peroxide and peroxynitrite but lack of reduction by hydrogen sulfide. *Arch Biochem Biophys* 514 (1–2):1–7.

214. Toledo, J. C. Jr., R. Audi, R. Ogusucu, G. Monteiro, L. E. Netto, and O. Augusto. 2011. Horseradish peroxidase compound I as a tool to investigate reactive protein-cysteine residues: From quantification to kinetics. *Free Radic Biol Med* 50 (9):1032–1038.

215. Horta, B. B., M. A. de Oliveira, K. F. Discola, J. R. Cussiol, and L. E. Netto. 2010. Structural and biochemical characterization of peroxiredoxin Qbeta from *Xylella fastidiosa*: Catalytic mechanism and high reactivity. *J Biol Chem* 285 (21):16051–16065.

216. Reeves, S. A., D. Parsonage, K. J. Nelson, and L. B. Poole. 2011. Kinetic and thermodynamic features reveal that *Escherichia coli* BCP is an unusually versatile peroxiredoxin. *Biochemistry* 50 (41):8970–8981.

217. Reyes, A. M., D. S. Vazquez, A. Zeida, M. Hugo, M. D. Pineyro, M. I. De Armas, D. Estrin, R. Radi, J. Santos, and M. Trujillo. 2016. PrxQ B from *Mycobacterium tuberculosis* is a monomeric, thioredoxin-dependent and highly efficient fatty acid hydroperoxide reductase. *Free Radic Biol Med* 101:249–260.

218. Jaeger, T., H. Budde, L. Flohe, U. Menge, M. Singh, M. Trujillo, and R. Radi. 2004. Multiple thioredoxin-mediated routes to detoxify hydroperoxides in *Mycobacterium tuberculosis*. *Arch Biochem Biophys* 423 (1):182–191.

219. Baker, L. M. and L. B. Poole. 2003. Catalytic mechanism of thiol peroxidase from *Escherichia coli*. Sulfenic acid formation and overoxidation of essential CYS61. *J Biol Chem* 278 (11):9203–9211.

220. Hillebrand, H., A. Schmidt, and R. L. Krauth-Siegel. 2003. A second class of peroxidases linked to the trypanothione metabolism. *J Biol Chem* 278 (9):6809–6815.

221. Pedre, B., I. Van Molle, A. F. Villadangos, K. Wahni, D. Vertommen, L. Turell, H. Erdogan, L. M. Mateos, and J. Messens. 2015. The *Corynebacterium glutamicum* mycothiol peroxidase is a reactive oxygen species-scavenging enzyme that shows promiscuity in thiol redox control. *Mol Microbiol* 96 (6):1176–1191.

222. Wang, L., L. Zhang, Y. Niu, R. Sitia, and C. C. Wang. 2014. Glutathione peroxidase 7 utilizes hydrogen peroxide generated by Ero1alpha to promote oxidative protein folding. *Antioxid Redox Signal* 20 (4):545–556.

223. Trujillo, M., P. Mauri, L. Benazzi, M. Comini, A. De Palma, L. Flohe, R. Radi et al. 2006. The mycobacterial thioredoxin peroxidase can act as a one-cysteine peroxiredoxin. *J Biol Chem* 281 (29):20555–20566.

224. Trujillo, M., H. Budde, M. D. Pineyro, M. Stehr, C. Robello, L. Flohe, and R. Radi. 2004. *Trypanosoma brucei* and *Trypanosoma cruzi* tryparedoxin peroxidases catalytically detoxify peroxynitrite via oxidation of fast reacting thiols. *J Biol Chem* 279 (33):34175–34182.

225. Crawford, M. A., T. Tapscott, L. F. Fitzsimmons, L. Liu, A. M. Reyes, S. J. Libby, M. Trujillo, F. C. Fang, R. Radi, and A. Vazquez-Torres. 2016. Redox-active sensing by bacterial DksA transcription factors is determined by cysteine and zinc content. *MBio* 7 (2):e02161–e02161-15.

226. Souza, J. M. and R. Radi. 1998. Glyceraldehyde-3-phosphate dehydrogenase inactivation by peroxynitrite. *Arch Biochem Biophys* 360 (2):187–194.
227. Peralta, D., A. K. Bronowska, B. Morgan, E. Doka, K. Van Laer, P. Nagy, F. Grater, and T. P. Dick. 2015. A proton relay enhances H_2O_2 sensitivity of GAPDH to facilitate metabolic adaptation. *Nat Chem Biol* 11 (2):156–163.
228. Winterbourn, C. C. and M. B. Hampton. 2008. Thiol chemistry and specificity in redox signaling. *Free Radic Biol Med* 45 (5):549–561.
229. Carballal, S., R. Radi, M. C. Kirk, S. Barnes, B. A. Freeman, and B. Alvarez. 2003. Sulfenic acid formation in human serum albumin by hydrogen peroxide and peroxynitrite. *Biochemistry* 42 (33):9906–9914.
230. Takakura, K., J. S. Beckman, L. A. MacMillan-Crow, and J. P. Crow. 1999. Rapid and irreversible inactivation of protein tyrosine phosphatases PTP1B, CD45, and LAR by peroxynitrite. *Arch Biochem Biophys* 369 (2):197–207.
231. Denu, J. M. and K. G. Tanner. 1998. Specific and reversible inactivation of protein tyrosine phosphatases by hydrogen peroxide: Evidence for a sulfenic acid intermediate and implications for redox regulation. *Biochemistry* 37 (16):5633–5642.
232. Sohn, J. and J. Rudolph. 2003. Catalytic and chemical competence of regulation of cdc25 phosphatase by oxidation/reduction. *Biochemistry* 42 (34):10060–10070.
233. Konorev, E. A., N. Hogg, and B. Kalyanaraman. 1998. Rapid and irreversible inhibition of creatine kinase by peroxynitrite. *FEBS Lett* 427 (2):171–174.
234. Li, C., S. Sun, D. Park, H. O. Jeong, H. Y. Chung, X. X. Liu, and H. M. Zhou. 2011. Hydrogen peroxide targets the cysteine at the active site and irreversibly inactivates creatine kinase. *Int J Biol Macromol* 49 (5):910–916.
235. Dairou, J., N. Atmane, F. Rodrigues-Lima, and J. M. Dupret. 2004. Peroxynitrite irreversibly inactivates the human xenobiotic-metabolizing enzyme arylamine N-acetyltransferase 1 (NAT1) in human breast cancer cells: A cellular and mechanistic study. *J Biol Chem* 279 (9):7708–7714.
236. Atmane, N., J. Dairou, A. Paul, J. M. Dupret, and F. Rodrigues-Lima. 2003. Redox regulation of the human xenobiotic metabolizing enzyme arylamine N-acetyltransferase 1 (NAT1). Reversible inactivation by hydrogen peroxide. *J Biol Chem* 278 (37):35086–35092.
237. Andres-Mateos, E., C. Perier, L. Zhang, B. Blanchard-Fillion, T. M. Greco, B. Thomas, H. S. Ko et al. 2007. DJ-1 gene deletion reveals that DJ-1 is an atypical peroxiredoxin-like peroxidase. *Proc Natl Acad Sci U S A* 104 (37):14807–14812.
238. Briviba, K., R. Kissner, W. H. Koppenol, and H. Sies. 1998. Kinetic study of the reaction of glutathione peroxidase with peroxynitrite. *Chem Res Toxicol* 11 (12):1398–1401.
239. Winterbourn, C. C. 2008. Reconciling the chemistry and biology of reactive oxygen species. *Nat Chem Biol* 4 (5):278–286.

4 Influence of CO_2 on Hydroperoxide Metabolism

Daniela Ramos Truzzi and Ohara Augusto

CONTENTS

INTRODUCTION

CO_2 is naturally present in the atmosphere (0.03%–0.06%) and is a normal constituent of the human body, which produces about 1.0 kg of CO_2/day through respiration. The level of atmospheric CO_2 is increasing and the best-known effect of such increase is global warming. However, there are a number of animal studies reporting that increased CO_2 levels (hypercapnia) leads to acute and long-term toxicity to the lungs, the cardiovascular system, and the nervous system of the exposed animals (for a review, see [1]). Such toxicity in organisms that are buffered mainly by bicarbonate/CO_2 is not surprising because increased CO_2 levels result in higher HCO_3^- and H^+ levels (Equations 4.1 and 4.2). These ions, in turn, modulate the activities of many of the enzymes involved in cellular processes through both pH-dependent and pH-independent mechanisms [1].

$$CO_{2(g)} \rightleftharpoons CO_{2(d)} \qquad (4.1)$$

$$CO_{2(d)} + H_2O \underset{k_r}{\overset{k_f}{\rightleftharpoons}} H_2CO_3 \overset{pK_a = 6.4/6.1\left(25^\circ C/37^\circ C\right)}{\rightleftharpoons} HCO_3^- + H^+ \qquad (4.2)$$

More surprising is the fact that the contribution of redox reactions to CO_2 toxicity has received limited attention in the literature despite early [2–4] and more recent studies reporting that the bicarbonate buffer increases biological oxidations, peroxidations, and nitrations [5–21]. The limited emphasis on the redox reactions of the bicarbonate buffer is likely due to many factors, including the difficulties in employing the buffer in routine *in vitro* experiments [22,23]. In fact, pH maintenance is hardly manageable under these conditions because of the dynamic equilibrium between HCO_3^-, H_2CO_3, dissolved CO_2 ($CO_{2(d)}$), and gaseous CO_2 ($CO_{2(g)}$) above the solution (Equations 4.1 and 4.2). Moreover, the detection and characterization of hypothetical oxidants derived from the bicarbonate buffer under physiological conditions are difficult [24].

The possibility of $CO_3^{\bullet-}$ mediating biological damage had been occasionally suggested prior to the 1990s [2–4,20]. The picture changed after Beckman and collaborators proposed that tissue injury associated with conditions of $NO^{\bullet}$ overproduction was due to the extremely rapid reaction of $NO^{\bullet}$ with $O_2^{\bullet-}$ ($k = (4.3–19) \times 10^9$ M^{-1} s^{-1}) to produce peroxynitrite* (Equation 4.3) [25]. This proposal greatly stimulated the study of peroxynitrite biochemistry. It soon became clear that one of the main biological targets of peroxynitrite was the physiologically ubiquitous CO_2 ($k = 2.6 \times 10^4$ M^{-1} s^{-1}; pH 7.4) [26]. Parallel studies by different groups established that this relatively rapid reaction produced NO_3^- (65%), $CO_3^{\bullet-}$ (35%), and $NO_2^{\bullet}$ (35%) in the specified yields (Equation 4.4) (reviewed in [27–30]). Paramount to this conclusion was the unequivocal characterization of $CO_3^{\bullet-}$ by continuous flow EPR of mixtures of peroxynitrite with HCO_3^- or $H^{13}CO_3^-$ [31].

$$NO^{\bullet} + O_2^{\bullet-} \rightarrow ONOO^- \tag{4.3}$$

$$ONOO^- + CO_2 \rightarrow [ONOOCO_2^-] \rightarrow 0.35NO_2^{\bullet} + 0.35CO_3^{\bullet-} + 0.65NO_3^{\bullet} + 0.65CO_2 \tag{4.4}$$

The demonstration that $CO_3^{\bullet-}$ existed as an independent species in aqueous solutions at physiological pH and temperature stimulated the interest in the pathophysiological roles of this radical. In recent years, $CO_3^{\bullet-}$ has been proposed to be an important mediator of the oxidative damage resulting from the production of peroxynitrite [5,27–29], the activity of xanthine oxidase (XO) [32], and the HCO_3^--dependent peroxidase activity of superoxide dismutase [11–14,33,34]. Additionally, the $CO_3^{\bullet-}$ has been proposed to be responsible for the stimulatory effects of the bicarbonate buffer on oxidations mediated by H_2O_2/transition metal ions [16–19,21]. Studies of the enzymatic pathways involved in the formation of $CO_3^{\bullet-}$ brought into focus peroxymonocarbonate ($HOOCO_2^-$; HCO_4^-), another oxidant derived from the bicarbonate buffer [33–37].

---

* The term peroxynitrite refers to the sum of peroxynitrite anion (ONOO⁻, oxoperoxonitrate (–1)) and peroxynitrous acid (ONOOH, hydrogen oxoperoxonitrate) unless otherwise specified. Other abbreviations are defined in the text.

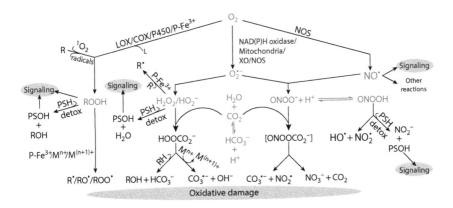

SCHEME 4.1 The reactions of the HCO$_3^-$/CO$_2$ pair with hydroperoxides and their possible biological consequences are schematically shown. For clarity, the reactions were not balanced and some intermediates were omitted. The following notations are used: LOX, lipoxygenase; COX, cyclooxygenase; P450, cytochrome P$_{450}$; XO, xanthine oxidase; L, Lipid; R, biomolecule; NOS, nitric oxide synthase; PSH, thiol protein; P-Fe^{3+}, iron(III) hemoproteins; and M^{n+}, transition metal ion.

Currently, the formation of CO$_3^{\bullet-}$ from peroxynitrite has been largely elucidated and the role of this radical in partially mediating the toxic effects of peroxynitrite has been widely accepted. The situation is not as clear in the case of the other biological routes for CO$_3^{\bullet-}$ production and for the pathophysiological roles of bicarbonate buffer-derived oxidants other than CO$_3^{\bullet-}$. Nevertheless, the majority of the studies reporting effects of the bicarbonate buffer on oxidative processes relate to hydroperoxides, that is, either peroxynitrous acid/peroxynitrite or H$_2$O$_2$. These and other hydroperoxides are important metabolites of molecular oxygen that exert a variety of biological functions. They can act as signaling molecules, as inflammation mediators, and as free radical generators, participating in cellular oxidative signaling and damage (Scheme 4.1). Therefore, it is worth examining the possible interactions between the HCO$_3^-$/CO$_2$ pair and biological hydroperoxides, the circumstances in which they may occur, and the physiological consequences they may have. To explore these interactions, it is helpful to summarize the chemical biology of the individual players (Scheme 4.1).

GENERAL PROPERTIES OF THE HCO$_3^-$/CO$_2$ PAIR AND OF BIOLOGICALLY RELEVANT HYDROPEROXIDES

HCO$_3^-$/CO$_2$ PAIR

The intracellular and serum concentrations of HCO$_3^-$ in humans are in the range of 14.7–25 mM and are in equilibrium with 5% CO$_2$ (1.3 mM) to render pH values of 7.2 and 7.4, respectively (Equations 4.1 and 4.2). Although it is usually considered as thermodynamically and kinetically stable, CO$_2$ has a strong affinity toward nucleophiles and electron-donating compounds due to the electron deficiency of carbonyl carbons [38]. Therefore, CO_2 is an acid anhydride that rapidly reacts with water, alkoxides, and amines. This property is responsible for CO_2 hydration and dehydration (Equation 4.2), which are relatively slow processes ($k_f = 3.0 \times 10^{-2}$ s^{-1}; $k_r = 1.16 \times 10^1$ s^{-1} at 25°C) [37] that are catalyzed by carbonic anhydrase in animal and many plant cells. Among the known enzymes, carbonic anhydrase has one of the highest turnover numbers, illustrating the importance of the HCO_3^-/CO_2 pair in animal physiology.

Under physiological conditions, the affinity of CO_2 for nucleophiles is responsible for its participation in metabolic reactions, in reversible binding to proteins, such as hemoglobin in blood, and in reactions with biological hydroperoxides. The latter reactions are usually initiated by the formation of an adduct product as a reversible reaction (Equation 4.5), whose equilibrium constant (K) varies linearly with the acid dissociation constant of the hydroperoxide (K_a) (Equation 4.6) [39]. Similarly, the fate of the produced adduct also depends on the intrinsic properties of the hydroperoxide, which are summarized here:

$$ROO^- + CO_2 \rightleftharpoons ROOCO_2^- \tag{4.5}$$

$$ROOH \rightleftharpoons ROO^- + H^+ \tag{4.6}$$

With regard to redox reactions, both HCO_3^- and CO_2 are not directly and easily oxidized by most biological oxidants. Even the most potent oxidant, which is HO$^\bullet$ ($E° = 2.3$ V), oxidizes HCO_3^- ($k = 8.5 \times 10^6$ M^{-1} s^{-1}) (Equation 4.7) relatively slowly as compared to most of its biological targets ($k \geq 10^9$ M^{-1} s^{-1}). The carbonate anion (CO_3^{2-}) reacts faster with HO$^\bullet$ ($k = 3 \times 10^8$ M^{-1} s^{-1}) (Equation 4.8) but the high pK_a of its conjugated acid ($pK_a = 10.32$) (Equation 4.9) makes this reaction of limited biological significance. The oxidation of CO_2 to $CO_2^{\bullet+}$ by an active Cu intermediate formed at the active site of the enzyme superoxide dismutase during its bicarbonate-dependent peroxidase activity has been proposed [13]; however, this reaction is unlikely to occur due to the extremely high reduction potential of $CO_2^{\bullet+}$ [40,41].

$$HCO_3^- + HO^\bullet \rightarrow CO_3^{\bullet-} + H_2O \tag{4.7}$$

$$CO_3^{2-} + HO^\bullet \rightarrow CO_3^{\bullet-} + OH \tag{4.8}$$

$$HCO_3^- \overset{pK_a = 10.32}{\rightleftharpoons} CO_3^{2-} + H^+ \tag{4.9}$$

It is worth mentioning that the product of CO_2 reduction, the carbon dioxide anion radical ($CO_2^{\bullet-}$), has been reported to be produced during hydralazine [42] and carbon tetrachloride [43] metabolism, but the radical source was not attributed to the bicarbonate buffer. Similarly, pyruvate decarboxylation promoted by H_2O_2/transition metal ions or peroxynitrite produces $CO_2^{\bullet-}$, which is derived from the substrate [44]. In contrast with $CO_3^{\bullet-}$, which is a strong one-electron oxidant ($E° = 1.78$ V, pH 7.0), $CO_2^{\bullet-}$ is a strong one-electron reductant ($E° = -1.8$ V, pH 7.0). It is capable of reducing molecular oxygen to $O_2^{\bullet-}$, triggering oxidative reactions, but it cannot directly promote oxidations.

ORGANIC HYDROPEROXIDES FROM BIOMOLECULES

Lipid hydroperoxides (LOOH) are enzymatically formed by reactions catalyzed by lipoxygenase, cyclooxygenase, cytochrome P_{450}, and heme peroxidases (Scheme 4.1). Additionally, several classes of organic hydroperoxides (ROOH) are produced upon oxidation of biomolecules (lipids, proteins and DNA) by radicals and by singlet oxygen ($^1O_2$) [45–48]. Many of these biomolecule-derived hydroperoxides have started to become fully characterized only recently because of the advances in mass spectrometry. Although ROOH are relatively stable, they participate in biological reactions that can increase or decrease their toxicity. The toxicity is increased when ROOH are converted to $RO^{\bullet}$ or $ROO^{\bullet}$ by transition metal ions, hemoproteins, and other one-electron oxidants. The reactivity and toxicity decrease by enzymes that reduce ROOH by two-electron mechanisms to their corresponding alcohols. These enzymes, which are normally present in cells, are glutathione peroxidase (GPx), glutathione S-transferases (GST), and peroxiredoxins (Prx) [45,49,50]. It is worth noting that kinetic studies of many of these enzymes with biomolecule-derived hydroperoxides are still limited [51–53].

To the best of our knowledge, the effect of the bicarbonate buffer on the reactions and metabolism of ROOH has yet to be examined. These compounds are expected to have considerably high pK_a values, making their efficient addition to CO_2 (Equations 4.5 and 4.6) unlikely under physiological conditions. The only pK_a value we found reported in the literature was the pK_a value of methyl hydroperoxide (11.5) [54]. It is also important to note that ROOH is a less efficient two-electron oxidant compared to H_2O_2 and peroxynitrite because RO^- is a worse leaving group than HO^- and NO_2^- [36,55,56].

PEROXYNITRITE

The diffusion-controlled reaction of $NO^{\bullet}$ with $O_2^{\bullet-}$ (Equation 4.4) is biologically relevant because it may limit $NO^{\bullet}$ availability interfering with physiological signaling mechanisms (Scheme 4.1). In addition, the reaction produces peroxynitrite ($ONOO^-$/$ONOOH$), which is a hydroperoxide with unusual properties. Compared to other hydroperoxides, peroxynitrite has a low pK_a (6.8) (Equation 4.10) and is present as a mixture of the anion and the acid ($ONOO^-$/$ONOOH$) at most physiological pHs. In contrast to ROOH ($pK_a \geq 11.5$) and H_2O_2 ($pK_a = 11.7$) [54], a considerable fraction of peroxynitrite is present as the anion form at pH 7.4, favoring its nucleophilic attack on CO_2 to produce the nitrosoperoxo-carboxylate adduct in equilibrium (as in Equation 4.5). The K value of this equilibrium has been estimated as 1 M^{-1} [39]. However, as the nitrosoperoxo-carboxylate adduct decomposes quite rapidly (k > 10^7 s^{-1}), the overall reaction is usually not shown as an equilibrium reaction (k = 2.6×10^4 M^{-1} s^{-1}) (Equation 4.4) [26]. Another difference between peroxynitrite and other hydroperoxides is that its O–O bond is prone to homolysis producing free radicals. The nature of the produced radicals varies with the agent that catalyzes it. Carbon dioxide-catalyzed homolysis produces $CO_3^{\bullet-}$ and $NO_2^{\bullet}$ (Equation 4.4) [31], whereas proton-catalyzed decomposition produces $HO^{\bullet}$ and $NO_2^{\bullet}$ (Equation 4.11) [57]; peroxynitrous acid decay is relatively rapid (k = 1.1 s⁻¹ at pH 7.4, 37°C) [58]. In contrast, the homolysis of other peroxides requires energy, such as heating and light or transition metal ions [59,60].

$$\text{ONOO}^- + \text{H}^+ \underset{\text{p}K_a = 6.8}{\rightleftharpoons} \text{ONOOH}$$
(4.10)

$$\text{ONOOH} \rightarrow 0.3\text{HO}^\bullet + 0.3\text{NO}_2^\bullet + 0.7\text{NO}_3^- + 0.7\text{H}^+$$
(4.11)

In summary, reaction with CO_2 and, thus, the production of $NO_2^\bullet$ and $CO_3^{\bullet-}$ radicals, is an important consequence of peroxynitrite production *in vivo* due to the high concentration of CO_2 in biological fluids and the considerable second-order rate constant of the reaction between CO_2 and peroxynitrite [26]. The radicals $NO_2^\bullet$ ($E° = 1.0$ V) and $CO_3^{\bullet-}$ (1.78 V) are moderately strong and strong one-electron oxidants, respectively, contributing to peroxynitrite-mediated toxicity [27,29,30]. The HCO_3^-/CO_2 pair can also decrease peroxynitrite-mediated oxidation of specific targets in determined cellular compartments because it greatly reduces the half-life of peroxynitrite, limiting its diffusion distance ($t_{1/2}$ values of 0.8 and 0.02 s in the absence and presence of 1.3 mM CO_2, 37°C, respectively). The HCO_3^-/CO_2 pair also decreases the total concentration of oxidizing species from peroxynitrite as one molecule of the oxidant is transformed into 0.7 radicals (0.35 $NO_2^\bullet$ and 0.35 $CO_3^{\bullet-}$) (Equation 4.4). The production of $HO^\bullet$ from peroxynitrite in physiological environments is less likely to occur than the formation of $CO_3^{\bullet-}$ because of the competition of the ubiquitous CO_2 for the oxidant. Even CO_2 may not compete with other biological targets, such as hemeproteins and thiol proteins, that quickly react with peroxynitrite (Scheme 4.1). In these reactions, peroxynitrite acts as a two-electron oxidant, although there are some exceptions in the case of hemeproteins [29]. The particularly high second-order rate constant of the reaction of peroxynitrite with thiol proteins from the peroxiredoxin family (Prx) (10^5–10^8 M⁻¹ s⁻¹) argues for a role of these proteins in peroxynitrite detoxification. The chemical biology of peroxynitrite is reviewed in this book and elsewhere [29,30]. We summarized it here because the influence of CO_2 on peroxynitrite toxicity is the better understood of the interactions of the bicarbonate buffer with hydroperoxides.

Hydrogen Peroxide

Hydrogen peroxide is continuously produced from the spontaneous or SOD-catalyzed dismutation of the $O_2^{\bullet-}$, which leaks from the mitochondrial electron transport chain during respiration. Additionally, H_2O_2 is produced from reactions catalyzed by several oxidases, such as NADPH oxidases (NOX), xanthine oxidase (XO), and monoamine oxidase (MAO) (Scheme 4.1), and from chemical and photochemical processes. Hydrogen peroxide is a powerful two-electron oxidant ($E° = 1.77$ V). However, its reactivity toward most of biological molecules is low because of the high activation energy of these oxidations [61]. The most damaging effects of H_2O_2 *in vivo* are considered to be mediated by either transition metal ions or enzymes, such as heme peroxidases [48]. These processes generate secondary species, which are more reactive than H_2O_2 and include radicals, such as $R^{\bullet}$ and $ROO^{\bullet}$, and nonradical species, such as $HOCl$ and related compounds. Reaction of H_2O_2 with reduced transition metal ions, such as copper (I) and iron (II), leads to the generation of $HO^{\bullet}$ [62]. Under physiological conditions, H_2O_2 rapidly decomposes to by catalase and by seleno-, heme-, and thiol- peroxidases (Prx). Thus, the steady-state concentration of H_2O_2 in cells and tissues are expected to be low, that is, in the submicromolar range [63–65]. Substantial cellular increases in H_2O_2 levels are expected to cause oxidative damage. In contrast, accumulating evidence indicates that transient and small increases in cellular H_2O_2 concentrations mediate cellular signaling through the reversible oxidation of catalytic thiol proteins, including Prx [64–66].

Despite the high pK_a of H_2O_2 (11.7), numerous studies have reported that the HCO_3^-/CO_2 pair increases H_2O_2-mediated oxidations at physiological pHs *in vitro* [16–19,21]. Since at these pHs the concentration of the deprotonated peroxide is quite low, it was not expected that its addition to CO_2 (Equations 4.5 and 4.6) would contribute to the activation of H_2O_2 by the bicarbonate buffer to a large extent. This apparent contradiction was resolved by a comprehensive study of the formation of peroxymonocarbonate (HCO_4^-) from the reaction between HCO_3^- and H_2O_2 by Richardson and coworkers [37] (Equation 4.12). These authors determined the second-order rate constants of the reaction of CO_2 with HOO^- and H_2O_2 (280 and 0.02 M^{-1} s^{-1}, respectively) by ^{13}C NMR, pH jump experiments, and kinetic simulations (Scheme 4.2). They concluded that the reaction of CO_2 with H_2O_2 (perhydration) is similar to the reaction of CO_2 with H_2O (hydration). The particularity of the perhydration reaction can explain the occurrence of the reaction between H_2O_2 and CO_2 at neutral pH, in contrast to organic hydroperoxides [39,54].

$$K = 0.33\ M^{-1}\left(25^{\circ}C\right)$$
$$HCO_3^- + H_2O_2 \quad \rightleftharpoons \quad HCO_4^- + H_2O \qquad (4.12)$$

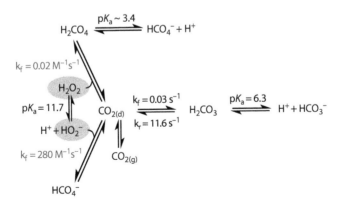

SCHEME 4.2 The mechanism of HCO_4^- formation from H_2O_2/HOO^- and the HCO_3^-/CO_2 pair as proposed by Richardson and coworkers (Bakhmutova-Albert et al. [37]).

Simple calculations based on the determined rate constants and the pK_a of H_2O_2 show that at pH 7.4, the H_2O_2 and HOO^- pathways contribute to approximately 59% and 41%, respectively, of HCO_4^- formation (Scheme 4.2) [37]. Therefore, the formation of HCO_4^- at pH 7.4 is slow (k = 0.034 M^{-1} s^{-1}) but has been shown to be accelerated by carbonic anhydrase and by protein- and lipid-enriched environments [36,37]. Whether these accelerating effects can facilitate HCO_4^- production *in vivo* remains an open question.

Up to this point, the determined rate constant values for HCO_4^- formation under different experimental conditions are several orders of magnitude lower than those determined for H_2O_2 reacting with catalase and with seleno-, heme-, and thiol- peroxidases (Prx). In addition, the reaction equilibrium is displaced to reagents and only approximately 1% of the H_2O_2 concentration is present as HCO_4^- at equilibrium (Equation 4.12). Although upon HCO_4^- consumption the equilibrium reestablishes, the process is slow. These facts are responsible for the limited attention received by HCO_4^- in the biological literature. However, HCO_4^- is produced at neutral pH, albeit slowly, and affects several biological oxidations *in vitro*. Additionally, HCO_4^- may also be produced at the active site of enzymes as evidenced during the catalytic activities of XO [32] and of SOD1 [33]. In the latter case, we showed that low steady-state concentration of H_2O_2 (4–10 µM), generated by glucose/glucose oxidase and controlled by catalase addition, promoted the continuous oxidation of dihydrorhodamine catalyzed by SOD1 in the presence of the pair HCO_3^-/CO_2 (Medinas et al. [33]). These results indicated that under a low and continuous flux of H_2O_2, HCO_4^- forms and exerts its effects. Therefore, HCO_4^- deserves attention as a hydroperoxide of potential biological importance.

A ROLE FOR PEROXYMONOCARBONATE IN REDOX BIOLOGY?

A NEWCOMER TO REDOX BIOLOGY

Peroxymonocarbonate is an oxidant whose existence in equilibrium with H_2O_2 and HCO_3^- has been known since the 1980s (Equation 4.12) [67,68]. Later, Richardson and coworkers studied HCO_4^- equilibrium in water and alcohol/water mixtures, as well as the ability of the oxidant to oxidize alkyl sulfides, alkenes, and methionine by a two-electron mechanism [69–71]. However, investigators in the field of biochemistry did not consider HCO_4^- until 2000. In 2002, we proposed HCO_4^- as a possible intermediate in the production of $CO_3^{\cdot-}$ during the oxidation of acetaldehyde catalyzed by xanthine oxidase (XO) in bicarbonate buffer [27]. This proposal was substantiated by several lines of experimental evidences and was published in 2004 [32]. A year prior, Elam and coworkers suggested that HCO_4^--bound to superoxide dismutase 1 (SOD1) was an intermediate produced during the bicarbonate-dependent peroxidase activity of SOD1 [72]. Although the mechanism they proposed did not hold after further studies, the participation of HCO_4^- in this enzymatic activity has been supported by other independent investigations [31,41,73,74]. This brief historical account is to emphasize that HCO_4^- is a newcomer in the field of redox biology and that its properties should be noted and compared with those of recognized biological hydroperoxides.

COMPARISON OF PEROXYMONOCARBONATE WITH OTHER HYDROPEROXIDES

Peroxymonocarbonate is a true hydroperoxide with the structure $HOOCO_2^-$, according to X-ray data [75] (Scheme 4.3). In contrast to H_2O_2, HCO_4^- is negatively charged at physiological pH because its lower pK_a is 3.4 [37,69]. Additionally, compared to H_2O_2 where both oxygens of the O–O bond have a similar tendency to undergo nucleophilic attack, in $HOOCO_2^-$, the oxygen of the hydroxyl group is the more susceptible oxygen [76] (Scheme 4.3). As a two-electron oxidant, peroxymonocarbonate ($E° = 1.80$ V) is as strong as H_2O_2 ($E° = 1.77$ V) and is stronger than ONOOH ($E° = 1.40$ V) [35,77]. However, the mechanism of the heterolytic oxidation by peroxides involves nucleophilic attack of the substrate on the electrophilic oxygen of the peroxide with displacement of the peroxide leaving group (Scheme 4.3). The second-order rate constants of these reactions are usually higher for peroxides with better leaving groups, a property that inversely correlates with the pK_a of the conjugated acid of the leaving group [55]. For ONOOH, HCO_4^-, H_2O_2, and ROOH, the corresponding leaving group and the pK_a value of the conjugated acid are NO_2^- ($pK_a = 3.15$), CO_3^{2-} ($pK_a = 10.3$), HO^- ($pK_a = 15.7$), and RO^- ($pK_a = 18$), respectively. Since the pK_a values of alkyl alkoxides from biologically relevant organic hydroperoxides are not known, the value listed here is the one of *tert*-butyl alkoxide [56]. From the pK_a value of the conjugated acid of the corresponding leaving group, it is possible to conclude that the reactivity of hydroperoxides in heterolytic oxidations follows the order: peroxynitrite > HCO_4^- > H_2O_2 > ROOH. For instance, this trend holds for the heterolytic oxidation of biothiols, although reactive thiol-proteins, such as Prx, increase the second-order rate constant of H_2O_2 reaction by several orders of magnitude as compared with that of low molecular thiols or BSA-SH, which are comparable (Table 4.1). Structural features of reactive thiol proteins that stabilize the transition state are likely involved in the extraordinary catalytic power of these enzymes [55,56,78]. The trend also holds for the heterolytic oxidation of biomolecules with different functional groups as becomes evident by comparing the second-order rate constant of the reactions of H_2O_2 with those of HCO_4^- (Table 4.1) (Figure 4.1).

In the case of the hydroperoxide-mediated oxidation of biomolecules by one-electron mechanisms, the hydroperoxides require previous activation or cleavage to radicals by hemeproteins, transition metal ions or energy (light/heat). The only exception is

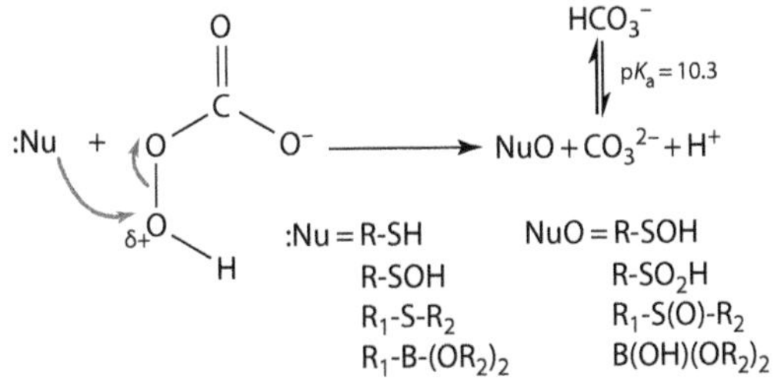

SCHEME 4.3 The mechanism of the two-electron heterolytic oxidation of biological nucleophilic targets by HCO_4^-.

TABLE 4.1
Comparison of Selected Second-Order Rate Constants for Two-Electron Oxidations Promoted by H_2O_2 or HCO_4^-

Substrate	H_2O_2 (M^{-1} s^{-1})	HCO_4^- (M^{-1} s^{-1})	pH	T (°C)
Thiol (-SH)				
GSH[a]	1.9	1.6×10^2	7.4	37
BSA[a]	1.2	2.3×10^2	7.4	37
PTP1B[b]	24	2.5×10^4	7.0	25
SHP-2[b]	15	2.1×10^4	7.0	25
Papain[b]	43	1.2×10^4	7.0	25
AhpE[c]	2.1×10^3	1.1×10^7	7.4	25
Sulfenic acid (-SOH)				
AhpE[c]	40	2.0×10^3	7.4	25
Thioether (R_1-S-R_2)				
Methionine[d]	7.5×10^{-3}	0.48	8.0	25
α1-P1[d]	1.4×10^{-2}	0.36	8.0	25
Boronate (R_1-B-$(OR_2)_2$)				
CBE	3.6	1.71×10^2	7.4	37

Notes: PTP1B and SHP-2 are members of the protein tyrosine phosphatase family; AhpE, alkyl hydroperoxide reductase E; α1-P1, α1 proteinase inhibitor; CBE, coumarin-7-boronic acid pinacolate ester.
[a] Trindade et al. [36].
[b] Zhou et al. [79].
[c] The data from Reyes et al. [56] were used to calculate the second-order rate constant as a function of HCO_4^- concentration.
[d] Richardson et al. [70].

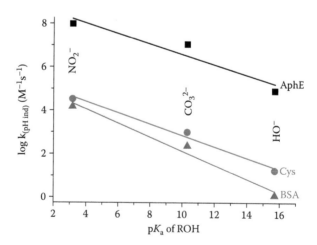

FIGURE 4.1 Log of pH-independent second-order rate constants for the thiol oxidation (k) of cysteine, BSA and AphE by ONOOH, HCO_4^-, and H_2O_2 plotted against the pK_a of the conjugate acid of the corresponding the leaving group (ROH). The second-order rate constants and the pK_a values were taken from the literature [36,56,80,81].

peroxynitrite, which decomposes to radicals by proton- and CO_2-catalyzed reactions (Scheme 4.1). NMR studies showed that HCO_4^- is stable in equilibrium with hydrogen peroxide and HCO_3^-/CO_2 solutions but it produces $CO_3^{\bullet-}$ upon reduction by transition metal ions (Fenton chemistry) (Equation 4.13). This type of reaction between HCO_4^- and metal ion complexes has been sparsely studied [14,16,19,82]. Nevertheless, it has been demonstrated to occur at the SOD-Cu(I) active site of the enzyme SOD1 during its bicarbonate-dependent peroxidase activity [33,34,41]. Using different strategies, these authors determined the second-order rate constant for the reaction of SOD-Cu(I) and HCO_4^- to be 2.0×10^3 and 1.6×10^3 $M^{-1} s^{-1}$ for the human and bovine enzyme, respectively. These values are two orders of magnitude higher than that estimated for the reaction between SODCu(I) with H_2O_2 ($k = 13$ $M^{-1} s^{-1}$) [11].

$$HOOCO_2^- + M^{n+} \rightarrow HO^- + CO_3^{\bullet-} + M^{(n+1)+} \qquad (4.13)$$

Among the hydroperoxide-derived radicals, $CO_3^{\bullet-}$ ($E° = 1.78$ V) is only less oxidizing than the $HO^{\bullet}$ ($E° = 2.31$ V). The latter reacts at the site of formation and oxidizes biomolecules by electron transfer and by addition to double bonds. In contrast, $CO_3^{\bullet-}$ diffuses longer distances and usually oxidizes biomolecules by electron transfer or by hydrogen abstraction mechanisms, producing radicals from the oxidized targets [27]. The produced radicals can engage in further radical reactions, including cross-linking reactions [83]. Indeed, $CO_3^{\bullet-}$ promotes protein–protein and DNA–DNA cross-links [74,84,85]. $CO_3^{\bullet-}$, produced from peroxynitrite or from HCO_4^-, is an important player in cell and tissue oxidative damage (Scheme 4.1).

CONSIDERATIONS ABOUT PEROXYMONOCARBONATE FORMATION IN CELLS

To discuss the possible HCO_4^- formation in cells, it is instructive to examine further the second-order rate constants of its two-electron oxidation of molecules containing different functional groups (Table 4.1). In general, the second-order rate constant of HCO_4^--mediated oxidations is two to three orders of magnitude higher than that of H_2O_2-mediated oxidations. As noted, the remarkable exception is oxidation of the catalytic thiol of AhpE, which is a one-cysteine Prx from *Mycobacterium tuberculosis* [56]. In contrast to the slow oxidation of the catalytic thiol of AhpE by H_2O_2 ($k = 2.1 \times 10^3$ $M^{-1} s^{-1}$), the catalytic thiols of mammalian Prx react with H_2O_2 with rate constants close to the diffusion limit ($k > 10^7 – 10^9$ $M^{-1} s^{-1}$) [86,87]. It is consequently unlikely that HCO_4^- could react considerably faster than H_2O_2 with the catalytic thiols of mammalian Prx. The picture changes in the case of Prx hyperoxidation, which is slower than catalytic thiol oxidation for both AhpE ($k = 4.0 \times 10^1$ $M^{-1} s^{-1}$) and mammalian Prx ($k = 1.2 \times 10^4$ $M^{-1} s^{-1}$) [88]. Prx hyperoxidation is the process by which the sulfenic acid resulting from Prx–SH oxidation (Prx–SOH) oxidizes further to sulfinic/sulfonic derivatives (Prx–SO_2H/–SO_3H). As shown in Table 4.1, AhpE hyperoxidation occurs approximately 100 times faster with HCO_4^- than with H_2O_2. Relevantly, Prx enzymes in the sulfinic/sulfonic forms are not recycled by cellular reductants and become inactivated [50,65,89]. Therefore, the possibility of HCO_4^- accelerating mammalian Prx hyperoxidation and inactivation is worth exploring. Preliminary studies from our laboratory with human Prx1 *in vitro* are indicating that HCO_4^- accelerates enzyme hyperoxidation and inactivation (ongoing work). If HCO_4^- contributes to the inactivation of Prx *in vivo*, the steady-state concentration of H_2O_2 is likely to increase. In parallel, the steady-state concentration of HCO_4^- is also likely to increase and these changes may affect cellular signaling and damaging pathways (Scheme 4.1).

With regard to cellular redox signaling, it is relevant noting that HCO_4^- oxidizes PTP1B and SHP-2, which are members of the protein tyrosine phosphatase (PTPs) family, three orders of magnitude faster than H_2O_2 (Table 4.1) [79]. Despite PTPs reacting slowly with H_2O_2 *in vitro*, several evidences indicate that they are important targets in H_2O_2-mediated signaling. The physiological bicarbonate buffer may be part of the solution of this enigma by increasing the rate of PTPs oxidation by H_2O_2 via HCO_4^-. As the increase in rate is small, PTPs cannot compete for H_2O_2 with Prx. Therefore, bicarbonate buffer may act in tandem with other biological characteristics, such as colocalization and compartmentalization [65,79].

The comparison of the second-order rate constants of the oxidation of boronates by H_2O_2 and HCO_4^- (Table 4.1) also has important implications. Boronate derivatives that become oxidized and fluorescent upon nucleophilic attack by peroxynitrite and H_2O_2 are being extensively employed to detect these oxidants in cells (for a review, see [90,91]). Different boronate derivatives have been synthesized and the second-order rate constants of their oxidation by peroxynitrite and H_2O_2 have been determined [92,93]. The bicarbonate buffer has been reported to increase boronate oxidation by H_2O_2; however, the data were not shown and the second-order rate constant was not reported [94]. Here, we show that the HCO_3^-/CO_2 pair increases the oxidation of coumarin-7-boronic acid pinacolate ester (CBE) by H_2O_2 (Figure 4.2). The determination of the second-order rate constants showed that it is higher for HCO_4^- (1.7×10^2 M^{-1} s^{-1}) than for H_2O_2 (3.6 M^{-1} s^{-1}) by two orders of magnitude, as expected from a S_N2 mechanism (Figure 4.2 inset, Table 4.1). Accordingly, the second-order rate constants of peroxynitrite-mediated oxidations are on the order of 10^6 M^{-1} s^{-1} [92,93]. It has been alleged that the fluorescence of boronate derivatives in cells that do not synthesize peroxynitrite can only be attributed to H_2O_2 because of the high pK_a of the conjugated acid of the organic hydroperoxide leaving group [90]. However, boronate fluorescence in these cells may well result from HCO_4^- formation (Figure 4.2). More recently, amino acid- and protein-derived hydroperoxides were shown to react with boronates, but the determined rate constants (k_{CBA} = 7–23 M^{-1} s^{-1}) were one to two orders of magnitude lower than those expected for HCO_4^- [95].

The data discussed here present new perspectives for the consideration of HCO_4^- as a relevant biological hydroperoxide. For instance, the well-controlled use of boronate derivatives in cells producing H_2O_2 levels that are insufficient to promote extensive biomolecule peroxidation in the presence and absence of the HCO_3^-/CO_2 pair may suggest HCO_4^- formation. Although these experiments have yet to be performed in controlled atmospheres, they may resolve the questioning of HCO_4^- production in cells.

HCO_4^- has been largely disregarded as an important biological oxidant because of the low second-order rate constant of the reaction between CO_2 and H_2O_2 at neutral pH and because of the low percentage of H_2O_2 present as HCO_4^- after equilibration. The available quantitative data were obtained in aqueous solutions, but the rate

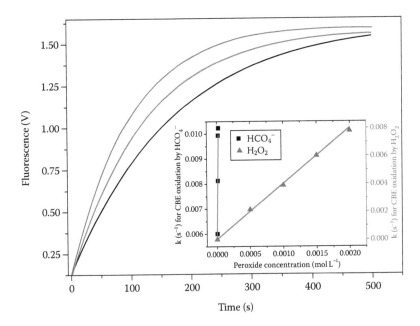

FIGURE 4.2 Representative kinetics of the oxidation of CBE (5 µM) by H_2O_2 (1.5 mM) in the absence (black trace) and in the presence of the HCO_3^-/CO_2 pair at 25 mM (red trace) or 45 mM (blue trace) monitored by the intrinsic fluorescence of the oxidation product of CBE in phosphate buffer 50 mM, pH 7.4, DTPA 100 µM at 37°C. The insert shows the plot of the k_{obs} of oxidation *versus* the HCO_4^- or H_2O_2 concentration as specified. The corresponding second-order rate constants were calculated from the slop of the plot. Bicarbonate was preincubated with H_2O_2 and the concentrations of HCO_4^- were calculated using Equation 4.12.

of HCO_4^- formation may increase in more physiological conditions, as indicated by the accelerating effect of carbonic anhydrase [37]. Unfortunately, there are no quantitative data about this accelerating effect. Likewise, limited quantitative data are available for HCO_4^- formation at the active site of XO [32] and of SOD1 [33]. It is certainly unexpected that in aqueous solutions, CO_2 can compete with enzymes, such as catalase and seleno-, heme-, and Prx that react with H_2O_2 with extremely high second-order rate constants. However, enzymes are compartmentalized in cells, which are constantly producing CO_2 from respiration. If the cells are producing H_2O_2 in parallel, then HCO_4^- formation could occur in certain compartments and promote oxidative damage. All of these considerations indicate that it is premature to exclude HCO_4^- as a possible mediator of CO_2 toxicity.

SUMMARY

The increasing levels of atmospheric CO_2 suggest that a better comprehension of how this increase may affect life on Earth is needed. It is known that high CO_2 levels (hypercapnia) cause acute and long-term toxicity to animals exposed to them. Levels of CO_2 above the physiological levels also occur in clinical situations, such as emphysema, respiratory muscle paralysis, and pulmonary fibrosis. These clinical conditions are usually associated with oxidative damage, but the knowledge regarding the role of the main physiological buffer, the HCO_3^-/CO_2 pair, in oxidative damage is partial and limited to the $CO_3^{\bullet-}$ radical generated from peroxynitrite. Here, the reactions of the HCO_3^-/CO_2 pair with biological hydroperoxides (peroxynitrite, H_2O_2, and organic hydroperoxides) are summarized. The possible biological consequences of these reactions are discussed. Peroxymonocarbonate (HCO_4^-), which is produced from the reaction between H_2O_2 and CO_2, was emphasized because its role in redox biology remains unproven. We conclude that it is premature to exclude the possibility of HCO_4^- formation *in vivo* and that further studies are required to advance the understanding of CO_2 toxicity.

ACKNOWLEDGMENTS

We thank Fernando R Coelho for technical assistance. The work in our laboratory is supported by Fundação de Amparo à Pesquisa do Estado de São Paulo (FAPESP) Grant 2013/07937-8 and Fellowship 2014/9518-5, Pro-Reitoria de Pesquisa da Universidade de São Paulo (PRPUSP) Grant 2011.1.9352.1.8 and Conselho Nacional de Desenvolvimento Científico Tecnológico (CNPq) Fellowship 303113/2014-6. The authors are members of the CEPID Redoxoma.

REFERENCES

1. Guais, A., G. Brand, L. Jacquot, M. Karrer, S. Dukan, G. Grevillot, T. J. Molina, J. Bonte, M. Regnier, and L. Schwartz. 2011. Toxicity of carbon dioxide: A review. *Chemical Research in Toxicology* 24 (12): 2061–2070.
2. Hodgson, E. K. and I. Fridovich. 1976. Mechanism of activity-dependent luminescence of xanthine-oxidase. *Archives of Biochemistry and Biophysics* 172 (1): 202–205.
3. Nakano, M. and K. Sugioka. 1977. Mechanism of chemiluminescence from the linoleate-lipoxygenase system. *Archives of Biochemistry and Biophysics* 181 (2): 371–383.
4. Michelson, A. M. and P. Durosay. 1977. Hemolysis of human erythrocytes by activated oxygen species. *Photochemistry and Photobiology* 25 (1): 55–63.
5. Dean, J. B. 2010. Hypercapnia causes cellular oxidation and nitrosation in addition to acidosis: Implications for CO_2 chemoreceptor function and dysfunction. *Journal of Applied Physiology* 108 (6): 1786–1795.
6. Queliconi, B. B., T. B. M. Marazzi, S. M. Vaz, P. S. Brookes, K. Nehrke, O. Augusto, and A. J. Kowaltowski. 2013. Bicarbonate modulates oxidative and functional damage in ischemia-reperfusion. *Free Radical Biology and Medicine* 55: 46–53.
7. Sankarapandi, S. and J. L. Zweier. 1999. Bicarbonate is required for the peroxidase function of Cu, Zn-superoxide dismutase at physiological pH. *Journal of Biological Chemistry* 274 (3): 1226–1232.
8. Zhang, H., J. Joseph, C. Felix, and B. Kalyanaraman. 2000. Bicarbonate enhances the hydroxylation, nitration, and peroxidation reactions catalyzed by copper, zinc superoxide dismutase—Intermediacy of carbonate anion radical. *Journal of Biological Chemistry* 275 (19): 14038–14045.
9. Santos, C. X., M. G. Bonini, and O. Augusto. 2000. Role of the carbonate radical anion in tyrosine nitration and hydroxylation by peroxynitrite. *Archives of Biochemistry and Biophysics* 377 (1): 146–152.

10. Bonini, M. G. and O. Augusto. 2001. Carbon dioxide stimulates the production of thiyl, sulfinyl, and disulfide radical anion from thiol oxidation by peroxynitrite. *The Journal of Biological Chemistry* 276 (13): 9749–9754.

11. Liochev, S. I. and I. Fridovich. 2002. Copper, zinc superoxide dismutase and H_2O_2—Effects of bicarbonate on inactivation and oxidations of NADPH and urate, and on consumption of H_2O_2. *Journal of Biological Chemistry* 277 (38): 34674–34678.

12. Zhang, H., C. Andrekopoulos, J. Joseph, K. Chandran, H. Karoui, J. P. Crow, and B. Kalyanaraman. 2003. Bicarbonate-dependent peroxidase activity of human Cu, Zn-superoxide dismutase induces covalent aggregation of protein—Intermediacy of tryptophan-derived oxidation products. *Journal of Biological Chemistry* 278 (26): 24078–24089.

13. Liochev, S. I. and I. Fridovich. 2004a. CO_2, Not HCO_3^-, facilitates oxidations by Cu, Zn superoxide dismutase plus H_2O_2. *Proceedings of the National Academy of Sciences of the United States of America* 101 (3): 743–744.

14. Liochev, S. I. and I. Fridovich. 2004b. Carbon dioxide mediates Mn(II)-catalyzed decomposition of hydrogen peroxide and peroxidation reactions. *Proceedings of the National Academy of Sciences of the United States of America* 101 (34): 12485–12490.

15. Bonini, M. G., D. C. Fernandes, and O. Augusto. 2004. Albumin oxidation to diverse radicals by the peroxidase activity of Cu, Zn-superoxide dismutase in the presence of bicarbonate or nitrite: Diffusible radicals produce cysteinyl and solvent-exposed and -unexposed tyrosyl radicals. *Biochemistry* 43 (2): 344–351.

16. Ramirez, D. C., S. E. G. Mejiba, and R. P. Mason. 2005. Copper-catalyzed protein oxidation and its modulation by carbon dioxide: Enhancement of protein radicals in cells. *The Journal of Biological Chemistry* 280 (29): 27402–27411.

17. Arai, H., B. S. Berlett, P. B. Chock, and E. R. Stadtman. 2005. Effect of bicarbonate on iron-mediated oxidation of low-density lipoprotein. *Proceedings of the National Academy of Sciences of the United States of America* 102 (30): 10472–10477.

18. Stadtman, E. R., H. Arai, and B. S. Berlett. 2005. Protein oxidation by the cytochrome P450 mixed-function oxidation system. *Biochemical and Biophysical Research Communications* 338 (1): 432–436.

19. Liochev, S. I. and I. Fridovich. 2005. The role of CO_2 in cobalt-catalyzed peroxidations. *Archives of Biochemistry and Biophysics* 439 (1): 99–104.

20. Berlett, B. S., P. B. Chock, M. B. Yim, and E. R. Stadtman. 1990. Manganese(II) catalyzes the bicarbonate-dependent oxidation of amino-acids by hydrogen-peroxide and the amino acid-facilitated dismutation of hydrogen-peroxide. *Proceedings of the National Academy of Sciences of the United States of America* 87 (1): 389–393.

21. Flemming, J. and J. Arnhold. 2007. Ferrous ion-induced strand breaks in the DNA plasmid pBR322 are not mediated by hydrogen peroxide. *European Biophysics Journal with Biophysics Letters* 36 (4–5): 377–384.

22. Wulff, R., G. M. Rappen, M. Koziolek, G. Garbacz, and C. S. Leopold. 2015. Controlled release of acidic drugs in compendial and physiological hydrogen carbonate buffer from polymer blend-coated oral Solid dosage forms. *European Journal of Pharmaceutical Sciences* 77: 246–253.

23. Garbacz, G., B. Kolodziej, M. Koziolek, W. Weitschies, and S. Klein. 2013. An automated system for monitoring and regulating the pH of bicarbonate buffers. *AAPS PharmSciTech* 14 (2): 517–522.

24. Medinas, D. B., G. Cerchiaro, D. F. Trindade, and O. Augusto. 2007. The carbonate radical and related oxidants derived from bicarbonate buffer. *IUBMB Life* 59 (4–5): 255–262.

25. Beckman, J. S., T. W. Beckman, J. Chen, P. A. Marshall, and B. A. Freeman. 1990. Apparent hydroxyl radical production by peroxynitrite—Implications for endothelial injury from nitric-oxide and superoxide. *Proceedings of the National Academy of Sciences of the United States of America* 87 (4): 1620–1624.

26. Lymar, S. V. and J. K. Hurst. 1995. Rapid reaction between peroxonitrite ion and carbon-dioxide—Implications for biological-activity. *Journal of the American Chemical Society* 117 (34): 8867–8868.

27. Augusto, O., M. G. Bonini, A. M. Amanso, E. Linares, C. C. X. Santos, and S. L. De Menezes. 2002. Nitrogen dioxide and carbonate radical anion: Two emerging radicals in biology. *Free Radical Biology and Medicine* 32 (9): 841–859.

28. Radi, R. 2004. Nitric oxide, oxidants, and protein tyrosine nitration. *Proceedings of the National Academy of Sciences of the United States of America* 101 (12): 4003–4008.

29. Ferrer-Sueta, G. and R. Radi. 2009. Chemical biology of peroxynitrite: Kinetics, diffusion, and radicals. *Chemical Biology* 4 (3): 161–177.

30. Radi, R. 2013. Peroxynitrite, a stealthy biological oxidant. *The Journal of Biological Chemistry* 288 (37): 26464–26472.

31. Bonini, M. G., R. Radi, G. Ferrer-Sueta, A. M. Ferreira, and O. Augusto. 1999. Direct EPR detection of the carbonate radical anion produced from peroxynitrite and carbon dioxide. *The Journal of Biological Chemistry* 274 (16): 10802–10806.

32. Bonini, M. G., S. Miyamoto, P. Di Mascio, and O. Augusto. 2004. Production of the carbonate radical anion during xanthine oxidase turnover in the presence of bicarbonate. *The Journal of Biological Chemistry* 279 (50): 51836–51843.

33. Medinas, D. B., J. C. Toledo Jr., G. Cerchiaro, A. T. Amaral, L. Rezende, A. Malvezzi, and O. Augusto. 2009. Peroxymonocarbonate and carbonate radical displace the hydroxyl-like oxidant in the Sod1 peroxidase activity under physiological conditions. *Chemical Research in Toxicology* 22 (4): 639–648.

34. Bonini, M. G., S. A. Gabel, K. Ranguelova, K. Stadler, E. F. DeRose, R. E. London, and R. P. Mason. 2009. Direct magnetic resonance evidence for peroxymonocarbonate involvement in the Cu, Zn-superoxide dismutase peroxidase catalytic cycle. *Journal of Biological Chemistry* 284 (21): 14618–14627.

35. Richardson, D. E., H. R. Yao, K. M. Frank, and D. A. Bennett. 2000. Equilibria, kinetics, and mechanism in the bicarbonate activation of hydrogen peroxide: Oxidation of sulfides by peroxymonocarbonate. *Journal of the American Chemical Society* 122 (8): 1729–1739.

36. Trindade, D. F., G. Cerchiaro, and O. Augusto. 2006. A role for peroxymonocarbonate in the stimulation of biothiol peroxidation by the bicarbonate/carbon dioxide pair. *Chemical Research in Toxicology* 19 (11): 1475–1482.

37. Bakhmutova-Albert, E. V., H. Yao, D. E. Denevan, and D. E. Richardson. 2010. Kinetics and mechanism of peroxymonocarbonate formation. *Inorganic Chemistry* 49 (24): 11287–11296.

38. Sakakura, T., J. C. Choi, and H. Yasuda. 2007. Transformation of carbon dioxide. *Chemical Reviews* 107 (6): 2365–2387.

39. Goldstein, S., J. Lind, and G. Merenyi. 2002. The reaction of ONOO$^-$ with carbonyls: Estimation of the half-lives of ONOOC(O)O$^-$ and O$_2$NOOC(O)O$^-$. *Journal of the Chemical Society-Dalton Transactions* 5: 808–810.

40. Smyth, H. D. and E. C. G. Stueckelberg. 1930. The ionization of carbon dioxide by electron impact. *Physical Review* 36 (3): 0472–0477.

41. Ranguelova, K., D. Ganini, M. G. Bonini, R. E. London, and R. P. Mason. 2012. Kinetics of the oxidation of reduced Cu, Zn-superoxide dismutase by peroxymonocarbonate. *Free Radical Biology and Medicine* 53 (3): 589–594.

42. Wong, P. K., J. L. Poyer, C. M. Dubose, and R. A. Floyd. 1988. Hydralazine-dependent carbon-dioxide free-radical formation by metabolizing mitochondria. *Journal of Biological Chemistry* 263 (23): 11296–11301.

43. Connor, H. D., R. G. Thurman, M. D. Galizi, and R. P. Mason. 1986. The formation of a novel free-radical metabolite from CCl4 in the perfused-rat-liver and *in vivo*. *Journal of Biological Chemistry* 261 (10): 4542–4548.

44. Vazquez, S., J. A. Aquilina, J. F. Jamie, M. M. Sheil, and R. J. W. Truscott. 2002. Novel protein modification by kynurenine in human lenses. *The Journal of Biological Chemistry* 277 (7): 4867–4873.

45. Girotti, A. W. 1998. Lipid hydroperoxide generation, turnover, and effector action in biological systems. *Journal of Lipid Research* 39 (8): 1529–1542.

46. Caro, A. A. and A. I. Cederbaum. 2006. Role of cytochrome P450 in phospholipase A2- and arachidonic acid-mediated cytotoxicity. *Free Radical Biology and Medicine* 40 (3): 364–375.

47. Miyamoto, S., G. E. Ronsein, F. M. Prado, M. Uemi, T. C. Corrêa, I. N. Toma, A. Bertolucci et al. 2007. Biological hydroperoxides and singlet molecular oxygen generation. *IUBMB Life* 59 (4–5): 322–331.

48. Davies, M. J., C. L. Hawkins, D. I. Pattison, and M. D. Rees. 2008. Mammalian heme peroxidases: From molecular mechanisms to health implications. *Antioxidants & Redox Signaling* 10 (7): 1199–1234.

49. Brigelius-Flohé, R. 1999. Tissue-specific functions of individual glutathione peroxidases. *Free Radical Biology and Medicine* 27 (9–10): 951–965.

50. Rhee, S. G., H. Z. Chae, and K. Kim. 2005. Peroxiredoxins: A historical overview and speculative preview of novel mechanisms and emerging concepts in cell signaling. *Free Radical Biology and Medicine* 38 (12): 1543–1552.

51. Davies, M. J. 2016. Protein oxidation and peroxidation. *Biochemical Journal* 473 (7): 805–825.

52. Alegria, T. G. P., D. A. Meireles, J. R. R. Cussiol, M. Hugo, M. Trujillo, M. A. de Oliveira, S. Miyamoto et al. 2017. Ohr plays a central role in bacterial responses against fatty acid hydroperoxides and peroxynitrite. *Proceedings of the National Academy of Sciences of the United States of America* 114 (2): E132–E141.

53. Peskin, A. V., A. G. Cox, P. Nagy, P. E. Morgan, M. B. Hampton, M. J. Davies, and C. C. Winterbourn. 2010. Removal of amino acid, peptide and protein hydroperoxides by reaction with peroxiredoxins 2 and 3. *Biochemical Journal* 432 (2): 313–321.

54. Jovanovic, S. V., I. Jankovic, and L. Josimovic. 1992. Electron-transfer reactions of alkylperoxy radicals. *Journal of the American Chemical Society* 114 (23): 9018–9021.

55. Ferrer-Sueta, G., B. Manta, H. Botti, R. Radi, M. Trujillo, and A. Denicola. 2011. Factors affecting protein thiol reactivity and specificity in peroxide reduction. *Chemical Research in Toxicology* 24 (4): 434–450.

56. Reyes, A. M., M. Hugo, A. Trostchansky, L. Capece, R. Radi, and M. Trujillo. 2011. Oxidizing substrate specificity of mycobacterium tuberculosis alkyl hydroperoxide reductase E: Kinetics and mechanisms of oxidation and overoxidation. *Free Radical Biology & Medicine* 51 (2): 464–473.

57. Augusto, O., R. M. Gatti, and R. Radi. 1994. Spin-trapping studies of peroxynitrite decomposition and of 3-morpholinosydnonimine N-ethylcarbamide autooxidation: Direct evidence for metal-independent formation of free radical intermediates. *Archives of Biochemistry and Biophysics* 310 (1): 118–125.

58. Molina, C., R. Kissner, and W. H. Koppenol. 2013. Decomposition kinetics of peroxynitrite: Influence of pH and buffer. *Dalton Transactions* 42 (27): 9898.

59. Vasquez-Vivar, J., A. Denicola, R. Radi, and O. Augusto. 1997. Peroxynitrite-mediated decarboxylation of pyruvate to both carbon dioxide and carbon dioxide radical anion. *Chemical Research in Toxicology* 10 (7): 786–794.

60. Halliwell, B. and J. M. C. Gutteridge. 2007. *Free Radicals in Biology and Medicine*, 4th edn. Oxford, U.K.: Oxford University Press.

61. Winterbourn, C. C. 2013. Chapter 1—The biological chemistry of hydrogen peroxide. In *Hydrogen Peroxide and Cell Signaling, Part C*, eds. E. Cadenas and L. Packer, Vol. 528, pp. 3–25. Methods in Enzymology, Amsterdam, The Netherlands: Academic Press, Elsevier.

62. Koppenol, W. H. 1994. Chemistry of iron and copper in radical reactions. In *New Comprehensive Biochemistry*, eds. C.A. Rice-Evans and R. H. Burdon, Vol. 28, pp. 3–24, Amsterdam, The Netherlands: Elsevier.

63. Antunes, F. and E. Cadenas. 2001. Cellular titration of apoptosis with steady state concentrations of H_2O_2: Submicromolar levels of H_2O_2 induce apoptosis through fenton chemistry independent of the cellular thiol state. *Free Radical Biology and Medicine* 30 (9): 1008–1018.

64. Stone, J. R. 2004. An assessment of proposed mechanisms for sensing hydrogen peroxide in mammalian systems. *Archives of Biochemistry and Biophysics* 422 (2): 119–124.

65. Winterbourn, C. C. and M. B. Hampton. 2008. Thiol chemistry and specificity in redox signaling. *Free Radical Biology & Medicine* 45 (5): 549–561.

66. Sobotta, M. C., W. Liou, S. Stöcker, D. Talwar, M. Oehler, T. Ruppert, A. N. D. Scharf, and T. P. Dick. 2015. Peroxiredoxin-2 and STAT3 form a redox relay for H_2O_2 signaling. *Nature Chemical Biology* 11 (1): 64–70.

67. Flanagan, J., D. P. Jones, W. P. Griffith, A. C. Skapski, and A. P. West. 1986. On the existence of peroxocarbonates in aqueous-solution. *Journal of the Chemical Society—Chemical Communications* 1: 20–21.

68. Jones, D. P. and W. P. Griffith. 1980. Alkali-metal peroxocarbonates, $M_2[CO_3] \cdot nH_2O_2$, $M_2[C_2O_6]$, $M[HCO_4] \cdot nH_2O$, and $Li_2[CO_4] \cdot H_2O$. *Journal of the Chemical Society, Dalton Transactions* 12: 2526–2532.

69. Bennett, D. A., H. Yao, and D. E. Richardson. 2001. Mechanism of sulfide oxidations by peroxymonocarbonate. *Inorganic Chemistry* 40 (13): 2996–3001.

70. Richardson, D. E., C. A. S. Regino, H. R. Yao, and J. V. Johnson. 2003. Methionine oxidation by peroxymonocarbonate, a reactive oxygen species formed from CO_2/bicarbonate and hydrogen peroxide. *Free Radical Biology and Medicine* 35 (12): 1538–1550.

71. Yao, H. R. and D. E. Richardson. 2003. Bicarbonate surfoxidants: Micellar oxidations of aryl sulfides with bicarbonate-activated hydrogen peroxide. *Journal of the American Chemical Society* 125 (20): 6211–6221.

72. Elam, J. S., K. Malek, J. A. Rodriguez, P. A. Doucette, A. B. Taylor, L. J. Hayward, D. E. Cabelli, J. S. Valentine, and P. J. Hart. 2003. An alternative mechanism of bicarbonate-mediated peroxidation by copper-zinc superoxide dismutase: Rates enhanced via proposed enzyme-associated peroxycarbonate intermediate. *The Journal of Biological Chemistry* 278 (23): 21032–21039.

73. Strange, R. W., M. A. Hough, S. V. Antonyuk, and S. S. Hasnain. 2012. Structural evidence for a copper-bound carbonate intermediate in the peroxidase and dismutase activities of superoxide dismutase. *PLoS One* 7 (9): e44811.

74. Medinas, D. B., F. C. Gozzo, L. F. A. Santos, A. H. Iglesias, and O. Augusto. 2010. A ditryptophan cross-link is responsible for the covalent dimerization of human superoxide dismutase 1 during its bicarbonate-dependent peroxidase activity. *Free Radical Biology and Medicine* 49 (6): 1046–1053.

75. Adam, A. and M. Mehta. 1998. $KH(O_2^-)CO_2 \cdot H_2O_2$—An oxygen-rich salt of monoperoxocarbonic acid. *Angewandte Chemie* 37 (10): 1387–1388.

76. Aparicio, F., R. Contreras, M. Galvan, and A. Cedillo. 2003. Global and local reactivity and activation patterns of HOOX (X = H, NO_2, CO_2^-, SO_3^-) peroxides with solvent effects. *Journal of Physical Chemistry A* 107 (47): 10098–10104.

77. Augusto, O. and S. Miyamoto. 2011. Chapter 2—Oxygen radicals and related species. In *Principles of Free Radical Biomedicine*, eds. K. Pantopoulos and H. M. Schipper, Vol. 1, pp. 19–41. Montreal, Quebec, Canada: Nova Science.

78. Hall, A., D. Parsonage, L. B. Poole, and P. A. Karplus. 2010. Structural evidence that peroxiredoxin catalytic power is based on transition-state stabilization. *Journal of Molecular Biology* 402 (1): 194–209.

79. Zhou, H., H. Singh, Z. D. Parsons, S. M. Lewis, S. Bhattacharya, D. R. Seiner, J. N. LaButti, T. J. Reilly, J. J. Tanner, and K. S. Gates. 2011. The biological buffer bicarbonate/CO$_2$ potentiates H$_2$O$_2$-mediated inactivation of protein tyrosine phosphatases. *Journal of the American Chemical Society* 133 (40): 15803–15805.
80. Radi, R., J. S. Beckman, K. M. Bush, and B. A. Freeman. 1991. Peroxynitrite oxidation of sulfhydryls—The cytotoxic potential of superoxide and nitric-oxide. *Journal of Biological Chemistry* 266 (7): 4244–4250.
81. Regino, C. A. S. and D. E. Richardson. 2007. Bicarbonate-catalyzed hydrogen peroxide oxidation of cysteine and related thiols. *Inorganica Chimica Acta* 360 (14): 3971–3977.
82. Carvalho do Lago, L. C., A. C. Matias, C. S. Nomura, and G. Cerchiaro. 2011. Radical production by hydrogen peroxide/bicarbonate and copper uptake in mammalian cells: Modulation by Cu(II) complexes. *Journal of Inorganic Biochemistry* 105 (2): 189–194.
83. Toledo, J. C. and O. Augusto. 2012. Connecting the chemical and biological properties of nitric oxide. *Chemical Research in Toxicology* 25 (5): 975–989.
84. Paviani, V., R. F. Queiroz, E. F. Marques, P. Di Mascio, and O. Augusto. 2015. Production of lysozyme and lysozyme-superoxide dismutase dimers bound by a ditryptophan cross-link in carbonate radical-treated lysozyme. *Free Radical Biology and Medicine* 89: 72–82.
85. Yun, B. H., N. E. Geacintov, and V. Shafirovich. 2011. Generation of guanine–thymidine cross-links in DNA by peroxynitrite/carbon dioxide. *Chemical Research in Toxicology* 24 (7): 1144–1152.
86. Cox, A. G., A. V. Peskin, L. N. Paton, C. C. Winterbourn, and M. B. Hampton. 2009. Redox potential and peroxide reactivity of human peroxiredoxin 3. *Biochemistry* 48 (27): 6495–6501.
87. Manta, B., M. Hugo, C. Ortiz, G. Ferrer-Sueta, M. Trujillo, and A. Denicola. 2009. The peroxidase and peroxynitrite reductase activity of human erythrocyte peroxiredoxin 2. *Archives of Biochemistry and Biophysics* 484 (2): 146–154.
88. Peskin, A. V., N. Dickerhof, R. A. Poynton, L. N. Paton, P. E. Pace, M. B. Hampton, and C. C. Winterbourn. 2013. Hyperoxidation of peroxiredoxins 2 and 3: Rate constants for the reactions of the sulfenic acid of the peroxidatic cysteine. *Journal of Biological Chemistry* 288 (20): 14170–14177.
89. Hall, A., P. A. Karplus, and L. B. Poole. 2009. Typical 2-cys peroxiredoxins: Structures, mechanisms and functions. *The FEBS Journal* 276 (9): 2469–2477.
90. Lippert, A. R., G. C. Van de Bittner, and C. J. Chang. 2011. Boronate oxidation as a bioorthogonal reaction approach for studying the chemistry of hydrogen peroxide in living systems. *Accounts of Chemical Research* 44 (9): 793–804.
91. Winterbourn, C. C. 2014. The challenges of using fluorescent probes to detect and quantify specific reactive oxygen species in living cells. *Biochimica et Biophysica Acta* 1840 (2): 730–738.
92. Sikora, A., J. Zielonka, M. Lopez, J. Joseph, and B. Kalyanaraman. 2009. Direct oxidation of boronates by peroxynitrite: Mechanism and implications in fluorescence imaging of peroxynitrite. *Free Radical Biology and Medicine* 47 (10): 1401–1407.
93. Zielonka, J., A. Sikora, J. Joseph, and B. Kalyanaraman. 2010. Peroxynitrite is the major species formed from different flux ratios of co-generated nitric oxide and superoxide direct reaction with boronate-based fluorescent probe. *Journal of Biological Chemistry* 285 (19): 14210–14216.
94. Zielonka, J., A. Sikora, M. Hardy, J. Joseph, B. P. Dranka, and B. Kalyanaraman. 2012. Boronate probes as diagnostic tools for real time monitoring of peroxynitrite and hydroperoxides. *Chemical Research in Toxicology* 25 (9): 1793–1799.
95. Michalski, R., J. Zielonka, E. Gapys, A. Marcinek, J. Joseph, and B. Kalyanaraman. 2014. Real-time measurements of amino acid and protein hydroperoxides using coumarin boronic acid. *Journal of Biological Chemistry* 289 (32): 22536–22553.

5 Peroxides and Protein Oxidation

Michael J. Davies

CONTENTS

INTRODUCTION

Biological systems are continually exposed to endogenous and exogenous free radicals and two-electron oxidants. The processes that give rise to these species have been reviewed (e.g., [1]). Usually the formation and reactions of these species are limited by defensive systems within cells and organisms, with these including low-molecular-mass scavengers (e.g., ascorbic acid, thiols, quinols, tocopherols, carotenoids, polyphenols, urate), enzymes that remove oxidants directly (e.g., superoxide dismutases), enzymes that remove oxidant precursors (e.g., peroxiredoxins, glutathione peroxidases, and catalases that remove peroxides), and enzyme systems that repair or remove damaged materials (methionine sulfoxide reductases, disulfide reductases/isomerases, sulfiredoxins, proteasomes, lysosomes, proteases, phospholipases, DNA repair enzymes) [1].

Despite this plethora of defense systems, elevated levels of oxidative damage have been detected in a wide range of human, animal, microbial, and plant systems (reviewed [1]). This may be due to increased oxidant levels, a decrease or failure of defense systems, or both. In many cases both are likely to be important, as defense systems are themselves subject to damage or depletion of critical cofactors required for activity. Aging is known to result in decline in enzyme activity, and lower levels of essential trace elements and metabolites, with this decline often accelerated by disease or environmental factors (reviewed [1]).

A large number of potential oxidants can be generated *in vivo*, and these species can vary markedly in their reactivity and specificity (reviewed: [2]). The resulting damage is therefore highly variable and complex. Some species, such as hydroxyl radicals (HO$^\bullet$), are relatively unselective and cause damage to nearly all biological targets, while other oxidants may react very slowly and with very high selectivity with a limited number of particular targets [2]. As HO$^\bullet$ is highly reactive and short-lived, it can only diffuse a very short distance before it reacts, whereas less chemically reactive species have longer biological half-lives, can diffuse long distances, and hence give rise to damage at remote locations [3]. Many oxidant reactions give rise to secondary oxidants, of different reactivity and lifetimes than the initial species, further complicating the analysis of damage. Some of these interconversion processes are illustrated in Figure 5.1. Understanding the nature and reactivity of potential oxidants and the patterns and extents of damage that they induce is therefore critical. This article will focus on the role of peroxides as initiators, intermediates, and products of protein modification.

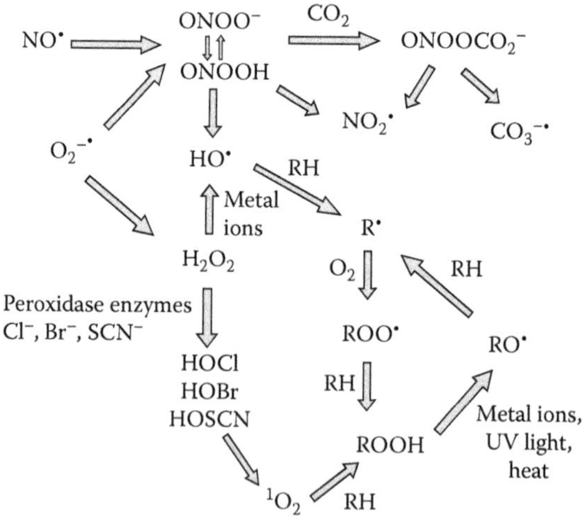

FIGURE 5.1   Interconversion pathways of some oxidants (two-electron species and radicals) of biological relevance. This is not exhaustive, and it should be noted that some of these processes occur in low yield (e.g., the hemolysis of ONOOH to radicals [154]).

SIGNIFICANCE OF PROTEINS AS OXIDATION TARGETS

As most oxidants react with multiple targets, there is the possibility of damage to all major classes of biological molecules. The extent of damage to these materials depends on multiple factors, including the absolute and relative concentrations of particular targets, the rate constants for reaction of a particular oxidant with different targets, the locations of the target and oxidant, the occurrence of secondary reactions including chain processes, intra- and inter-molecular transfer reactions, and the occurrence of repair and oxidant scavenging reactions [2]. Although each of these factors can be important, the initial sites of damage appear to be predominantly driven by the first two—concentrations and rate constants, and specifically the *product* of these two parameters. Proteins are the major (non water) components of most biological systems with concentrations in plasma of 1–3 mM, and 5–10 mM in cells (calculated assuming an average protein molecular mass of 25–50 kDa), and consequently these are likely to be major targets [2,4], especially when the rate constants for reaction of oxidants with proteins are high, as they are in many cases [2,5–9]. Rate constant and abundance data have allowed "guesstimates" of the likely fate of different oxidants in cells, and in most cases, proteins have been predicted to be major targets (e.g., values for oxidant consumption by proteins of 65%–70% for HO$^{\bullet}$ and singlet oxygen [10,11]). The *extent of damage* and its *importance* are not however necessarily equivalent, as limited damage to a critical target may be of much greater importance than a high extent of modification of redundant or unimportant targets.

PEROXIDES AS INITIATORS OF PROTEIN OXIDATION

Peroxides can act as both direct and indirect initiating agents in protein damage. Direct reaction of peroxides with proteins is often limited in extent and occurs with slow rate constants, though there are exceptions to this.

DIRECT OXIDATION REACTIONS

Direct reaction of H_2O_2 and other peroxides with protein side chains is predominantly limited to specific side chains and particularly to sulfur and selenium residues (i.e., cysteine [Cys], cystine, methionine [Met], and the rare amino acids selenocysteine [Sec] and selenomethionine [SeMet]). Reaction with other amino acid chains is negligible under most physiologically relevant conditions. Of these susceptible sites, reaction with the selenol function (RSeH) of Sec is the most rapid, as selenium is more readily oxidized and is a better nucleophile than sulfur, particularly when the selenol group is present in its ionized (RSe$^-$) form as is usually the case at physiological pH values (cf. a pK_a of ~4.8 for Sec [12]). The occurrence of Sec in biological samples is limited, with these primarily found in selenoproteins with specific functions (e.g., glutathione peroxidases, thioredoxin reductases, and some isoforms of methionine sulfoxide reductase) with these residues typically involved in catalytic activity, including reaction with H_2O_2 in the case of glutathione peroxidases (and lipid hydroperoxides for GPx4) [13,14]. Reaction of peroxides with selenols is believed to generate transient selenenic acids (RSeOH), but these are highly reactive and typically react with a thiol molecule to give a mixed RSeSR' adduct. In (rare) cases where the concentration of the selenol is high, a diselenide can be formed. In situations where reaction with a thiol (or another selenol) is not rapid, further oxidation of the selenium center can occur to give seleninic ($RSeO_2H$) and selenonic acids ($RSeO_3H$). The latter undergo rapid β-elimination to give dehydroalanine, a known precursor of protein cross-links [15,16].

Direct reaction of thiols with peroxides can be a significant reaction, but the rate constants for these reactions vary dramatically. With neutral Cys residues (i.e., RSH), reaction with H_2O_2 is relatively slow [17,18] (cf. k_2 for reaction with the Cys34 residue of BSA of ~34 M^{-1} s^{-1} [19]), but ionization of a Cys residue to give the corresponding thiolate (RS⁻) enhances the rate of reaction [6]. Thus proteins with low pK_a thiol residues can be significant targets for H_2O_2. The protein structure and environment also affect the rate of these reactions, with the presence of suitable groups that can promote weakening of the oxygen–oxygen bond in the peroxide (e.g., by acting as a proton source to protonate the leaving group) can dramatically enhance the rate constant for reaction [20]. Thus peroxiredoxins have rate constants for reaction with H_2O_2 of ~10^7 M^{-1} s^{-1} [21]. There is therefore a wide range of rate constants for reaction of peroxides with thiol groups present on proteins. These reactions yield short-lived sulfenic acids (RSOH), which undergo further reactions, with a major route being formation of cystine or mixed disulfides (RSSR') [22]. These types of reactions play critical roles in the enzymatic cycles of peroxide removing enzymes such as the peroxiredoxins, where the second (resolving) Cys is present within the same protein (or on another molecules) [20]. Other reactions can also occur, including over oxidation to give sulfinic (RSO_2H) and sulfonic (RSO_3H) acids, and reaction with amines and amides to give sulfenamides and sulfonamides, amongst others [23]. Similar reactions occur with a wide range of other peroxides, as well as H_2O_2, and are important redox switches that control enzyme activity and cell-signaling processes.

Evidence has been presented for slow reaction of H_2O_2 with Met residues on proteins, with this giving rise to the corresponding sulfoxide. The rate constants for such reactions are poorly defined, but appear to be in the range $10^{-2} \sim 10^{-3}$ M^{-1} s^{-1} [24]. Similar processes occur with the rare amino acid residue selenomethionine [25,26]. Such reactions would be expected to be faster, as the selenium center is a better nucleophile, but these are not well defined. Given the slow rate of reaction of peroxides with Met, this type of process is unlikely to be of major significance in viable cells, where rapid and effective removal of H_2O_2 can occur [27]. However, in other situations, this type of process may be a significant source of protein damage (e.g., with protein- or peptide-based medicines on prolonged storage [28–30]).

Limited evidence has been presented for reaction of cystine (i.e., the disulfide) with H_2O_2 [31]. This is typically a slow process, but may be significant in some circumstances when favorable conformations or environments are present [32]. Thus reaction with the strained 5-membered ring disulfides, lipoic acid and lipoamide, has been characterized with the initial product being the monooxide (disulfide S-oxide), but further reaction to give a dioxide and sulfonic acids (i.e., cleavage of the disulfide bond) can occur [33–36]. The disulfide S-oxides are moderate oxidants and can react further with thiols, resulting in the formation of new disulfide bonds and sulfenic acids [36]. Whether these types of reactions are of biological significance *in vivo* remains to be established.

Direct reactions of peroxides with other residues at physiological pH values and temperatures do not appear to be significant processes, though some evidence has been reported for Trp oxidation at elevated temperatures [37,38].

INDIRECT PEROXIDE-MEDIATED PROTEIN OXIDATION REACTIONS

The most widely studied, and probably most important, indirect mechanism through which peroxides can initiate protein damage is metal-ion catalyzed decomposition of the peroxide to give $HO^{\bullet}$ (from H_2O_2) and $RO^{\bullet}$ (from alkyl, aryl, and lipid hydroperoxides and peroxides). These reactions are usually termed Fenton, or pseudo-Fenton, reactions, with the resulting radicals undergoing hydrogen atom abstraction, electron abstraction, and addition reactions with protein side chains and the backbone ([39,40], see also [41]). These types of reactions have been widely reviewed (for a recent discussion, see [2]). Formation of carbon-centered radicals on proteins under conditions where O_2 is present results in the generation of peroxyl radicals ($ROO^{\bullet}$), which can then undergo multiple complex further reactions that can give more H_2O_2 and hydroperoxides. These reactions have been recently reviewed [2]. Thus peroxides can be initiators of protein damage, be formed as intermediates during oxidation processes on proteins, and also be products of such damage.

A second pathway to indirect peroxide-mediated protein damage is via the action of H_2O_2 as a cofactor for peroxidase and oxidase enzymes (e.g., myeloperoxidase, eosinophil peroxidase, lactoperoxidase, peroxidasins) and other heme proteins (e.g., myoglobin and hemoglobin) [42]. Reaction of H_2O_2 with the iron heme centers of such enzymes generates high-oxidation state species (e.g., Compound I from the Fe^{3+} form, Compound II from the Fe^{2+} form) that can initiate electron transfer reactions both within proteins (reviewed in [43]) and with other target species [44]. Thus Compound I of myeloperoxidase is known to oxidize Tyr residues to the corresponding phenoxyl radical ($TyrO^{\bullet}$) and indole derivatives (e.g., Trp, and species related to Trp) to indolyl radicals [45–48]. Related species formed from myoglobin and hemoglobin have been shown to induce protein–protein radical transfer (probably via electron transfer), with this resulting in the formation Tyr phenoxyl, Trp indolyl, and Cys-derived thiyl radicals on target proteins [49–51]. Subsequent reactions of Tyr phenoxyl and Trp indolyl radicals, if these are not repaired by reaction with ascorbate, urate, or GSH [52–55], can give to peroxides on these residues. Thus both Tyr phenoxyl and Trp indolyl radicals can give rise to rapid reaction with superoxide radicals ($O_2^{\bullet-}$) to give hydroperoxides [56–59]. Trp indolyl radicals and related species undergo reaction with O_2 to give peroxyl radicals, which can then undergo hydrogen atom abstraction reactions to yield peroxides [59–61]. The corresponding reaction of O_2 with Tyr phenoxyl radicals is however slow, $k < 10^3$ M^{-1} s^{-1} [62], and does not appear to be a major source of peroxyl radicals and hydroperoxides.

FORMATION OF PEROXIDES DURING
PROTEIN MODIFICATION

As most carbon-centered radicals react rapidly with O_2 (with rate constants at or close to the diffusion limit, i.e., $k \sim 10^9$ M^{-1} s^{-1} [9], though reaction with Trp indolyl and Tyr phenoxyl radicals are much slower—see earlier in Indirect Peroxide-Mediated Protein Oxidation Reactions), and there is an abundance of targets with which the resulting ROO• can react by hydrogen atom abstraction, it is not surprising that hydroperoxides can be generated in high yields (see Table 5.1). However, as these species often undergo subsequent reactions, they are both major intermediates and products formed on protein oxidation.

The formation of peroxides on proteins as a result of oxidation is not a new discovery, with the first report on these being from 1942 [63], though subsequent research was sporadic for several decades. Alexander and colleagues [64] showed that BSA which had been subjected to irradiation, could induce polymerization of methacrylate, with this being ascribed to the presence of peroxides on the irradiated protein. Occasional reports on protein peroxides appeared in the radiation chemistry/biology literature up to the 1980s, but in-depth research on this topic only began to appear in the 1990s; early work is reviewed in [10] and recent data in [2].

Steady-state γ-irradiation experiments have shown that amino acid, peptide, and protein hydroperoxides can be formed in high yield when these species are exposed to HO• in the presence of O_2 [10] (Table 5.1). Early studies were confounded by the presence of both H_2O_2 and hydroperoxides in the irradiated samples, but the use of catalase that efficiently removes the former but not the latter has allowed quantification of amino acid and protein hydroperoxide yields [10,65]. Considerable evidence has now been presented to show that this type of chemistry is not limited to HO•/O_2 systems, with many oxidizing systems involving radicals, and other oxidants (e.g., $^1O_2$), now known to give these species when molecular oxygen is present [2]. The hydroperoxide yields vary with the oxidant, the reaction conditions, and the targets present, due to the occurrence of other competing processes, but there are limited examples where these are *not* detected [2,10]. Examples of some of the systems that are known to give rise to hydroperoxides are illustrated in Figure 5.2.

Exposure of *free amino acids* to γ-irradiation in the presence of O_2 yields hydroperoxide in a radiation dose-dependent manner [65–70]. The yields are highly dependent on the structure of the amino acids [67], with the maximum levels corresponding to ~40% of the initial HO• generated [65,67]. This is likely to be an underestimate of the true level, as hydroperoxides are known to undergo decomposition during continued irradiation (via electron-induced cleavage of the peroxide bond) and via thermal degradation processes. Some amino acids (Cys, cystine, Ser, Thr) give negligible yields [67], which have been rationalized in terms of the chemistry of the radicals formed on the amino acid [2,71]. In the case of Ser and Thr, this is due to alternative reactions of the ROO• formed during the initial oxidation, with these undergoing elimination reactions to give H_2O_2 rather than amino acid hydroperoxides [72]. Aromatic amino acids give moderate yields, due to the occurrence of alternative reactions at the aromatic rings, though this is dependent on the radical and reaction conditions. High yields are detected from amino acids with large

TABLE 5.1

Efficiency of Hydroperoxide Formation (Number of Hydroperoxide Groups Formed per HO$^{\bullet}$ × 100) on Free Amino Acids Exposed to γ-Radiation at pH 7.4)

Amino Acid	Peroxidation Efficiency
Valine	49
Leucine	44
Proline	44
Isoleucine	43
Lysine	34
Glutamic acid	28
Tryptophan	18
Glutamine	16
Arginine	13
Alanine	11
Aspartic acid	6
Phenylalanine	5
Histidine	4
Glycine	3
Tyrosine	3
Asparagine	2
Hydroxyproline	2
Cysteine	0.4
Methionine	0
Serine	0
Threonine	0

Source: Gebicki, S. and Gebicki, J.M., *Biochem. J.*, 289, 743, 1993.

numbers of aliphatic C–H bonds (Val, Leu, Ile, Glu, Lys, Pro) from which hydrogen atom abstraction and subsequent ROO$^{\bullet}$ formation can occur [65,67]. Some of these hydroperoxides have been characterized in detail (e.g., from Leu, Val, and Lys [73–75]), but in other cases these are less well defined. For free amino acids, the yield of α-carbon hydroperoxides is low (e.g., only low levels are detected with free Gly) consistent with a slow rate of hydrogen abstraction from this site as a result of the influence of the protonated amine function [2,10,68].

For Tyr, a major route to hydroperoxide formation is via reaction of phenoxyl radicals formed by ring oxidation and subsequent deprotonation at the hydroxyl group, with $O_2^{\bullet-}$. This radical–radical reaction is very rapid, and near the diffusion limit, as a result of the low energy barrier, and is an efficient pathway to peroxide formation due to the long lifetimes (and hence higher steady-state concentrations) of the parent radicals. Multiple isomers can be formed via addition of $O_2^{\bullet-}$ at the different ring positions, but reaction at C1 (the site of –CH$_2$- attachment) and C3 (*ortho* to the –OH group) predominate, as these are the sites of the highest spin density in the

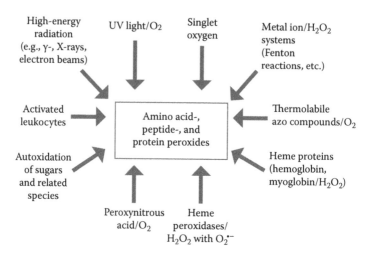

FIGURE 5.2 Biological processes known to generate amino acid, peptide, and protein hydroperoxides. (From Davies, M.J., *Biochem. J.*, 473, 805, 2016.)

parent phenoxyl radical. The C1 species is the more stable species due to its tertiary structure [56–58,76–80]. The C1 hydroperoxide has also been detected with HO$^\bullet$ from Fe^{2+}/H$_2$O$_2$ [57]. Analogous processes generate hydroperoxides from Trp indolyl radicals [59], though these are less well characterized.

High yields of endo- and hydroperoxides are formed from Tyr and Trp, and also His, via cycloaddition reactions of singlet oxygen ($^1\Delta g$; ^{1}O$_2$) to the aromatic rings (reviewed [11,81,82]. ^{1}O$_2$ can be generated in biological systems by multiple routes (reviewed [11,81,82]) including light-mediated reactions (Type 2 photochemical processes), enzymatic (peroxidase-, lipoxygenase-, and cyclooxygenase- and heme-mediated reactions), and chemical reactions (e.g., termination reactions of peroxyl radicals, reaction of H$_2$O$_2$ with HOCl, via ozone-mediated reactions, and from some ONOO$^-$/ONOOH reactions). The endoperoxides formed from Tyr, Trp, and His can undergo ring-opening reactions to give hydroperoxides at ring positions, with these including hydroperoxides at C1 and C4 for Tyr; C3 for Trp; and C2, C4, and C5 for His. The structures of some of these species have been elucidated by MS and NMR [59,83–91]. ^{1}O$_2$ reacts rapidly, by addition, with the sulfur atoms of Cys and Met residues to give peroxidic species [92–94]. The subsequent reactions of these species are not fully understood, though Cys gives cystine as a major product, but also thiosulfinates and oxy acids, whereas reaction with Met gives variable yields of the sulfoxide and H$_2$O$_2$ [92–94].

With *peptides*, side-chain hydroperoxide formation remains an important process, but increased yields of backbone hydroperoxides are also formed, due to the increased generation of initial α-carbon radicals as a result of the absence of the deactivating protonated amine function, and capto-dative stabilization of these radicals. Thus the yield of backbone peroxide increases with increasing chain length for small peptides (e.g., [Ala]$_n$ and related species [68,95,96]). With large peptides, and particularly those with defined secondary structure (e.g., β-sheets), access to the α-carbon sites appears to be a significant barrier to the generation of these species, and consequently the yield of peroxides formed on *proteins* is variable. However there are few, if any, proteins that *do not* give rise to peroxides on exposure to a wide range of radicals (or radical-generating systems) in the presence of O_2, or $^1O_2$ [2].

Protein hydroperoxides are typically formed at multiple sites on proteins (e.g., with $HO^\bullet/O_2$) [2], but examples are known where hydroperoxide formation is highly specific, both with regard to amino acid type and location. Site-specific formation of Tyr phenoxyl and also Trp indolyl radicals can result in hydroperoxide generation at specific residues via reaction with $O_2^{\bullet-}$ [77–79,97], or for Trp residues by the addition of O_2 at C3 of indolyl radicals, to give a C3 peroxyl radical, and hence a C3 hydroperoxide. This last pathway may be protein specific, as O_2 does not affect the rate of decay of *free* Trp indolyl radicals, indicating that k_2 for O_2 addition must be $< \sim 10^5$ M^{-1} s^{-1} and a lot slower than radical dimerization (k_2 for free Trp radicals $\sim 7.3 \times 10^8$ M^{-1} s^{-1} [59]). Whether related chemistry occurs with His-derived radicals is unknown.

While many of the peroxides formed on proteins are similar to those formed on amino acids and small peptides, the decreased rate of radical–radical termination and radical–molecule repair reactions for protein radicals, for steric and electronic reasons, appears to increase hydroperoxide yields as these will result in an increased extent of O_2 addition to carbon-centered radicals to give $ROO^\bullet$ and hence hydroperoxides, and $O_2^{\bullet-}$ addition to Tyr phenoxyl and Trp indolyl radicals.

PEROXIDES AS PRODUCTS OF PROTEIN OXIDATION

Amino acid, peptide, and protein hydroperoxides can have lifetimes of hours–weeks when kept under appropriate conditions (low temperatures, absence of light and metal ions, reductants, and other reactive species such as enzymes [65]). For some proteins, ~30% of the initial hydroperoxide remained after one week at room temperature [65], but a half-life of ~1.5 days has been estimated for others [67]. Elevated temperatures, redox active metal ions (e.g., Cu^+, Fe^{2+} [65,68,69,98,99]), UV light [70], or reductants (e.g., dithionite, triphenylphosphine, ascorbate, GSH, sodium borohydride, but *not* NADH and NADPH [65,67]) can induce rapid decay. Those formed at tertiary carbon sites or at sterically isolated sites on proteins appear to have the greatest stability [10,67–69]. Some protein hydroperoxides can survive enzymatic digestion by proteolytic enzymes such as pronase [100], thereby aiding analysis.

The hydro- and endoperoxides formed on amino acids, peptides, and proteins can be detected by multiple methods (reviewed [2]), including classical titration methods (e.g., using $KMnO_4$, iodometric, or Ti^{3+} [101]); by spectrophotometric reactions (e.g., the ferric iron–xylenol orange, FOX, assay, where the hydroperoxide oxidizes Fe^{2+} to Fe^{3+}, which forms a complex with xylenol orange that absorbs strongly at 560 nm [102,103]); by chemiluminescence using microperoxidase and luminol [70]; by reaction with (nonfluorescent) coumarin boronic acid probes to give fluorescent products (e.g., 7-hydroxycoumarin); and by mass spectroscopic (MS) approaches, as peroxides give distinctive m/z +32 peaks (e.g., [84–88,95,96]). Peroxide stability is a significant constraint for some of these assays (e.g., the elevated temperatures used in some MS protocols), as is the requirement for the pretreatment to remove H_2O_2 in others (e.g., titrations and FOX assays). This makes absolute quantification difficult, as does the absence of appropriate standards. MS can be valuable in determining the *sites*, and *identities* of both initial peroxides and the sites (and yields) of alcohols (m/z +16) arising from hydroperoxide decomposition/reduction [57,73–75,95].

AMINO ACID, PEPTIDE, AND PROTEIN HYDROPEROXIDE FORMATION *IN VITRO* AND *IN VIVO*

There is increasing direct and indirect evidence for hydroperoxide formation in complex systems, but the exact identity of these species is less clear. Incubation of isolated low-density lipoproteins with $O_2^{\cdot-}$ [104], or human macrophage–like THP-1 cells, gives hydroperoxides on both the apolipoprotein B100 protein and lipids [105]. Exposure of a number of different cells (e.g., mouse myeloma, monocyte-derived U937, and HL-60 cells) to HO$^{\cdot}$ generated by γ-irradiation generates hydroperoxides at concentrations in the low micromolar to millimolar range [106–108]. This occurred under conditions where no lipid or DNA damage could be detected (though this may be method related), and have been interpreted as indicating that proteins are both the major and initial targets for radiation-generated radicals [106,107]. As the peroxide concentration was independent of the medium used, only radicals generated in close proximity to or within cells appeared to be the source of the peroxides in accordance with the very high reactivity of HO$^{\cdot}$ [107]. In other studies (HL-60 cells [108]) protein hydroperoxide formation occurred after a lag phase, which has been ascribed to the action of endogenous antioxidants (e.g., GSH and ascorbate) in preventing hydroperoxide formation, or removing them once formed [108]; this interpretation is supported by data from supplementation and depletion studies [108].

ROO$^{\cdot}$ (from the thermally labile azo compound AAPH in the presence of O_2) also generates protein peroxides in HL-60, U937, and human monocyte-derived macrophage cells [109,110], and visible light photolysis in the presence of the sensitizer Rose Bengal formed protein peroxides in murine macrophage-like (J774A.1) or human monocyte (THP-1) cells or cell lysates [96,111,112]. The peroxide yields in these latter experiments were enhanced in D_2O, and decreased by sodium azide, consistent with the involvement of $^1O_2$ [111,112].

Evidence has also been provided for peroxide formation on endogenous or exogenous proteins by appropriately stimulated cells. Stimulated neutrophils can generate hydroperoxides on free Tyr [113] or enkephalins [97] via myeloperoxidase-mediated reactions involving Tyr phenoxyl radical formation and reaction with $O_2^{\cdot-}$ [97]. Hydroperoxides have also been detected on fresh human plasma proteins exposed to $^1O_2$ generated by photosensitizers ([96], Silvester, unpublished data); HPLC fractionation indicated that the majority of the peroxides coeluted with human serum albumin.

Less direct evidence is available for hydroperoxides in intact normal or diseased tissues, probably reflecting the short half-lives and/or reactivity of these species [2]. However, considerable *indirect* evidence is available via the detection of (presumed) decomposition products such as carbonyls and alcohols. While there is good evidence that hydroperoxide decomposition gives these products, the quantitative significance of these pathways is unknown (though this can be near quantitative *in vitro* [73–75,114]). The detection and quantification of proteins carbonyls in healthy and diseased tissues have been reviewed in multiple articles (e.g., [1,115–122]), but whether these reflect (wholly or partially) hydroperoxide formation is unclear. Elevated levels of protein-derived alcohols, which can be formed from hydroperoxides by reduction, have been detected in multiple tissues, including atherosclerotic lesions [123], cataractous and normal-aged lenses [124], and diabetic materials [125]. Quantitative data have been obtained for particular Leu- and Val-derived species in human atherosclerotic lesions (50–100 μmol/mol parent amino acid, 1–4 pmol/mg wet mass of intimal tissue) [123] and in advanced human lens cataracts (200–400 μmol oxidized amino acid per mole parent amino acid, 60–120 picomol oxidized amino acid/mg of dried lens tissue) [124]. On the basis of these numbers, and estimates of protein concentrations in cells and their composition, the total flux of protein hydroperoxide formed are 50–100 μmol hydroperoxide for the proteins in advanced atherosclerotic lesions, and 200–400 μmol hydroperoxide for cataractous lens proteins. The *overall* concentration of protein hydroperoxides is likely to be higher than this, as this calculation considers products from only two amino acids on which hydroperoxides are generated.

PROTEIN HYDROPEROXIDES AS INITIATORS
OF SECONDARY DAMAGE

One-electron reduction (e.g., by transition metal ions via pseudo-Fenton reactions) of hydroperoxides gives alkoxyl radicals ($RO^{\bullet}$), while exposure to short wavelength UV light or high temperatures induces thermal homolysis to give both $HO^{\bullet}$ and $RO^{\bullet}$. These processes may therefore enhance or propagate damage via fragmentation reactions, electron transfer, or hydrogen atom abstraction.

$RO^{\bullet}$ formed at primary or secondary carbon atom sites can undergo rapid (formally 1,2-) hydrogen shift reactions to give α-hydroxyalkyl (carbon-centered) radicals [126,127]. These carbon-centered radicals undergo rapid reaction with O_2 to give peroxyl radicals that then rapidly eliminate $HOO^{\bullet}$ (and hence give additional H_2O_2 via dismutation). Tertiary $RO^{\bullet}$, such as species formed at the tertiary carbons in Val, Leu, Ile, cannot undergo 1,2-shift reactions and hence undergo rapid β-fragmentation to give a released carbon-centered radical (e.g., methyl radicals from Val and Leu) and a carbonyl (aldehydes/ketones) [128–131]. Side-chain (β-carbon) $RO^{\bullet}$, formed from hydroperoxides, can also fragment to give both a backbone α-carbon radical with the release of a low-molecular-mass carbonyl from the side chain [132,133]. These reactions result in the transfer of damage between sites in an amino acid or protein, the release of further radicals, and formation of additional peroxide or carbonyl species. These reactions constitute chain-carrying reactions in protein peroxidation reactions [2]. Other fragmentation reactions are also known, including β-scission of $RO^{\bullet}$ on Glu side chains, which results in the loss of the side-chain carboxyl group (as $CO_2^{\bullet-}$) and formation of an aldehyde [69]. Other hydrogen atom shift reactions can transfer damage from side-chain sites to α-carbon (backbone) sites, or other side chains.

Fragmentation reactions of $RO^{\bullet}$ formed from backbone α-carbon hydroperoxides may be of particular significance as these can result in backbone cleavage [68,69], with β-scission of the $RO^{\bullet}$ giving a carbonyl group and a secondary acyl radical ($^{\bullet}$C(O)NHR) when the hydroperoxide is present within a peptide chain [68]. Carboxyl-terminal hydroperoxides give RO$^{\bullet}$ that release CO$_2^{\bullet-}$ (or $^{\bullet}$C(O)NH$_2$ from C-terminal amides). This hydroperoxide/RO$^{\bullet}$ pathway competes with an alternative pathway to backbone fragmentation involving initial formation of backbone carbon radicals, ROO$^{\bullet}$ formation, elimination of HOO$^{\bullet}$ (which then yields H$_2$O$_2$), and hydrolysis of the resulting imine [4,66,71].

These secondary reactions may rationalize, at least in part, the observation that more amino acids are lost during γ-irradiation of proteins, than initial radicals supplied, an observation interpreted in terms of the occurrence of *chain reactions* [134]. Although the chain lengths are modest (10–15) when compared to lipid peroxidation (where values of >100 have been reported), protein chain oxidation reactions appear to occur under certain circumstances, and both with isolated proteins and in cells [108,134].

For some radical transfer reactions (e.g., of initial R$^{\bullet}$), no additional amino acid loss occurs—only *transfer* of damage—but this may be of major significance if these reactions change the amino acid *stereochemistry* (i.e., conversion of L-amino acids to the D-form). This can occur by reaction at either face of a planar radical intermediate. Such stereochemical inversion has been reported [135–137], and may be a common process, with this resulting in perturbation of protein structures.

REMOVAL OR DETOXIFICATION OF HYDROPEROXIDES ON PROTEINS

In addition to undergoing one-electron reduction to RO$^{\bullet}$, as described above, hydroperoxides on proteins can also undergo two-electron reduction. This is a detoxification pathway as it removes the oxidizing capacity and limits further damage, but it does not *repair* the initial lesion, as an alcohol is formed from the hydroperoxide. Rapid reduction occurs with thiols (RSH) and particularly thiolate anions (RS^{-}) [65,67,100]. Such reactions are likely to be the major detoxification route in cells, and *in vivo* [114]. Disulfides react less rapidly, with lipoic acid being a possible exception [100]. Reduction by thioethers (e.g., free Met) is inefficient in most cases, though it can occur if the hydroperoxide and Met oxidation are in close proximity (e.g., in the same peptide) [97,100].

Ascorbate is an effective hydroperoxide reductant [65,67,100], as are some (low abundance) selenium-containing compounds [26,100]. With selenomethionine, which is oxidized to the selenoxide, reduction can be catalytic, as GSH and various enzymes can rapidly reduce the selenoxide.

The rate of hydroperoxide removal appears to depend on the structure on which the hydroperoxide is present, with more rapid loss seen with low mass materials than with proteins probably for steric and electronic reasons [100]. The peroxides detected in cells may therefore be present on large proteins, and/or inaccessible to reductants such as GSH and ascorbate [2]. Phenols do not remove these hydroperoxides, but may scavenge radicals derived from them [100].

As might be expected from these data, thiol-dependent enzymes can also react with amino acid, peptide, and protein hydroperoxides, and this can either be a detoxification process (when the enzyme can be readily re-reduced) or result in damage to the target protein. Catalase does not remove these species readily, and affords a convenient means of distinguishing between hydroperoxides and H_2O_2 [65,67,100]. No reaction occurs with superoxide dismutases, or peroxidases such as horserad-ish peroxidase, lactoperoxidase, and myeloperoxidase [100,138], and reaction with ferric (met)myoglobin and hemoglobin is slow [100]. Rapid reaction occurs with Oxymyoglobin and oxyhemoglobin [100], including the latter in red blood cells [139]. These reactions occur readily with amino acid hydroperoxides but at slower rates with peptide hydroperoxides, and protein hydroperoxides are essentially inert. The oxy-to-Met conversion is faster than with H_2O_2 and this is a general phenome-non: many amino acid and peptide hydroperoxides are both more reactive than H_2O_2, and poorly removed by protective systems [100].

With glutathione peroxidase (GPx1), a selenocysteine-dependent enzyme, a similar pattern of reactivity is detected: Rapid catalytic removal (at the expense of GSH) is seen with amino acid hydroperoxides and some peptide and small pro-tein hydroperoxides (e.g., insulin hydroperoxide [138], but slow, or no, enzyme-mediated reaction is detected with large hydroperoxides [100]. The related enzyme GPx4 (phospholipid-GPx) does not appear to stimulate decay of any protein hydro-peroxides compared to GSH alone [138]. In contrast, peroxiredoxins 2 and 3, which are abundant Cys-containing cytosolic and mitochondrial enzymes [140], rapidly reduce hydroperoxides with the proteins being re-reduced by the NADPH/thio-redoxin/thioredoxin reductase system [141]. Thus, addition of leucine and BSA hydroperoxides to erythrocyte lysates results in Prx2 oxidation without significant GSH loss, indicating that Prxs are major intracellular targets for these hydroper-oxides and can detoxify these species in cells [141]. These reactions are not *repair* reactions, as they only convert the hydroperoxide to the alcohol, but would limit secondary damage.

BIOLOGICAL CONSEQUENCES OF AMINO ACID, PEPTIDE, AND PROTEIN HYDROPEROXIDE FORMATION

As hydroperoxides readily oxidize Cys residues, these can induce loss of activity or function of Cys-dependent proteins, and particularly those with highly reactive (low pK_a) thiolate anion (RS$^-$) residues. Thus inactivation has been detected with glyceraldehyde-3-phosphate dehydrogenase (GAPDH) [142], glutathione reduc-tase [142,143], papain, and some cathepsin isoforms (the Cys-dependent B and L isoforms; but not the non-Cys-dependent D and G isoforms) [144], caspases [145], protein tyrosine phosphatases (PTPs) (both isolated and in cell lysates) [146], and the Ca^{2+} pump SERCA (sarco/endoplasmic reticulum Ca-ATPase) [147]. Evidence has also been presented for a decreased rate of turnover of both native and modified proteins as a result of oxidation of critical residues present on particular subunits of the 26S proteasome hydroperoxide-mediated, with the chymotryptic and tryptic (and also probably the caspase-like) activities affected by amino acid, peptide, and protein hydroperoxides, in both isolated systems and cell lysate preparations [148]. For purified human 26S proteasomes, inhibition appears to be associated, at least partly, with oxidation of a critical Cys on the S6 ATPase control subunits [148].

INDUCTION OF DNA DAMAGE

In addition to modifying other proteins, decomposition of protein hydroperoxides and related species can induce DNA damage. Histone proteins appear to give particularly high hydroperoxide yields, accounting for up to 70% of the initial radicals, when exposed to radiation in the presence of O_2, possibly as a result of their high Lys and Arg content and low levels of Cys, Met, and aromatic residues that might act as alternative targets [149,150]. Decomposition of these peroxides by Cu^+ and Fe^{2+} results in protein radicals that react with pyrimidine bases and nucleosides to give protein-DNA base adducts [149–151], the mutagenic base 7,8-dihydro-8-oxo-2′-deoxyguanosine [149,150], and strand cleavage [152].

Little work has been carried out on the initiation of lipid oxidation by protein hydroperoxide–derived species, though it has been shown that protein hydroperoxide formation can precede lipid oxidation [106,109]. It is therefore likely that radicals formed from amino acid, peptide, and protein hydroperoxides can contribute to the initiation of lipid oxidation.

INDUCTION OF SECONDARY DAMAGE IN CELLS

In the light of the data mentioned, it is not surprising that exposure of cells to amino acid, peptide, and protein hydroperoxides results in a complex pattern of changes. Visible light photo-oxidation (using Rose Bengal as a sensitizer) has been employed to generate peroxides *in situ*, or prior to addition to cells, with the effects of the hydroperoxides examined after the cessation of illumination. With murine macrophage-like (J774A.1) cells, exposure to preformed peptide hydroperoxides resulted in a loss of total cellular thiols and GSH that occurred concurrently with hydroperoxide consumption, and prior to decreased cell viability [153]. GSH loss was more rapid than the total thiol pool, suggesting that protein thiols are spared by GSH. Inhibition of cathepsins B and L, and caspases 3/7 was detected, but not for other (non-thiol dependent) cathepsins or aryl sulfatase. These effects on lysosomal enzymes are consistent with hydroperoxide uptake and accumulation in the endo-lysosomal compartment [153].

When hydroperoxides were generated *in situ* in cells by limited photooxidation, a wide range of effects were detected that may arise from either direct reactions of transient intermediates (radicals or excited state) or peroxides. These effects included loss of GSH and total thiols, inhibition of multiple Cys-dependent (but not non-thiol) enzymes (e.g., GAPDH, thioredoxin, protein tyrosine phosphatases, creatine kinase, and cathepsins B and L), increases in NADPH levels and enhanced activity of glutathione reductase, glutathione peroxidase and thioredoxin reductase, which are key oxidative defense enzymes [111]. After cessation of peroxide formation (i.e., illumination) a limited recovery of the thiols was detected, compared to the concentrations detected immediately after cessation of illumination, but these decreased with further incubation. Increases in the activity of GAPDH and protein tyrosine phosphatases were also detected, together with a marked increase in caspase 3/7 activity, and decreased cell viability. These observations are probably due to limited repair and recovery, followed by the induction of apoptosis [111].

SUMMARY

The data reviewed here indicate that peroxides are key players in the development of oxidative stress and are important mediators of damage. Both direct peroxide oxidation and damage induced by radicals derived from peroxides can be important in altering protein side chains and the protein backbone. Direct molecular oxidation is a relatively slow and specific process in many cases, but rapid reaction can occur with specific Cys and Sec residues on some proteins. In contrast, radicals generated from peroxides (e.g., $HO^{\bullet}$ and $RO^{\bullet}$) react very rapidly (often at diffusion or near diffusion controlled rates) with many protein side-chain and backbone sites, to give wide range of radical species. Many of these species react with molecular oxygen to give peroxyl radicals ($ROO^{\bullet}$) and hence hydroperoxides or H_2O_2. Thus hydroperoxides and H_2O_2 can be initiators of damage, intermediates, and also products of protein oxidation reactions. The yield of hydroperoxides formed on proteins can be very high (40%–70%), and these values may be underestimated due to the instability of these species and problems with accurate quantification. High concentrations of hydroperoxides can also be generated by species such as $^1O_2$. These hydroperoxides can be long-lived in the absence of other reagents and are readily detected on isolated proteins, lipoproteins, in plasma, in cell lysates and some intact cells. Data from cell studies indicate that micromolar concentrations can be present, possibly as a result of the slow rate of removal of some of these species by cellular protective systems, with these being much less efficient and effective than with H_2O_2. The hydroperoxides react at highly variable rates with one- and two-electron reducing agents. One-electron reduction can generate additional radicals that contribute to protein chain oxidation reactions, cleavage of the protein backbone, and modification of side chains via complex reactions, some of which generate carbonyls and alcohols. These are well-established biomarkers of protein oxidation and are known to accumulate in aged and some diseased tissues. Competing reactions of these protein-derived radicals can damage DNA, including formation of 7,8-dihydro-8-oxo-2'-deoxyguanosine (8-oxodG), protein-DNA base adducts and strand cleavage. Two-electron reactions of protein hydroperoxides occur predominantly with Cys, selenium species, and (to a lesser extent) Met, with these process resulting in inactivation of multiple cellular enzymes including those involved in calcium handling, phosphorylation, energy metabolism, apoptosis and protein turnover.

ACKNOWLEDGMENTS

Financial support from the Novo Nordisk Foundation (Laureate grant: NNF13OC0004294) and the Australian Research Council (through the Centres of Excellence Scheme, CE0561607, and Discovery Programs DP140103116 and DP160102063) is gratefully acknowledged.

DECLARATIONS OF INTERESTS

The author has no competing interests to declare.

REFERENCES

1. Halliwell, B. and J. M. C. Gutteridge. 2015. *Free Radicals in Biology & Medicine*, 5th ed. Oxford, U.K.: Oxford University Press.
2. Davies, M. J. 2016. Protein oxidation and peroxidation. *Biochem. J.* 473:805–825. doi: 10.1042/BJ20151227.
3. Winterbourn, C. C. 2008. Reconciling the chemistry and biology of reactive oxygen species. *Nat. Chem. Biol.* 4 (5):278–286 doi: 10.1038/nchembio.85.
4. Davies, M. J. 2005. The oxidative environment and protein damage. *Biochim. Biophys. Acta* 1703 (2):93–109.
5. Buxton, G. V., C. L. Greenstock, W. P. Helman, and Ross, A. B. 1988. Critical review of rate constants for reactions of hydrated electrons, hydrogen atoms, and hydroxyl radicals ($^{\bullet}$OH/$^{\bullet}$O-) in aqueous solution. *J. Phys. Chem. Ref. Data* 17:513–886.
6. Ferrer-Sueta, G. and R. Radi. 2009. Chemical biology of peroxynitrite: Kinetics, diffusion, and radicals. *ACS Chem. Biol.* 4 (3):161–177.
7. Pattison, D. I. and M. J. Davies. 2006. Reactions of myeloperoxidase-derived oxidants with biological substrates: Gaining insight into human inflammatory diseases. *Curr. Med. Chem.* 13:3271–3290.
8. Pattison, D. I., M. J. Davies, and C. L. Hawkins. 2012. Reactions and reactivity of myeloperoxidase-derived oxidants: Differential biological effects of hypochlorous and hypothiocyanous acids. *Free Radic. Res.* 46:975–995.
9. Ross, A. B., B. H. J. Bielski, G. V. Buxton, D. E. Cabelli, C. L. Greenstock, W. P. Helman, R. E. Huie, J. Grodkowski, and P. Neta. 1994. NDRL-NIST Solution Kinetics Database: Ver. 2. National Institute of Standards and Technology, Gaithersburg, MD. http://kinetics.nist.gov/solution/.
10. Gebicki, J. M. 1997. Protein hydroperoxides as new reactive oxygen species. *Redox Rep.* 3:99–110.
11. Davies, M. J. 2004. Reactive species formed on proteins exposed to singlet oxygen. *Photochem. Photobiol. Sci.* 3:17–25.
12. Stadtman, T. C. 1996. Selenocysteine. *Annu. Rev. Biochem* 65:83–100. doi: 10.1146/annurev.bi.65.070196.000503.
13. Brigelius-Flohe, R. 2015. The evolving versatility of selenium in biology. *Antioxid. Redox Signal.* 23 (10):757–760. doi: 10.1089/ars.2015.6469.
14. Brigelius-Flohe, R. and M. Maiorino. 2013. Glutathione peroxidases. *Biochim. et Biophys. Acta* 1830 (5):3289–3303. doi: 10.1016/j.bbagen.2012.11.020.
15. Cho, C. S., S. Lee, G. T. Lee, H. A. Woo, E. J. Choi, and S. G. Rhee. 2010. Irreversible inactivation of glutathione peroxidase 1 and reversible inactivation of peroxiredoxin II by H_2O_2 in red blood cells. *Antioxid. Redox Signal.* 12 (11):1235–1246. doi: 10.1089/ars.2009.2701.
16. Rhee, S. G. and C. S. Cho. 2010. Blot-based detection of dehydroalanine-containing glutathione peroxidase with the use of biotin-conjugated cysteamine. *Methods Enzymol.* 474:23–34. doi: 10.1016/S0076-6879(10)74002-7.
17. Winterbourn, C. C. and D. Metodiewa. 1999. Reactivity of biologically important thiol compounds with superoxide and hydrogen peroxide. *Free Radic. Biol. Med.* 27 (3–4):322–328.
18. Luo, D. Y., S. W. Smith, and B. D. Anderson. 2005. Kinetics and mechanism of the reaction of cysteine and hydrogen peroxide in aqueous solution. *J. Pharmaceut. Sci.* 94 (2):304–316. doi: 10.1002/jps.20253.
19. Carballal, S., R. Radi, M. C. Kirk, S. Barnes, B. A. Freeman, and B. Alvarez. 2003. Sulfenic acid formation in human serum albumin by hydrogen peroxide and peroxynitrite. *Biochemistry* 42:9906–9914.

20. Nagy, P., A. Karton, A. Betz, A. V. Peskin, P. Pace, R. J. O'Reilly, M. B. Hampton, L. Radom, and C. C. Winterbourn. 2011. Model for the exceptional reactivity of peroxiredoxins 2 and 3 with hydrogen peroxide: A kinetic and computational study. *J. Biol. Chem.* 286 (20):18048–18055. doi: 10.1074/jbc.M111.232355.

21. Ogusucu, R., D. Rettori, D. C. Munhoz, L. E. Netto, and O. Augusto. 2007. Reactions of yeast thioredoxin peroxidases I and II with hydrogen peroxide and peroxynitrite: Rate constants by competitive kinetics. *Free Radic. Biol. Med.* 42 (3):326–334. doi: 10.1016/j.freeradbiomed.2006.10.042.

22. Turell, L., H. Botti, S. Carballal, G. Ferrer-Sueta, J. M. Souza, R. Duran, B. A. Freeman, R. Radi, and B. Alvarez. 2008. Reactivity of sulfenic acid in human serum albumin. *Biochemistry* 47 (1):358–367. doi: 10.1021/bi701520y.

23. Trujillo, M., B. Alvarez, and R. Radi. 2016. One- and two-electron oxidation of thiols: Mechanisms, kinetics and biological fates. *Free Radic. Res.* 50:150–171.

24. Carruthers, N. J. and P. M. Stemmer. 2008. Methionine oxidation in the calmodulin-binding domain of calcineurin disrupts calmodulin binding and calcineurin activation. *Biochemistry* 47 (10):3085–3095. doi: 10.1021/bi702044x.

25. Rahmanto, A. S. and M. J. Davies. 2012. Selenium-containing amino acids as direct and indirect antioxidants. *IUBMB Life* 64 (11):863–871. doi: 10.1002/iub.1084.

26. Suryo Rahmanto, A. and M. J. Davies. 2011. Catalytic activity of selenomethionine in removing amino acid, peptide, and protein hydroperoxides. *Free Radic. Biol. Med.* 51 (12):2288–2299. doi: 10.1016/j.freeradbiomed.2011.09.027.

27. Chance, B., H. Sies, and A. Boveris. 1979. Hydroperoxide metabolism in mammalian organs. *Physiol. Rev.* 59 (3):527–605.

28. Manning, M. C., K. Patel, and R. T. Borchardt. 1989. Stability of protein pharmaceuticals. *Pharm. Res.* 6 (11):903–918.

29. Jiskoot, W., T. W. Randolph, D. B. Volkin, C. R. Middaugh, C. Schoneich, G. Winter, W. Friess, D. J. Crommelin, and J. F. Carpenter. 2012. Protein instability and immunogenicity: Roadblocks to clinical application of injectable protein delivery systems for sustained release. *J. Pharmaceut. Sci.* 101 (3):946–954. doi: 10.1002/jps.23018.

30. Torosantucci, R., C. Schoneich, and W. Jiskoot. 2014. Oxidation of therapeutic proteins and peptides: Structural and biological consequences. *Pharmaceut. Res.* 31 (3):541–553. doi: 10.1007/s11095-013-1199-9.

31. Lipton, S. H., C. E. Bodwell, and A. H. Coleman Jr. 1977. Amino acid analyzer studies of the products of peroxide oxidation of cystine, lanthionine, and homocystine. *J. Agric. Food Chem.* 25 (3):624–628.

32. Stary, F. E., S. J. Jindal, and R. W. Murray. 1975. Oxidation of alpha-lipoic acid. *J. Org. Chem.* 40 (1):58–62. doi: 10.1021/jo00889a013.

33. Packer, L., E. H. Witt, and H. J. Tritschler. 1995. Alpha-Lipoic acid as a biological antioxidant. *Free Radic. Biol. Med.* 19:227–250.

34. Giles, G. I. and C. Jacob. 2002. Reactive sulfur species: An emerging concept in oxidative stress. *Biol. Chem.* 383 (3–4):375–388.

35. Giles, G. I., K. M. Tasker, C. Collins, N. M. Giles, E. O'Rourke, and C. Jacob. 2002. Reactive sulphur species: An *in vitro* investigation of the oxidation properties of disulphide S-oxides. *Biochem. J.* 364 (Pt 2):579–585.

36. Giles, G. I., K. M. Tasker, and C. Jacob. 2002. Oxidation of biological thiols by highly reactive disulfide-S-oxides. *Gen. Physiol. Biophys.* 21 (1):65–72.

37. Deweck, D., H. K. Nielsen, and P. A. Finot. 1987. Oxidation rate of free and protein-bound tryptophan by hydrogen-peroxide and the bioavailability of the oxidation-products. *J. Sci. Food Agric.* 41 (2):179–185. doi: 10.1002/Jsfa.2740410211.

38. Kramer, A. C., P. W. Thulstrup, M. N. Lund, and M. J. Davies. 2016. Key role of cysteine residues and sulfenic acids in thermal- and H_2O_2-mediated modification of beta-lactoglobulin. *Free Radic. Biol. Med.* 97:544–555. doi: 10.1016/j.freeradbiomed.2016.07.010.

39. Fenton, H. J. H. 1894. Oxidation of tartaric acid in the presence of iron. *J. Chem. Soc.* 65:899–910.

40. Fenton, H. J. H. and H. Jackson. 1899. The oxidation of polyhydric alcohols in the presence of iron. *J. Chem. Soc. Trans. (Lond.)* 75:1–11.

41. Koppenol, W. H. 1993. The centennial of the Fenton reaction. *Free Radic. Biol. Med.* 15 (6):645–651.

42. Dunford, H. B. 1982. Peroxidases. *Advances in the Inorganic Biochemistry*, eds. G. L. Eichorn, and L. G. Marzilli, Elsevier, New York, 41–68.

43. Prutz, W. A., J. Butler, E. J. Land, and A. J. Swallow. 1989. The role of sulphur peptide functions in free radical transfer: A pulse radiolysis study. *Int. J. Radiat. Biol.* 55 (4):539–556.

44. Ostdal, H., H. J. Andersen, and M. J. Davies. 1999. Formation of long-lived radicals on proteins by radical transfer from heme enzymes—A common process? *Arch. Biochem. Biophys.* 362:105–112.

45. Marquez, L. A. and H. B. Dunford. 1995. Kinetics of oxidation of tyrosine and dityrosine by myeloperoxidase compounds I and II. Implications for lipoprotein peroxidation studies. *J. Biol. Chem.* 270 (51):30434–30440.

46. Kettle, A. J. and L. P. Candaeis. 2000. Oxidation of tryptophan by redox intermediates of myeloperoxidase and inhibition of hypochlorous acid production. *Redox Rep.* 5 (4):179–184.

47. Galijasevic, S., I. Abdulhamid, and H. M. Abu-Soud. 2008. Potential role of tryptophan and chloride in the inhibition of human myeloperoxidase. *Free Radic. Biol. Med.* 44 (8):1570–1577 doi: 10.1016/j.freeradbiomed.2008.01.003.

48. Ximenes, V. F., S. O. Silva, M. R. Rodrigues, L. H. Catalani, G. J. Maghzal, A. J. Kettle, and A. Campa. 2005. Superoxide-dependent oxidation of melatonin by myeloperoxidase. *J. Biol. Chem.* 280 (46):38160–38169. doi: 10.1074/jbc.M506384200.

49. Irwin, J. A., H. Ostdal, and M. J. Davies. 1999. Myoglobin-induced oxidative damage: Evidence for radical transfer from oxidized myoglobin to other proteins and antioxidants. *Arch. Biochem. Biophys.* 362 (1):94–104.

50. Ostdal, H., M. J. Bjerrum, J. A. Pedersen, and H. J. Andersen. 2000. Lactoperoxidase-induced protein oxidation in milk. *J. Agric. Food Chem.* 48 (9):3939–3944.

51. Ostdal, H., B. Daneshvar, and L. H. Skibsted. 1996. Reduction of ferrylmyoglobin by β-lactoglobulin. *Free Radic. Res.* 24:429–438.

52. Domazou, A. S., W. H. Koppenol, and J. M. Gebicki. 2009. Efficient repair of protein radicals by ascorbate. *Free Radic. Biol. Med.* 46 (8):1049–1057. doi: 10.1016/j.freeradbiomed.2009.01.001.

53. Domazou, A. S., H. Zhu, and W. H. Koppenol. 2012. Fast repair of protein radicals by urate. *Free Radic. Biol. Med.* 52 (9):1929–1936. doi: 10.1016/j.freeradbiomed.2012.02.045.

54. Gebicki, J. M., T. Nauser, A. Domazou, D. Steinmann, P. L. Bounds, and W. H. Koppenol. 2010. Reduction of protein radicals by GSH and ascorbate: Potential biological significance. *Amino Acids* 39 (5):1131–1137. doi: 10.1007/s00726-010-0610-7.

55. Domazou, A. S., J. M. Gebicki, T. Nauser, and W. H. Koppenol. 2014. Repair of protein radicals by antioxidants. *Isr. J. Chem.* 54 (3):254–264. doi: 10.1002/ijch.201300117.

56. Jin, F., J. Leitich, and C. von Sonntag. 1993. The superoxide radical reacts with tyrosine-derived phenoxyl radicals by addition rather than by electron transfer. *J. Chem. Soc. Perkin Trans.* 2:1583–1588.

57. Moller, M. N., D. M. Hatch, H. Y. Kim, and N. A. Porter. 2012. Superoxide reaction with tyrosyl radicals generates para-hydroperoxy and para-hydroxy derivatives of tyrosine. *J. Am. Chem. Soc.* 134 (40):16773–16780. doi: 10.1021/ja307215z.

58. Winterbourn, C. C., H. N. Parsons-Mair, S. Gebicki, J. M. Gebicki, and M. J. Davies. 2004. Requirements for superoxide-dependent tyrosine hydroperoxide formation in peptides. *Biochem. J.* 381:241–248.

59. Fang, X., F. Jin, H. Jin, and C. von Sonntag. 1998. Reaction of the superoxide radical with the N-centred radical derived from N-acetyltryptophan methyl ester. *J. Chem. Soc. Perkin Trans.* 2:259–263.

60. Gunther, M. R., D. J. Kelman, J. T. Corbett, and R. P. Mason. 1995. Self-peroxidation of metmyoglobin results in formation of an oxygen-reactive tryptophan-centered radical. *J. Biol. Chem.* 270 (27):16075–16081.

61. Tafazoli, S. and P. J. O'Brien. 2004. Prooxidant activity and cytotoxic effects of indole-3-acetic acid derivative radicals. *Chem. Res. Toxicol.* 17 (10):1350–1355.

62. Hunter, E. P., M. F. Desrosiers, and M. G. Simic. 1989. The effect of oxygen, antioxidants, and superoxide radical on tyrosine phenoxyl radical dimerization. *Free Radic. Biol. Med.* 6:581–585.

63. Latarjet, R. and J. Loiseleur. 1942. Modalités de la Fixation de L'Oxygène en Radiobiologie. *Societe de Biologie Comptes Rendue* 136:60–63.

64. Alexander, P., M. Fox, K. A. Stacey, and D. Rosen. 1956. Comparison of some direct and indirect effects of ionising radiation in proteins. *Nature* 178:846–849.

65. Simpson, J. A., S. Narita, S. Gieseg, S. Gebicki, J. M. Gebicki, and R. T. Dean. 1992. Long-lived reactive species on free-radical-damaged proteins. *Biochem. J.* 282:621–624.

66. Garrison, W. M. 1987. Reaction mechanisms in the radiolysis of peptides, polypeptides, and proteins. *Chem. Rev.* 87:381–398.

67. Gebicki, S. and J. M. Gebicki. 1993. Formation of peroxides in amino acids and proteins exposed to oxygen free radicals. *Biochem. J.* 289:743–749.

68. Davies, M. J. 1996. Protein and peptide alkoxyl radicals can give rise to C-terminal decarboxylation and backbone cleavage. *Arch. Biochem. Biophys.* 336:163–172.

69. Davies, M. J., S. Fu, and R. T. Dean. 1995. Protein hydroperoxides can give rise to reactive free radicals. *Biochem. J.* 305:643–649.

70. Robinson, S., R. Bevan, J. Lunec, and H. Griffiths. 1998. Chemiluminescence determination of hydroperoxides following radiolysis and photolysis of free amino acids. *FEBS Lett.* 430 (3):297–300.

71. Hawkins, C. L. and M. J. Davies. 2001. Generation and propagation of radical reactions on proteins. *Biochim. Biophys. Acta* 1504:196–219.

72. von Sonntag, C. 1987. *The Chemical Basis of Radiation Biology.* London, U.K.: Taylor & Francis.

73. Fu, S., L. A. Hick, M. M. Sheil, and R. T. Dean. 1995. Structural identification of valine hydroperoxides and hydroxides on radical-damaged amino acid, peptide, and protein molecules. *Free Radic. Biol. Med.* 19 (3):281–292.

74. Fu, S. L. and R. T. Dean. 1997. Structural characterization of the products of hydroxyl-radical damage to leucine and their detection on proteins. *Biochem. J.* 324 (Pt 1):41–48.

75. Morin, B., W. A. Bubb, M. J. Davies, R. T. Dean, and S. Fu. 1998. 3-hydroxylysine, a potential marker for studying radical-induced protein oxidation. *Chem. Res. Toxicol.* 11:1265–1273.

76. d'Alessandro, N., G. Bianchi, X. Fang, F. Jin, H. P. Schuchmann, and C. von Sonntag. 2000. Reaction of superoxide with phenoxyl-type radicals. *J. Chem. Soc. Perkin. Trans.* 2:1862–1867.

77. Das, A. B., P. Nagy, H. F. Abbott, C. C. Winterbourn, and A. J. Kettle. 2010. Reactions of superoxide with the myoglobin tyrosyl radical. *Free Radic. Biol. Med.* 48 (11):1540–1547. doi: 10.1016/j.freeradbiomed.2010.02.039.

78. Das, A. B., T. Nauser, W. H. Koppenol, A. J. Kettle, C. C. Winterbourn, and P. Nagy. 2014. Rapid reaction of superoxide with insulin-tyrosyl radicals to generate a hydroperoxide with subsequent glutathione addition. *Free Radic. Biol. Med.* 70:86–95. doi: 10.1016/j.freeradbiomed.2014.02.006.

79. Nagy, P., A. J. Kettle, and C. C. Winterbourn. 2009. Superoxide-mediated formation of tyrosine hydroperoxides and methionine sulfoxide in peptides through radical addition and intramolecular oxygen transfer. *J. Biol. Chem.* 284 (22):14723–14733. doi: 10.1074/jbc.M809396200.

80. Pichorner, H., D. Metodiewa, and C. C. Winterbourn. 1995. Generation of superoxide and tyrosine peroxide as a result of tyrosyl radical scavenging by glutathione. *Arch. Biochem. Biophys.* 323 (2):429–437.

81. Davies, M. J. 2003. Singlet oxygen-mediated damage to proteins and its consequences. *Biochem. Biophys. Res. Commun.* 305:761–770.

82. Pattison, D. I., A. S. Rahmanto, and M. J. Davies. 2012. Photo-oxidation of proteins. *Photochem. Photobiol. Sci.* 11 (1):38–53. doi: 10.1039/c1pp05164d.

83. Jin, F. M., J. Leitich, and C. von Sonntag. 1995. The photolysis (λ = 254 nm) of tyrosine in aqueous solutions in the absence and presence of oxygen—The reaction of tyrosine with singlet oxygen. *J. Photochem. Photobiol. A Chem.* 92 (3):147–153.

84. Agon, V. V., W. A. Bubb, A. Wright, C. L. Hawkins, and M. J. Davies. 2006. Sensitizer-mediated photooxidation of histidine residues: Evidence for the formation of reactive side-chain peroxides. *Free Radic. Biol. Med.* 40 (4):698–710.

85. Gracanin, M., C. L. Hawkins, D. I. Pattison, and M. J. Davies. 2009. Singlet oxygen-mediated amino acid and protein oxidation: Formation of tryptophan peroxides and decomposition products. *Free Radic. Biol. Med.* 47:92–102.

86. Ronsein, G. E., M. C. de Oliveira, M. H. de Medeiros, and P. Di Mascio. 2009. Characterization of O(2) ((1)delta(g))-derived oxidation products of tryptophan: A combination of tandem mass spectrometry analyses and isotopic labeling studies. *J. Am. Soc. Mass Spectrom.* 20 (2):188–197.

87. Ronsein, G. E., M. C. Oliveira, S. Miyamoto, M. H. Medeiros, and P. Di Mascio. 2008. Tryptophan oxidation by singlet molecular oxygen [O2(1Deltag)]: Mechanistic studies using 18O-labeled hydroperoxides, mass spectrometry, and light emission measurements. *Chem. Res. Toxicol.* 21 (6):1271–1283.

88. Wright, A., W. A. Bubb, C. L. Hawkins, and M. J. Davies. 2002. Singlet oxygen-mediated protein oxidation: Evidence for the formation of reactive side-chain peroxides on tyrosine residues. *Photochem. Photobiol.* 76:35–46.

89. Wright, A., C. L. Hawkins, and M. J. Davies. 2000. Singlet oxygen-mediated protein oxidation: Evidence for the formation of reactive peroxides. *Redox Rep.* 5:159–161.

90. Saito, I., T. Matsuura, M. Nakagawa, and T. Hino. 1977. Peroxidic intermediates in photosensitized oxygenation of tryptophan derivatives. *Acc. Chem. Res.* 10:346–352.

91. Santus, R., L. K. Patterson, G. L. Hug, M. Bazin, J. C. Maziere, and P. Morliere. 2000. Interactions of superoxide anion with enzyme radicals: Kinetics of reaction with lysozyme tryptophan radicals and corresponding effects on tyrosine electron transfer. *Free Radic. Res.* 33 (4):383–391.

92. Foote, C. S. and J. V. Peters. 1971. Chemistry of singlet oxygen 14. A reactive intermediate in sulfide photooxidation. *J. Am. Chem. Soc.* 93 (15):3795–3796.

93. Gu, C. L., C. S. Foote, and M. L. Kacher. 1981. Chemistry of singlet oxygen 35. Nature of intermediates in the photooxygenation of sulfides. *J. Am. Chem. Soc.* 1981 (103):5949–5951.

94. Sysak, P. K., C. S. Foote, and T.-Y. Ching. 1977. Chemistry of singlet oxygen—XXV. Photooxygenation of methionine. *Photochem. Photobiol.* 26:19–27.

95. Morgan, P. E., D. I. Pattison, and M. J. Davies. 2012. Quantification of hydroxyl radical-derived oxidation products in peptides containing glycine, alanine, valine, and proline. *Free Radic. Biol. Med.* 52 (2):328–339. doi: 10.1016/j.freeradbiomed.2011.10.448.

96. Morgan, P. E., D. I. Pattison, C. L. Hawkins, and M. J. Davies. 2008. Separation, detection and quantification of hydroperoxides formed at side-chain and backbone sites on amino acids, peptides and proteins. *Free Radic. Biol. Med.* 45:1279–1289.

97. Nagy, P., A. J. Kettle, and C. C. Winterbourn. 2010. Neutrophil-mediated oxidation of enkephalins via myeloperoxidase-dependent addition of superoxide. *Free Radic. Biol. Med.* 49 (5):792–799. doi: 10.1016/j.freeradbiomed.2010.05.033.

98. Davies, M. J. 1997. Radicals derived from amino acid and protein hydroperoxides—Key mediators in protein damage? In *Free Radicals in Biology and the Environment*, ed. F. Minisci, pp. 251–262. Dordrecht, the Netherlands: Kluwer Academic Publishers.

99. Gebicki, S., G. Bartosz, and J. M. Gebicki. 1995. The action of iron on amino acid and protein peroxides. *Biochem. Soc. Trans.* 23 (2):2495.

100. Morgan, P. E., R. T. Dean, and M. J. Davies. 2004. Protective mechanisms against peptide and protein peroxides generated by singlet oxygen. *Free Radic. Biol. Med.* 36:484–496.

101. Jessup, W., R. T. Dean, and J. M. Gebicki. 1994. Iodometric determination of hydroperoxides in lipids and proteins. *Methods Enzymol.* 233:289–303.

102. Wolff, S. P. 1994. Ferrous ion oxidation in the presence of ferric ion indicator xylenol orange for measurement of hydroperoxides. *Methods Enzymol.* 233:182–189.

103. Gay, C., J. Collins, and J. M. Gebicki. 1999. Hydroperoxide assay with the ferric-xylenol orange complex. *Anal. Biochem.* 273:149–155.

104. Babiy, A. V., S. Gebicki, and J. M. Gebicki. 1992. Protein peroxides: Formation by superoxide-generating systems and during oxidation of low density lipoprotein. In *Free Radicals: From Basic Science to Clinical Medicine*, eds. G. Poli, E. Albano, and M. U. Dianzani. Basel, Switzerland: Birkhauser.

105. Firth, C. A., E. M. Crone, E. A. Flavall, J. A. Roake, and S. P. Gieseg. 2008. Macrophage mediated protein hydroperoxide formation and lipid oxidation in low density lipoprotein are inhibited by the inflammation marker 7,8-dihydroneopterin. *Biochim. et Biophys. Acta* 1783 (6):1095–1101. doi: 10.1016/j.bbamcr.2008.02.010.

106. Gebicki, J., J. Du, J. Collins, and H. Tweeddale. 2000. Peroxidation of proteins and lipids in suspensions of liposomes, in blood serum, and in mouse myeloma cells. *Acta Biochim. Polonica* 47 (4):901–911.

107. Du, J. and J. M. Gebicki. 2004. Proteins are major initial cell targets of hydroxyl free radicals. *Int. J. Biochem. Cell Biol.* 36 (11):2334–2343.

108. Liu, C. C. and J. M. Gebicki. 2012. Intracellular GSH and ascorbate inhibit radical-induced protein chain peroxidation in HL-60 cells. *Free Radic. Biol. Med.* 52 (2):420–426. doi: 10.1016/j.freeradbiomed.2011.10.450.

109. Gieseg, S., S. Duggan, and J. M. Gebicki. 2000. Peroxidation of proteins before lipids in U937 cells exposed to peroxyl radicals. *Biochem. J.* 350 (Pt 1):215–218.

110. Firth, C. A., Y. T. Yang, and S. P. Gieseg. 2007. Lipid oxidation predominates over protein hydroperoxide formation in human monocyte-derived macrophages exposed to aqueous peroxyl radicals. *Free Radic. Res.* 41 (7):839–848. doi: 10.1080/10715760701416442.

111. Rahmanto, A. S., P. E. Morgan, C. L. Hawkins, and M. J. Davies. 2010. Cellular effects of photogenerated oxidants and long-lived, reactive, hydroperoxide photoproducts. *Free Radic. Biol. Med.* 49 (10):1505–1515. doi: 10.1016/j.freeradbiomed.2010.08.006.

112. Wright, A., C. L. Hawkins, and M. J. Davies. 2003. Photo-oxidation of cells generates long-lived intracellular protein peroxides. *Free Radic. Biol. Med.* 34:637–647.

113. Winterbourn, C. C., H. Pichorner, and A. J. Kettle. 1997. Myeloperoxidase-dependent generation of a tyrosine peroxide by neutrophils. *Arch. Biochem. Biophys.* 338:15–21.

114. Fu, S., S. Gebicki, W. Jessup, J. M. Gebicki, and R. T. Dean. 1995. Biological fate of amino acid, peptide and protein hydroperoxides. *Biochem. J.* 311:821–827.

115. Dalle-Donne, I., R. Rossib, D. Giustarinib, A. Milzania, and R. Colomboa. 2003. Protein carbonyl groups as biomarkers of oxidative stress. *Clin. Chim. Acta* 329:23–38.

116. Dalle-Donne, I., G. Aldini, M. Carini, R. Colombo, R. Rossi, and A. Milzani. 2006. Protein carbonylation, cellular dysfunction, and disease progression. *J. Cell. Mol. Med.* 10 (2):389–406.

117. Winterbourn, C. C., M. J. Bonham, H. Buss, F. M. Abu-Zidan, and J. A. Windsor. 2003. Elevated protein carbonyls as plasma markers of oxidative stress in acute pancreatitis. *Pancreatology* 3 (5):375–382. doi: 10.1159/000073652.

118. Winterbourn, C. C., I. H. Buss, T. P. Chan, L. D. Plank, M. A. Clark, and J. A. Windsor. 2000. Protein carbonyl measurements show evidence of early oxidative stress in critically ill patients. *Crit. Care Med.* 28 (1):143–149.

119. Requena, J. R., R. L. Levine, and E. R. Stadtman. 2003. Recent advances in the analysis of oxidized proteins. *Amino Acids* 25 (3–4):221–226.

120. Oh-Ishi, M., T. Ueno, and T. Maeda. 2003. Proteomic method detects oxidatively induced protein carbonyls in muscles of a diabetes model Otsuka Long-Evans Tokushima Fatty (OLETF) rat. *Free Radic. Biol. Med.* 34 (1):11–22.

121. Buss, H., T. P. Chan, K. B. Sluis, N. M. Domigan, and C. C. Winterbourn. 1997. Protein carbonyl measurement by a sensitive ELISA method. *Free Radic. Biol. Med.* 23 (3):361–366.

122. Brown, R. K. and F. J. Kelly. 1994. Evidence for increased oxidative damage in patients with cystic fibrosis. *Pediatr. Res.* 36 (4):487–493.

123. Fu, S., M. J. Davies, R. Stocker, and R. T. Dean. 1998. Evidence for roles of radicals in protein oxidation in advanced human atherosclerotic plaque. *Biochem. J.* 333:519–525.

124. Fu, S., R. Dean, M. Southan, and R. Truscott. 1998. The hydroxyl radical in lens nuclear cataractogenesis. *J. Biol. Chem.* 273:28603–28609.

125. Fu, S., M.-X. Fu, J. W. Baynes, S. R. Thorpe, and R. T. Dean. 1998. Presence of dopa and amino acid hydroperoxides in proteins modified with advanced glycation end products : Amino acid oxidation products as possible source of oxidative stress induced by age proteins. *Biochem. J.* 330:233–239.

126. Berdnikov, V. M., N. M. Bazhin, V. K. Federov, and O. V. Polyakov. 1972. Isomerization of the ethoxyl radical to the α-hydroxyethyl radical in aqueous solution. *Kinet. Catal. (Engl. Trans.)* 13:986–987.

127. Gilbert, B. C., R. G. G. Holmes, H. A. H. Laue, and R. O. C. Norman. 1976. Electron spin resonance studies. Part L. Reactions of alkoxyl radicals generated from hydroperoxides and titanium(III) ion in aqueous solution. *J. Chem. Soc. Perkin Trans.* 2:1047–1052.

128. Gilbert, B. C., P. D. R. Marshall, R. O. C. Norman, N. Pineda, and P. S. Williams. 1981. Electron spin resonance studies. Part 61. The generation and reactions of the t-butoxyl radical in aqueous solution. *J. Chem. Soc. Perkin Trans.* 2:1392–1400.

129. Bors, W., D. Tait, C. Michel, M. Saran, and M. Erben-Russ. 1984. Reactions of alkoxy radicals in aqueous solution. *Isr. J. Chem.* 24:17–24.

130. Neta, P., M. Dizdaroglu, and M. G. Simic. 1984. Radiolytic studies of the cumyloxyl radical in aqueous solutions. *Isr. J. Chem.* 24:25–28.

131. Erben-Russ, M., C. Michel, W. Bors, and M. Saran. 1987. Absolute rate constants of alkoxyl radical reactions in aqueous solution. *J. Phys. Chem.* 91:2362–2365.

132. Headlam, H. A. and M. J. Davies. 2002. Beta-scission of side-chain alkoxyl radicals on peptides and proteins results in the loss of side chains as aldehydes and ketones. *Free Radic. Biol. Med.* 32:1171–1184.

133. Headlam, H. A., A. Mortimer, C. J. Easton, and M. J. Davies. 2000. β-Scission of C-3 (β-carbon) alkoxyl radicals on peptides and proteins: A novel pathway which results in the formation of α-carbon radicals and the loss of amino acid side chains. *Chem. Res. Toxicol.* 13:1087–1095.

134. Neuzil, J., J. M. Gebicki, and R. Stocker. 1993. Radical-induced chain oxidation of proteins and its inhibition by chain-breaking antioxidants. *Biochem. J.* 293:601–606.

135. Mozziconacci, O., B. A. Kerwin, and C. Schoneich. 2010. Reversible hydrogen transfer between cysteine thiyl radical and glycine and alanine in model peptides: Covalent H/D exchange, radical-radical reactions, and L- to D-Ala conversion. *J. Phys. Chem. B* 114 (19):6751–6762. doi: 10.1021/jp101508b.

136. Mozziconacci, O. and C. Schoneich. 2014. Sequence-specific formation of d-amino acids in a monoclonal antibody during light exposure. *Mol. Pharmaceut.* 11 (11):4291–4297. doi: 10.1021/mp500508w.

137. Fradkin, A. H., O. Mozziconacci, C. Schoneich, J. F. Carpenter, and T. W. Randolph. 2014. UV photodegradation of murine growth hormone: Chemical analysis and immunogenicity consequences. *Eur. J. Pharmaceut. Biopharmaceut.* 87 (2):395–402. doi: 10.1016/j.ejpb.2014.04.005.

138. Gebicki, S., K. H. Gill, R. T. Dean, and J. M. Gebicki. 2002. Action of peroxidases on protein hydroperoxides. *Redox Rep.* 7 (4):235–242.

139. Soszynski, M., A. Filipiak, G. Bartosz, and J. M. Gebicki. 1996. Effect of amino acid peroxides on the erythrocyte. *Free Rad. Biol. Med.* 20:45–51.

140. Rhee, S. G., S. W. Kang, L. E. Netto, M. S. Seo, and E. R. Stadtman. 1999. A family of novel peroxidases, peroxiredoxins. *Biofactors* 10 (2-3):207–209.

141. Peskin, A. V., A. G. Cox, P. Nagy, P. E. Morgan, M. B. Hampton, M. J. Davies, and C. C. Winterbourn. 2010. Removal of amino acid, peptide and protein hydroperoxides by reaction with peroxiredoxins 2 and 3. *Biochem. J.* 432 (2):313–321. doi: 10.1042/BJ20101156.

142. Morgan, P. E., R. T. Dean, and M. J. Davies. 2002. Inhibition of glyceraldehyde-3-phosphate dehydrogenase by peptide and protein peroxides generated by singlet oxygen attack. *Eur. J. Biochem.* 269:1916–1925.

143. Gebicki, S., R. T. Dean, and J. M. Gebicki. 1996. Inactivation of glutathione reductase by protein and amino acid peroxides. In *Oxidative Stress and Redox Regulation: Cellular Signalling, AIDS, Cancer and Other Diseases*, p. 139. Paris, France: Institute Pasteur.

144. Headlam, H. A., M. Gracanin, K. J. Rodgers, and M. J. Davies. 2006. Inhibition of cathepsins and related proteases by amino acid, peptide, and protein hydroperoxides. *Free Radic. Biol. Med.* 40 (9):1539–1548.

145. Hampton, M. B., P. E. Morgan, and M. J. Davies. 2002. Inactivation of cellular caspases by peptide-derived tryptophan and tyrosine peroxides. *FEBS Lett.* 527 (1–3):289–292.

146. Gracanin, M. and M. J. Davies. 2007. Inhibition of protein tyrosine phosphatases by amino acid, peptide and protein hydroperoxides: Potential modulation of cell signaling by protein oxidation products. *Free Radic. Biol. Med.* 42 (10):1543–1551.

147. Dremina, E. S., V. S. Sharov, M. J. Davies, and C. Schoneich. 2007. Oxidation and inactivation of SERCA by selective reaction of cysteine residues with amino acid peroxides. *Chem. Res. Toxicol.* 20 (10):1462–1469.

148. Gracanin, M., M. A. Lam, P. E. Morgan, K. J. Rodgers, C. L. Hawkins, and M. J. Davies. 2011. Amino acid, peptide and protein hydroperoxides, and their decomposition products, modify the activity of the 26S proteasome. *Free Radic. Biol. Med.* 50:389–399.

149. Luxford, C., B. Morin, R. T. Dean, and M. J. Davies. 1999. Histone H1- and other protein- and amino acid-hydroperoxides can give rise to free radicals which oxidize DNA. *Biochem. J.* 344:125–134.

150. Luxford, C., R. T. Dean, and M. J. Davies. 2000. Radicals derived from histone hydroperoxides damage nucleobases in RNA and DNA. *Chem. Res. Toxicol.* 13:665–672.

151. Gebicki, S. and J. M. Gebicki. 1999. Crosslinking of DNA and proteins induced by protein hydroperoxides. *Biochem. J.* 338:629–636.

152. Luxford, C., R. T. Dean, and M. J. Davies. 2002. Induction of DNA damage by oxidised amino acids and proteins. *Biogerentology* 3:95–102.

153. Rahmanto, A. S., P. E. Morgan, C. L. Hawkins, and M. J. Davies. 2010. Cellular effects of peptide and protein hydroperoxides. *Free Radic. Biol. Med.* 48 (8):1071–1078. doi: 10.1016/j.freeradbiomed.2010.01.025.

154. Koppenol, W. H., P. L. Bounds, T. Nauser, R. Kissner, and H. Ruegger. 2012. Peroxynitrous acid: Controversy and consensus surrounding an enigmatic oxidant. *Dalton Trans.* 41 (45):13779–13787. doi: 10.1039/c2dt31526b.

6 Catalysis by Peroxiredoxins at High Temporal and Structural Resolution

Leslie B. Poole and P. Andrew Karplus

CONTENTS

INTRODUCTION

Peroxiredoxins (Prxs) are cysteine-dependent hydroperoxide reductases with varying specificities for their substrates, and typically very high catalytic efficiencies, with their best substrates on the order of 10^6–10^8 M^{-1} s^{-1} [1–5]. This implies a very specialized active site since a typical cysteine residue, even in its fully ionized state as a thiolate (e.g., with a low pK_a or measured at high pH), exhibits rate constants of only up to ~20 M^{-1} s^{-1} with H_2O_2 [6,7]. Prxs are also remarkable, given the considerable conformational rearrangements that they must go through during the various phases of the catalytic cycle.

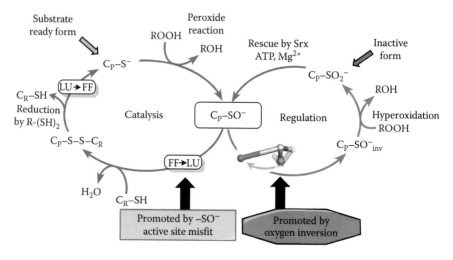

FIGURE 6.1 Catalytic cycle of 2-Cys peroxiredoxins. The catalytic cycle (left ellipse) is shown along with the peroxide-mediated inactivation pathway (right ellipse). Reaction of the hydroperoxide substrate (including H_2O_2, lipid hydroperoxide, or peroxynitrite) with the active, substrate-ready enzyme in its thiolate form (C_P–S^-) generates the sulfenate (C_P–SO^-), which can then either proceed through the catalytic cycle or become hyperoxidized. The conformational change necessary for Prx catalysis (lower part of left cycle) is shown as a transition from fully folded (FF) to locally unfolded (LU) enzyme, after which the disulfide bond between C_P and C_R (the resolving Cys) is formed. After reduction and release of C_R–SH, the opposite transition occurs (LU → FF) and C_P–S^- is regenerated, ready for another cycle. Blue arrows indicate pathways supporting catalysis, whereas red pathways lead toward inactivating hyperoxidation. The ball-and-stick structure depicts the C_P–SO^- side chain rotating into an inverted position to allow for binding and oxidation of C_P by a second peroxide molecule to form the inactive sulfinate (C_P–SO_2^-). Recovery of activity for typical 2-Cys Prxs is catalyzed by the enzyme sulfiredoxin (Srx) in an ATP-dependent manner. The text boxes summarize new insights into structural features that can independently promote either catalysis (in green) or hyperoxidation (in red).

To initiate the reaction, the absolutely conserved cysteine poised within the highly organized active site architecture is oxidized by the peroxide substrate to a sulfenic acid (R-SOH, or R-SO⁻ for the sulfenate form). Then it must leave that environment (termed the fully folded or FF conformation) to become locally unfolded (termed the LU conformation) to undergo protein disulfide and/or mixed disulfide formation and then reduction to restore the reaction-ready thiolate form of the active site cysteine (Figure 6.1). In the course of these reactions, the active site cysteine sulfenate can sometimes react with a second hydroperoxide, yielding the sulfinate (R–SO_2^-) and inactivating the protein. The degree of sensitivity of individual Prxs toward hyperoxidation by their substrates is significantly controlled by structural and dynamic elements of these proteins, allowing tuning of this sensitivity to match the requirements for the biological function [8]. Much current research focuses on gaining a better understanding of the forces at play in modulating catalytic power and hyperoxidation sensitivity of Prxs that link to their biological function, as elaborated on further in this chapter.

As inactivation of these efficient peroxide scavengers can serve to elevate levels of H_2O_2 that may serve signaling roles, in 2003 we posited in our floodgate hypothesis that one biological function of inactivation sensitivity may be to promote localized, H_2O_2-mediated cell signaling processes [9]. Other proposed roles for Prx hyperoxidation in some settings include promotion of a chaperone-like holdase activity [10–12], involvement in the maintenance of circadian rhythms in cells and/or organisms [13,14], and serving as a direct signaling molecule itself, although the prevalence and significance of each of these is as yet largely unclear. The existence of a repair enzyme (sulfiredoxin; [15,16]) in many eukaryotes with hyperoxidation-sensitive Prxs apparently dedicated to the recovery of activity in hyperoxidized Prxs, as well as the presence of evolutionarily conserved structural features that impart sensitivity to specific Prxs [8], argue strongly that hyperoxidation sensitivity is a feature of biological significance rather than an unavoidable Achilles heel for these enzymes, and much remains to be discovered in this important area.

Another mode in which Prxs can participate in H_2O_2 signal transduction that would not be enhanced by hyperoxidation is as direct sensor and transducer of the signal to effector proteins. In this mode, Prxs serve as a redox mediator in which oxidation by H_2O_2 to yield either sulfenic acid or the disulfide form of Prx is followed by a subsequent relaying of the redox equivalents to a signaling protein such as a transcription factor or kinase. Evidence for specific cases in which such a redox relay is operative are few but are increasing (for recent reviews, see [17,18]).

It is also clear that a number of types of posttranslational modifications, such as phosphorylation, acetylation, glutathionylation, nitrosylation, and nitrosation, can modulate Prx catalytic activity and/or hyperoxidation sensitivity, but only inactivating phosphorylation by Src-family kinases or cyclin-dependent kinases has so far been clearly linked to biologically important outputs [18]. With roles in regulation of cell growth, metabolism, and immunity, Prxs are found to be implicated as regulators or protectants in pathological settings including cancer, neurodegenerative diseases, and inflammatory diseases, and may represent viable therapeutic targets in particular disease settings [8,19]. Reflecting possible roles in influencing diverse processes, prior to them being named "peroxiredoxin," the abundantly expressed human PrxI and PrxII enzymes were variously identified as proliferation-associated gene, calpromotin, torin, heme binding protein 23, macrophage stress protein 23, osteoblast specific factor 3, and natural killer cell enhancing factor [9].

As the interplay between structure, dynamics, and catalysis in Prxs is intimately associated with chemical and biological functions of these enzymes, our research team has focused on better understanding the chemical and structural mechanisms governing the diverse catalytic and regulatory features of this large and ubiquitous family of proteins. In this chapter, we elaborate on recent studies of two bacterial Prxs we have developed as model systems: AhpC from *Salmonella typhimurium* (*St*AhpC) and PrxQ from *Xanthomonas campestris* (*Xc*PrxQ). Specifically, rapid reaction kinetics studies of *St*AhpC and high-resolution crystallographic studies of *Xc*PrxQ have revealed mechanistic details not before obtainable in such systems. As additional features involved in tuning of Prx hypersensitivity toward H_2O_2 emerge, the question of biological roles for this sensitivity remains. We close this chapter with a perspective on mechanisms involved in Prx roles as modulators of redox signaling.

RAPID REACTION KINETIC STUDIES OF
Salmonella typhimurium AhpC (*St*AhpC)

Bacterial AhpCs are representative of a large group of Prxs, now designated as the Prx1 group [20]. For these highly efficient Prxs, the active site, peroxidatic Cys residue (termed C_P) during recycling forms an intersubunit disulfide bond with a resolving Cys (termed C_R) located in the C-terminus of an adjacent subunit, covalently tying the two subunits together in two places if both active sites are oxidized. The Prx1 group is arguably the most widely distributed of the six subgroups of Prxs identifiable by structural and bioinformatics approaches [20,21]. This group includes human PrxI, PrxII, PrxIII, and PrxIV, and is sometimes referred to as the typical 2-Cys Prxs for historical reasons [22]. While the minimal functional structure is a homodimer, which stays intact whether or not the disulfide linkage is present, the assembly of five such dimers into doughnut-shaped decamers (or less commonly dodecamers formed from six dimers) is observed in solution and in structural studies, with these higher-order oligomers stabilized by reduction [22,23]. Steady-state kinetic studies have revealed low to submicromolar K_m values for preferred peroxide substrates and turnover rates as high as ~55 s^{-1} for *St*AhpC with H_2O_2 [3,24].

Rapid Reaction Kinetic Results with *St*AhpC

To gain a better understanding of the steps involved in peroxide binding and reduction, we conducted pre-steady-state, rapid reaction kinetic analyses of *St*AhpC using spectroscopic data collected on the millisecond time scale. One fortuitous feature of *St*AhpC that greatly enabled this research is the presence of Trp residues in the protein that undergo detectable fluorescence changes upon mixing reduced *St*AhpC with H_2O_2. Further study of variants having the three individual Trp residues in *St*AhpC converted to Phe indicated that two of these residues, one proximal to C_P (Trp81) and the other proximal to C_R (Trp169), served as sensitive reporters of catalytic processes occurring in important regions of this dynamic protein [25].

As shown in Figure 6.2, upon the mixing of hydroperoxide substrates with reduced *St*AhpC in a stopped-flow spectrofluorometer, fluorescence changes ensue, which involve an initial, peroxide-dependent decrease in fluorescence, followed by an increase that levels off at a fluorescence value only slightly higher than the starting fluorescence of the reduced protein. The Trp mutagenesis studies mentioned here indicated that the peroxide-dependent first phase reflected primarily fluorescence changes involving Trp81, whereas the peroxide-independent changes in fluorescence observed in the second phase were largely a result of changes occurring in the environment of Trp169. Each of the fluorescence traces obtained at various H_2O_2 concentrations can be fitted to a double exponential rate equation. The first, fastest rate constant measured for each trace increases linearly with H_2O_2 concentrations up to about 8 µM (then becomes too rapid to monitor due to the technical limitation of a 1.5 ms deadtime in the instrument). Notably, the second rate constant remains at a steady value of approximately 80 s^{-1} across all peroxide concentrations (Figure 6.2, inset). This initial, simplistic data analysis therefore supports a minimum of two phases in the reaction of *St*AhpC with H_2O_2. Very similar fluorescence profiles were

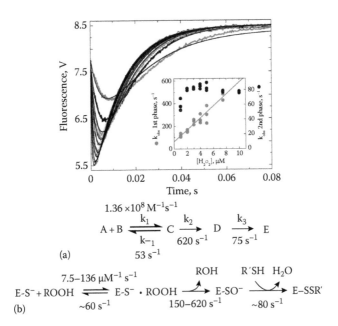

$$A + B \underset{\substack{k_{-1} \\ 53\ \text{s}^{-1}}}{\overset{\substack{1.36 \times 10^8\ \text{M}^{-1}\text{s}^{-1} \\ k_1}}{\rightleftharpoons}} C \overset{\substack{k_2 \\ 620\ \text{s}^{-1}}}{\longrightarrow} D \overset{\substack{k_3 \\ 75\ \text{s}^{-1}}}{\longrightarrow} E$$

(a)

$$\text{E-S}^- + \text{ROOH} \underset{\substack{\sim 60\ \text{s}^{-1}}}{\overset{\substack{7.5-136\ \mu\text{M}^{-1}\text{s}^{-1}}}{\rightleftharpoons}} \text{E-S}^- \cdot \text{ROOH} \overset{\overset{\text{ROH}}{\nearrow}}{\underset{150-620\ \text{s}^{-1}}{\longrightarrow}} \text{E-SO}^- \overset{\overset{\text{R'SH}\quad\text{H}_2\text{O}}{\nwarrow\nearrow}}{\underset{\sim 80\ \text{s}^{-1}}{\longrightarrow}} \text{E-SSR'}$$

(b)

FIGURE 6.2 Kinetic profiles of intrinsic Trp fluorescence changes along with fits to a three-step kinetic model for reduced, wild type *Salmonella typhimurium* AhpC (*St*AhpC) reacting with H_2O_2. (a) One µM reduced AhpC in reaction buffer (25 mM potassium phosphate at pH 7, with 1 mM EDTA and 100 mM ammonium sulfate) at 25°C was mixed in a stopped-flow spectrofluorometer with increasing concentrations of H_2O_2, with representative traces shown in red (1 µM), black (2 µM), green (3 µM), blue (5 µM), magenta (10 µM), and cyan (25 µM). The kinetic model to which these data fit best (using KinTek Global Kinetic Explorer, fit lines in black) and the values for the individual rate constants are shown below the plot. The inset depicts the rate constants derived from simple double exponential fits of the data at each H_2O_2 concentration, with the data in red reflecting the rate constants for the first phase (decreasing fluorescence) and data in black indicating those for the second phase (increasing fluorescence), plotted *versus* $[H_2O_2]$. (b) Our interpretation of the reaction steps detected and the range of values for the fitted kinetic constants using both H_2O_2 and *t*-butyl hydroperoxide are shown. (Reprinted with permission from Parsonage, D., Nelson, K.J., Ferrer-Sueta, G., Alley, S., Karplus, P.A., Furdui, C.M., and Poole, L.B., Dissecting peroxiredoxin catalysis: Separating binding, peroxidation, and resolution for a bacterial AhpC, *Biochemistry*, 54(7), 1567–1575. Copyright 2015 American Chemical Society.)

also observed upon mixing reduced *St*AhpC with *t*-butyl hydroperoxide, a bulky substrate, but the concentrations needed to reproduce data like those of Figure 6.2 were much higher, fitting with the higher K_m for this substrate (238 µM versus 1.4 µM for H_2O_2) [24,25].

GLOBAL FITTING OF *St*AhpC STOPPED-FLOW DATA TO A KINETIC MODEL

To develop a kinetic model consistent with all the data, we carried out a global fitting of the data obtained at multiple concentrations of the same peroxide. We started with a two-step model (A + B ↔ C → D) that could in theory reflect the binding of peroxide in the first step, followed by a second step representing all of the subsequent reactions, namely oxygen transfer to C_P to form C_P–SOH, the conformational change (FF → LU) to expose C_P–SOH, then disulfide bond formation between C_P and C_R. Such treatment of kinetic data can often work even with multiple steps, as the rate of the C → D conversion could be dominated kinetically by a single slowest rate-limiting step. In this case, however, the simple, two-step kinetic model did not yield a sufficiently satisfactory global fit to the data, so an additional step was added (A + B ↔ C → D → E). Fitting of the data to this model yielded much better results (black lines in Figure 6.2).

There are eight variables in this 3-step model to be determined during the fitting process (four kinetic rate constants, k_1, $k_{-1,}$ k_2, and k_3 (Figure 6.2), and four fluorescence scaling factors, for species A, C, D, and E) for which only the initial and final fluorescence values are easy to estimate (to start the fit). Thus we took advantage of our knowledge of the rate constant for the unchanging value in the last step that we interpret as disulfide bond formation (from the data shown in black for the inset of Figure 6.2). Holding k_3 constant at 75 s⁻¹, all other values were readily fit to give relatively well determined values. Of these, k_{-1} was the least well determined, and as we assign this rate constant to the dissociation of H_2O_2 from the enzyme in the first step, we conducted a second, independent fitting of the data by using as an estimate for this value the y-intercept of the H_2O_2-dependent first phase shown in red in the inset of Figure 6.2 (estimated as 53 s⁻¹). This yielded nearly identical values compared to the first fit for all other parameters (kinetic constants and fluorescence scaling factors), with the final results shown at the bottom of Figure 6.2.

INTERPRETATION OF THE KINETIC MODEL FOR *St*AhpC

Given the global fitting results described, the putative reaction mechanism corresponding to the observed kinetic profile involves the initial reversible binding of peroxide (k_1 and k_{-1}), followed by the chemical step leading to sulfenic acid formation (k_2), then by structural rearrangements that allow for disulfide bond formation between C_P and C_R (k_3). Interestingly, when this exercise was repeated for the data obtained with *t*-butyl hydroperoxide, only two of the four kinetic constants were affected. k_1 (the forward step of binding) changed from 1.36×10^8 to 7.5×10^6 M⁻¹ s⁻¹ going from H_2O_2 to the bulkier substrate, and the next (chemical) step of oxygen transfer from the substrate to the enzyme slowed from 620 to 150 s⁻¹ [25].

In an effort to collect complementary kinetic information about chemical changes in *St*AhpC during catalysis, we also monitored absorbance changes in the stopped-flow spectrophotometer at 240 nm, a wavelength where thiolate anion (R–S⁻) is known to contribute to the absorbance. In these experiments, the observed decrease in A_{240} was coincident with the peroxide-independent phase of fluorescence increase, interpreted here as disulfide bond formation. While we might have expected the loss of thiolate to occur during step two of the chemical mechanism shown, an alternative possibility is that both the thiolate, R–S⁻, and the sulfenate anions, R–SO⁻, absorb at this wavelength, and it is the sulfenate that is decreasing as observed at 240 nm, in line with step three of the model.

Regarding the spectral changes observed, in general, we note that a detectable change in fluorescence upon the binding of peroxide substrate is not necessarily expected, particularly with the very small H_2O_2 molecule used as substrate. However, the noncovalent binding of H_2O_2 to the enzyme may invoke a change in conformation of the active site that yields a change in the fluorescence of Trp81, which is very close to the β-carbon of the active site Cys in FF. Interestingly, if k_1 and k_{-1} simply represent reversible binding of peroxide to the enzyme, then the K_d for that binding would be about 400 nM for H_2O_2, a value that would seem surprisingly low, given that this tiny molecule can form, at most, only six hydrogen bonds (with the two hydrogens serving as H-bond donors, and the two lone pairs on each of the two oxygens serving as H-bond acceptors).

One possibility is that k_{-1} (the so-called off rate, at 53–70 s^{-1}) represents the rate of FF → LU in this enzyme, whether or not the active site contains a substrate or product (although the apparent rapid exchange between conformations suggests that such rates may be faster than this would represent). Consistent with this interpretation is the similar rate constant obtained for disulfide bond formation (75–88 s^{-1}), which clearly involves a transition of FF to LU (of the C_P–SO$^-$ form of the enzyme) and a potentially more rapid rate of disulfide bond formation that would follow. Also consistent with this potential interpretation is the fact that neither of these rates, ranging overall from 53 to 88 s^{-1}, change even when very different substrates are used. It will be interesting to develop a better understanding of the events, including conformational changes underlying the kinetic observations here with additional future experimental work with Prxs.

DISULFIDE BOND FORMATION AND HYPEROXIDATION IN MULTIPLE Prxs

Among Prxs that have both a C_P and a C_R and so form an intraprotein disulfide during catalysis (either within or between subunits), StAhpC exhibits the fastest rate of disulfide bond formation observed to date, at ~75 s^{-1}. For two human Prxs, PrxIII (in the same subgroup as AhpC) and PrxV (in a distinct subgroup, with C_R in a different location and forming an intrasubunit disulfide bond during catalysis), disulfide bond formation rates are 22 and 15 s^{-1}, respectively [26,27]. For human PrxII (also in the Prx1 subgroup like AhpC and PrxIII), this rate is particularly slow, at 1.7 s^{-1}, and this protein is also particularly sensitive toward inactivation by H_2O_2 during turnover [26]. This emphasizes the principle that the sulfenic acid group, when formed, can either become reoriented by local unfolding to react with the C_R and form the disulfide bond (i.e., promoting continuation through the catalytic cycle), or it can persist longer within the active site and be subject to further oxidation by a second peroxide molecule (i.e., promoting the hyperoxidation pathway; Figure 6.1).

Further details of the structural features of these chemical transformations during catalysis and hyperoxidation were subsequently obtained by our research team through studies of XcPrxQ, for which the monomeric protein could be crystallized in a crystal form that diffracted to ultrahigh resolution and that retained catalytic activity [28]. As is described next, this allowed us to obtain high-resolution data for not only oxidized and reduced forms of the protein, but also for intermediates along the pathway of peroxide reaction.

HIGH-RESOLUTION STRUCTURAL STUDIES OF
Xanthomonas campestris PrxQ (*Xc*PrxQ) CATALYSIS

To elucidate structural details important for catalysis and hyperoxidation, we selected a model Prx protein from the PrxQ group, specifically the PrxQ from the bacterial plant pathogen *X. campestris* (*Xc*PrxQ). We selected it in large part because, in contrast to the overwhelming majority of Prxs, it was monomeric and thus would potentially allow us to carry out detailed NMR studies that could give much more insight into the dynamic properties of the protein than can be discovered by crystal structure analyses [29]. Also, we knew that this protein could grow crystals that diffracted to high resolution as it had previously yielded structures of both the FF and LU disulfide forms at resolutions of 1.5–1.8 Å [30]. Furthermore, we anticipated that these crystals could be suitable for studies of ligand binding and catalysis, because they had an accessible active site pocket. As the Liao et al. study of the *Xc*PrxQ FF conformation was carried out using a catalytically inactive double mutant with Ser in place of both C_P and C_R, to carry out in-crystal studies of catalysis we needed to grow similar crystals of the authentic wild type enzyme.

ESTABLISHING THE *Xc*PrxQ MODEL SYSTEM

To develop the model system, we initially expressed and purified recombinant wild type *Xc*PrxQ along with the individual $C_P \rightarrow$ Ser (C48S) and $C_R \rightarrow$ Ser (C84S) mutants, and assessed their catalytic activities in solution by conducting bisubstrate kinetic analyses varying both the reductant concentration (using *E. coli* thioredoxin) and the hydroperoxide substrate concentration. We studied both the smallest possible substrate (hydrogen peroxide) and a bulky substrate (cumene hydroperoxide). Similar to our previous steady-state kinetics results with the PrxQ from *E. coli* (previously known as *E. coli* BCP) [31], the k_{cat}/K_m values with H_2O_2 or cumene hydroperoxide at multiple Trx concentrations were about the same, at $2–4 \times 10^4$ M^{-1} s^{-1} (Figure 6.3). It has been reported that PrxQ B from *Mycobacterium tuberculosis* has ~100-fold higher activity with lipid hydroperoxides compared with H_2O_2 [5], but we have not yet tested this with *Xc*PrxQ. *Xc*PrxQ was also relatively resistant to hyperoxidation during turnover with H_2O_2 (about 20-fold more robust than PrxI and 7-fold less robust than *St*AhpC), but as expected became highly sensitized to hyperoxidation in the C84S variant in which C_R was absent [28]. Thus *Xc*PrxQ is an active, broad spectrum peroxide reductase with activities representative of this group of Prxs.

Both the dithiol and disulfide forms of the protein provide excellent NMR spectra, and we have completed backbone assignments for both forms [29]. Already, at this stage we have evidence the protein will have interesting dynamics, as while most all residues had clear resonances in the dithiol form, the disulfide form was missing about one-third of the backbone resonances and these were all from residues near parts of the protein that shift during the FF–LU conformational change. It is interesting that the disulfide bond formation, which is usually thought of as a rigidifying event, seems to increase the mobility in this Prx. We were also able to grow crystals of all three recombinant enzymes, including both the dithiol and disulfide forms for

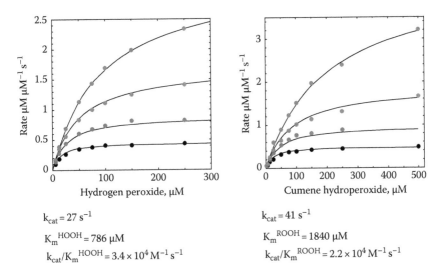

$k_{cat} = 27 \text{ s}^{-1}$

$K_m^{HOOH} = 786 \text{ } \mu M$

$k_{cat}/K_m^{HOOH} = 3.4 \times 10^4 \text{ M}^{-1}\text{s}^{-1}$

$k_{cat} = 41 \text{ s}^{-1}$

$K_m^{ROOH} = 1840 \text{ } \mu M$

$k_{cat}/K_m^{ROOH} = 2.2 \times 10^4 \text{ M}^{-1}\text{s}^{-1}$

FIGURE 6.3 In solution kinetics of *Xanthomonas campestris* PrxQ (*Xc*PrxQ). Shown on the left are bisubstrate kinetic data of *Xc*PrxQ (0.2 μM) turning over with hydrogen peroxide at various concentrations, including *Ec*TrxA at 5 μM (black), 10 μM (red), 20 μM (green), and 40 μM (blue). Curves in black are the results from the global fitting of all four data sets. Data on the right reflected similar data gathered for reaction with cumene hydroperoxide, including *Ec*TrxA at 5 μM (black), 10 μM (red), 20 μM (green), and 50 μM (blue). Kinetic parameters obtained from the global fits are shown beneath each plot. (Reprinted from *Structure*, 24(10), Perkins, A., Parsonage, D., Nelson, K.J., Ogba, O.M., Cheong, P.H., Poole, L.B., and Karplus, P.A., Peroxiredoxin catalysis at atomic resolution, 1668–1678. Copyright 2016, with permission from Elsevier.)

the wild type protein. To our delight, the wild type dithiol crystals yielded data to 1.05 Å resolution and the other forms diffracted to between 1.2 and 1.3 Å resolution.

ANALYSIS OF FF AND LU STRUCTURES OF *Xc*PrxQ

The FF active site structure of the wild type enzyme in the dithiol form gave a clearly defined view of the FF dithiol wild type active site at the highest resolution so far for any Prx. The comparison of this form with its LU counterpart also illustrated the significant conformational changes that occur upon oxidation (Figure 6.4). Briefly, in this subgroup of the PrxQ class, the C_R is located in the α3 helix that follows the C_P-containing α2 helix. After reaction of C_P with the peroxide substrate, the α3 helix undergoes substantial conformational rearrangements, which move the C_R into a new position; unfolding of the first turn of the α2 helix similarly moves the C_P out of the active site to allow formation of a disulfide bond between C_R and C_P [28,30]. Although details vary for Prxs from different classes, particularly with C_R in various other parts of the structure or missing altogether in some Prxs, some kind of a local unfolding process of the C_P-loop is necessary in all Prxs for regeneration of the reduced C_P for another round of catalysis with peroxide.

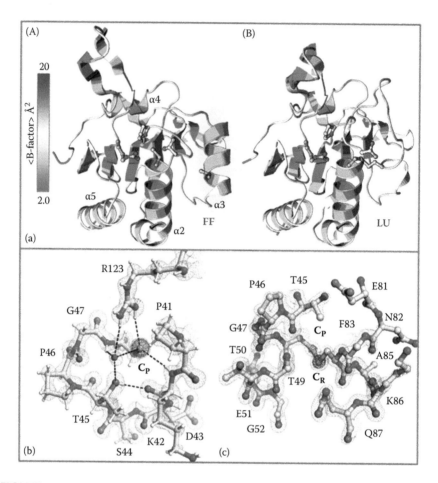

FIGURE 6.4 Fully folded (FF) and locally unfolded (LU) conformations of *Xc*PrxQ. (a) The active FF conformation (A) and the LU disulfide conformation (B) are shown as cartoons colored by local mobility as indicated by B-factor, and with sticks for C_P, C_R and the conserved active site Pro, Thr and Arg. The largest conformation changes involve the helix 3 region that is highlighted in yellow. This structural rearrangement moves C_P away from the active site pocket, disrupting all of its hydrogen bonding interactions, as described by Liao et al. [30]. (b) The active site of the *Xc*PrxQ FF structure with $2F_O–F_C$ density (thin blue mesh $1.5\rho_{rms}$ and pink mesh $3.5\rho_{rms}$). (c) Same as (b) but for the LU $C_P–C_R$ disulfide structure. (Reprinted from *Structure*, 24(10), Perkins, A., Parsonage, D., Nelson, K.J., Ogba, O.M., Cheong, P.H., Poole, L.B., and Karplus, P.A., Peroxiredoxin catalysis at atomic resolution, 1668–1678. Copyright 2016, with permission from Elsevier.)

Focusing in on the question of how well the active site architecture of Prxs is conserved, we compare in Figure 6.5 the FF active site of *Xc*PrxQ, captured at ~1 Å resolution, with the well-defined high-resolution active site structures from representatives of two other groups of Prxs, Prx5 (*Homo sapiens* PrxV, 3MNG at 1.5 Å resolution), and Prx6 (*Aeropyrum pernix* Tpx, 3A2V at 1.7 Å resolution).

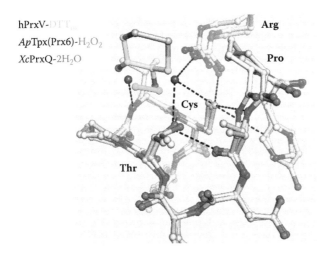

FIGURE 6.5 Overlay of active sites of DTT-bound human PrxV (light blue) and peroxide-bound *Aeropyrum pernix* Tpx (green) with water-bound wild type *Xc*PrxQ (white), with ligands colored cyan (DTT), lime (H_2O_2), and dark gray (waters), respectively. Dashed lines indicate active site interactions for *Xc*PrxQ, including a contact between the CδH of Tyr40 with the thiolate that would also contribute weak electrostatic stabilization to the thiolate. This appears to be a conserved interaction in FF Prx structures and involves either a Tyr or His. (Reprinted from *Structure*, 24(10), Perkins, A., Parsonage, D., Nelson, K.J., Ogba, O.M., Cheong, P.H., Poole, L.B., and Karplus, P.A., Peroxiredoxin catalysis at atomic resolution, 1668–1678. Copyright 2016, with permission from Elsevier.)

Remarkably, despite only having sequence identities of ~20%–30%, all three proteins have all of their active site side chains at nearly identical positions (Figure 6.5). The C_p sulfur is hydrogen bonding to a backbone NH (of Lys42) and the conserved active site Arg (Arg123 in *Xc*PrxQ), while the conserved Thr (Thr45) is donating a hydrogen bond to a backbone carbonyl (also from Lys42) as well as (presumably) accepting a hydrogen bond from a bound water molecule observable in the active site of this structure. This and a second active site water (gray spheres in Figure 6.5) occupy positions similar to the observed binding sites of the oxygen atoms of bound H_2O_2 and the diol of oxidized 1,4-dithiothreitol captured in the structures of *Ap*Tpx and *Hs*PrxV, respectively.

These bound atoms help identify locations and interactions with active site side chains that represent substrate and product locations, including in this set the bona fide H_2O_2 substrate of *Ap*Tpx. The contributions of the conserved Arg and Thr interacting with the peroxide substrate, as well as the Arg-to-C_p-S$^-$ interaction, have previously been described by our group as promoting catalysis in Prxs by stabilization of the transition state (as the O–O bond of the peroxide is breaking and the O–S bond of the C_p-sulfenate is forming) [32]. We further noted that the geometries of these H-bonding interactions appear to be such that they would become optimized in the transition state [32]. Comparing the structure of the authentic dithiol FF active site with that seen for the C48S mutant of *Xc*PrxQ shows small shifts just as were seen for the equivalent structures of *St*AhpC [33], reinforcing the conclusions we made then that even such a subtle mutation as $C_P \rightarrow$ Ser changes the thermodynamic stability of the FF active site, and so it is important to realize that $C_P \rightarrow$ Ser mutants are not perfect models for the native FF active site.

HIGH-RESOLUTION STRUCTURES ALONG THE REACTION PATH OF XcPrxQ

In order to obtain snapshots of the protein structure at various stages of catalysis, crystals of wild type XcPrxQ were soaked with hydrogen peroxide, cumene hydroperoxide, or t-butyl hydroperoxide for various times and in various pH buffers, then plunged into liquid nitrogen to stop the reaction and observe partially reacted active sites. From a large set of solved structures, a series of nine ~1 Å resolution structures were refined (named FF0 through FF8, Figure 6.6a) that provide detailed information about the transition from the substrate-ready form (the thiolate) through reaction to yield first the sulfenate (SO⁻, from the first catalytic step), then the sulfinate form (SO₂⁻, from hyperoxidation) of C_P within the active site. A tenth structure, designated FF9, represents the fully occupied sulfinate form at pH 7. With these structures, important new insights could be obtained because even the partially occupied representatives (from ~0.1 to 0.5 occupancy) are well defined at this resolution, with coordinate uncertainties <0.05 Å.

With increasing electron density up to ~0.5 occupancy for the sulfenate oxygen, represented by structures FF1 through FF3, the first surprise was the large $C\beta$–$S\gamma$–$O\delta$ bond angle (153°) and short $S\gamma$–$O\delta$ bond length (1.38 Å) observed for the sulfenate (Figure 6.6b). These differ significantly from the expected bond angle of ~110° [34] and a bond length that one would assume would be between that of a disulfide (S-S distance at ~2 Å) and the O–O bond of a peroxide (at ~1.5 Å) [32]. This location for the sulfenate oxygen also perturbs the geometry of the nearby conserved Pro ring (Pro41) creating a C–H hydrogen bond between Pro-CδH and the SO⁻ oxygen. We inferred from this very high energy geometry that a relaxed C_P–SO⁻ does not fit well into the XcPrxQ FF active site.

Based on this observation, we propose that in solution, C_P–SO⁻ formation results in an incompatibility with the FF active site and this shifts the thermodynamics of the C_P-loop to strongly favor the LU conformation, which in turn facilitates disulfide bond formation and promotes progression through the catalytic cycle. We use the phrase *shifts the thermodynamic equilibrium* rather than *triggers the conformational change* because multiple lines of evidence indicate that the FF–LU conformations are in rapid equilibrium at all times (e.g., see [23,35]). According to this proposal, an FF conformation of the C_P–SO⁻ form would never occur to a high degree in solution as it is very unstable; it is only observed in the crystal because the crystal packing interactions provide enough energy to block the local unfolding and stabilize the high energy C_P–SO⁻ configuration.

To better understand the unusual C–S–O⁻ bond angle and length, computational chemistry calculations in the gas phase were conducted with several model methyl-SO(H) species to evaluate the energetics of bond angle contortions (Figure 6.6c). While the energetic penalty for adopting the 153° bond angle was quite high for just the methyl sulfenate (~40–50 kcal/mol), related species with additional hydrogens (e.g., methyl-SH₂O⁻) dropped this penalty to ~9 kcal/mol. This suggests that the

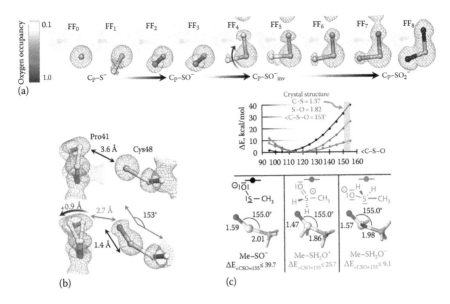

FIGURE 6.6 Crystallographic snapshots of *Xc*PrxQ undergoing peroxide-mediated oxidation and depictions of unusual Cys-sulfenate geometry. (a) C_P is shown with occupancies of its oxygen adducts indicated by color (from light to dark red). $2F_O\text{–}F_C$ density is shown (blue contoured at $1.0\rho_{rms}$ and for the sulfenate oxygen in FF1 olive contoured at $0.3\rho_{rms}$). Density near the top of the image in structures FF4–FF7 is from a partially occupied Arg conformation that occurs as the protein converts from the sulfenate to sulfinate state. (b) Interaction of the C_P-thiolate with Pro41 (upper image) compared with that of the C_P-sulfenate (at 0.5 occupancy and labeled with its unusual bond angle and distance reported; lower image). A brown arrow shows the shift of Pro41-Cγ accommodating the sulfenate formation by relieving a steric clash (red double-headed arrow) to increase the Pro41-Cγ to sulfenate oxygen distance to 3.6 Å. (c) Gas phase quantum mechanical calculations of methyl SO⁻, SH₂O⁺, and SH₂O⁻ species showing their energies as a function of their C–S–O bond angle (upper panel) as well as the optimized geometries and energies for the structures constrained to have a C–S–O angle of 155° (lower panel). (Reprinted from *Structure*, 24(10), Perkins, A., Parsonage, D., Nelson, K.J., Ogba, O.M., Cheong, P.H., Poole, L.B., and Karplus, P.A., Peroxiredoxin catalysis at atomic resolution, 1668–1678. Copyright 2016, with permission from Elsevier.)

universally conserved hydrogen bond interactions of the sulfenate with Arg123-NH and a backbone amide, described here, may serve in a similar way as the hydrogens of methyl-SH₂O⁻ to lower the energetic penalty expected for the Cys-sulfenate to adopt such an unusual structure.

HIGH-RESOLUTION STRUCTURES ON THE PATH TOWARD HYPEROXIDATION OF *Xc*PrxQ

As noted, our FF crystals of *Xc*PrxQ cannot undergo the transitions required for disulfide bond formation. In this sense the crystal is like a large, noncovalent, conformation-specific inhibitor that binds to the FF conformation and stabilizes it.

This has the effect of strongly promoting hyperoxidation, and in the crystals the sulfenate form is fully converted to sulfinate upon further incubation with peroxide. Structures FF4 through FF8 have therefore fortuitously provided insights into hyperoxidation of XcPrxQ.

As shown in Figure 6.6a, structures FF4 through FF7 provide evidence that, after the initial oxidation, the oxygen of the sulfenate rotates in (toward the Arg), providing room for the binding of another molecule of peroxide. This inverted position of the oxygen, which had been suggested as a necessary intermediate [36] but not yet observed, allows for the relaxation of the Pro side chain back to the exo rotamer but also shifts the conserved Arg side chain out of the active site pocket. Once hyperoxidation is complete, placing two oxygens on the $C_P-S\gamma$, the Arg side chain swings into a third location where it forms a strong, bidentate salt bridge with the negatively charged sulfinate oxygens, similar to the structure at the active site originally reported for the hyperoxidized form of human PrxII, with pdb identifier 1QMV [37]. Although further oxidation to a sulfonic acid is theoretically possible, the active site would need to be able to rotate the oxygens of $-SO_2^-$ yet again. This apparently does not happen in these XcPrxQ crystals as no evidence of sulfonic acid was observed even after extensive peroxide soaks, and this transition may also be uncommon even in solution for most Prxs [36].

TUNING OF HYPERSENSITIVITY THROUGH STRUCTURAL FEATURES

In the case presented, trapping the FF conformation in a crystal form that effectively lowered the rate of C_P-C_R disulfide bond formation to zero yielded a protein for which hyperoxidation was 100% favored. In solution, the local unfolding and hyperoxidation pathways compete in Prxs, with hyperoxidation being enhanced kinetically by increasing peroxide concentrations (since it is a bimolecular reaction). What is clear at this point is that each individual Prx has evolved its own characteristic sensitivity toward hyperoxidation that reflects the intersection of structural and dynamic factors with active site chemistry. For example, hyperoxidation sensitivity among the human PrxI, PrxII, and PrxIII varies a great deal, with the order of sensitivity of PrxII > PrxI >> PrxIII ([26,38] and personal communication with Jesalyn Bolduc and Todd Lowther), even though their sequences are quite similar to one another. As noted here, the Winterbourn group contributed clear evidence that modulation of the rate of disulfide bond formation is the main factor influencing hyperoxidation sensitivity, rather than variations in the rate constant for the hyperoxidation step. For instance, human PrxII and PrxIII both have rate constants for hyperoxidation of ~12,000 M^{-1} s^{-1}, yet their hyperoxidation sensitivities during turnover differ by more than a factor of 10 [26].

A few features that influence sensitivity have already been described in the literature. The first of these was a conserved GGLG motif and a C-terminal α-helix containing a conserved YF motif, which pack around C_P in the FF conformation in eukaryotic Prx1 group proteins. These were proposed to promote sensitivity of these Prxs toward hyperoxidation by H_2O_2 by globally stabilizing the FF conformation and minimizing the concentration of the locally unfolded population [9]. This proposal was experimentally confirmed by swapping of the C-terminal regions of one sensitive and one robust Prx from *Schistosoma mansoni*, which showed that the susceptibility of these proteins to hyperoxidation was indeed conferred by the C-terminal tail [39]. The importance of the C-terminal region was further confirmed as additional interconversions of internal and C-terminal tail residues between human PrxII and PrxIII [38], and of other residues between PrxI and PrxII (Bolduc and Lowther, personal communication), did yield some predicted changes in sensitivity, with the effects as expected being more modest than was caused by the wholesale removal of the C-terminal tail. Furthermore, as noted, studies with a number of mutants of *St*AhpC show that the FF–LU equilibrium can be perturbed by even single, relatively conservative amino acid substitutions such as $C_P \rightarrow$ Ser or $C_R \rightarrow$ Ala or Ser [33] and others [40].

What this study adds to our understanding is two additional more subtle factors that can strongly influence the sensitivity to hyperoxidation. The first of these is the ability of the FF active site to accommodate the C_P–SO^- form of the enzyme. If the C_P–SO^- cannot be well-accommodated in the FF active site (as we infer is the case for *Xc*PrxQ), then its formation will promote local unfolding and the enzyme will (all other things being equal) be less sensitive to hyperoxidation. The second independent factor is the ability of the FF active site to accommodate an inverted C_P–SO^- side chain conformation in which the oxygen is rotated out of the active site pocket to a more buried position in which it interacts with the conserved active site Arg. Prx active sites that accommodate this inverted conformation well, without the Arg being displaced, will in this regard be more susceptible to hyperoxidation, as a second peroxide can then bind in the active site and be activated for attack. But Prx active sites that do not accommodate the inverted conformation without disruption of the FF active site interactions will be less sensitive to hyperoxidation.

PERSPECTIVES ON Prx REGULATION AND REDOX SIGNALING

A much discussed issue in redox biology is the role that peroxidases, and especially abundant and widespread Prxs, play in modulating or transducing signals from oxidants including H_2O_2 and lipoxygenase products [6,17,41–43]. H_2O_2 is especially key since it can come from several sources, including damaged Fe–S centers or mitochondrial electron transport chain leakage, and is the major product generated (directly or through spontaneous or catalyzed dismutation of superoxide) from NADPH oxidase complexes that are assembled and/or activated in response to extracellular signals [18]. The second order rate constants for the reaction of H_2O_2 with specific cysteine residues on target proteins (i.e., signaling related) do not appear to be sufficient to allow for their oxidation in the context of an environment replete with reactive Prx proteins, creating a conundrum that has been especially well described by Christine Winterbourn in her seminal paper regarding how the chemistry and biology of reactive oxygen species might be reconciled [6].

Yet if sites or hotspots of H_2O_2 generation in cells exist near which target proteins are localized, for example, in signaling complexes organized by adapter or scaffolding proteins, the global kinetic competition can be outstripped by proximity influences [44]. Moreover, Prx activity is not always fully on. This can, in the most simple case, be due to the accumulation of the Prx disulfide form when the rapid turnover of Prxs in the presence of substantial amounts of peroxide cannot be matched by the rate of regeneration of reduced Prxs through the reductant pool (primarily reduced thioredoxin). If one adds to that the propensity, à la floodgate hypothesis, for Prx activity to be turned down or off by post-translational modifications such as phosphorylation or hyperoxidation, then it can be seen that the presence of Prxs as chief competitors for H_2O_2 may not always be a barrier to the direct oxidation of other proteins during signaling.

In fact, concrete examples also exist for which, as part of normal non-stress-related physiology, the direct oxidation of downstream targets appears to be promoted by localized H_2O_2 buildup due to similarly localized Prx inactivation. One is the phosphorylation of PrxI (on Tyr194) by Src-family kinases localized near membrane-bound signaling complexes formed as a result of growth factor stimulation (observed in both platelet-derived and epithelial growth factor signaling), which provides requisite buildup of H_2O_2 to cause inactivation of associated protein tyrosine phosphatases and thus produce a stronger downstream signal [45]. Another is the phosphorylation of centrosome-associated PrxI (on Thr90) by cyclin-dependent kinase 1 during mitosis, an inactivating modification required to allow localized buildup of H_2O_2 at the centrosome to inactivate centrosome-bound phosphatases such as Cdc14B, enabling cell cycle progression [46]. A third is a report from the Rhee lab of a physiological role for PrxIII hyperoxidation in a negative feedback loop that coordinates diurnal cycles of corticosteroid synthesis [47,48]. They demonstrated a choreographed dance whereby a leaky cytochrome P450 (CYP11B1) generates H_2O_2 during corticosterone production. This H_2O_2 inactivates PrxIII via hyperoxidation and builds up sufficiently to escape the mitochondria and trigger stress-linked pathways that turn off steroidogenesis and promote import of sulfiredoxin into mitochondria to resurrect the PrxIII to be functional until the next circadian cycle of corticosteroid synthesis. Interesting about this scenario is that PrxIII, as noted, is actually more resilient toward hyperoxidation than PrxI and PrxII.

Also, another proven signaling mode in which highly reactive Prxs take center stage is when they serve as the conduit through which oxidizing equivalents from H_2O_2 flow [49]. The best demonstration of this redox-relay phenomenon so far has been the inhibitory oxidation in engineered HEK-293 cells of STAT3 through direct interaction with PrxII. In this case, exogenously applied H_2O_2 or the immunomodulatory molecules oncostatin M or interleukin-6 stimulate the formation of transient Prx2-STAT3 complexes to generate high molecular weight, inactive forms of STAT3 [50]. The scenario of having the oxidation of the target protein mediated by an enzyme is of course very attractive in that it can bring a high degree of specificity to the process in terms of both the proteins that will undergo oxidation and the site where that oxidation will occur [42,43,50,51].

Clearly much is still to be learned about the roles Prxs play in peroxide-related signaling, especially in normal non-stress-related physiology [52]. But, in closing, one interesting contrast between the floodgate and redox relay roles of Prxs in promoting peroxide signaling is that hyperoxidation impacts them in completely opposite ways. In the floodgate model, hyperoxidation promotes signal transduction, and indeed the hypothesis was generated to provide a rationale for why hyperoxidation appears to have been selected for in eukaryotes. In contrast, redox relay activity of Prxs would be completely shut off by hyperoxidation as the passing on of redox equivalents requires the Prx to be in the sulfenic acid or disulfide form. Thus, in pathways in which Prx are functioning as redox relays, the hypersensitivity of Prxs could play a role as an off switch. In this respect, it is worth noting that while PrxII is the Prx expected to be most active in redox relays according to the many disulfide-linked complexes it makes (compared with PrxI, [50]), it is also most hyperoxidation sensitive and will thus be most readily switched off under conditions not only of exposure to exogenous or non-natural stressors, but also with aging and disease states (see, e.g., [53]).

The extent to which Prx hyperoxidation enhances or hinders signaling has yet to be elucidated, but it is apparent that Prxs play important roles in signaling. Indeed, one of the most powerful pieces of evidence that Prxs play important roles in signaling and are not simply redundant antioxidants is that the PrxI knockout mouse has the dramatic phenotype of being riddled with cancer at six to nine months of age [8,54]. Given the contrasting predictions for how hyperoxidation sensitivity will impact signaling through floodgate-like or relay-like mechanisms, insights may ultimately be gained by examining the phenotype of transgenic mice in which the wild type sensitive PrxI and/or PrxII are genetically exchanged for robust versions of these enzymes. Addressing these questions in the future will not be a simple task; as Prxs appear to be hubs of signaling events, straightforward conclusions may be difficult to draw from studies of such complex biological systems. Thus, this is likely to be an active area of investigation well into the future requiring innovative and sensitive approaches to further clarify the significance of each of these mechanisms.

SUMMARY

Rapid reaction kinetic studies and high-resolution crystallographic analyses of model Prxs are bringing enhanced clarity to the atomic-level structural and dynamic features of Prxs that make them such effective catalysts as antioxidants, and help tune their regulatory properties in terms of hyperoxidation sensitivity. The new, high-resolution structures will be particularly ideal as starting models for future computational studies of peroxidation and hyperoxidative regulation, and for further exploration of dynamic properties of this monomeric PrxQ model protein by NMR. As Prxs act as hubs of peroxide reactivity modulating H_2O_2-associated cell signaling, a better understanding of these proteins will be critical to gaining a clearer view of the myriad pathways modulated by this important small messenger molecule.

ACKNOWLEDGMENTS

Thanks to all the researchers who contributed their skills and efforts to this work, including (but not limited to) Arden Perkins, Derek Parsonage, Kimberly Nelson, Andrea Hall, Zachary Wood, and Holly Ellis, who as graduate students and research associates have been lead authors helping drive the projects forward. For being a wonderfully collegial and encouraging collaborator and friendly competitor helping to keep each other honest and challenging each other with questions that drive us all to dig deeper and sort out complexities and get to answers and explanations that will stand the test of time, we are indebted to Christine Winterbourn. She continues to be an inspiration to us all to aspire to contributing high quality work in this field.

142

Hydrogen Peroxide Metabolism in Health and Disease

REFERENCES

1. Dubuisson, M., D. Vander Stricht, A. Clippe, F. Etienne, T. Nauser, R. Kissner, W. H. Koppenol, J. F. Rees, and B. Knoops. 2004. Human peroxiredoxin 5 is a peroxynitrite reductase. *FEBS Lett* 571 (1–3):161–165. doi: 10.1016/j.febslet.2004.06.080.
2. Manta, B., M. Hugo, C. Ortiz, G. Ferrer-Sueta, M. Trujillo, and A. Denicola. 2009. The peroxidase and peroxynitrite reductase activity of human erythrocyte peroxiredoxin 2. *Arch Biochem Biophys* 484 (2):146–154. doi: 10.1016/j.abb.2008.11.017.
3. Parsonage, D., D. S. Youngblood, G. N. Sarma, Z. A. Wood, P. A. Karplus, and L. B. Poole. 2005. Analysis of the link between enzymatic activity and oligomeric state in AhpC, a bacterial peroxiredoxin. *Biochemistry* 44 (31):10583–10592. doi: 10.1021/bi050448i.
4. Peskin, A. V., F. M. Low, L. N. Paton, G. J. Maghzal, M. B. Hampton, and C. C. Winterbourn. 2007. The high reactivity of peroxiredoxin 2 with H(2)O(2) is not reflected in its reaction with other oxidants and thiol reagents. *J Biol Chem* 282 (16):11885–11892. doi: 10.1074/jbc.M700339200.
5. Reyes, A. M., D. S. Vazquez, A. Zeida, M. Hugo, M. D. Pineyro, M. I. De Armas, D. Estrin, R. Radi, J. Santos, and M. Trujillo. 2016. PrxQ B from Mycobacterium tuberculosis is a monomeric, thioredoxin-dependent and highly efficient fatty acid hydroperoxide reductase. *Free Radic Biol Med* 101:249–260. doi: 10.1016/j.freeradbiomed.2016.10.005.
6. Winterbourn, C. C. 2008. Reconciling the chemistry and biology of reactive oxygen species. *Nat Chem Biol* 4 (5):278–286. doi: 10.1038/nchembio.85.
7. Winterbourn, C. C. and D. Metodiewa. 1999. Reactivity of biologically important thiol compounds with superoxide and hydrogen peroxide. *Free Radic Biol Med* 27 (3–4):322–328.
8. Perkins, A., L. B. Poole, and P. A. Karplus. 2014. Tuning of peroxiredoxin catalysis for various physiological roles. *Biochemistry* 53 (49):7693–7705. doi: 10.1021/bi5013222.
9. Wood, Z. A., L. B. Poole, and P. A. Karplus. 2003. Peroxiredoxin evolution and the regulation of hydrogen peroxide signaling. *Science* 300 (5619):650–653. doi: 10.1126/science.1080405.
10. Jang, H. H., S. Y. Kim, S. K. Park, H. S. Jeon, Y. M. Lee, J. H. Jung, S. Y. Lee et al. 2006. Phosphorylation and concomitant structural changes in human 2-Cys peroxiredoxin isotype I differentially regulate its peroxidase and molecular chaperone functions. *FEBS Lett* 580 (1):351–355. doi: 10.1016/j.febslet.2005.12.030.
11. Lee, W., K. S. Choi, J. Riddell, C. Ip, D. Ghosh, J. H. Park, and Y. M. Park. 2007. Human peroxiredoxin 1 and 2 are not duplicate proteins: The unique presence of CYS83 in Prx1 underscores the structural and functional differences between Prx1 and Prx2. *J Biol Chem* 282 (30):22011–22022. doi: 10.1074/jbc.M610330200.
12. Saccoccia, F., P. Di Micco, G. Boumis, M. Brunori, I. Koutris, A. E. Miele, V. Morea et al. 2012. Moonlighting by different stressors: Crystal structure of the chaperone species of a 2-Cys peroxiredoxin. *Structure* 20 (3):429–439. doi: 10.1016/j.str.2012.01.004.
13. Hoyle, N. P. and J. S. O'Neill. 2015. Oxidation-reduction cycles of peroxiredoxin proteins and nontranscriptional aspects of timekeeping. *Biochemistry* 54 (2):184–193. doi: 10.1021/bi5008386.
14. Putker, M. and J. S. O'Neill. 2016 Reciprocal control of the circadian clock and cellular redox state—A critical appraisal. *Mol Cells* 39 (1):6–19. doi: 10.14348/molcells.2016.2323.
15. Biteau, B., J. Labarre, and M. B. Toledano. 2003. ATP-dependent reduction of cysteine-sulphinic acid by *S. cerevisiae* sulphiredoxin. *Nature* 425 (6961):980–984. doi: 10.1038/nature02075.
16. Lowther, W. T. and A. C. Haynes. 2011. Reduction of cysteine sulfinic acid in eukaryotic, typical 2-Cys peroxiredoxins by sulfiredoxin. *Antioxid Redox Signal* 15 (1):99–109. doi: 10.1089/ars.2010.3564.

17. Latimer, H. R. and E. A. Veal. 2016. Peroxiredoxins in regulation of MAPK signalling pathways; sensors and barriers to signal transduction *Mol Cells* 39 (1):40–45. doi: 10.14348/molcells.2016.2327.
18. Rhee, S. G., H. A. Woo, and D. Kang 2017. Signaling via oxidation of proteinaceous cysteine by H_2O_2 and the role of peroxiredoxin in the local control of H_2O_2 concentration and transduction of the H_2O_2 signal. *Antiox Red Sign.* (Epub ahead of print. June 6. doi: 10.1089/ars.2017.7167.)
19. Park, M. H., M. Jo, Y. R. Kim, C. K. Lee, and J. T. Hong. 2016. Roles of peroxiredoxins in cancer, neurodegenerative diseases and inflammatory diseases. *Pharmacol Ther* 163:1–23. doi: 10.1016/j.pharmthera.2016.03.018.
20. Nelson, K. J., S. T. Knutson, L. Soito, C. Klomsiri, L. B. Poole, and J. S. Fetrow. 2011. Analysis of the peroxiredoxin family: Using active-site structure and sequence information for global classification and residue analysis. *Proteins* 79 (3):947–964. doi: 10.1002/prot.22936.
21. Poole, L. B. and K. J. Nelson. 2016. Distribution and features of the six classes of peroxiredoxins. *Mol Cells* 39 (1):53–59. doi: 10.14348/molcells.2016.2330.
22. Perkins, A., K. J. Nelson, D. Parsonage, L. B. Poole, and P. A. Karplus. 2015. Peroxiredoxins: Guardians against oxidative stress and modulators of peroxide signaling. *Trends Biochem Sci* 40 (8):435–445. doi: 10.1016/j.tibs.2015.05.001.
23. Wood, Z. A., L. B. Poole, R. R. Hantgan, and P. A. Karplus. 2002. Dimers to doughnuts: Redox-sensitive oligomerization of 2-cysteine peroxiredoxins. *Biochemistry* 41 (17):5493–5504.
24. Parsonage, D., P. A. Karplus, and L. B. Poole. 2008. Substrate specificity and redox potential of AhpC, a bacterial peroxiredoxin. *Proc Natl Acad Sci USA* 105 (24):8209–8214. doi: 10.1073/pnas.0708308105.
25. Parsonage, D., K. J. Nelson, G. Ferrer-Sueta, S. Alley, P. A. Karplus, C. M. Furdui, and L. B. Poole. 2015. Dissecting peroxiredoxin catalysis: Separating binding, peroxidation, and resolution for a bacterial AhpC. *Biochemistry* 54 (7):1567–1575. doi: 10.1021/bi501515w.
26. Peskin, A. V., N. Dickerhof, R. A. Poynton, L. N. Paton, P. E. Pace, M. B. Hampton, and C. C. Winterbourn. 2013. Hyperoxidation of peroxiredoxins 2 and 3: Rate constants for the reactions of the sulfenic acid of the peroxidatic cysteine. *J Biol Chem* 288 (20):14170–14177. doi: 10.1074/jbc.M113.460881.
27. Trujillo, M., A. Clippe, B. Manta, G. Ferrer-Sueta, A. Smeets, J. P. Declercq, B. Knoops, and R. Radi. 2007. Pre-steady state kinetic characterization of human peroxiredoxin 5: Taking advantage of Trp84 fluorescence increase upon oxidation. *Arch Biochem Biophys* 467 (1):95–106. doi: 10.1016/j.abb.2007.08.008.
28. Perkins, A., D. Parsonage, K. J. Nelson, O. M. Ogba, P. H. Cheong, L. B. Poole, and P. A. Karplus. 2016. Peroxiredoxin catalysis at atomic resolution. *Structure* 24 (10):1668–1678. doi: 10.1016/j.str.2016.07.012.
29. Buchko, G. W., A. Perkins, D. Parsonage, L. B. Poole, and P. A. Karplus. 2016. Backbone chemical shift assignments for *Xanthomonas campestris* peroxiredoxin Q in the reduced and oxidized states: A dramatic change in backbone dynamics. *Biomol NMR Assign* 10 (1):57–61. doi: 10.1007/s12104-015-9637-8.
30. Liao, S. J., C. Y. Yang, K. H. Chin, A. H. Wang, and S. H. Chou. 2009. Insights into the alkyl peroxide reduction pathway of *Xanthomonas campestris* bacterioferritin comigratory protein from the trapped intermediate-ligand complex structures. *J Mol Biol* 390 (5):951–966. doi: 10.1016/j.jmb.2009.05.030.
31. Reeves, S. A., D. Parsonage, K. J. Nelson, and L. B. Poole. 2011. Kinetic and thermodynamic features reveal that *Escherichia coli* BCP is an unusually versatile peroxiredoxin. *Biochemistry* 50 (41):8970–8981. doi: 10.1021/bi200935d.
32. Hall, A., D. Parsonage, L. B. Poole, and P. A. Karplus. 2010. Structural evidence that peroxiredoxin catalytic power is based on transition-state stabilization. *J Mol Biol* 402 (1):194–209. doi: 10.1016/j.jmb.2010.07.022.

33. Perkins, A., K. J. Nelson, J. R. Williams, D. Parsonage, L. B. Poole, and P. A. Karplus. 2013. The sensitive balance between the fully folded and locally unfolded conformations of a model peroxiredoxin. *Biochemistry* 52 (48):8708–8721. doi: 10.1021/bi4011573.

34. Engh, R. A. and R. Huber. 1991. Accurate bond and angle parameters for x-ray protein structure refinement. *Acta Crystallogr Sect A Found Crystallogr* 47:392–400.

35. Perkins, A., M. C. Gretes, K. J. Nelson, L. B. Poole, and P. A. Karplus. 2012. Mapping the active site helix-to-strand conversion of CxxxxC peroxiredoxin Q enzymes. *Biochemistry* 51 (38):7638–7650. doi: 10.1021/bi301017s.

36. Sarma, G. N., C. Nickel, S. Rahlfs, M. Fischer, K. Becker, and P. A. Karplus. 2005. Crystal structure of a novel Plasmodium falciparum 1-Cys peroxiredoxin. *J Mol Biol* 346 (4):1021–1034. doi: 10.1016/j.jmb.2004.12.022.

37. Schroder, E., J. A. Littlechild, A. A. Lebedev, N. Errington, A. A. Vagin, and M. N. Isupov. 2000. Crystal structure of decameric 2-Cys peroxiredoxin from human erythrocytes at 1.7 A resolution. *Structure* 8 (6):605–615.

38. Haynes, A. C., J. Qian, J. A. Reisz, C. M. Furdui, and W. T. Lowther. 2013. Molecular basis for the resistance of human mitochondrial 2-Cys peroxiredoxin 3 to hyperoxidation. *J Biol Chem* 288 (41):29714–29723. doi: 10.1074/jbc.M113.473470.

39. Sayed, A. A. and D. L. Williams. 2004. Biochemical characterization of 2-Cys peroxiredoxins from *Schistosoma mansoni*. *J Biol Chem* 279 (25):26159–26166. doi: 10.1074/jbc.M401748200.

40. Nelson, K. J., A. Perkins, A. E. D. Van Swearingen, S. Hartman, A. E. Brereton, D. Parsonage, F. R. Salsbury Jr., P. A. Karplus, and L. B. Poole. 2017. Experimentally dissecting the origins of peroxiredoxin catalysis. *Antiox Red Sign* (Epub ahead of print. April 4. doi: 10.1089/ars.2016.6922.)

41. Cordray, P., K. Doyle, K. Edes, P. J. Moos, and F. A. Fitzpatrick. 2007. Oxidation of 2-Cys-peroxiredoxins by arachidonic acid peroxide metabolites of lipoxygenases and cyclooxygenase-2. *J Biol Chem* 282 (45):32623–32629. doi: 10.1074/jbc.M704369200.

42. Forman, H. J., M. Maiorino, and F. Ursini. 2010. Signaling functions of reactive oxygen species. *Biochemistry* 49 (5):835–842. doi: 10.1021/bi9020378.

43. Randall, L. M., G. Ferrer-Sueta, and A. Denicola. 2013. Peroxiredoxins as preferential targets in H_2O_2-induced signaling. *Methods Enzymol* 527:41–63. doi: 10.1016/b978-0-12-405882-8.00003-9.

44. Heppner, D. E., Y. M. Janssen-Heininger, and A. van der Vliet. 2017. The role of sulfenic acids in cellular redox signaling: Reconciling chemical kinetics and molecular detection strategies. *Arch Biochem Biophys* 616:40–46. doi: 10.1016/j.abb.2017.01.008.

45. Woo, H. A., S. H. Yim, D. H. Shin, D. Kang, D. Y. Yu, and S. G. Rhee. 2010. Inactivation of peroxiredoxin I by phosphorylation allows localized H(2)O(2) accumulation for cell signaling. *Cell* 140 (4):517–528. doi: 10.1016/j.cell.2010.01.009.

46. Lim, J. M., K. S. Lee, H. A. Woo, D. Kang, and S. G. Rhee. 2015. Control of the pericentrosomal H_2O_2 level by peroxiredoxin I is critical for mitotic progression. *J Cell Biol* 210 (1):23–33. doi: 10.1083/jcb.201412068.

47. Kil, I. S., S. K. Lee, K. W. Ryu, H. A. Woo, M. C. Hu, S. H. Bae, and S. G. Rhee. 2012. Feedback control of adrenal steroidogenesis via H_2O_2-dependent, reversible inactivation of peroxiredoxin III in mitochondria. *Mol Cell* 46 (5):584–594. doi: 10.1016/j.molcel.2012.05.030.

48. Kil, I. S., K. W. Ryu, S. K. Lee, J. Y. Kim, S. Y. Chu, J. H. Kim, S. Park, and S. G. Rhee. 2015. Circadian oscillation of sulfiredoxin in the mitochondria. *Mol Cell* 59 (4):651–663. doi: 10.1016/j.molcel.2015.06.031.

49. Herrmann, J. M., K. Becker, and T. P. Dick. 2015. Highlight: Dynamics of thiol-based redox switches. *Biol Chem* 396 (5):385–387. doi: 10.1515/hsz-2015-0135.

50. Sobotta, M. C., W. Liou, S. Stocker, D. Talwar, M. Oehler, T. Ruppert, A. N. Scharf, and T. P. Dick. 2015. Peroxiredoxin-2 and STAT3 form a redox relay for H2O2 signaling. *Nat Chem Biol* 11 (1):64–70. doi: 10.1038/nchembio.1695.
51. Jarvis, R. M., S. M. Hughes, and E. C. Ledgerwood. 2012. Peroxiredoxin 1 functions as a signal peroxidase to receive, transduce, and transmit peroxide signals in mammalian cells. *Free Radic Biol Med* 53 (7):1522–1530. doi: 10.1016/j.freeradbiomed.2012.08.001.
52. Karplus, P. A. and L. B. Poole. 2012. Peroxiredoxins as molecular triage agents, sacrificing themselves to enhance cell survival during a peroxide attack. *Mol Cell* 45 (3):275–278. doi: 10.1016/j.molcel.2012.01.012.
53. Collins, J. A., S. T. Wood, K. J. Nelson, M. A. Rowe, C. S. Carlson, S. Chubinskaya, L. B. Poole, C. M. Furdui, and R. F. Loeser. 2016. Oxidative stress promotes peroxiredoxin hyperoxidation and attenuates pro-survival signaling in aging chondrocytes. *J Biol Chem* 291 (13):6641–6654. doi: 10.1074/jbc.M115.693523.
54. Neumann, C. A., D. S. Krause, C. V. Carman, S. Das, D. P. Dubey, J. L. Abraham, R. T. Bronson, Y. Fujiwara, S. H. Orkin, and R. A. Van Etten. 2003. Essential role for the peroxiredoxin Prdx1 in erythrocyte antioxidant defence and tumour suppression. *Nature* 424 (6948):561–565. doi: 10.1038/nature01819.

Section II

Biological Sources of H_2O_2

7 Mitochondrial Hydrogen Peroxide as a Redox Signal

Elizabeth C. Hinchy and Michael P. Murphy

CONTENTS

INTRODUCTION

Mitochondria are at the heart of cell metabolism and contribute to many aspects of the life and death of the cell [1–3]. Mitochondria are the site of oxidative phosphorylation; consequently, they are the major source of ATP in most eukaryotic cells and hence any defects to these processes impact cell survival [1,2] (Figure 7.1). The metabolic pathways, notably Krebs cycle and fatty acid oxidation, that provide electrons to the respiratory chain are vital for oxidative phosphorylation, but they affect many other processes beyond the supply of ATP [1–5]. The metabolic roles of mitochondria also include many other biosynthetic processes, such as the assembly of iron-sulfur centers and heme biosynthesis [6,7], and may also have regulatory roles in the cytosol, such as in hypoxia sensing and in generating epigenetic modifications to the nuclear genome [8–10].

The many roles carried out by the mitochondrion are closely integrated into the function of the cell. For example, the mitochondrion is intimately associated with the endoplasmic reticulum [11], which assists in determining the fission of the organelle, and also the movement of calcium from the endoplasmic reticulum to the cytosol and from there into the mitochondrial matrix as a way of modulating mitochondrial ATP production [12,13]. Another aspect of the cell where mitochondria are intimately involved is in cell death [2,3,12]. This is because the mitochondrial pathway of apoptosis involves the release of factors such as cytochrome c from the intermembrane

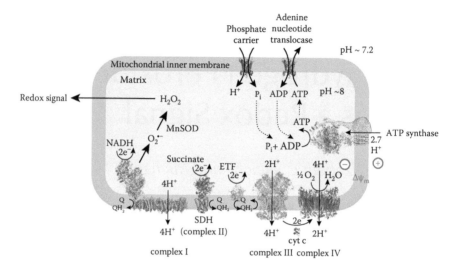

FIGURE 7.1   This figure shows a mitochondrion with the respiratory chain generating a mitochondrial membrane potential ($\Delta\psi_m$). The $\Delta\psi_m$ is used to synthesize ATP. The production of superoxide ($O_2^{\cdot-}$) by complex I is converted to hydrogen peroxide by Mn superoxide dismutase (MnSOD) and can then diffuse from the mitochondria as a redox signal. Cyt c, cytochrome c; ETF, electron transfer flavoprotein; SDH, succinate dehydrogenase.

space as a critical step in committing the cell to activating the apoptotic cell death program [14]. As well as regulated cell death, the central role of mitochondria in ATP production means that damage to the organelle will lead on to necrotic cell death, due to the lack of ATP preventing the cell from sustaining ion gradients. During necrotic cell death, the induction of the mitochondrial permeability transition pore (MPTP) is a major player, committing the cell to a rapid death [15–17].

Mitochondria are also a major source of reactive oxygen species (ROS) within the cell [4,18,19]. These ROS come from the respiratory chain, primarily in the form of superoxide, which then goes on to form hydrogen peroxide (H_2O_2) [4] (Figure 7.1). These ROS can overwhelm the multitude of antioxidant defenses within the mitochondrial matrix and thereby cause extensive oxidative damage to mitochondria that contributes to a wide range of pathologies [3]. More interesting is the growing view that the production of ROS from mitochondria can act as a redox signal to the rest of the cell, suggesting that the production of ROS by mitochondria may be a way in which the mitochondria "talks" to the rest of the cell, coordinating the function of the mitochondria with that of the cell [4,18–20] (Figure 7.1). One situation in which mitochondrial ROS signaling seems to be particularly important is in the activation of cells such as macrophages during inflammation [21,22], and also when mitochondria act as signaling hubs in response to viral infections [23,24].

Thus there is a lot of interest in how ROS production by mitochondria might act as a potential redox signal (Figure 7.1). However, there are considerable uncertainties about the mechanisms and significance. Here we outline how these signals may be generated in mitochondria and how they can be passed on to the rest of the cell.

HOW DO MITOCHONDRIA GENERATE H_2O_2 AS A REDOX SIGNAL?

Mitochondria can generate superoxide within the matrix from a number of sources, and this superoxide is then dismutated to H_2O_2 by the high concentration of MnSOD present [4]. While there are other locations within mitochondria, such as the intermembrane space and the outer membrane, that can also produce superoxide and H_2O_2 and that may be important in redox signaling [4], the focus here will be on redox signals generated from the mitochondrial matrix. This is because such signals may most directly enable the organelle to feed back to the rest of the cell about its metabolic status [25]. Within the matrix, the respiratory chain complexes I and III are usually considered to be the sources of superoxide, although some matrix dehydrogenases may also contribute [4]. However, in many situations this ROS production is in response to damage to the complexes, leading to a backup of electrons that spill out as superoxide [4]. As a potential redox signal it is likely that the ROS generation would be more regulated [20]. Therefore, the focus is on two sites in the respiratory chain that have the potential to produce superoxide in a regulated way: these are complexes I and III [4]. The production of superoxide by complex III occurs in response to inhibition at the internal ubiquinone binding site by inhibitors such as antimycin, which is unlikely to be physiological [4]. However, there are a number of reports that complex III may be a source of superoxide as a signal during hypoxia [26,27], but the mechanistic details of how this occurs are obscure. This leaves the generation of superoxide by complex I through the process of reverse electron transport (RET) as the best understood potential source of redox signaling from mitochondria (Figure 7.2) [25]. This process of RET occurs when electrons are forced backward through complex I from the ubiquinone (CoQ) pool, which requires a high proton motive force and a very reduced CoQ pool [4,25]. The main attraction of RET as a physiological redox signal is that it can occur robustly in response to two factors: the CoQ pool and the membrane potential, both of which are physiologically accessible and do not require artificial intervention [4].

The site of superoxide production from complex I is not certain, although we favor production from the flavin site in complex I [4,25,28]. Therefore, our current view is that RET at complex I is likely to be the major physiological source of superoxide within the mitochondrial matrix and that once formed it is then converted rapidly to H_2O_2.

HOW IS THE H_2O_2 REDOX SIGNAL MODULATED?

Our working hypothesis is that the major mitochondrial source of H_2O_2 is RET at complex I, generating superoxide that is then converted to H_2O_2 by MnSOD [4]. For mitochondrial H_2O_2 to be a redox signal, it is important that it be responsive to mitochondrial status [20,29,30]. The major way in which this is likely to occur is in the alteration to the generation of superoxide at complex I in response to changes in the proton motive force and the redox state of the CoQ pool [4,25]. This mechanism has a lot of appeal as RET is exquisitely sensitive to these factors, both of which are central components of mitochondrial function and which vary in direct response to mitochondrial activity. Thus the redox signal would respond rapidly to mitochondrial

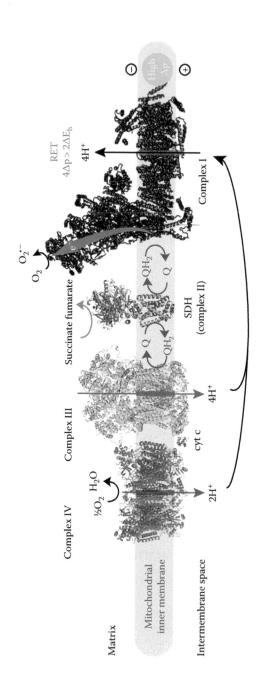

FIGURE 7.2 Reverse electron transport (RET) by complex I. For forward electron transport to occur the difference in reduction potential between the NAD⁺/NADH and the Coenzyme Q (CoQ) pool across complex I (ΔE_h) has to be sufficient to pump protons across the mitochondrial inner membrane against the proton motive force (Δp). As 4 protons are pumped for every 2 electrons that pass through complex I, $2\Delta E_h > 4\Delta p$ is the requirement for the forward reaction to occur. When the Δp is high and/or the ΔE_h across complex I is low such that $4\Delta p > 2\Delta E_h$, electrons can be driven backward from the CoQ pool onto the flavin mononucleotide (FMN) of complex I, reducing the FMN, which can donate a pair of electrons to NAD⁺ to form NADH, or pass one electron to oxygen to generate superoxide. The red arrow in complex I indicates reverse electron transport. Cyt c, cytochrome c; SDH, succinate dehydrogenase.

status and would signal this situation to the rest of the cell. For example, during isch-emia reperfusion injury, where superoxide production by RET has been found to play a key role in pathology, the rapid oxidation of the succinate that accumulates during ischemia favors reduction of the CoQ pool. The reduced CoQ pool favors proton pumping by complexes III and IV helping maintain a large membrane potential upon reperfusion, and the degradation of adenine nucleotides during ischemia limits ADP availability upon reperfusion that would otherwise diminish membrane potential by stimulating mitochondrial ATP synthesis [25]. Similarly, succinate oxidation and the metabolic switch to glycolysis in activated macrophages during inflammation provide the reduced CoQ pool and high membrane potential required for increased superoxide production by RET [21].

In addition to modulation by the factors that affect membrane potential and CoQ redox state, it could be that RET could be affected by other changes to complex I, for example, by post-translational modifications (PTMs) such as phosphorylation [31], but in the absence of evidence this possibility is only theoretical at this stage. The rate of RET by complex I is also affected by the proportion of complex I that is in the active or the deactive state, for example, during ischemia, complex I adopts the deac-tive state [32,33], but whether such transitions are used to regulate RET is not known. Finally, complex I can exist as an isolated complex, or as a supercomplex with other respiratory chain complexes [34]. The physiological role, if any, of supercomplex formation is not clear at present [35]. However, as changes in the extent of complex I incorporation into supercomplexes correlates with changes in mitochondrial ROS formation, it may be that these processes are interrelated [34,36]. Therefore, RET at complex I is determined by factors that alter the membrane potential and the CoQ redox state, with the possibility that PTMs, conformation changes, or incorporation into supercomplexes may also affect RET.

Once formed at the flavin of complex I, superoxide is dismutated spontaneously or enzymatically by the large excess of MnSOD, forming H_2O_2 [4]. Although superox-ide is a negatively charged molecule that will not readily diffuse through membranes, it is possible that some superoxide that escapes MnSOD may also act independently as a signaling molecule within the matrix, possibly by acting on the iron-sulfur center in aconitase [37,38].

It is important to note that there are effects of MnSOD level of expression on cell fate that are poorly understood [39], perhaps suggesting that the balance between superoxide and H_2O_2 level may modulate signaling pathways.

Once formed within the mitochondrial matrix, the H_2O_2 level can then be regu-lated by its degradation (Figure 7.3). This is done by a series of peroxidases within mitochondria, including glutathione peroxidases and peroxiredoxins (Prx) 3 and 5 [40,41]. Most studies indicate that the Prx 3 is the major peroxidase in mitochondria [41,42]. Thus, regulating the activity of Prx 3 may be a way of modulating H_2O_2 levels within mitochondria. This could be done in a number of ways. The activity of Prx 3 is modulated by the ratio of its active to reversibly inactive form, due to disulfide bond formation [41,43]. In addition, Prx 3 can be oxidized to a sulfinic acid form that can be slowly reactivated by sulfiredoxin [44]. These alterations may alter the flux of H_2O_2 from the mitochondria. There is also the possibility that Prx 3 activity may be altered by PTMs, or by the formation of higher-order structures of Prx 3 [45], but the details

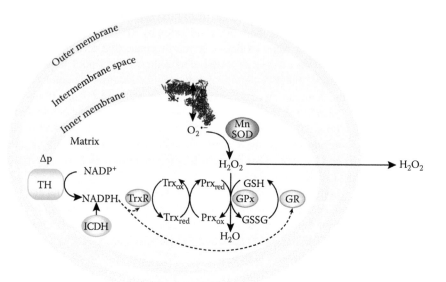

FIGURE 7.3 The production of $O_2^{\bullet-}$ within the mitochondrial matrix leads to the formation of H_2O_2 from SOD-catalyzed dismutation. Within mitochondria, H_2O_2 is degraded by glutathione peroxidases (GPx) or peroxiredoxins (Prx), which depend for their reduction on glutathione (GSH) and thioredoxin-2 (Trx), respectively. Glutathione disulfide (GSSG) is reduced back to GSH by glutathione reductase (GR). Trx is reduced by thioredoxin reductase-2 (TrxR). Both enzymes receive reducing equivalents from the NADPH pool, which is kept reduced by the Δp-dependent transhydrogenase (TH), and by isocitrate dehydrogenase (ICDH).

of these are not clear yet. Finally, the activity of Prx 3, and other peroxidases in mitochondria, is determined by the redox state of the mitochondrial NADPH pool. The NADPH pool itself can respond to a number of factors (the activity of Δp-dependent transhydrogenase (TH), isocitrate dehydrogenase (ICDH), and malic enzyme), which maintain NADPH, and the redox state of the GSH and thioredoxin systems, which are reduced by NAPDH [40,46,47], potentially adding a further way of regulating the steady state level of H_2O_2 level in mitochondria.

HOW DOES THE MITOCHONDRIAL REDOX SIGNAL MOVE FROM MITOCHONDRIA TO THE CYTOSOL?

For H_2O_2 itself to act as a redox signal it has to leave the mitochondria and enter the cytosol. It is known that isolated mitochondria can produce a flux of H_2O_2 from both the intermembrane space and the matrix [4] and this indicates that H_2O_2 itself can leave the mitochondrion and pass to the cytosol. The passage of H_2O_2 through the plasma membrane, for example, in response to its generation outside the cell by NADPH oxidases, is mediated by aquaporins [48,49]. There have been reports of aquaporin 8 in the mitochondrial inner membrane [50,51]; however, these have been disputed [52] and no aquaporins are found in databases of mitochondrial proteins, such as MitoCarta2. The very large surface-to-volume area of the mitochondrial inner membrane most probably enables the rapid diffusion of H_2O_2 through the phospholipid bilayer in the absence of aquaporins [52]. Therefore, we currently favor a model in which H_2O_2 efflux from mitochondria is unmediated, but we cannot exclude the possibility that other proteins in the mitochondrial inner membrane may facilitate H_2O_2 movement out of the mitochondria. Once in the intermembrane space, the H_2O_2 should then be able to diffuse out of that compartment by movement through porins in the outer membrane.

This scenario indicates how H_2O_2 itself can move from the matrix to the rest of the cell. Of course, it is also possible that the H_2O_2 in the mitochondrial matrix can alter mitochondrial metabolism, perhaps by altering the activity of metal-center proteins [42], and thus lead to the generation of another signal that can pass on to the rest of the cell, for example, by generating an electrophile.

WHAT ARE THE CYTOSOLIC TARGETS OF THE MITOCHONDRIAL H_2O_2 REDOX SIGNAL?

The efflux of H_2O_2 from the mitochondria to the cytosol can thus act as a potential redox signal. The way in which H_2O_2 acts as a signal requires that it affect the activity or location of a protein in some way [20,53,54]. This is usually thought to happen by the redox modification of a target cysteine residue. This can be done by the direct reaction of H_2O_2 with the target and can be reversible (e.g., reversible inactivation of tyrosine phosphatases through oxidation of an active site cysteine thiol [55]) or irreversible (e.g., thiol alkylation of Kelch-like ECH-associated protein 1 (KEAP1), inducing nuclear translocation of nuclear factor erythroid 2-related factor 2 (NRF2) [56]). However, an alternative mechanism has also been described in some cases such that it is not H_2O_2 itself that acts as a direct signal but that it is picked up by a disulfide relay such that this then passes on the redox signal to the target protein [57,58]. It is likely that in the cell the hydrogen peroxide interacts with such a similar relay, but whether this is specific for the H_2O_2 emanating from the mitochondria or not is currently unclear.

Whether the signal emanates from mitochondria as a H_2O_2 signal itself, a redox relay or another signaling molecule, this will ultimately have to act on a target protein and lead to a biological effect, notable examples including cytosolic stabilization of hypoxia-inducible factor-1α (HIF-1α) in response to mitochondrial ROS during hypoxia (leading to transcriptional adaptation to hypoxia) [59], and mitochondrial ROS–induced oligomerization of mitochondrial antiviral signaling (MAVS) protein in immune cells in response to viral infection, enhancing production of proinflammatory cytokines [23]. Many proteins have been claimed to be targets for mitochondrial redox signals, but often the evidence is circumstantial or indirect. The development of redox proteomic methods should enable theses targets to be picked up more definitively [60–62]. However, it may be that a recurring problem is that these changes are a series of reversible redox changes that have multiple small effects, and thus may be challenging to detect definitively.

CONCLUSION

This survey of the potential roles of H_2O_2 in mitochondrial redox signaling focused on its generation by RET in mitochondria. This is because to us it seems the most physiologically tractable method for mitochondria to generate a redox signal. However, many other possibilities are conceivable, and time will tell the extent to which RET is a significant player in mitochondrial redox signaling. Even so, at the moment it seems that RET is moving away from being an *in vitro* curiosity to a potential contributor to *in vivo* signaling in a number of scenarios [21,25,34,63]. However, there are still huge gaps in our understanding. We are only starting to learn the web of interactions that link mitochondrial function to the rest of the cell.

NOTES

Competing interests: M.P. Murphy holds patents in the development of mitochondrial therapies.

Acknowledgments: We thank our many colleagues with whom we have worked on the area over the past years.

Funding: This work was supported by the Medical Research Council UK (MC_U105663142) and by a Wellcome Trust Investigator award (110159/Z/15/Z) to MPM.

Authors' contributions: Both authors prepared the manuscript and gave final approval for publication.

REFERENCES

1. Duchen, M. R. and G. Szabadkai. 2010. Roles of mitochondria in human disease. *Essays Biochem* 47:115–137. doi: 10.1042/bse0470115.
2. Wallace, D. C., W. Fan, and V. Procaccio. 2010. Mitochondrial energetics and therapeutics. *Annu Rev Pathol* 5:297–348. doi: 10.1146/annurev.pathol.4.110807.092314.
3. Smith, R. A. J., R. C. Hartley, H. M. Cocheme, and M. P. Murphy. 2012. Mitochondrial pharmacology. *Trends Pharmacol Sci* 33 (6):341–352. doi: 10.1016/j.tips.2012.03.010.
4. Murphy, M. P. 2009. How mitochondria produce reactive oxygen species. *Biochem J* 417 (1):1–13. doi: 10.1042/BJ20081386.
5. Smith, R. A. J., R. C. Hartley, and M. P. Murphy. 2011. Mitochondria-targeted small molecule therapeutics and probes. *Antioxid Redox Signal* 15 (12):3021–3038. doi: 10.1089/ars.2011.3969.
6. Paul, V. D. and R. Lill. 2015. Biogenesis of cytosolic and nuclear iron-sulfur proteins and their role in genome stability. *Biochim Biophys Acta* 1853 (6):1528–1539. doi: 10.1016/j.bbamcr.2014.12.018.
7. Ye, H. and T. A. Rouault. 2010. Human iron-sulfur cluster assembly, cellular iron homeostasis, and disease. *Biochemistry* 49 (24):4945–4956. doi: 10.1021/bi1004798.
8. Chouchani, E. T., V. R. Pell, A. M. James, L. M. Work, K. Saeb-Parsy, C. Frezza, T. Krieg, and M. P. Murphy. 2016. A unifying mechanism for mitochondrial superoxide production during ischemia-reperfusion injury. *Cell Metab* 23 (2):254–263. doi: 10.1016/j.cmet.2015.12.009.

9. Sciacovelli, M. and C. Frezza. 2016. Oncometabolites: Unconventional triggers of oncogenic signalling cascades. *Free Radic Biol Med* 100:175–181. doi: 10.1016/j. freeradbiomed.2016.04.025.

10. Wallace, D. C. 2012. Mitochondria and cancer. *Nat Rev Cancer* 12 (10):685–698. doi: 10.1038/nrc3365.

11. Murley, A. and J. Nunnari. 2016. The emerging network of mitochondria-organelle contacts. *Mol Cell* 61 (5):648–653. doi: 10.1016/j.molcel.2016.01.031.

12. Mammucari, C., M. Patron, V. Granatiero, and R. Rizzuto. 2011. Molecules and roles of mitochondrial calcium signaling. *Biofactors* 37 (3):219–227. doi: 10.1002/biof.160.

13. Brand, M. D. and M. P. Murphy. 1987. Control of electron flux through the respiratory chain in mitochondria and cells. *Biol Rev Camb Philos Soc* 62 (2):141–193.

14. Tait, S. W. and D. R. Green. 2010. Mitochondria and cell death: Outer membrane permeabilization and beyond. *Nat Rev Mol Cell Biol* 11 (9):621–632. doi: 10.1038/nrm2952.

15. Halestrap, A. 2005. Biochemistry: A pore way to die. *Nature* 434 (7033):578–579. doi: 10.1038/434578a.

16. Murphy, E. and C. Steenbergen. 2011. What makes the mitochondria a killer? Can we condition them to be less destructive? *Biochim Biophys Acta* 1813 (7):1302–1308. doi: 10.1016/j.bbamcr.2010.09.003.

17. Rasola, A. and P. Bernardi. 2011. Mitochondrial permeability transition in Ca(2+)-dependent apoptosis and necrosis. *Cell Calcium* 50 (3):222–233. doi: 10.1016/j. ceca.2011.04.007.

18. Arnoult, D., F. Soares, I. Tattoli, and S. E. Girardin. 2011. Mitochondria in innate immunity. *EMBO Rep* 12 (9):901–910. doi: 10.1038/embor.2011.157.

19. Tormos, K. V., E. Anso, R. B. Hamanaka, J. Eisenbart, J. Joseph, B. Kalyanaraman, and N. S. Chandel. 2011. Mitochondrial complex III ROS regulate adipocyte differentiation. *Cell Metab* 14 (4):537–544. doi: 10.1016/j.cmet.2011.08.007.

20. Holmstrom, K. M. and T. Finkel. 2014. Cellular mechanisms and physiological consequences of redox-dependent signalling. *Nat Rev Mol Cell Biol* 15 (6):411–421. doi: 10.1038/nrm3801.

21. Mills, E. L., B. Kelly, A. Logan, A. S. Costa, M. Varma, C. E. Bryant, P. Tourlomousis et al. 2016. Succinate dehydrogenase supports metabolic repurposing of mitochondria to drive inflammatory macrophages. *Cell* 167 (2):457–470.e13. doi: 10.1016/j. cell.2016.08.064.

22. Green, D. R., L. Galluzzi, and G. Kroemer. 2011. Mitochondria and the autophagy-inflammation-cell death axis in organismal aging. *Science* 333 (6046):1109–1112. doi: 10.1126/science.1201940.

23. Buskiewicz, I. A., T. Montgomery, E. C. Yasewicz, S. A. Huber, M. P. Murphy, R. C. Hartley, R. Kelly et al. 2016. Reactive oxygen species induce virus-independent MAVS oligomerization in systemic lupus erythematosus. *Sci Signal* 9 (456):ra115. doi: 10.1126/scisignal.aaf1933.

24. Zhou, R., A. S. Yazdi, P. Menu, and J. Tschopp. 2011. A role for mitochondria in NLRP3 inflammasome activation. *Nature* 469 (7329):221–225. doi: 10.1038/nature09663.

25. Chouchani, E. T., V. R. Pell, E. Gaude, D. Aksentijevic, S. Y. Sundier, E. L. Robb, A. Logan et al. 2014. Ischaemic accumulation of succinate controls reperfusion injury through mitochondrial ROS. *Nature* 515 (7527):431–435. doi: 10.1038/nature13909.

26. Bell, E. L. and N. S. Chandel. 2007. Mitochondrial oxygen sensing: Regulation of hypoxia-inducible factor by mitochondrial generated reactive oxygen species. *Essays Biochem* 43:17–27. doi: 10.1042/BSE0430017.

27. Chandel, N. S., E. Maltepe, E. Goldwasser, C. E. Mathieu, M. C. Simon, and P. T. Schumacker. 1998. Mitochondrial reactive oxygen species trigger hypoxia-induced transcription. *Proc Natl Acad Sci USA* 95 (20):11715–11720.

28. Pryde, K. R. and J. Hirst. 2011. Superoxide is produced by the reduced flavin in mitochondrial complex I: A single, unified mechanism that applies during both forward and reverse electron transfer. *J Biol Chem* 286 (20):18056–18065. doi: 10.1074/jbc.M110.186841.

29. Finkel, T. 2012. Signal transduction by mitochondrial oxidants. *J Biol Chem* 287 (7):4434–4440. doi: 10.1074/jbc.R111.271999.

30. Janssen-Heininger, Y. M., B. T. Mossman, N. H. Heintz, H. J. Forman, B. Kalyanaraman, T. Finkel, J. S. Stamler, S. G. Rhee, and A. van der Vliet. 2008. Redox-based regulation of signal transduction: Principles, pitfalls, and promises. *Free Radic Biol Med* 45 (1):1–17. doi: 10.1016/j.freeradbiomed.2008.03.011.

31. Covian, R. and R. S. Balaban. 2012. Cardiac mitochondrial matrix and respiratory complex protein phosphorylation. *Am J Physiol Heart Circ Physiol* 303 (8):H940–H966. doi: 10.1152/ajpheart.00077.2012.

32. Gorenkova, N., E. Robinson, D. J. Grieve, and A. Galkin. 2013. Conformational change of mitochondrial complex I increases ROS sensitivity during ischemia. *Antioxid Redox Signal* 19 (13):1459–1468. doi: 10.1089/ars.2012.4698.

33. Drose, S., A. Stepanova, and A. Galkin. 2016. Ischemic A/D transition of mitochondrial complex I and its role in ROS generation. *Biochim Biophys Acta* 1857 (7):946–957. doi: 10.1016/j.bbabio.2015.12.013.

34. Moreno-Loshuertos, R. and J. A. Enriquez. 2016. Respiratory supercomplexes and the functional segmentation of the CoQ pool. *Free Radic Biol Med* 100:5–13. doi: 10.1016/j.freeradbiomed.2016.04.018.

35. Blaza, J. N., R. Serreli, A. J. Jones, K. Mohammed, and J. Hirst. 2014. Kinetic evidence against partitioning of the ubiquinone pool and the catalytic relevance of respiratory-chain supercomplexes. *Proc Natl Acad Sci USA* 111 (44):15735–15740. doi: 10.1073/pnas.1413855111.

36. Lopez-Fabuel, I., J. Le Douce, A. Logan, A. M. James, G. Bonvento, M. P. Murphy, A. Almeida, and J. P. Bolanos. 2016. Complex I assembly into supercomplexes determines differential mitochondrial ROS production in neurons and astrocytes. *Proc Natl Acad Sci USA* 113 (46):13063–13068. doi: 10.1073/pnas.1613701113.

37. Gardner, P. R. 2002. Aconitase: Sensitive target and measure of superoxide. *Methods Enzymol* 349:9–23.

38. James, A. M., Y. Collins, A. Logan, and M. P. Murphy. 2012. Mitochondrial oxidative stress and the metabolic syndrome. *Trends Endocrinol Metab* 23 (9):429–434. doi: 10.1016/j.tem.2012.06.008.

39. Hart, P. C., M. Mao, A. L. de Abreu, K. Ansenberger-Fricano, D. N. Ekoue, D. Ganini, A. Kajdacsy-Balla et al. 2015. MnSOD upregulation sustains the Warburg effect via mitochondrial ROS and AMPK-dependent signalling in cancer. *Nat Commun* 6:6053. doi: 10.1038/ncomms7053.

40. Murphy, M. P. 2012. Mitochondrial thiols in antioxidant protection and redox signaling: Distinct roles for glutathionylation and other thiol modifications. *Antioxid Redox Signal* 16 (6):476–495. doi: 10.1089/ars.2011.4289.

41. Cox, A. G., C. C. Winterbourn, and M. B. Hampton. 2009. Mitochondrial peroxiredoxin involvement in antioxidant defence and redox signalling. *Biochem J* 425 (2):313–325. doi: 10.1042/BJ20091541.

42. Winterbourn, C. C. and M. B. Hampton. 2008. Thiol chemistry and specificity in redox signaling. *Free Radic Biol Med* 45 (5):549–561. doi: 10.1016/j.freeradbiomed.2008.05.004.

43. Cox, A. G., J. M. Pullar, G. Hughes, E. C. Ledgerwood, and M. B. Hampton. 2008. Oxidation of mitochondrial peroxiredoxin 3 during the initiation of receptor-mediated apoptosis. *Free Radic Biol Med* 44 (6):1001–1109. doi: 10.1016/j.freeradbiomed.2007.11.017.

44. Noh, Y. H., J. Y. Baek, W. Jeong, S. G. Rhee, and T. S. Chang. 2009. Sulfiredoxin translocation into mitochondria plays a crucial role in reducing hyperoxidized peroxiredoxin III. *J Biol Chem* 284 (13):8470–8477. doi: 10.1074/jbc.M808981200.

45. Barranco-Medina, S., J. J. Lazaro, and K. J. Dietz. 2009. The oligomeric conformation of peroxiredoxins links redox state to function. *FEBS Lett* 583 (12):1809–1816. doi: 10.1016/j.febslet.2009.05.029.

46. Murphy, M. P. 2015. Redox modulation by reversal of the mitochondrial nicotinamide nucleotide transhydrogenase. *Cell Metab* 22 (3):363–365. doi: 10.1016/j.cmet.2015.08.012.

47. Nickel, A. G., A. von Hardenberg, M. Hohl, J. R. Loffler, M. Kohlhaas, J. Becker, J. C. Reil et al. 2015. Reversal of mitochondrial transhydrogenase causes oxidative stress in heart failure. *Cell Metab* 22 (3):472–484. doi: 10.1016/j.cmet.2015.07.008.

48. Bienert, G. P., A. L. Moller, K. A. Kristiansen, A. Schulz, I. M. Moller, J. K. Schjoerring, and T. P. Jahn. 2007. Specific aquaporins facilitate the diffusion of hydrogen peroxide across membranes. *J Biol Chem* 282 (2):1183–1192. doi: 10.1074/jbc.M603761200.

49. Miller, E. W., B. C. Dickinson, and C. J. Chang. 2010. Aquaporin-3 mediates hydrogen peroxide uptake to regulate downstream intracellular signaling. *Proc Natl Acad Sci USA* 107 (36):15681–15686. doi: 10.1073/pnas.1005776107.

50. Calamita, G., D. Ferri, P. Gena, G. E. Liquori, A. Cavalier, D. Thomas, and M. Svelto. 2005. The inner mitochondrial membrane has aquaporin-8 water channels and is highly permeable to water. *J Biol Chem* 280 (17):17149–17153. doi: 10.1074/jbc.C400595200.

51. Marchissio, M. J., D. E. Frances, C. E. Carnovale, and R. A. Marinelli. 2012. Mitochondrial aquaporin-8 knockdown in human hepatoma HepG2 cells causes ROS-induced mitochondrial depolarization and loss of viability. *Toxicol Appl Pharmacol* 264 (2):246–254. doi: 10.1016/j.taap.2012.08.005.

52. Yang, B., D. Zhao, and A. S. Verkman. 2006. Evidence against functionally significant aquaporin expression in mitochondria. *J Biol Chem* 281 (24):16202–16206. doi: 10.1074/jbc.M601864200.

53. Collins, Y., E. T. Chouchani, A. M. James, K. E. Menger, H. M. Cocheme, and M. P. Murphy. 2012. Mitochondrial redox signalling at a glance. *J Cell Sci* 125 (Pt 4):801–806. doi: 10.1242/jcs.098475.

54. Gough, D. R. and T. G. Cotter. 2011. Hydrogen peroxide: A Jekyll and Hyde signalling molecule. *Cell Death Dis* 2:e213. doi: 10.1038/cddis.2011.96.

55. Meng, T. C., T. Fukada, and N. K. Tonks. 2002. Reversible oxidation and inactivation of protein tyrosine phosphatases *in vivo*. *Mol Cell* 9 (2):387–399.

56. Kobayashi, M. and M. Yamamoto. 2006. Nrf2-Keap1 regulation of cellular defense mechanisms against electrophiles and reactive oxygen species. *Adv Enzyme Regul* 46:113–140.

57. Sobotta, M. C., W. Liou, S. Stocker, D. Talwar, M. Oehler, T. Ruppert, A. N. Scharf, and T. P. Dick. 2015. Peroxiredoxin-2 and STAT3 form a redox relay for H2O2 signaling. *Nat Chem Biol* 11 (1):64–70. doi: 10.1038/nchembio.1695.

58. Azevedo, D., F. Tacnet, A. Delaunay, C. Rodrigues-Pousada, and M. B. Toledano. 2003. Two redox centers within Yap1 for H_2O_2 and thiol-reactive chemicals signaling. *Free Radic Biol Med* 35 (8):889–900.

59. Sanjuan-Pla, A., A. M. Cervera, N. Apostolova, R. Garcia-Bou, V. M. Victor, M. P. Murphy, and K. J. McCreath. 2005. A targeted antioxidant reveals the importance of mitochondrial reactive oxygen species in the hypoxic signaling of HIF-1alpha. *FEBS Lett* 579 (12):2669–2674. doi: 10.1016/j.febslet.2005.03.088.

60. Rinalducci, S., L. Murgiano, and L. Zolla. 2008. Redox proteomics: Basic principles and future perspectives for the detection of protein oxidation in plants. *J Exp Bot* 59 (14):3781–3801. doi: 10.1093/jxb/ern252.

61. Kumar, V., T. Kleffmann, M. B. Hampton, M. B. Cannell, and C. C. Winterbourn. 2013. Redox proteomics of thiol proteins in mouse heart during ischemia/reperfusion using ICAT reagents and mass spectrometry. *Free Radic Biol Med* 58:109–117. doi: 10.1016/j.freeradbiomed.2013.01.021.

62. Menger, K. E., A. M. James, H. M. Cocheme, M. E. Harbour, E. T. Chouchani, S. Ding, I. M. Fearnley, L. Partridge, and M. P. Murphy. 2015. Fasting, but not aging, dramatically alters the redox status of cysteine residues on proteins in Drosophila melanogaster. *Cell Rep* 11 (12):1856–1865. doi: 10.1016/j.celrep.2015.05.033.

63. Scialo, F., A. Sriram, D. Fernandez-Ayala, N. Gubina, M. Lohmus, G. Nelson, A. Logan et al. 2016. Mitochondrial ROS produced via reverse electron transport extend animal lifespan. *Cell Metab* 23 (4):725–734. doi: 10.1016/j.cmet.2016.03.009.

8 Quinone Toxicity
Involvement of Reactive Oxygen Species

*Rex Munday**

CONTENTS

INTRODUCTION

Quinones have been detected in interstellar dust and are thus among the oldest organic chemicals in the universe [1]. Quinones and their reduction products, hydroquinones, are also widely distributed on earth, as metabolites of fungi, bacteria, lichen, plants,

---

* Sadly, Rex Munday passed away unexpectedly on July 20, 2017. The editors acknowledge his contribution to the field of Free Radical Biology over many years and are thankful that he accepted our invitation to contribute this chapter on quinones in biology.

insects, worms, spiders, and mammals [2]. Ubiquinones are components of the electron transport chain, and vitamin K_1, which is essential for the synthesis of blood clotting factors, is a naphthoquinone derivative. Quinones are found in tea, coffee, fruit, wine, vegetables, wheat-based cereals, bread, and drinking water [3,4]. Henna, which has been used since antiquity as a dye for the hair and skin, owes its activity to the presence of 2-hydroxy-1,4-naphthoquinone [5], and the anthraquinone derivative, carminic acid, is the major component of cochineal, which is widely used as a dye for food and lipstick. Creams containing hydroquinone are used as cosmetics in many parts of the world, particularly in the Middle East, sub-Saharan Africa, and Asia, where fair skin is considered beautiful and an indicator of high socioeconomic class [6]. 2,3-Dichloro-1,4-naphthoquinone has been used for many years as a fungicide and the anthraquinone derivatives emodin and chrysazin are, or have been, used as laxatives. Quinones are also produced during combustion, and are important environmental pollutants. Plant-derived quinones have been widely used in traditional medicine, and quinones comprise one of the largest classes of anticancer drugs [7] and are in use for therapy of malaria [8]. Other quinones are under investigation for possible use in the treatment of schistosomiasis and trypanosomiasis.

We are thus exposed to a wide range of quinones and hydroquinones, which may have beneficial or adverse effects upon our health. In this review, the focus will be on natural and synthetic derivatives of 1,2-benzoquinone (**1**), 1,4-benzoquinone (**2**), 1,2-naphthoquinone (**3**) and 1,4-naphthoquinone (**4**), 9,10-phenanthrenequinone (**5**) and 9,10-anthraquinone (**6**), including the azirinidylquinones, such as diaziquone (2,5-diaziridinyl-3,6-bis(carboethoxyamino)-1,4-benzoquinone (**7**), mitomycin C (**8**), β-lapachone (**9**) streptonigrin (**10**), anthracyclines, such as doxorubicin (**11**, R = CH$_2$OH) and daunorubicin (**11**, R = CH$_3$), 17-(allylamino)-17-demethoxygeldanamycin (**12**), deoxynyboquinone (**13**), gossypol (**14**), mitoxantrone (**15**), and chrysazin (**16**). The effects of certain derivatives of hydroquinone and catechol, the reduction products of 1,4- and 1,2-benzoquinone respectively, are also discussed.

1

2

3

4

5

6

7

8

9

10

11

12

13

14

15

16

REDUCTION OF QUINONES

Quinones and hydroquinones constitute a 2-electron redox system, with intermediacy of the one-electron reduction/oxidation product, the semiquinone (Reaction 8.1).

(8.1)

The reduction of quinones is mediated by several enzyme systems. Reduction may be a one-electron reaction, yielding the semiquinone, as mediated by NADPH-cytochrome P450 reductase, cytochrome b5 reductase, mitochondrial NADH:ubiquinone oxidoreductase (Complex I) [9], sepiapterin reductase [10], and neuronal nitric oxide synthase [11]. Alternatively, reduction may be a two-electron reaction, yielding the hydroquinone, as catalyzed by NAD(P)H:quinone acceptor oxidoreductase (NQO1) [12] and other carbonyl reductases [13] or may proceed via both one- and two-electron reactions, as seen with xanthine oxidase [9], glutathione reductase [14], thioredoxin reductase [15], trypanothione-disulfide reductase [16], and lipoamide dehydrogenase [17].

Quinones with relatively high reduction potentials, such as benzoquinone, methylbenzoquinone, and halogenated benzoquinones, are reduced nonenzymatically by NAD(P)H [18] and by GSH [19]. Quinones are reduced by ascorbate in two one-electron steps, forming the semiquinone and hydroquinone [20]. They are also reduced to the semiquinone by reaction with oxyhemoglobin (Hb^{II}O$_2$), with concomitant formation of methemoglobin (Hb^{III}) [21] (Reaction 8.2).

$$\text{(benzoquinone)} + Hb^{II}O_2 \rightleftharpoons \text{(semiquinone)} + Hb^{III} + O_2 \qquad (8.2)$$

The ease of reduction of a particular quinone is related to its reduction potential. The more positive the reduction potential, the more easily the quinone or semiquinone is reduced. Conversely, the more negative the potential of a quinone, the more difficult it is to reduce [1].

AUTOXIDATION OF HYDROQUINONES

Hydroquinones undergo autoxidation. Reaction of the monoanion of a hydroquinone with molecular oxygen produces the semiquinone and superoxide radical (Reaction 8.3).

$$\text{(hydroquinone monoanion)} + O_2 \rightarrow \text{(semiquinone)} + O_2^{\bullet -} + H^+ \qquad (8.3)$$

The same products are formed by the reaction of the monoanion of a hydroquinone with a transition metal in its higher oxidation state (Reaction 8.4), followed by autoxidation of the reduced metal (Reaction 8.5).

$$\text{(hydroquinone monoanion)} + M^{(n+1)+} \longrightarrow \text{(semiquinone)} + M^{n+} + H^+ \qquad (8.4)$$

$$M^{n+} + O_2 \longrightarrow M^{(n+1)+} + O_2^{\bullet-} \qquad (8.5)$$

The semiquinone reacts with molecular oxygen to form the quinone and superoxide radical (Reaction 8.6). This reaction is reversible:

$$\text{(semiquinone)} + O_2 \rightleftharpoons \text{(quinone)} + O_2^{\bullet-} \qquad (8.6)$$

More semiquinone is formed by comproportionation between the quinone and hydroquinone, a reaction that is in equilibrium with disproportionation of the semiquinone (Reaction 8.7), and by oxidation of the hydroquinone by superoxide (Reaction 8.8).

$$\text{(quinone)} + \text{(hydroquinone)} \rightleftharpoons 2\,\text{(semiquinone)} + H^+ \qquad (8.7)$$

The rate of autoxidation of a hydroquinone depends upon its reduction potential. The more negative the reduction potential, the more easily the hydroquinone is oxidized [1]. Substitution with electron-donating groups in the aromatic ring decreases reduction potentials, thereby increasing autoxidation rates [22]. Conversely, compounds substituted with electron-withdrawing groups have more positive reduction potentials, and a lower rate of oxidation would be expected [22]. Since the initial step of oxidation involves the hydroquinone anion, autoxidation rates will also be dependent upon ionization potentials. This is again influenced by substituent groups, with electron-donating groups leading to a decrease in the degree of ionization, while electron-withdrawing groups increase it [1]. The effect of these parameters is shown by the rapid autoxidation of hydroquinones [23–25] and naphthohydroquinones [26,27] containing electron-donating groups. Halogenated hydroquinones [1], halogenated 1,4-naphthohydroquinones [27], and 5-hydroxy-1,4-naphthohydroquinone (juglone) [27] also undergo rapid autoxidation at neutral pH. These compounds have very high reduction potentials, and for this reason low oxidation rates would be expected. However, they are highly ionized at neutral pH, and the high concentration of the anion in solution compensates for the effect of the reduction potential.

The position of the equilibrium in the reaction between the semiquinone and oxygen and the reaction between the quinone and superoxide radical (Reaction 8.6) is also of crucial importance, since the forward reaction initiates a radical chain reaction for oxidation of the hydroquinone via Reaction 8.8. If the reduction potential of a quinone is lower than the one-electron reduction potential of oxygen (i.e., < -180 mV), the equilibrium position for Reaction 8.6 will lie to the right, favoring formation of superoxide. For quinones with a potential greater than -180 mV, however, the generation of superoxide will not be thermodynamically favorable [1]. In this case, a lag phase in the autoxidation reaction will be observed, with the rate increasing as the level of quinone increases, facilitating semiquinone formation via Reaction 8.7. The lag phase would be abolished by the addition of catalytic amounts of quinone. Such effects have been observed with hydroquinone [28], phenylhydroquinone [29], and 1,4-naphthohydroquinone [30].

This reaction will also determine the effect of superoxide dismutase on the rate of autoxidation. If the equilibrium lies to the left, superoxide dismutase, by destroying superoxide radical, will shift the equilibrium to the right, increasing the formation of the quinone, which can then participate in the comproportionation reaction (Reaction 8.7). In this way, the rate of autoxidation will be increased. Such an effect has been seen with hydroquinone [28], phenylhydroquinone [31], chlorophenylhydroquinones [32], and 1,2-naphthohydroquinone [33]. Conversely, if the equilibrium lies to the right, superoxide dismutase will decrease the rate of autoxidation by eliminating the superoxide-driven radical chain reaction, as seen with 1,2,3-trihydroxybenzene (pyrogallol) and 1,2-4-trihydroxybenzene [34,35], 1,2-dihydroxy-4-(1-hydroxy-2-(methylamino)ethyl)benzene (epinephrine) [36], 1,2,4-trihydroxy-5-(2-aminoethyl)benzene (6-hydroxydopamine) [37], and alkyl-, alkoxy-, hydroxy-, and amino-1,4-naphthohydroquinones [26,27].

REDOX CYCLING OF QUINONES

The reduction of quinones by cellular reducing agents and the autoxidation of the hydroquinones and semiquinones leads to redox cycling, in which a single molecule of the quinone may generate many molecules of ROS. Redox cycling of quinones has been observed in mitochondria [38,39] and in microsomes [40,41], and when these compounds are incubated with NAD(P)H [42,43], ascorbate [44,45], NQO1 [46], or with monothiols [47] or dithiols [48].

Thiols are nucleophiles as well as reducing agents, and thiolated hydroquinones as well as unsubstituted hydroquinones have been observed as products of the reaction between thiols and quinones [49]. Since the thiol moiety is electron donating, the redox potential of the hydroquinone thiol conjugates will be lower than that of the parent hydroquinone, and will therefore undergo more rapid autoxidation [50]. The reaction with thiols may therefore constitute an activation reaction, and redox cycling of quinones has been observed in the presence of thiols, with the generation of ROS. This may occur not only with low molecular weight thiols [51], but also with protein thiols [52]. It has been suggested, however, that conjugation of quinones with N-acetylcysteine could constitute a detoxification reaction [53].

CYTOTOXICITY AND ROS PRODUCTION BY QUINONES AND HYDROQUINONES *IN VITRO*

The toxicity of quinones and hydroquinones to cell lines *in vitro* has been extensively studied. Cancer cell lines have been most extensively used, mainly with a view to identifying compounds that could possibly be useful in the therapy of cancer *in vivo*. More than a thousand compounds have been tested, in a wide range of cell lines. In general, cell death by quinones involves apoptosis via the intrinsic pathway, although necrosis is sometimes seen in cells exposed to very high levels of these substances.

In many cases, increased intracellular levels of ROS have been reported, generally attributed to redox cycling of the test material after uptake by the cells, and such species have been suggested to be responsible for the observed toxic effects. In most instances, fluorescent probes have been used to detect ROS in cells, although there are significant problems with the specificity of such probes and the interpretation of the results derived from their use [54]. Another problem arises in studies on hydroquinones due to the possibility that they may undergo autoxidation in the culture medium [55,56], generating ROS extracellularly. Superoxide radical is unable to cross cell membranes, but its decomposition to hydrogen peroxide, which is able to enter cells, could lead to increased intracellular levels of this substance, as detected by the fluorescent dyes. Extracellular SOD was shown to protect against the cytotoxicity of gallic acid [57], 6-hydroxydopamine [58], gossypol [59], and 4-allyl-catechol [60], the autoxidation of which is inhibited by this enzyme. In contrast, the cytotoxicity of phenylhydroquinone was increased by addition of SOD to the culture medium [31], consistent with the fact that the rate of autoxidation of this substance is increased by SOD [31,61]. It is possible that extracellular autoxidation and generation of ROS contributes to the cytotoxic activity of these compounds.

The cytotoxicity of 2-methyl-1,4-naphthoquinone (menadione) [62,63], 2,3-dimethoxy-1,4-naphthoquinone [64] and β-lapachone [65] was also decreased by addition of SOD to the culture medium, again suggesting extracellular production of ROS. With these compounds, such an effect could reflect intracellular formation of the hydroquinone with subsequent release into the culture medium, although in view of the instability of naphthohydroquinones, this appears unlikely. Alternatively, the quinone may be reduced by membranal quinone reductase [64] or by components of the culture medium, such as the thiol groups of proteins. Ascorbate has been shown to increase the cytotoxicity of menadione [66], juglone and 2,3-dichloro-1,4-naphthoquinone [67], and 2-(4-hydroxyanilino)- and 2-(4-methoxyanilino)-1,4-naphthoquinone [68], again most likely due to extracellular redox cycling.

While high levels of ROS may kill cancer cells *in vitro*, low concentrations of such species stimulate cancer cell growth [69,70], and such an effect has been observed with several quinones [71,72].

ROS production and oxidative damage by quinones in erythrocytes has been extensively studied. In these cells, generation of hydrogen peroxide has been investigated by the rather old-fashioned, but apparently specific, technique involving inhibition of cellular catalase activity in the presence of 3-aminotriazole [73]. Oxidative damage to erythrocytes, occurring via Reaction 8.2, is indicated by methemoglobin formation and by further irreversible oxidation of hemoglobin leading to the formation of intracellular precipitates (Heinz bodies). Such changes have been observed in erythrocytes incubated with benzoquinones [74], catechols [75], 1,2- and 1,4-naphthoquinone [76], alkyl-1,4-naphthoquinones [77], dialkyl-1,4-naphthoquinones [78], alkoxy-1,4-naphthoquinones [79], chloro-1,4-naphthoquinones [80], juglone [81], β-lapachone [82], and daunorubicin [83]. No effects were observed in rodent erythrocytes incubated with 2-hydroxy-1,4-naphthoquinone [81] or in erythrocytes from normal individuals. Oxidative damage was seen, however, in erythrocytes from individuals deficient in glucose-6-phosphate dehydrogenase [84], an enzyme of the pentose phosphate pathway, which is the sole route for NADPH generation in erythrocytes. Low levels of this enzyme compromise the ability of erythrocytes to withstand oxidative stress and such cells are particularly susceptible to oxidative damage induced by quinones [85].

EFFECT OF QUINONES ON CHEMICALLY-INDUCED CANCER IN ANIMALS

Protection against chemically induced cancer has been observed in animals dosed with a number of quinones. With 2,6-dimethoxy-1,4-benzoquinone [86], plumbagin and juglone [87], 5,8-dihydroxy-2-(1-hydroxy-4-methyl-3-pentenyl)-1,4-naphthoquinone (shikonin) [88], and β-lapachone [89], protection was given when the quinone was given before, during, and after the carcinogen. In comparative studies, 2-isopropyl-5-methyl-benzoquinone (thymoquinone) [90] and gallic acid [91] were much more effective when given before, rather than after, the carcinogen. 3,4-Dihydroxybenzoic acid was equally effective in either situation [92,93].

EFFECT OF QUINONES AND HYDROQUINONES ON TUMOR GROWTH *IN VIVO*

Thymoquinone [94], mitomycin C [95], gossypol [96], juglone [97], plumbagin [98,99], shikonin [100], doxorubicin [101], and β-lapachone [102] have been shown to decrease the growth of cancer cell transplants in immunocompromised mice or in syngeneic mouse models.

ANTIPARASITIC ACTION OF QUINONES

Protozoans of the genus *Plasmodium* are responsible for malaria, a major problem in countries within a broad band around the equator. Organisms of the genera *Trypanosoma* and *Leishmania* are responsible for African trypanosomiasis (sleeping sickness), American trypanosomiasis (Chagas' disease), and leishmaniasis, which are major health problems found almost exclusively in low-income populations of developing countries in tropical and subtropical areas of the world [103]. Drugs to treat these diseases are available, but they are expensive and are associated with severe side effects. Furthermore, resistance is growing to such drugs. New, preferably cheap, drugs are needed, but development of such drugs is of little interest to pharmaceutical companies [104]. Research is being undertaken in academic institutions, however, particularly in South America. The main focus of this research has been on naphthoquinones, in view of the traditional use of lapachol (2-hydroxy-3-[3-methylbut-2-enyl]-1,4-naphthoquinone) in South America, which was originally isolated from the Brazilian tree *Tabebuia avellanedae*. Many naturally occurring and synthetic naphthoquinone derivatives have been tested, and some have shown high activity against various stages of the life cycle of *Trypanosoma* [105] and of *Plasmodium* [106] *in vitro*.

INDUCTION OF PHASE 2 ENZYMES BY QUINONES AND HYDROQUINONES IN ANIMALS

t-Butylhydroquinone increased the activities of NQO1 and glutathione *S*-transferase in the liver, lungs, kidneys, forestomach, glandular stomach, and upper small intestine of mice [107] while 3,4-dihydroxybenzoic acid increased hepatic glutathione *S*-transferase activity in rats [108]. 2-Amino-, 2-methylamino-, 2-dimethylamino-, 2-amino-3-methyl-, 2-amino-3-hydroxy-, 2,3-dichloro, 2-bromo- and 2-amino-3-chloro-1,4-naphthoquinone increased NQO1 activity in the livers and kidneys of rats [79]. Juglone and 5-hydroxy-2-methyl-1,4-naphthoquinone (plumbagin) increased the activity of NQO1 and glutathione *S*-transferase and NQO1 in the forestomach, glandular stomach, duodenum, jejunum, caecum, and colon of rats, but they had no significant effect on the activities of these enzymes in the liver, spleen, heart, lungs, or urinary bladder of these animals [109].

TOXICITY OF QUINONES AND HYDROQUINONES TO ANIMALS

HEMOLYTIC ANEMIA

Oxidative damage to erythrocytes *in vivo* is reflected by methemoglobinemia and by the presence of Heinz bodies within the cells. Such damaged erythrocytes are removed from the circulation by phagocytic cells. In rodents, the primary site of erythrocyte destruction is the spleen, although in severe cases, hepatic Kupffer cells are also involved [110,111]. The spleen plays a major role in erythrocyte regeneration following hemolysis [112], and splenic erythropoiesis is associated with engorgement of red pulp sinusoids, recognized macroscopically as splenic enlargement and darkening. Iron released from phagocytized cells is stored as hemosiderin, predominantly in the spleen, with lesser amounts in the liver and kidneys. When the rate of erythro-clasis exceeds that of compensatory erythropoiesis, blood packed cell volumes and hemoglobin levels decline. In severe hemolytic anemia, immature erythrocytes are released into the circulation, recognized as reticulocytosis. The hematological, histological, and organ weight changes associated with oxidative hemolysis are thus quite characteristic, and easily distinguished from other forms of anemia.

Many quinones have been shown to induce oxidative hemolysis in animals. Among monocyclic compounds, benzoquinone, hydroquinone, and *t*-butylhydroquinone are relatively weak hemolytic agents in animals. The addition of a third hydroxy group increases activity, and 1,2,4-trihydroxybenzene, gallic acid, and pyrogallol are potent hemolytic agents *in vivo* [113,114]. The last-named compound, which is found in acorns and oak leaves, has caused severe toxicity in farm animals grazing in fields containing oak trees. Thousands of animals are poisoned each year [115]. Severe, and sometimes fatal, methemoglobinemia, Heinz body formation, and hemolytic anemia have been recorded in horses eating the leaves of red maple (*Acer rubrum*), and poisoning by this plant is a serious problem for horse owners in the eastern United States and Canada [116]. Again, pyrogallol is held responsible for these toxic effects [117]. Mitomycin C [118] and 17-(allylamino)-17-demethoxygeldanamycin [119] are also hemolytic agents in animals.

Similarly, 1,2-naphthoquinone [76] and its alkyl, alkoxy, and amino derivatives [79] induce oxidative hemolysis in rats, as does β-lapachone [120]. 1,4-Naphthoquinone is also a hemolytic agent in animals, as are 2-monoalkyl- [77], 2,3-dialkyl- [121], 2-amino- and 2-alkylamino- [79], 2,3-dichloro- [122], and 2-hydroxy-1,4-naphthoquinone [77]. 9,10-Anthraquinone and its 2-amino- [123] and 1-amino-2,4-dibromo derivatives [124] are also hemolytic agents in rodents. The toxic effects of a series of 2-hydroxy-3-alkyl-4-naphthoquinones, with alkyl substitution from methyl to pentyl, have been examined in rats. Hemolytic activity decreased with increasing size of the alkyl group, possibly reflecting a steric effect of the substituent [125]. Steric effects cannot be the whole answer, however, since substitution at the 3-position with relatively small groups (chloro or amino) also decreased the severity of hemolysis. Similar effects were observed with 2-amino-1,4-naphthoquinones substituted at the 3-position [79].

When fed at excessive levels, gossypol-containing cottonseed causes hemolysis in farm animals [126], and at times, gossypol poisoning has caused major problems in agricultural production, such as the death of 1600 calves in a single incident [127].

HEPATOTOXICITY

Centrilobular hepatic necrosis was induced in mice after intraperitoneal injection of tetrachloro-1,4-benzoquinone [128] and there is evidence for gossypol-induced hepato-toxicity in rats [129]. Liver damage was also observed in animals after acute or chronic administration of anthracyclines [130,131]. Hepatic necrosis was observed in rats injected with 17-[(dimethylaminoethyl)amino]-17-demethoxygeldanamycin [132], and liver damage, in addition to hemolytic anemia, has been observed in farm animals eating cottonseed or acorns and oak leaves [126,133].

NEPHROTOXICITY

Oral administration of hydroquinone induces renal tubular necrosis in male F344 rats, but not in female F344 rats, Sprague-Dawley rats, or male or female mice [134]. F344 rats, particularly males, are known to be very susceptible to renal damage, and the relevance of this strain of rat in the evaluation of nephrotoxic agents is questionable [135].

Unlike the mono-alkyl derivatives, 2,3-dimethyl-1,4-naphthoquinone not only caused hemolysis in Sprague-Dawley rats but also renal tubular necrosis in the distal segment of the proximal convoluted tubules [121]. Mitoxantrone [136] and amino-anthraquinones [137] have also been shown to be nephrotoxic in mice.

2-Hydroxy-1,4-naphthoquinone is a potent nephrotoxin in rats [138] and both hemolytic anemia and nephrotoxicity were reported in a dog that ate a large amount of a hair dye containing henna [139].

In none of these cases was nephrotoxicity due to urinary excretion of hemoglobin or methemoglobin, since there was no evidence of intravascular hemolysis in the animals.

CARDIOTOXICITY

Although effective broad-spectrum anticancer drugs, the clinical use of doxorubicin, daunorubicin, and other anthracyclines is restricted by their toxicity, particularly irreversible cardiomyopathy. Acute effects on the heart are rare with currently-employed doses of anthracyclines, but cumulative toxicity, resembling congestive heart failure, occurs in a significant proportion of patients, with occurrence proportional to dose. Myocytic vacuolation, accompanied by areas of interstitial fibrosis, are seen in patients, although necrosis is rarely seen [140]. Mitoxantrone, another anthraquinone anticancer drug, has been shown to be cardiotoxic in mice, although the severity of the cardiac effects induced by this compound was lower than those induced by doxorubicin [136]. Replacement of one of the quinone groups of doxorubicin with an imine function, to yield 5-iminodaunorubicin, diminishes the rate of redox cycling in mitochondria and cardiotoxic activity [141,142].

EFFECTS ON REPRODUCTION

Oral administration of plumbagin decreased spermatogenesis in male rats [143], and in dogs dosed with this compound by intraperitoneal injection. In the latter animals, testicular atrophy and damage to seminiferous tubules were also observed [143]. In female rats, plumbagin prolonged the duration of the estrus cycle, inhibited fetal implantation, and induced abortion [144]. Decreased spermatogenesis was observed in rodents after acute [131] or chronic [145] administration of anthracyclines, and gossypol decreased spermatogenesis and decreased spermatozoal motility in rats [146].

CARCINOGENICITY

In a two-year feeding study with hydroquinone in F344 rats and B6C3F$_1$ mice, an increase in the incidence of renal tubular adenomas were seen in male rats, but not in female rats or in either sex of mice, and it was concluded that this compound is a carcinogen [134]. The interpretation of the results of this study has been criticized, however, on the basis of the association between the spontaneous nephropathy seen in F344 rats (mainly in males) and the observed increase in the incidence of renal adenomas [147]. In short-term studies, hydroquinone administered by gavage induced degeneration and increased cell proliferation in the renal tubules of male F344 rats, but not in female F344 rats or in male Sprague-Dawley rats. It was concluded that chemically-induced cell proliferation secondary to toxicity may be important in the pathogenesis of the benign renal tumors observed in male F344 rats dosed with hydroquinone, which may be associated with chronic regenerative activity [148].

In long-term feeding studies in rats, catechol induced adenocarcinoma formation and adenomatous hyperplasia in the glandular stomach of rats and mice. Papillomas and squamous cell carcinomas were occasionally observed in the forestomachs of these rats [92]. Similar studies with 9,10-anthraquinone showed an increased incidence of hepatic, renal, and bladder cancer [149], while 1-amino-2,4-dibromoanthraquinone induced cancer at various sites in both mice and rats [150]. 1-Hydroxy-9,10-anthraquinone, 1-amino-2-methylanthraquinone [137], and chrysazin [151] were shown to be carcinogenic in the large intestine and liver of animals, and formation of 1,2-quinones is possibly involved in the carcinogenic action of polycyclic hydrocarbons [152].

CATARACT FORMATION

1,2- and 1,4-naphthoquinone [153] and doxorubicin [154] are cataractogenic in rats.

TOXICITY OF QUINONES AND HYDROQUINONES TO HUMANS

Hydroquinone appears to be of low oral toxicity in humans. There is no evidence for adverse effects in individuals employed in the manufacture or use of this compound [155] and oral administration of hydroquinone at 300–500 mg/day, in 3 divided doses, to volunteers for three months showed no adverse effects [156]. 17-(Allylamino)-17-demethoxygeldanamycin causes anemia in humans. It is also hepatotoxic [157]. Azirinidylquinones have also been shown to induce anemia in cancer patients, together with myelosuppression and renal damage [158].

Hykinone (menadione sodium bisulphite) and Synkavit (menadiol sodium phosphate) are drugs that were once routinely administered to babies in order to prevent hemorrhagic disease of the newborn, an uncommon, but very serious, event. These compounds are rapidly metabolized to menadione *in vivo*, and it was first noticed in 1953 that some infants treated with these drugs suffered oxidative hemolysis [159]. The association between the use of these drugs and hemolytic anemia was later confirmed, with some deaths being reported [160]. A decrease in the recommended dose levels, and the use of vitamin K$_1$ rather than menadione or its derivatives, appears to have resolved this problem. A Phase 1 trial of Synkavit in patients with advanced cancer showed a high incidence of hemolytic anemia [161] and trials in cancer patients with menadione in addition to other cancer chemotherapeutics revealed rate-limiting hemolytic anemia [162]. 1,4-Naphthoquinone is a metabolite of naphthalene, which may contribute to the hemolytic anemia seen in humans after ingestion of this substance, which is a common household item used as a moth deterrent [163].

Application of henna extract to the skin of newborn babies is a traditional practice in some cultures. In individuals with glucose-6-phosphate dehydrogenase deficiency, such application may be associated with hemolytic anemia and even death [164]. Fatal hemolytic anemia and renal failure also occurred in a 27-day-old boy treated with henna for nappy rash [165]. Acute renal failure and hemolysis occurred after accidental ingestion of henna [166] and in an individual who consumed a large amount of henna as a presumed therapeutic [167]. Severe hemolytic anemia occurred in a young woman who ingested a henna decoction as an abortifacient [168]. A complex of β-lapachone with hydroxypropyl-β-cyclodextrin, code-named ARQ 501, underwent unsuccessful clinical trials as a therapeutic for a variety of cancers. The reasons for failure included dose-limiting toxicity in the form of hemolytic anemia [169].

The cardiotoxicity and myelosuppression of anthracyclines are important factors in the use of these compounds in cancer therapy.

In the 1930s and 1940s, the culinary use of unrefined cottonseed oil, containing high levels of gossypol, in certain areas of China was associated with a pronounced decrease in birth rate, and purified gossypol has been evaluated as a male contraceptive [170].

No information on the chronic oral or inhalation toxicity of phenanthrene-9,10-quinone is available. Since this substance is a common atmospheric pollutant, and in view of the observation that individuals exposed to diesel exhaust fumes, of which phenanthrene-9,10-quinone is a major component [171], show oxidative DNA damage, such studies would be of considerable value.

ROLE OF NQO1 IN QUINONE TOXICITY

The enzyme NQO1 (then named "DT-diaphorase") was first reported by Ernster et al. in 1962 [12]. They showed that NQO1 reduced a number of quinones to the corresponding hydroquinones with either NADH or NADPH as cofactor, and that the enzyme was strongly inhibited by dicoumarol. In later publications by Ernster and colleagues, it was suggested that NQO1 could protect against the toxic effect of quinones by producing "relatively stable" hydroquinones, thus avoiding one-electron reduction of the quinones to the highly unstable semiquinones, which can react with molecular oxygen to produce ROS [172]. The concept that NQO1 protected against quinone toxicity became dogma [173], even though later publications indicated that two-electron reduction of quinones does not necessarily lead to detoxification [174,175], and it is still regularly stated in the literature that NQO1 detoxifies quinones.

In fact, NQO1 may either detoxify or activate quinones, and which process occurs depends upon the rate of reduction of the quinone, the rate and mechanism of autoxidation of the hydroquinone, and the concentration of NQO1 employed. These factors are clearly shown in studies with quinones and purified NQO1. With menadione, redox cycling occurred at low levels of NQO1, indicating activation. As the level of NQO1 was increased, however, redox cycling was progressively inhibited, indicating detoxication. Menadione is rapidly reduced by NQO1. At low levels, only a proportion of the quinone will be reduced to the hydroquinone, so that semiquinone formation via the comproportionation reaction (Reaction 8.7) will proceed, leading to redox cycling and ROS production. Furthermore, with menadione, the equilibrium position of Reaction 8.6 lies to the right, so the superoxide radical will be available for further semiquinone formation via Reaction 8.8. As the level of NQO1 is increased, the concentration of quinone will decrease, and oxidation via the comproportionation reaction will be inhibited, thereby leading to a decrease in oxidation rate and ROS production. Similar effects were seen with 2,3-dimethyl- and 2,3-dimethoxy-1,4-naphthohydroquinone, although with these compounds, a higher concentration of NQO1 was required for inhibition than that for menadione, possibly reflecting the relatively low rates of reduction of these compounds by NQO1 [46].

In contrast, the rates of redox cycling of 2-hydroxy- and 2-amino-1,4-naphthoquinone and streptonigrin increased with increasing levels of NQO1, and no inhibition was seen even at high levels of this enzyme. These compounds are but slowly reduced by NQO1, so that it is likely that levels of the hydroquinone were never low enough to inhibit Reaction 8.7 [30,46]. Redox cycling with juglone in the presence of NQO1 was very fast, and again no inhibition was observed in the presence of high levels of this enzyme [46]. Juglone is reduced relatively rapidly by NQO1, and in this case it is likely that the failure of inhibition is due not to slow reduction but to the fact that the comproportionation reaction is unimportant in the mechanism of oxidation of this compound [46]. Similar results are to be expected with 2,3-dichloro- and 2-bromo-1,4-naphthoquinone, since the comproportionation reaction likewise plays little part in the autoxidation of the hydroquinones derived from these substances.

CONCLUSIONS

It has been concluded in many cases that the ability of quinones to undergo redox cycling, with the generation of ROS, is responsible for the biological effects of the compounds. Certainly, redox cycling occurs during the interaction of these compounds with biological reducing agents at physiological pH, making such a conclusion feasible.

With regard to effects in cells *in vitro*, several criteria for the involvement of ROS in toxicity may be considered:

1. The demonstration of ROS within the cells incubated with the test compounds
2. The demonstration of a protective effect of scavengers of ROS
3. The demonstration that factors that increase or decrease the production of ROS by quinones *in vitro* increase or decrease the cytotoxicity of these substances

ROS formation has consistently been reported in cells incubated with quinones and hydroquinones. There is a question with regard to the site at which ROS production occurs, particularly with hydroquinones, which may undergo autoxidation in the culture medium, thereby generating ROS extracellularly.

N-Acetylcysteine has regularly been described as a scavenger of ROS, and addition of this substance to the culture medium has been shown to protect against the cytotoxicity of benzoquinones [176], 6-hydroxydopamine [177], naphthoquinones [178,179], anthraquinones [180], and phenanthrene-9,10-quinone [181]. The rate of reaction of *N*-acetylcysteine with superoxide radical and hydrogen peroxide is low, however [182,183], and it has been suggested that protection is given not by direct scavenging of ROS but by the ability of this thiol to increase intracellular levels of NADPH, thereby maintaining intracellular antioxidant defenses [184].

As discussed, ROS production by menadione and 2,3-dimethoxy-1,4-naphthoquinone was inhibited by high levels of NQO1. It would be expected that cytotoxicity would be decreased by increasing cellular levels of NQO1, while it would be increased by inhibiting this enzyme. In accord with these expectations, the toxic effects of menadione [185] and 2,3-dimethoxy-1,4-naphthoquinone [186] in cells *in vitro* have been shown to be inversely proportional to cellular NQO1 levels and the cytotoxicity of these substances was increased by inhibition of NQO1 [187,188]. Similar effects have been observed with benzoquinone [189], doxorubicin [190], 2,6-dimethoxy-1,4-benzoquinone [187], and mitoxantrone [191].

In contrast, high levels of NQO1 stimulated ROS production by streptonigrin, and it would therefore be expected that the cytotoxicity of this compound would be increased by increasing cellular levels of this enzyme and protection would be afforded by its inhibition. Again, such expectation has been fulfilled [192]. Similar effects have been recorded in cells exposed to 17-(allylamino)-17-demethoxygeldanamycin [193], 2,3-dimethoxy-1,4-naphthoquinone [186], and β-lapachone [102]. Overall, there is good evidence for the involvement of ROS in quinone toxicity to cells *in vitro*.

For consideration of the involvement of ROS in the *in vivo* toxicity of quinones and hydroquinones, one may consider the following criteria:

1. Demonstration of ROS production, or of the products of oxidation by ROS, *in vivo* at the sites at which toxicity has been observed
2. Demonstration of ROS production in the target cells *in vivo*
3. Proportionality between the efficacy of derivatives in generating ROS *in vitro* and the severity of the toxic effects that they induce
4. Demonstration of a protective effect of scavengers of ROS
5. Effects of modulators of ROS production on toxicity *in vivo*
6. The induction of toxic effects similar to those observed with the test compound by other ROS generators

Some of these criteria have been met with regard to the hemolytic activity of quinones and hydroquinones. ROS production was observed with several of the hemolytic quinones *in vitro*, and hydrogen peroxide was detected in erythrocytes of mice injected with menadione [74]. Hemolysis by quinones and hydroquinones was of the oxidative type, as indicated by methemoglobinemia and Heinz body formation, in erythrocytes *in vivo*. Erythrocytes are particularly susceptible to oxidative damage, and oxidative hemolysis is seen with other compounds that are known to generate ROS, such as polysulfides [194] and aromatic amines [195]. In the case of mono-alkyl-1,4-naphthoquinones, hemolytic activity in rats was correlated with their ability to generate superoxide radical through reaction with oxyhemoglobin, and with their ability to generate hydrogen peroxide and induce oxidative damage in erythro-cytes *in vitro*, indicating that no extra-erythrocytic activation is needed for their *in vivo* hemolytic action [77].

Preadministration of compounds that increase the activity of Phase 2 enzymes, including NQO1, would be expected to decrease the toxicity of compounds that are detoxified by this enzyme while toxicity would be increased by concurrent administration of an NQO1 inhibitor. The severity of hemolytic anemia induced in rats by menadione was indeed decreased by increasing tissue levels of Phase 2 enzymes, and the degree of protection was correlated with the extent of NQO1 induction in the livers of the animals. Inhibition of NQO1 increased the severity of hemolysis [196]. Similarly, the hemolytic action of 2,3-dimethyl-1,4-naphthoquinone was decreased in rats by induction of Phase 2 enzymes [121]. Conversely, pretreatment of rats with the Phase 2 enzyme inducers butylated hydroxyanisole, butylated hydroxytoluene, dimethyl fumarate, or disulfiram increased the severity of the hemolytic anemia induced by 2-hydroxy- and 2-amino-1,4-naphthoquinone, for which high levels of NQO1 were shown to increase ROS production. The nephrotoxicity of the latter compound was also decreased by all the inducers, but that of 2-hydroxy-1,4-naphthoquinone was decreased only by dimethyl fumarate or disulfiram [196,197]. The reason for this disparity is unknown.

The increased susceptibility of individuals with glucose-6-phosphate dehydrogenase deficiency, in which the inability of erythrocytes to maintain intracellular levels of GSH makes them particularly susceptible to oxidative damage, is also consistent with the involvement of ROS in quinone-induced hemolysis. Although high levels of doxorubicin were shown to generate ROS in erythrocytes *in vitro*, accompanied by methemoglobin formation [83], no hemolysis has been reported in animals dosed with these compounds. In view of the susceptibility of erythrocytes from glucose-6-phosphate dehydrogenase–deficient individuals to oxidative damage, it was suggested in 1980 that these drugs should be used with caution in patients with this disorder [198]. Three years later, severe oxidative hemolysis was reported in such an individual, following administration of doxorubicin [199].

Thus, there is evidence for the involvement of ROS in the hemolytic activity of quinones *in vivo*.

The involvement of ROS in the nephrotoxicity of quinones is not established. The nephrotoxic action of 2,3-dimethyl-1,4-naphthoquinone, which is detoxified by high levels of NQO1, was abolished by induction of Phase 2 enzymes, while inhibition of NQO1 caused a massive increase in the severity of the renal tubular necrosis induced by this compound [121]. The nephrotoxicity of 2-amino-1,4-naphthoquinone was decreased by butylated hydroxyanisole, butylated hydroxytoluene, dimethyl fumarate, or disulfiram, but that of 2-hydroxy-1,4-naphthoquinone was decreased only by dimethyl fumarate or disulfiram [196,197]. The reason for this disparity is unknown. It is therefore not possible to generalize with regard to the effects of modulators of Phase 2 enzyme induction on the nephrotoxicity of naphthoquinones *in vivo*. Structure–activity relationships indicate the importance of electron-donating groups for nephrotoxic activity. Compounds with a relatively weak electron-donating effect, such as an alkyl group, or with an electron-withdrawing group, are not nephrotoxic. Dialkyl-1,4-naphthoquinones contain two electron-donating groups, and these are nephrotoxic. 2-Hydroxy- and 2-amino-1,4-naphthoquinone are potent nephrotoxins, and both these substituents are strongly electron donating. Furthermore, methylation of the amino group, which produces a greater electron-donating effect, increases the nephrotoxic activity of 2-aminonaphthoquinones. However, substitution with either electron-donating or electron-withdrawing groups at the 3-position of 2-hydroxy- or 2-amino-naphthoquinones decreases nephrotoxic activity. It must be concluded that a free 3-position in hydroxy and amino-1,4-naphthoquinones is important for renal toxicity, although the role that this may play in the nephrotoxic action is unclear, and further work on the mechanism of the renal damage induced by these compounds is required.

Although several other ROS-generating compounds, such as diquat [200], paracetamol [201], and sodium selenite [202], have been shown to be cataractogenic in animals, there is no direct evidence that the cataract formation by quinones is attributable to ROS production. Similarly, testicular damage and decreased spermatogenesis have been observed not only with quinones, but also with other ROS producers, including arsenic trioxide [203], aromatic nitro compounds [204], sodium selenite [205], quinacrine [206], 1,4-diaminobenzene [207], and paraquat [208], but again there is no direct evidence for the involvement of ROS in the induction of such lesions by quinones.

Innumerable studies on the toxicity of quinones to cells *in vitro* have been conducted, and the results of experiments on new compounds and on the effects of previously studied compounds on different cell lines continue to be published.

Detailed investigations on the mechanisms of cytotoxicity of quinones and the signaling pathways involved have been conducted, but no clear structure–activity relationships have been identified. Although many reports on the *in vitro* effects of quinones end with a statement along the lines of "this compound is a promising drug for the therapy of such-and-such cancer," few studies have progressed to the next steps of the evaluation process, namely, demonstration of growth inhibition effects in xenografts and assessment of toxicity.

There has been much interest in the use of quinones in cancer therapy, and the use of anthracyclines is an outstanding example of the value of such compounds. Results with other quinones have, however, generally been rather disappointing. In a Phase I clinical trial in cancer patients, administration of 17-(allylamino)-17-de-methoxygeldanamycin was associated with dose-limiting hepatotoxicity and anemia [157]. A small Phase II trial involving administration of this compound to cancer patients showed no beneficial effects, but good effects in the intravesicle therapy of nonmuscle invasive bladder cancer were observed [209]. Clinical development of other geldanamycin derivatives is continuing [210]. Several azirinidylquinones have been used clinically as anticancer drugs, although further development of some compounds of this class was discontinued because of toxicity issues [211]. Mitomycin C has been used as a cancer therapeutic agent in Japan since the early 1960s, although it was not approved for use in North America until 1974. It is effective in the treatment of a variety of solid tumors, although its toxicity has restricted its use. Menadione and derivatives were tested in clinical trials, but no beneficial effects were seen, and hemolytic anemia was a serious side effect. In the 1960s and 1970s, there was considerable interest in the potential use of streptonigrin for cancer chemotherapy. In early studies in patients with advanced cancer at various sites [212,213], streptonigrin was administered at various doses and by different routes of administration. A positive response in terms of objective tumor regression and clinical benefit was claimed in some cases. These findings were not substantiated in later studies, however, and streptonigrin was found to cause severe bone marrow depression in patients [214], and human trials were discontinued. A complex of β-lapachone with hydroxypropyl-β-cyclodextrin, code-named ARQ 501, underwent unsuccessful clinical trials as a therapeutic for a variety of cancers. The reasons for failure included dose-limiting toxicity in the form of hemolytic anemia [49,169]. Research on derivatives of β-lapachone is continuing.

Many cancer cells contain high levels of NQO1, and it was suggested that this property could be exploited by anticancer drugs that redox cycle with this enzyme leading to the generation of ROS [215]. However, this conclusion does not take into account the fact that while many quinones undergo redox cycling with NQO1 at relatively low concentrations, redox cycling and ROS production by some such compounds are inhibited at high levels of the enzyme. Such compounds may therefore be ineffective in cancer cells with high activities of NQO1. In this situation, compounds with which redox cycling with NQO1 is not inhibited at high levels of the enzyme may be the agents of choice. This property is seen with some dihydroxy-1,4-naphtho-quinones and dihydroxyanthraquinones, with halogenated 1,4-naphthoquinones and with many 1,2-naphthoquinones.

The situation with regard to the use of quinones in the therapy of malaria, schistosomiasis, and trypanosomiasis is in some respects similar to that regarding their use as anticancer agents. Atovaquone (*trans*-2-[4-(4-chlorophenyl) cyclohexyl]-3-hydroxy-1,4-naphthoquinone) in combination with proguanil hydrochloride is used in the prophylaxis and therapy of malaria under the trade name "Malarone" [8] and many new hydroxynaphthoquinones have been tested for their effects on the parasites *in vitro*, and some have been shown to be effective in mouse models [216], while others, while highly effective *in vitro*, showed no benefit in such models [217]. A mixture of menadione and ascorbate was shown to be highly toxic to the epimastigote, trypomastigote, and amastigote forms of *Trypanosoma cruzi* [218], suggesting that further work with this preparation would be worthwhile, although the possibility of hemolytic anemia as a side effect must be considered.

Several quinones have been shown to protect against chemically induced cancer, and such effects may reflect intervention at either the initiation or subsequent phases of carcinogenesis. Quinones have been shown to increase tissue activities of Phase 2 enzymes in animal tissues, and since these enzymes are known to facilitate the detoxification of carcinogens by conversion to water-soluble metabolites that can be excreted in urine [219,220], it is possible that enzyme induction could be responsible for the observed protection against chemically-induced carcinogenesis. In this situation, the test material would have to be given before exposure to the carcinogen. Alternatively, the toxic effects of quinones and hydroquinones on cancer cells, as demonstrated *in vitro*, could, if translated to the *in vivo* situation, lead to a decrease in the promotion and/or progression of chemically induced tumors. In this case, the test compounds would be effective when administered after of the carcinogen. These possibilities have not been sufficiently explored, with the test compounds in most experiments being administered before, during and after the carcinogen. In the few comparative studies that have been conducted, some compounds were shown to be more effective when given before the carcinogen, suggesting an effect at the initiation phase, while one compound, 3,4-dihydroxy-benzoic acid, was effective when given either before or after the carcinogen. How such effects could be translated into the human situation for protection against cancer is unclear.

There is evidence for the involvement of reactive oxygen species in the toxicity of quinones *in vitro* and *in vivo*, and their ability to generate such species may be exploited for protection against cancer, and for therapy of cancer and of parasitic diseases. Substantial progress has been made on the possibility of such use via *in vitro* experiments and studies in animals, and it is to be hoped that some of the more active compounds identified in these studies will progress toward clinical use.

ABBREVIATIONS

ROS   reactive oxygen species
GSH   reduced glutathione
SOD   superoxide dismutase
NQO1 NAD(P)H:quinone acceptor oxidoreductase

183

REFERENCES

1. Song, Y. and G. R. Buettner. 2010. Thermodynamic and kinetic considerations for the reaction of semiquinone radicals to form superoxide and hydrogen peroxide. *Free Radical Biology and Medicine* 49: 919–962.
2. Thomson, R. H. 1987. *Naturally Occurring Quinones III: Recent Advances.* London, U.K.: Chapman & Hall.
3. Deisinger, P. J., T. S. Hill, and J. C. English. 1996. Human exposure to naturally occurring hydroquinone. *Journal of Toxicology and Environmental Health* 47: 31–46.
4. Zhao, Y., J. Anichina, X. Lu et al. 2012. Occurrence and formation of chloro- and bromo-benzoquinones during drinking water disinfection. *Water Research* 46: 4351–4360.
5. Semwal, R. B., D. K. Semwal, S. Combrinck, C. Cartwright-Jones, and A. Viljoen. 2014. *Lawsonia inermis* L.(henna): Ethnobotanical, phytochemical and pharmacological aspects. *Journal of Ethnopharmacology* 155: 80–103.
6. AlGhamdi, K. M. 2010. The use of topical bleaching agents among women: A cross-sectional study of knowledge, attitude and practices. *Journal of the European Academy of Dermatology and Venereology* 24: 1214–1219.
7. Asche, C. 2005. Antitumour quinones. *Mini-Reviews in Medicinal Chemistry* 5: 449–467.
8. Schuck, D. C., S. B. Ferreira, L. N. Cruz et al. 2013. Biological evaluation of hydroxy-napthoquinones as anti-malarials. *Malaria Journal* 12: 234.
9. Siegel, D., P. Reigan, and D. Ross. 2008. One-and two-electron-mediated reduction of quinones: Enzymology and toxicological implications. In *Advances in Bioactivation Research*, ed. A. A. Elfarra. Springer, Berlin, Germany, pp. 169–200.
10. Yang, S., Y.-H. Jan, J. P. Gray et al. 2013. Sepiapterin reductase mediates chemical redox cycling in lung epithelial cells. *Journal of Biological Chemistry* 288: 19221–19237.
11. Matsuda, H., S. Kimura, and T. Iyanagi. 2000. One-electron reduction of quinones by the neuronal nitric-oxide synthase reductase domain. *Biochimica et Biophysica Acta* 1359: 106–116.
12. Ernster, L., L. Danielson, and M. Ljunggren. 1962. DT diaphorase I. Purification from the soluble fraction of rat-liver cytoplasm, and properties. *Biochimica et Biophysica Acta* 58: 171–188.
13. Oppermann, U. 2007. Carbonyl reductases: The complex relationships of mammalian carbonyl-and quinone-reducing enzymes and their role in physiology. *Annual Review of Pharmacology and Toxicology* 47: 293–322.
14. Čénas, N. K., G. A. Rakauskiené, and J. J. Kulys. 1989. One-and two-electron reduction of quinones by glutathione reductase. *Biochimica et Biophysica Acta* 973: 399–404.
15. Čénas, N., H. Nivinskas, Z. Anusevicius, J. Sarlauskas, F. Lederer, and E. S. J. Arnér. 2004. Interactions of quinones with thioredoxin reductase: A challenge to the antioxidant role of the mammalian selenoprotein. *Journal of Biological Chemistry* 279: 2583–2592.
16. Belorgey, D., D. A. Lanfranchi, and E. Davioud-Charvet. 2013. 1,4-Naphthoquinones and other NADPH-dependent glutathione reductase-catalyzed redox cyclers as anti-malarial agents. *Current Pharmaceutical Design* 19: 2512–2528.
17. Vienožinskis, J., A. Butkus, N. Čénas, and J. Kulys. 1990. The mechanism of the quinone reductase reaction of pig heart lipoamide dehydrogenase. *Biochemical Journal* 269: 101–105.
18. Carlson, B. W. and L. L. Miller. 1985. Mechanism of the oxidation of NADH by quinones Energetics of one-electron and hydride routes. *Journal of the American Chemical Society* 107: 479–485.
19. Wilson, I., P. Wardman, T.-S. Lin, and A. C. Sartorelli. 1987. Reactivity of thiols towards derivatives of 2- and 6-methyl-1,4-naphthoquinone bioreductive alkylating agents. *Chemico-Biological Interactions* 61: 229–240.

20. Isaacs, N. S. and R. van Eldik. 1997. A mechanistic study of the reduction of quinones by ascorbic acid. *Journal of the Chemical Society, Perkin Transactions* 2: 1465–1468.
21. Winterbourn, C. C., J. K. French, and R. F. Claridge. 1979. The reaction of menadione with haemoglobin. Mechanism and effect of superoxide dismutase. *Biochemical Journal* 179: 665–673.
22. Monks, T. J., R. P. Hanzlik, G. M. Cohen, D. Ross, and D. G. Graham. 1992. Quinone chemistry and toxicity. *Toxicology and Applied Pharmacology* 112: 2–16.
23. Gao, R., Z. Yuan, Z. Zhao, and X. Gao. 1998. Mechanism of pyrogallol autoxidation and determination of superoxide dismutase enzyme activity. *Bioelectrochemistry and Bioenergetics* 45: 41–45.
24. Lewis, J. G., W. Stewart, and D. O. Adams. 1988. Role of oxygen radicals in induction of DNA damage by metabolites of benzene. *Cancer Research* 48: 4762–4765.
25. Varela, E. and M. Tien. 2003. Effect of pH and oxalate on hydroquinone-derived hydroxyl radical formation during brown rot wood degradation. *Applied and Environmental Microbiology* 69: 6025–6031.
26. Munday, R. 1997. Inhibition of naphthohydroquinone autoxidation by DT-diaphorase (NAD(P)H:[quinone acceptor]oxidoreductase). *Redox Report* 3: 189–196.
27. Munday, R. 2000. Autoxidation of naphthohydroquinones: Effects of pH, naphthoquinones and superoxide dismutase. *Free Radical Research* 32: 245–253.
28. Eyer, P. 1991. Effects of superoxide dismutase on the autoxidation of 1,4-hydroquinone. *Chemico-Biological Interactions* 80: 159–176.
29. Kwok, E. S. C. and D. A. Eastmond. 1997. Effects of pH on nonenzymatic oxidation of phenylhydroquinone: Potential role in urinary bladder carcinogenesis induced by *o*-phenylphenol in Fischer 344 rats. *Chemical Research in Toxicology* 10: 742–749.
30. Munday, R. 2004. Activation and detoxification of naphthoquinones by NAD(P)H:quinone oxidoreductase. *Methods in Enzymology* 382: 364–380.
31. Tayama, S. and Y. Nakagawa. 1994. Effect of scavengers of active oxygen species on cell damage caused in CHO-K1 cells by phenylhydroquinone, an *o*-phenylphenol metabolite. *Mutation Research* 324: 121–131.
32. Amaro, A. R., G. G. Oakley, U. Bauer, H. P. Spielmann, and L. W. Robertson. 1996. Metabolic activation of PCBs to quinones: Reactivity toward nitrogen and sulfur nucleophiles and influence of superoxide dismutase. *Chemical Research in Toxicology* 9: 623–629.
33. Cadenas, E., D. Mira, A. Brunmark, C. Lind, J. Segura-Aguilar, and L. Ernster. 1988. Effect of superoxide dismutase on the autoxidation of various hydroquinones—A possible role of superoxide dismutase as a superoxide: semiquinone oxidoreductase. *Free Radical Biology and Medicine* 5: 71–79.
34. Hiramoto, K., R. Mochizuki, and K. Kikugawa. 2001. Generation of hydrogen peroxide from hydroxyhydroquinone and its inhibition by superoxide dismutase. *Journal of Oleo Science* 50: 21–28.
35. Marklund, S. and G. Marklund. 1974. Involvement of the superoxide anion radical in the autoxidation of pyrogallol and a convenient assay for superoxide dismutase. *European Journal of Biochemistry* 47: 469–474.
36. Misra, H. P. and I. Fridovich. 1972. The role of superoxide anion in the autoxidation of epinephrine and a simple assay for superoxide dismutase. *Journal of Biological Chemistry* 247: 3170–3175.
37. Heikkila, R. E. and F. Cabbat. 1976. A sensitive assay for superoxide dismutase based on the autoxidation of 6-hydroxydopamine. *Analytical Biochemistry* 75: 356–362.
38. Frei, B., K. H. Winterhalter, and C. Richter. 1986. Menadione- (2-methyl-1,4-naphthoquinone)-dependent enzymic redox cycling and calcium release by mitochondria. *Biochemistry* 25: 4438–4443.

39. Pritsos, C. A., D. E. Jensen, D. Pisani, and R. S. Pardini. 1982. Involvement of super-oxide in the interaction of 2,3-dichloro-1,4-naphthoquinone with mitochondrial membranes. *Archives of Biochemistry and Biophysics* 217: 98–109.
40. Bergmann, B., J. K. Dohrmann, and R. Kahl. 1992. Formation of the semiquinone anion radical from *tert*-butylquinone and from *tert*-butylhydroquinone in rat liver microsomes. *Toxicology* 74: 127–133.
41. Kumagai, Y., Y. Tsurutani, M. Shinyashiki et al. 1997. Bioactivation of lapachol responsible for DNA scission by NADPH-cytochrome P450 reductase. *Environmental Toxicology and Pharmacology* 3: 245–250.
42. Cone, R., S. K. Hasan, J. W. Lown, and A. R. Morgan. 1976. The mechanism of the degradation of DNA by streptonigrin. *Canadian Journal of Biochemistry* 54: 219–223.
43. McLean, M. R., T. P. Twaroski, and L. W. Robertson. 2000. Redox cycling of 2-(x′-mono,-di,-trichlorophenyl)-1,4-benzoquinones, oxidation products of polychlorinated biphenyls. *Archives of Biochemistry and Biophysics* 376: 449–455.
44. Kviecinski, M. R., R. C. Pedrosa, K. B. Felipe et al. 2012. Inhibition of cell proliferation and migration by oxidative stress from ascorbate-driven juglone redox cycling in human bladder-derived T24 cells. *Biochemical and Biophysical Research Communications* 421: 268–273.
45. Shang, Y., C. Chen, Y. Li, J. Zhao, and T. Zhu. 2012. Hydroxyl radical generation mechanism during the redox cycling process of 1,4-naphthoquinone. *Environmental Science & Technology* 46: 2935–2942.
46. Munday, R. 2001. Concerted action of DT-diaphorase and superoxide dismutase in preventing redox cycling of naphthoquinones: An evaluation. *Free Radical Research* 35: 145–158.
47. Ross, D., J. K. Kepa, S. L. Winski, H. D. Beall, A. Anwar, and D. Siegel. 2000. NAD(P) H:quinone oxidoreductase 1 (NQO1): Chemoprotection, bioactivation, gene regulation and genetic polymorphisms. *Chemico-Biological Interactions* 129: 77–97.
48. Molina Portela, M. P. and A. O. M. Stoppani. 1996. Redox cycling of β-lapachone and related *o*-naphthoquinones in the presence of dihydrolipoamide and oxygen. *Biochemical Pharmacology* 51: 275–283.
49. Wilson, W. L., C. Labra, and E. Barrist. 1961. Preliminary observations on the use of streptonigrin as an antitumor agent in human beings. *Antibiotics & Chemotherapy* 11: 147–150.
50. Monks, T. J. and S. S. Lau. 1992. Toxicology of quinone-thioethers. *Critical Reviews in Toxicology* 22: 243–270.
51. Sun, Y., K. Taguchi, D. Sumi, S. Yamano, and Y. Kumagai. 2006. Inhibition of endothelial nitric oxide synthase activity and suppression of endothelium-dependent vaso-relaxation by 1,2-naphthoquinone, a component of diesel exhaust particles. *Archives of Toxicology* 80: 280–285.
52. Chung, S.-M., J.-Y. Lee, M.-Y. Lee, O.-N. Bae, and J.-H. Chung. 2001. Adverse consequences of erythrocyte exposure to menadione: Involvement of reactive oxygen species generation in plasma. *Journal of Toxicology and Environmental Health* 63A: 617–629.
53. Mlejnek, P. and P. Dolezel. 2014. N-acetylcysteine prevents the geldanamycin cytotoxicity by forming geldanamycin–N-acetylcysteine adduct. *Chemico-Biological Interactions* 220: 248–254.
54. Winterbourn, C. C. 2014. The challenges of using fluorescent probes to detect and quantify specific reactive oxygen species in living cells. *Biochimica et Biophysica Acta* 1840: 730–738.
55. Passi, S., M. Picardo, and M. Nazzaro-Porro. 1987. Comparative cytotoxicity of phenols in vitro. *Biochemical Journal* 245: 537–542.
56. Saito, Y., K. Nishio, Y. Ogawa et al. 2007. Molecular mechanisms of 6-hydroxydopamine-induced cytotoxicity in PC12 cells: Involvement of hydrogen peroxide-dependent and-independent action. *Free Radical Biology and Medicine* 42: 675–685.

57. Nose, M., T. Koide, K. Morikawa et al. 1998. Formation of reactive oxygen interme-diates might be involved in the trypanocidal activity of gallic acid. *Biological and Pharmaceutical Bulletin* 21: 583–587.

58. Tiffany-Castiglioni, E., R. P. Saneto, P. H. Proctor, and J. R. Perez-Polo. 1982. Participation of active oxygen species in 6-hydroxydopamine toxicity to a human neu-roblastoma cell line. *Biochemical Pharmacology* 31: 181–188.

59. Grankvist, K. 1989. Gossypol-induced free radical toxicity to isolated islet cells. *International Journal of Biochemistry* 21: 853–856.

60. Jeng, J.-H., Y.-J. Wang, W. H. Chang et al. 2004. Reactive oxygen species are crucial for hydroxychavicol toxicity toward KB epithelial cells. *Cellular and Molecular Life Sciences* 61: 83–96.

61. Inoue, S., K. I. Yamamoto, and S. Kawanishi. 1990. DNA damage induced by metabolites of o-phenylphenol in the presence of copper(II) ion. *Chemical Research in Toxicology* 3: 144–149.

62. Abe, K. and H. Saito. 1996. Menadione toxicity in cultured rat cortical astrocytes. *Japanese Journal of Pharmacology* 72: 299–306.

63. Sun, J.-S., Y.-H. Tsuang, W.-C. Huang, L.-T. Chen, Y.-S. Hang, and F.-J. Lu. 1997. Menadione-induced cytotoxicity to rat osteoblasts. *Cellular and Molecular Life Sciences* 53: 967–976.

64. Tan, A. S. and M. V. Berridge. 2010. Evidence for NAD(P)H:quinone oxidoreductase 1 (NQO1)-mediated quinone-dependent redox cycling via plasma membrane electron transport: A sensitive cellular assay for NQO1. *Free Radical Biology and Medicine* 48: 421–429.

65. Bey, E. A., K. E. Reinicke, M. C. Srougi, M. Varnes, V. E. Anderson, J. J. Pink, L. S. Li, M. Patel, L. Cao, and Z. Moore. 2013. Catalase abrogates β-lapachone-induced PARP1 hyperactivation–directed programmed necrosis in NQO1-positive breast can-cers. *Molecular Cancer Therapeutics* 12: 2110–2120.

66. McGuire, K., J. M. Jamison, J. Gilloteaux, and J. L. Summers. 2013. Synergistic anti-tumor activity of vitamins C and K_3 on human bladder cancer cell lines. *Journal of Cancer Therapy* 4: 7–19.

67. Verrax, J., M. Delvaux, N. Beghein, H. Taper, B. Gallez, and P. Buc Calderon. 2005. Enhancement of quinone redox cycling by ascorbate induces a caspase-3 indepen-dent cell death in human leukaemia cells. An *in vitro* comparative study. *Free Radical Research* 39: 649–657.

68. Felipe, K. B., J. Benites, C. Glorieux et al. 2013. Antiproliferative effects of phenyl-aminonaphthoquinones are increased by ascorbate and associated with the appearance of a senescent phenotype in human bladder cancer cells. *Biochemical and Biophysical Research Communications* 433: 573–578.

69. Burdon, R. H. 1995. Superoxide and hydrogen peroxide in relation to mammalian cell proliferation. *Free Radical Biology and Medicine* 18: 775–794.

70. Halliwell, B. 2007. Oxidative stress and cancer: Have we moved forward? *Biochemical Journal* 401: 1–11.

71. Kimura, K., M. Takada, T. Ishii, K. Tsuji-Naito, and M. Akagawa. 2012. Pyrroloquinoline quinone stimulates epithelial cell proliferation by activating epider-mal growth factor receptor through redox cycling. *Free Radical Biology and Medicine* 53: 1239–1251.

72. Matsunaga, T., Y. Morikawa, M. Haga et al. 2014. Exposure to 9,10-phenanthrene-quinone accelerates malignant progression of lung cancer cells through up-regulation of aldo-keto reductase 1B10. *Toxicology and Applied Pharmacology* 278: 180–189.

73. Margoliash, E. and A. Novogrodsky. 1958. A study of the inhibition of catalase by 3-amino-1:2:4-triazole. *Biochemical Journal* 68: 468–475.

74. Cohen, G. and P. Hochstein. 1964. Generation of hydrogen peroxide in erythrocytes by hemolytic agents. *Biochemistry* 3: 895–900.

75. Kusumoto, S. and T. Nakajima. 1964. Methemoglobin formation by aminophenol and diphenol in rabbits. *Industrial Health* 2: 133–138.

76. Harley, J. D. and H. Robin. 1963. Adaptive mechanisms in erythrocytes exposed to naphthoquinones. *Australian Journal of Experimental Biology* 41: 281–292.

77. Munday, R., E. A. Fowke, B. L. Smith, and C. M. Munday. 1994. Comparative toxicity of alkyl-1,4-naphthoquinones in rats: Relationship to free radical production *in vitro*. *Free Radical Biology and Medicine* 16: 725–731.

78. Munday, R., B. L. Smith, and C. M. Munday. 1995. Comparative toxicity of 2-hydroxy-3-alkyl-1,4-naphthoquinones in rats. *Chemico-Biological Interactions* 98: 185–192.

79. Munday, R., B. L. Smith, and C. M. Munday. 2007. Structure-activity relationships in the haemolytic activity and nephrotoxicity of derivatives of 1,2- and 1,4-naphthoquinone. *Journal of Applied Toxicology* 27: 262–269.

80. Sikka, H. C., E. H. Schwartzel, J. Saxena, and G. Zweig. 1974. Interaction of dichlone with human erythrocytes. I. Changes in cell permeability. *Chemico-Biological Interactions* 9: 261–272.

81. Hart, L. A., N. A. A. van der Wal, A. S. Koster, and R. P. Labadie. 1989. A rapid and sensitive micro-assay to determine the capacity of quinones to undergo redox cycling. *Toxicology Letters* 48: 151–157.

82. Lopes, J. N., F. S. Cruz, R. Docampo et al. 1978. *In vitro* and *in vivo* evaluation of the toxicity of 1,4-naphthoquinone and 1,2-naphthoquinone derivatives against *Trypanosoma cruzi*. *Annals of Tropical Medicine and Parasitology* 72: 523–531.

83. Pedersen, J. Z., L. Marcocci, L. Rossi, I. Mavelli, and G. Rotilio. 1988. First electron spin resonance evidence for the generation of the daunomycin free radical and superoxide by red blood cell membranes. *Annals of the New York Academy of Sciences* 551: 121–127.

84. Zinkham, W. H. and F. A. Oski. 1996. Henna: A potential cause of oxidative hemolysis and neonatal hyperbilirubinemia. *Pediatrics* 97: 707–709.

85. Bashan, N., O. Makover, A. Livne, and S. Moses. 1980. Effect of oxidant agents on normal and G6PD-deficient erythrocytes. *Israel Journal of Medical Sciences* 16: 351–356.

86. Zalatnai, A., K. Lapis, B. Szende et al. 2001. Wheat germ extract inhibits experimental colon carcinogenesis in F-344 rats. *Carcinogenesis* 22: 1649–1652.

87. Sugie, S., K. Okamoto, K. M. W. Rahman et al. 1998. Inhibitory effects of plumbagin and juglone on azoxymethane-induced intestinal carcinogenesis in rats. *Cancer Letters* 127: 177–183.

88. Yoshimi, N., A. Wang, Y. Morishita et al. 1992. Modifying effects of fungal and herb metabolites on azoxymethane-induced intestinal carcinogenesis in rats. *Japanese Journal of Cancer Research* 83: 1273–1278.

89. Higa, R. A., R. D. Aydos, I. S. Silva, R. T. Ramalho, and A. S. de Souza. 2011. Study of the antineoplastic action of *Tabebuia avellanedae* in carcinogenesis induced by azoxymethane in mice. *Acta Cirurgica Brasileira* 26: 125–128.

90. Raghunandhakumar, S., A. Paramasivam, S. Senthilraja et al. 2013. Thymoquinone inhibits cell proliferation through regulation of G1/S phase cell cycle transition in N-nitrosodiethylamine-induced experimental rat hepatocellular carcinoma. *Toxicology Letters* 223: 60–72.

91. Jagan, S., G. Ramakrishnan, P. Anandakumar, S. Kamaraj, and T. Devaki. 2008. Antiproliferative potential of gallic acid against diethylnitrosamine-induced rat hepatocellular carcinoma. *Molecular and Cellular Biochemistry* 319: 51–59.

92. Tanaka, T., T. Kojima, T. Kawamori, and H. Mori. 1995. Chemoprevention of digestive organs carcinogenesis by natural product protocatechuic acid. *Cancer* 75: 1433–1439.

93. Tanaka, T., T. Kojima, T. Kawamori, N. Yoshimi, and H. Mori. 1993. Chemoprevention of diethylnitrosamine-induced hepatocarcinogenesis by a simple phenolic acid protocatechuic acid in rats. *Cancer Research* 53: 2775–2779.

94. Zhu, W.-Q., J. Wang, X.-F. Guo, Z. Liu, and W.-G. Dong. 2016. Thymoquinone inhibits proliferation in gastric cancer via the STAT3 pathway *in vivo* and *in vitro*. *World Journal of Gastroenterology* 22: 4149–4159.

95. Phillips, R. M., A. M. Burger, P. M. Loadman, C. M. Jarrett, D. J. Swaine, and H.-H. Fiebig. 2000. Predicting tumor responses to mitomycin C on the basis of DT-diaphorase activity or drug metabolism by tumor homogenates: Implications for enzyme-directed bioreductive drug development. *Cancer Research* 60: 6384–6390.

96. Ko, C.-H., S.-C. Shen, L.-Y. Yang, C.-W. Lin, and Y.-C. Chen. 2007. Gossypol reduction of tumor growth through ROS-dependent mitochondria pathway in human colorectal carcinoma cells. *International Journal of Cancer* 121: 1670–1679.

97. Aithal, B. K., S. M. R. Kumar, B. N. Rao et al. 2011. Evaluation of pharmacokinetic, biodistribution, pharmacodynamic, and toxicity profile of free juglone and its sterically stabilized liposomes. *Journal of Pharmaceutical Sciences* 100: 3517–3528.

98. Hafeez, B. B., W. Zhong, J. W. Fischer et al. 2013. Plumbagin, a medicinal plant (*Plumbago zeylanica*)-derived 1,4-naphthoquinone, inhibits growth and metastasis of human prostate cancer PC-3M-luciferase cells in an orthotopic xenograft mouse model. *Molecular Oncology* 7: 428–439.

99. Niu, M., W. Cai, H. Liu et al. 2015. Plumbagin inhibits growth of gliomas *in vivo* via suppression of FOXM1 expression. *Journal of Pharmacological Sciences* 128: 131–136.

100. Wang, Y., Y. Zhou, G. Jia et al. 2014. Shikonin suppresses tumor growth and synergizes with gemcitabine in a pancreatic cancer xenograft model: Involvement of NF-κB signaling pathway. *Biochemical Pharmacology* 88: 322–333.

101. Giuliani, F. C., K. A. Zirvi, and N. O. Kaplan. 1981. Therapeutic response of human tumor xenografts in athymic mice to doxorubicin. *Cancer Research* 41: 325–335.

102. Li, L. S., E. A. Bey, Y. Dong et al. 2011. Modulating endogenous NQO1 levels identifies key regulatory mechanisms of action of β-lapachone for pancreatic cancer therapy. *Clinical Cancer Research* 17: 275–285.

103. Fidalgo, L. M. and L. Gille. 2011. Mitochondria and trypanosomatids: Targets and drugs. *Pharmaceutical Research* 28: 2758–2770.

104. de Castro, S. L. 1993. The challenge of Chagas' disease chemotherapy: An update of drugs assayed against *Trypanosoma cruzi*. *Acta Tropica* 53: 83–98.

105. Diogo, E. B. T., G. G. Dias, B. L. Rodrigues et al. 2013. Synthesis and anti-*Trypanosoma cruzi* activity of naphthoquinone-containing triazoles: Electrochemical studies on the effects of the quinoidal moiety. *Bioorganic & Medicinal Chemistry* 21: 6337–6348.

106. Ehrhardt, K., C. Deregnaucourt, A.-A. Goetz et al. 2016. The redox cycler plasmodione is a fast-acting antimalarial lead compound with pronounced activity against sexual and early asexual blood-stage parasites. *Antimicrobial Agents and Chemotherapy* 60: 5146–5158.

107. De Long, M. J., H. J. Prochaska, and P. Talalay. 1985. Tissue-specific induction patterns of cancer-protective enzymes in mice by *tert*-butyl-4-hydroxyanisole and related substituted phenols. *Cancer Research* 45: 546–551.

108. Hung, M.-Y., T. Y.-C. Fu, P.-H. Shih, C.-P. Lee, and G.-C. Yen. 2006. Du-Zhong (*Eucommia ulmoides* Oliv.) leaves inhibits CCl₄-induced hepatic damage in rats. *Food and Chemical Toxicology* 44: 1424–1431.

109. Munday, R. and C. M. Munday. 2000. Induction of quinone reductase and glutathione transferase in rat tissues by juglone and plumbagin. *Planta Medica* 66: 399–402.

110. Azen, E. A. and R. F. Schilling. 1964. Extravascular destruction of acetylphenylhydrazine-damaged erythrocytes in the rat. *Journal of Laboratory and Clinical Medicine* 63: 122–136.

111. Rifkind, R. A. and D. Danon. 1965. Heinz body anemia- an ultrastructural study. I. Heinz body formation. *Blood* 25: 885–896.

112. Jenkins, F. P., J. A. Robinson, J. B. M. Gellatly, and G. W. A. Salmond. 1972. The no-effect dose of aniline in human subjects and a comparison of aniline toxicity in man and the rat. *Food and Cosmetics Toxicology* 10: 671–679.

113. Jung, F. and P. Witt. 1947. Studien über methämoglobinbildung XXX Mitteilung. Versuche mit Polyphenolen. *Naunyn-Schmiedebergs Archiv für experimentelle Pathologie und Pharmakologie* 204: 426–438.

114. Niho, N., M. Shibutani, T. Tamura, K. Toyoda, C. Uneyama, N. Takahashi, and M. Hirose. 2001. Subchronic toxicity study of gallic acid by oral administration in F344 rats. *Food and Chemical Toxicology* 39: 1063–1070.

115. Plumlee, K. H., B. Johnson, and F. D. Galey. 1998. Comparison of disease in calves dosed orally with oak or commercial tannic acid. *Journal of Veterinary Diagnostic Investigation* 10: 263–267.

116. Tennant, B., S. G. Dill, L. T. Glickman et al. 1981. Acute hemolytic anemia, methemo-globinemia, and Heinz body formation associated with ingestion of red maple leaves by horses. *Journal of the American Veterinary Medical Association* 179: 143–150.

117. Agrawal, K., J. G. Ebel, C. Altier, and K. Bischoff. 2012. Identification of protoxins and a microbial basis for red maple (*Acer rubrum*) toxicosis in equines. *Journal of Veterinary Diagnostic Investigation* 25: 112–119.

118. Adikesavan, A. K., R. Barrios, and A. K. Jaiswal. 2007. *In vivo* role of NAD(P)H:quinone oxidoreductase 1 in metabolic activation of mitomycin C and bone marrow cytotoxicity. *Cancer Research* 67: 7966–7971.

119. Solit, D. B., F. F. Zheng, M. Drobnjak et al. 2002. 17-Allylamino-17-demethoxygeldanamycin induces the degradation of androgen receptor and HER-2/*neu* and inhibits the growth of prostate cancer xenografts. *Clinical Cancer Research* 8: 986–993.

120. Noh, J.-Y., J.-S. Park, K.-M. Lim et al. 2010. A naphthoquinone derivative can induce anemia through phosphatidylserine exposure-mediated erythrophagocytosis. *Journal of Pharmacology and Experimental Therapeutics* 333: 414–420.

121. Munday, R., B. L. Smith, and C. M. Munday. 2001. Effects of modulation of tissue activities of DT-diaphorase on the toxicity of 2,3-dimethyl-1,4-naphthoquinone to rats. *Chemico-Biological Interactions* 134: 87–100.

122. Ivanov, B. G. and E. I. Makovskaya. 1970. Effect of dichloronaphthoquinone on animals. *Gigiena Primeneniya, Toksikologiya Pestitsidov i Klinika Osravlenii* 8: 160–166.

123. Baker, J. R., E. R. Smith, Y. H. Yoon, G. G. Wade, H. Rosenkrantz, B. Schmall, J. H. Weisburger, and E. K. Weisburger. 1975. Nephrotoxic effect of 2-aminoanthraquinone in Fischer rats. *Journal of Toxicology and Environmental Health* 1: 1–11.

124. Fleischman, R. W., H. J. Esber, M. Hagopian, H. S. Lilja, and J. E. Huff. 1986. Thirteen-week toxicology studies of 1-amino-2, 4-dibromoanthraquinone in Fischer 344/N rats and B6C3F$_1$ mice. *Toxicology and Applied Pharmacology* 82: 389–404.

125. Munday, R., B. L. Smith, and C. M. Munday. 1995. Toxicity of 2,3-dialkyl-1,4-naph-thoquinones in rats: Comparison with cytotoxicity *in vitro*. *Free Radical Biology and Medicine* 19: 759–765.

126. Rogers, G. M., M. H. Poore, and J. C. Paschal. 2002. Feeding cotton products to cattle. *Veterinary Clinics of North America: Food Animal Practice* 18: 267–294.

127. Morgan, S. E. 1997. Stocker and feedlot toxicology: Six investigations. *Compendium on Continuing Education for the Practicing Veterinarian* 19: S166–S173.

128. Xu, D., L. Hu, C. Su et al. 2014. Tetrachloro-*p*-benzoquinone induces hepatic oxidative damage and inflammatory response, but not apoptosis in mouse: The prevention of cur-cumin. *Toxicology and Applied Pharmacology* 280: 305–313.

129. Wang, Y. and H.-P. Lei. 1987. Hepatotoxicity of gossypol in rats. *Journal of Ethnopharmacology* 20: 53–64.

130. Hiona, A., A. S. Lee, J. Nagendran et al. 2011. Pre-treatment with ACE inhibitor attenuates doxorubicin induced cardiomyopathy via preservation of mitochondrial function. *Journal of Thoracic and Cardiovascular Surgery* 142: 396–403.

131. Park, E.-S., S.-D. Kim, M.-H. Lee et al. 2003. Protective effects of N-acetylcysteine and selenium against doxorubicin toxicity in rats. *Journal of Veterinary Science* 4: 129–136.

132. Glaze, E. R., A. L. Lambert, A. C. Smith et al. 2005. Preclinical toxicity of a geldanamycin analog, 17-(dimethylaminoethylamino)-17-demethoxygeldanamycin (17-DMAG), in rats and dogs: Potential clinical relevance. *Cancer Chemotherapy and Pharmacology* 56: 637–647.

133. Basden, K. W. and R. R. Dalvi. 1987. Determination of total phenolics in acorns from different species of oak trees in conjunction with acorn poisoning in cattle. *Veterinary and Human Toxicology* 29: 305–306.

134. Kari, F. W., J. Bucher, S. L. Eustis, J. K. Haseman, and J. E. Huff. 1992. Toxicity and carcinogenicity of hydroquinone in F344/N rats and B6C3F$_1$ mice. *Food and Chemical Toxicology* 30: 737–747.

135. English, J. C., L. G. Perry, M. Vlaovic, C. Moyer, and J. L. O'Donoghue. 1994. Measurement of cell proliferation in the kidneys of Fischer 344 and Sprague-Dawley rats after gavage administration of hydroquinone. *Fundamental and Applied Toxicology* 23: 397–406.

136. Alderton, P. M., J. Gross, and M. D. Green. 1992. Comparative study of doxorubicin, mitoxantrone, and epirubicin in combination with ICRF-187 (ADR-529) in a chronic cardiotoxicity animal model. *Cancer Research* 52: 194–201.

137. Murthy, A. S. K., A. B. Russfield, M. Hagopian, R. Monson, J. M. Snell, and E. K. Weisburger. 1979. Carcinogenicity and nephrotoxicity of 2-amino-, 1-amino-2-methyl-, and 2-methyl-1-nitro-anthraquinone. *Toxicology Letters* 4: 71–78.

138. Munday, R., B. L. Smith, and E. A. Fowke. 1991. Haemolytic activity and nephrotoxicity of 2-hydroxy-1,4-naphthoquinone in rats. *Journal of Applied Toxicology* 11: 85–90.

139. Jardes, D. J., L. A. Ross, and J. E. Markovich. 2013. Hemolytic anemia after ingestion of the natural hair dye *Lawsonia inermis* (henna) in a dog. *Journal of Veterinary Emergency and Critical Care* 23: 648–651.

140. Takemura, G. and H. Fujiwara. 2007. Doxorubicin-induced cardiomyopathy: From the cardiotoxic mechanisms to management. *Progress in Cardiovascular Diseases* 49: 330–352.

141. Tong, G. L., D. W. Henry, and E. M. Acton. 1979. 5-Iminodaunorubicin reduced cardiotoxic properties in an antitumor anthracycline. *Journal of Medicinal Chemistry* 22: 36–39.

142. Doroshow, J. H. and K. J. A. Davies. 1986. Redox cycling of anthracyclines by cardiac mitochondria. II. Formation of superoxide anion, hydrogen peroxide, and hydroxyl radical. *Journal of Biological Chemistry* 261: 3068–3074.

143. Bhargava, S. K. 1984. Effects of plumbagin on reproductive function of male dog. *Indian Journal of Experimental Biology* 22: 153–156.

144. Premakumari, P., K. Rathinam, and G. Santhakumari. 1977. Antifertility activity of plumbagin. *Indian Journal of Medical Research* 65: 829–838.

145. Desai, V. G., E. H. Herman, C. L. Moland et al. 2013. Development of doxorubicin-induced chronic cardiotoxicity in the B6C3F$_1$ mouse model. *Toxicology and Applied Pharmacology* 266: 109–121.

146. Gadelha, I. C. N., N. B. S. Fonseca, S. C. S. Oloris, M. M. Melo, and B. Soto-Blanco. 2014. Gossypol toxicity from cottonseed products. *The Scientific World Journal* 2014: Article ID 231635.

147. O'Donoghue, J. L. and J. C. English. 1994. Some comments on the potential effects of hydroquinone exposure. *Food and Chemical Toxicology* 32: 863–866.

148. Hard, G. C., J. Whysner, J. C. English, E. Zang, and G. M. Williams. 1997. Relationship of hydroquinone-associated rat renal tumors with spontaneous chronic progressive nephropathy. *Toxicologic Pathology* 25: 132–143.

149. National Toxicology Program. 2005. NTP technical report on the toxicology and carcinogenesis studies of anthraquinone in F344/N rats and B6C3F1 mice (feed studies). U.S. Department of Health and Human Services, Washington, DC.

150. National Toxicology Program. 1996. NTP technical report on the toxicology and carcinogenesis studies of 1-amino-2,4-dibromoanthraquinone in F344/N rats and B6C3F1 mice (feed studies). U.S. Department of Health and Human Services, Washington, DC.

151. Mori, H., S. Sugie, K. Niwa, N. Yoshimi, T. Tanaka, and I. Hirono. 1986. Carcinogenicity of chrysazin in large intestine and liver of mice. *Japanese Journal of Cancer Research* 77: 871–876.

152. Xue, W. and D. Warshawsky. 2005. Metabolic activation of polycyclic and heterocyclic aromatic hydrocarbons and DNA damage: A review. *Toxicology and Applied Pharmacology* 206: 73–93.

153. Wells, P. G., B. Wilson, and B. M. Lubek. 1989. *In vivo* murine studies on the biochemical mechanism of naphthalene cataractogenesis. *Toxicology and Applied Pharmacology* 99: 466–473.

154. Bayer, A., C. Evereklioglu, E. Demirkaya, S. Altun, Y. Karslioglu, and G. Sobaci. 2005. Doxorubicin-induced cataract formation in rats and the inhibitory effects of hazelnut, a natural antioxidant: A histopathological study. *Medical Science Monitor* 11: BR300–BR304.

155. Pifer, J. W., F. T. Hearne, F. A. Swanson, and J. L. O'Donoghue. 1995. Mortality study of employees engaged in the manufacture and use of hydroquinone. *International Archives of Occupational and Environmental Health* 67: 267–280.

156. Carlson, A. J. and N. R. Brewer. 1953. Toxicity studies on hydroquinone. *Experimental Biology and Medicine* 84: 684–687.

157. Sausville, E. A., J. E. Tomaszewski, and P. Ivy. 2003. Clinical development of 17-allylamino,17-demethoxygeldanamycin. *Current Cancer Drug Targets* 3: 377–383.

158. Danson, S. J., P. Johnson, T. H. Ward et al. 2011. Phase I pharmacokinetic and pharmacodynamic study of the bioreductive drug RH1. *Annals of Oncology* 22: 1653–1660.

159. Gasser, C. 1953. Die hämolytische Frühgeburtenanämie mit spontaner Innenkörperbildung. *Helvetica Paediatrica Acta* 8: 491–529.

160. Laurance, B. 1955. Danger of vitamin-K analogues to newborn. *Lancet* 265: 819.

161. Lim, D., R. J. Morgan, S. Akman et al. 2005. Phase I trial of menadiol diphosphate (vitamin K_3) in advanced malignancy. *Investigational New Drugs* 23: 235–239.

162. Tetef, M., K. Margolin, C. Ahn et al. 1995. Mitomycin C and menadione for the treatment of advanced gastrointestinal cancers: A phase II trial. *Journal of Cancer Research and Clinical Oncology* 121: 103–106.

163. Zuelzer, W. W. and L. Apt. 1949. Acute hemolytic anemia due to naphthalene poisoning: A clinical and experimental study. *Journal of the American Medical Association* 141: 185–190.

164. Raupp, P., J. Al Hassan, M. Varughese, and B. Kristiansson. 2001. Henna causes life threatening haemolysis in glucose-6-phosphate dehydrogenase deficiency. *Archives of Disease in Childhood* 85: 411–412.

165. Devecioğlu, C., S. Katar, O. Doğru, and M. A. Taş. 2000. Henna-induced hemolytic anemia and acute renal failure. *Turkish Journal of Pediatrics* 43: 65–66.

166. Rund, D., T. Schaap, N. Da'as, D. B. Yehuda, and J. Kalish. 2007. Plasma exchange as treatment for lawsone (henna) intoxication. *Journal of Clinical Apheresis* 22: 243–245.

167. Qurashi, H. E. A., A. A. Qumqumji, and Y. Zacharia. 2013. Acute renal failure and intravascular hemolysis following henna ingestion. *Saudi Journal of Kidney Diseases and Transplantation* 24: 553–556.

168. Perinet, I., E. Lioson, L. Tichadou, M. Glaizal, and L. de Haro. 2011. Ingestion volontaire de décoction de henné (*Lawsonia inermis*) à l'origine d'une anémie hémolytique chez une patiente atteinte d'un déficit en G6PD. *Médecine Tropicale* 71: 292–294.

169. Blanco, E., E. A. Bey, C. Khemtong, S.-G. Yang, J. Setti-Guthi, H. Chen, C. W. Kessinger, K. A. Carnevale, W. G. Bornmann, and D. A. Boothman. 2010. β-Lapachone micellar nanotherapeutics for non–small cell lung cancer therapy. *Cancer Research* 70: 3896–3904.

170. Wang, X., C. P. Howell, F. Chen, J. Yin, and Y. Jiang. 2009. Gossypol—A polyphenolic compound from cotton plant. *Advances in Food and Nutrition Research* 58: 215–263.

171. Schuetzle, D. 1983. Sampling of vehicle emissions for chemical analysis and biological testing. *Environmental Health Perspectives* 47: 65–80.

172. Lind, C., P. Hochstein, and L. Ernster. 1982. DT-diaphorase as a quinone reductase: A cellular control device against semiquinone and superoxide radical formation. *Archives of Biochemistry and Biophysics* 216: 178–185.

173. Cadenas, E. 1995. Antioxidant and prooxidant functions of DT-diaphorase in quinone metabolism. *Biochemical Pharmacology* 49: 127–140.

174. Lind, C., E. Cadenas, P. Hochstein, and L. Ernster. 1989. DT-diaphorase: Purification, properties, and function. *Methods in Enzymology* 186: 287–301.

175. Ross, D., H. Thor, S. Orrenius, and P. Moldeus. 1985. Interaction of menadione (2-methyl-1,4-naphthoquinone) with glutathione. *Chemico-Biological Interactions* 55: 177–184.

176. Dong, H., D. Xu, L. Hu, L. S. Li, E. Song, and Y. Song. 2014. Evaluation of N-acetylcysteine against tetrachlorobenzoquinone-induced genotoxicity and oxidative stress in HepG2 cells. *Food and Chemical Toxicology* 64: 291–297.

177. Wanpen, S., P. Govitrapong, S. Shavali, P. Sangchot, and M. Ebadi. 2004. Salsolinol, a dopamine-derived tetrahydroisoquinoline, induces cell death by causing oxidative stress in dopaminergic SH-SY5Y cells, and the said effect is attenuated by metallothionein. *Brain Research* 1005: 67–76.

178. Gaascht, F., M.-H. Teiten, C. Cerella, M. Dicato, D. Bagrel, and M. Diederich. 2014. Plumbagin modulates leukemia cell redox status. *Molecules* 19: 10011–10032.

179. Xu, H. L., X. F. Yu, S. C. Qu, X. R. Qu, and Y. F. Jiang. 2012. Juglone, from *Juglans mandshruica Maxim*, inhibits growth and induces apoptosis in human leukemia cell HL-60 through a reactive oxygen species-dependent mechanism. *Food and Chemical Toxicology* 50: 590–596.

180. Bair, J. S., R. Palchaudhuri, and P. J. Hergenrother. 2010. Chemistry and biology of deoxynyboquinone, a potent inducer of cancer cell death. *Journal of the American Chemical Society* 132: 5469–5478.

181. Shang, Y., L. Zhang, Y. Jiang, Y. Li, and P. Lu. 2014. Airborne quinones induce cytotoxicity and DNA damage in human lung epithelial A549 cells: The role of reactive oxygen species. *Chemosphere* 100: 42–49.

182. Aruoma, O. I., B. Halliwell, B. M. Hoey, and J. Butler. 1989. The antioxidant action of N-acetylcysteine: Its reaction with hydrogen peroxide, hydroxyl radical, superoxide, and hypochlorous acid. *Free Radical Biology and Medicine* 6: 593–597.

183. Winterbourn, C. C. and D. Metodiewa. 1999. Reactivity of biologically important thiol compounds with superoxide and hydrogen peroxide. *Free Radical Biology and Medicine* 27: 322–328.

184. Rushworth, G. F. and I. L. Megson. 2014. Existing and potential therapeutic uses for N-acetylcysteine: The need for conversion to intracellular glutathione for antioxidant benefits. *Pharmacology & Therapeutics* 141: 150–159.

185. Chiou, T.-J., Y.-T. Wang, and W.-F. Tzeng. 1999. DT-diaphorase protects against menadione-induced oxidative stress. *Toxicology* 139: 103–110.

186. Karczewski, J. M., J. G. P. Peters, and J. Noordhoek. 1999. Quinone toxicity in DT-diaphorase-efficient and-deficient colon carcinoma cell lines. *Biochemical Pharmacology* 57: 27–37.
187. Helinska, A., T. Belej, and P. J. O'Brien. 1996. Cytotoxic mechanisms of anti-tumour quinones in parental and resistant lymphoblasts, *British Journal of Cancer* 74 (Supplement 27): S23–S27.
188. Thor, H., M. T. Smith, P. Hartzell, G. Bellomo, S. A. Jewell, and S. Orrenius. 1982. The metabolism of menadione (2-methyl-1,4-naphthoquinone) by isolated hepatocytes. A study of the implications of oxidative stress in intact cells. *Journal of Biological Chemistry* 257: 12419–12425.
189. Rubio, V., J. Zhang, M. Valverde, E. Rojas, and Z.-Z. Shi. 2011. Essential role of Nrf2 in protection against hydroquinone- and benzoquinone-induced cytotoxicity. *Toxicology in Vitro* 25: 521–529.
190. Wang, X.-J., Z. Sun, N. F. Villeneuve et al. 2008. Nrf2 enhances resistance of cancer cells to chemotherapeutic drugs, the dark side of Nrf2. *Carcinogenesis* 29: 1235–1243.
191. Duthie, S. J. and M. H. Grant. 1989. The role of reductive and oxidative metabolism in the toxicity of mitoxantrone, adriamycin and menadione in human liver derived Hep G2 hepatoma cells. *British Journal of Cancer* 60: 566–571.
192. Beall, H. D., Y. Liu, D. Siegel, E. M. Bolton, N. W. Gibson, and D. Ross. 1996. Role of NAD(P)H:quinone oxidoreductase (DT-diaphorase) in cytotoxicity and induction of DNA damage by streptonigrin. *Biochemical Pharmacology* 51: 645–652.
193. Siegel, D., B. Shieh, C. Yan, J. K. Kepa, and D. Ross. 2011. Role for NAD(P)H:quinone oxidoreductase 1 and manganese-dependent superoxide dismutase in 17-(allylamino)-17-demethoxygeldanamycin-induced heat shock protein 90 inhibition in pancreatic cancer cells. *Journal of Pharmacology and Experimental Therapeutics* 336: 874–880.
194. Munday, R. 2012. Harmful and beneficial effects of organic monosulfides, disulfides, and polysulfides in animals and humans. *Chemical Research in Toxicology* 25: 47–60.
195. Khan, M. F., X. Wu, B. S. Kaphalia, P. J. Boor, and G. A. S. Ansari. 1997. Acute hematopoietic toxicity of aniline in rats. *Toxicology Letters* 92: 31–37.
196. Munday, R., B. L. Smith, and C. M. Munday. 1999. Effect of inducers of DT-diaphorase on the toxicity of 2-methyl- and 2-hydroxy-1,4-naphthoquinone to rats. *Chemico-Biological Interactions* 123: 219–237.
197. Munday, R., B. L. Smith, and C. M. Munday. 1998. Effects of butylated hydroxyanisole and dicoumarol on the toxicity of menadione to rats. *Chemico-Biological Interactions* 108: 155–170.
198. Shinohara, K. and K. R. Tanaka. 1980. The effects of adriamycin (doxorubicin HCl) on human red blood cells. *Hemoglobin* 4: 735–745.
199. Doll, D. C. 1983. Oxidative haemolysis after administration of doxorubicin. *British Medical Journal* 287: 180–181.
200. Clark, D. G. and E. W. Hurst. 1970. The toxicity of diquat. *British Journal of Industrial Medicine* 27: 51–55.
201. Lubek, B. M., P. K. Basu, and P. G. Wells. 1988. Metabolic evidence for the involvement of enzymatic bioactivation in the cataractogenicity of acetaminophen in genetically susceptible (C57BL6) and resistant (DBA2) murine strains. *Toxicology and Applied Pharmacology* 94: 487–495.
202. Huang, L.-L., C.-Y. Zhang, J. L. Hess, and G. E. Bunce. 1992. Biochemical changes and cataract formation in lenses from rats receiving multiple, low doses of sodium selenite. *Experimental Eye Research* 55: 671–678.
203. da Silva, R. F., C. S. dos Santos Borges, P. Villela e Silva et al. 2016. The coadministration of N-acetylcysteine ameliorates the effects of arsenic trioxide on the male mouse genital system. *Oxidative Medicine and Cellular Longevity* 2016: Article ID 4257498.

204. Matsumoto, M., A. Hirose, and M. Ema. 2008. Review of testicular toxicity of dinitro-phenolic compounds, 2-sec-butyl-4,6-dinitrophenol, 4,6-dinitro-o-cresol and 2,4-dini-trophenol. *Reproductive Toxicology* 26: 185–190.
205. Cabaj, M., R. Toman, M. Adamkovičová, P. Massányi, S. Hluchý, N. Lukáč, and J. Golian. 2012. Structural changes of the testis and changes in semen quality parameters caused by intraperitoneal and peroral administration of selenium in rats. *Scientific Papers: Animal Science and Biotechnologies* 45: 125–131.
206. Siegel, H. and C. W. Mushett. 1944. Structural changes following administration of quinacrine hydrochloride. *AMA Archives of Pathology* 38: 63–70.
207. Bharali, M. K. and K. Dutta. 2012. Testicular toxicity of para-phenylenediamine after subchronic topical application in rat. *International Journal of Environmental Health Research* 22: 270–278.
208. Clark, D. G., T. F. McElligott, and E. W. Hurst. 1966. The toxicity of paraquat. *British Journal of Industrial Medicine* 23: 126–132.
209. Anastasiadis, A. and T. M. de Reijke. 2012. Best practice in the treatment of nonmuscle invasive bladder cancer. *Therapeutic Advances in Urology* 4: 13–32.
210. Fukuyo, T., C. R. Hunt, and N. Horikoshi. 2010. Geldanamycin and its anti-cancer activities. *Cancer Letters* 290: 25–35.
211. Miliukienė, V., H. Nivinskas, and N. Čėnas. 2014. Cytotoxicity of anticancer aziridinyl-substituted benzoquinones in primary mice splenocytes. *Acta Biochimica Polonica* 61: 833–836.
212. Harris, M. N., T. J. Medrek, F. M. Golomb, S. L. Gumport, A. H. Postel, and J. C. Wright. 1965. Chemotherapy with streptonigrin in advanced cancer. *Cancer* 18: 49–57.
213. Sullivan, R. D., E. Miller, W. Z. Zurek, and F. R. Rodriguez. 1963. Clinical effects of prolonged (continuous) infusion of streptonigrin (NSC-45383) in advanced cancer. *Cancer Chemotherapy Reports* 33: 27–40.
214. Smith, G. M. R., J. A. Gordon, I. A. Sewell, and H. Ellis. 1967. A trial of streptonigrin in the treatment of advanced malignant disease. *British Journal of Cancer* 21: 295–301.
215. Leinonen, H. N., E. Kansanen, P. Pölönen, M. Heinäniemi, and A.-L. Levonen. 2014. Role of the Keap1-Nrf2 pathway in cancer. *Advances in Cancer Research* 122: 281–320.
216. de Rezende, L. C. D., F. Fumagalli, M. S. Bortolin et al. 2013. In vivo antimalarial activity of novel 2-hydroxy-3-anilino-1,4-naphthoquinones obtained by epoxide ring-opening reaction. *Bioorganic and Medicinal Chemistry Letters* 23: 4583–4586.
217. Docampo, R. 1990. Sensitivity of parasites to free radical damage by antiparasitic drugs. *Chemico-Biological Interactions* 73: 1–27.
218. Desoti, V. C., D. Lazarin-Bidóia, F. M. Ribeiro et al. 2015. The combination of vitamin K_3 and vitamin C has synergic activity against forms of *Trypanosoma cruzi* through a redox imbalance process. *PloS One* 10: e0144033.
219. Slocum, S. L. and T. W. Kensler. 2011. Nrf2: Control of sensitivity to carcinogens. *Archives of Toxicology* 85: 273–284.
220. Talalay, P., A. T. Dinkova-Kostova, and W. D. Holtzclaw. 2003. Importance of phase 2 gene regulation in protection against electrophile and reactive oxygen toxicity and carcinogenesis. *Advances in Enzyme Regulation* 43: 121–134.

9 NADPH Oxidases and Reactive Oxygen in Viral Infections, with Emphasis on Influenza

Shivaprakash Gangappa, Amelia R. Hofstetter, Becky A. Diebold, and J. David Lambeth

CONTENTS

196	Hydrogen Peroxide Metabolism in Health and Disease

ROS AND NADPH OXIDASES

REACTIVE OXYGEN SPECIES AND THEIR PRODUCTION

Reactive oxygen species (ROS) are short-lived, electrophilic molecules generated by the partial reduction of oxygen to form superoxide (O_2^-), hydrogen peroxide (H_2O_2), and hydroxyl radical ($HO^•$) as well as secondary metabolites including lipid peroxides, peroxynitrite ($ONOO^-$), and hypochlorous acid ($HOCl$) [1]. ROS in biological systems were originally thought to be produced mainly by accidental mechanisms such as by-products of metabolism, or from ingestion of toxins or chemicals that can induce redox cycling [2]. The herbicide paraquat is an example of a class of toxic bipyridinium derivatives of 4,4′-bipyridyl (viologens) that is easily reduced in cells to a radical mono cation that produces superoxide upon reaction with molecular oxygen, regenerating the parent molecule and leading to redox cycling. Likewise, some quinoidal chemical food additives, for example, the vitamin K analog menadione, undergo redox cycling reacting with dioxygen to form superoxide [3,4]. Also in this category is the initiation of free radical chain reactions by radiation that can result in formation of superoxide, hydroxyl radical, and peroxyl radicals (Table 9.1).

In early studies, mitochondria were proposed as a major source of ROS since it was estimated that 1% of respired oxygen resulted in ROS generation by mitochondria [5,6]. However, this estimate has been called into question since these experiments were performed on isolated mitochondria that may have undergone structural and functional alterations during the isolation procedure. Recent estimates derived using methods that permit measurements in living cells indicate that normally respiring mitochondria release very low levels of ROS [7]. Nevertheless, inherited or acquired mutations in DNA encoding mitochondrial respiratory chain components can perturb electron transfer reactions increasing basal ROS generation. Thus, normal mitochondria likely provide a low basal level of ROS in many cells, and this level may increase during aging or mitochondrial damage.

Early studies also pointed to reactive oxygen generation as a side reaction during catalysis by a variety of redox-active enzymes. These include xanthine oxidase (XO), cytochrome p450, cyclooxygenase, lipoxygenase, and nitric oxide synthase (NOS). Oxidative destruction of tetrahydrobiopterin, a cofactor of the L-arginine-dependent NOS, results in an L-arginine-dependent increase in superoxide generation with a concomitant decrease in NOS [8]. In this manner, oxidative stress initiated from other sources may be amplified. Nevertheless, relatively few studies support the idea that these enzymes are a primary source of most ROS under normal or most pathological conditions.

TABLE 9.1
Reactive Oxygen Species and Their Properties and Reactivities

Molecular Species	Properties	Molecular Reactivity
Superoxide (O_2^-)	Weak oxidant, weak reductant Membrane impermeable	Iron sulfur centers Reacts with NO to form $ONOO^-$ Reacts with H_2O_2 to form $HO^\cdot$
Hydrogen peroxide (H_2O_2)	Moderate oxidant Membrane permeable	Proteins with low-pKa cysteine residues Reacts with Cl^- to form HOCl (catalyzed by myeloperoxidase) Peroxidases, unsaturated lipids
Hydroxyl radical ($HO^\cdot$)	Highly reactive Produces secondary radicals	Protein, DNA, lipids
Peroxynitrite ($ONOO^-$)	Highly reactive Produces secondary radicals	Protein, DNA, lipids
Hypochlorous acid (HOCl)	Highly reactive Produced in areas of inflammation by reaction of Cl^- and H_2O_2 with myeloperoxidase	Chlorination of tyrosine Oxidation of thiols, methionine, amines

Source: Adapted from Winterbourn, C.C. and Hampton, M.B., *Free Radic. Biol. Med.*, 45(5), 549, 2008.

Purpose-Driven ROS Production: The NADPH Oxidase Family

Unlike the mentioned enzymes that produce ROS as a side reaction during the course of catalysis, the NADPH oxidase enzymes (NOX) (see Figure 9.1) use NADPH to reduce molecular oxygen to form ROS as their sole known enzymatic function, and hence are termed "professional" generators of ROS. The phagocytic NADPH oxidase (a.k.a., respiratory burst oxidase), reviewed more thoroughly by Nauseef et al. in Chapter 10 of this book, is the earliest example of a professional ROS-generating enzyme, reviewed in [9,10]. Its function in neutrophils is to produce large amounts of microbicidal ROS during the "respiratory burst," which is characterized by consumption of O_2 and production of superoxide as well as of secondary ROS products shown in Figure 9.1. The catalytic subunit gp91^{phox}—now referred to as NOX2—is a membrane-associated flavocytochrome (i.e., it contains both flavin adenine nucleotide and heme) that utilizes electrons from cytosolic NADPH to reduce molecular oxygen to form superoxide. NOX2 partners in the membrane with p22^{phox}, a subunit that serves as a docking partner for the cytosolic regulatory subunits (Figure 9.2). NOX2, predominantly but not exclusively expressed in phagocytic cells of the innate immune system, is dormant in circulating neutrophils but becomes activated upon exposure to microbes, microbial products, or inflammatory mediators, producing large amounts of ROS. Owing to the reactivity and cytotoxicity of high levels of ROS (particularly hypochlorous acid; see Figure 9.1), this process is tightly regulated. Upon phagocytosis of a microbe, membranes from specific granules and secretory vesicles in which

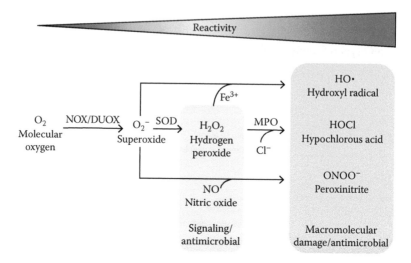

FIGURE 9.1   ROS generated by NADPH oxidases. Superoxide (O_2^-) is generated from various sources, which include NADPH oxidase (NOX) and dual oxidases (DUOX) enzymes. Two molecules of superoxide can react to generate hydrogen peroxide (H_2O_2) plus O_2 in a reaction known as dismutation, which is accelerated by the enzyme superoxide dismutase (SOD). In the presence of iron, superoxide and hydrogen peroxide react in a metal-catalyzed reaction to generate hydroxyl radical (HO·). In addition to superoxide and hydrogen peroxide, and hydroxyl radical, other ROS occur in biological systems, including hypochlorous acid (HOCl), formed from hydrogen peroxide and chloride by the phagocytic enzyme myeloperoxidase (MPO). In addition to producing hypochlorous acid as a major product, MPO can also oxidize a variety of other biological molecules, and can react in a multifaceted manner with superoxide. Nitric oxide (NO) can react with superoxide to form peroxynitrite ($ONOO^-$). The gradient indicates the relative reactivity of individual molecules. Shaded boxes indicate functional and pathological roles of individual ROS, such as signal transduction and reactions with macromolecules that can participate in microbial killing and inflammation.

NOX2 and p22^{phox} are embedded fuse with the plasma membrane of phagosomes. There, the NOX2 becomes activated as a result of assembly of the NOX2-p22^{phox} complex with cytosolic regulatory subunits that include p47^{phox}, p40^{phox}, p67^{phox}, and the small GTPase Rac (see Figure 9.2). In addition to their microbicidal role in host defense, NOX2-generated ROS can also damage normal tissues and initiate an inflammatory response, a side effect (the proverbial double-edged sword) of its function.

Tissue NADPH Oxidases

In addition to phagocytes, ROS are produced by a variety of cell types and tissues, often in response to hormones and growth factors including insulin, angiotensin II, epidermal growth factor, and platelet-derived growth factor. ROS levels in these tissues are typically much lower than those in neutrophils, and have been proposed to play a variety of roles, notably in signal transduction. While the ROS-generating

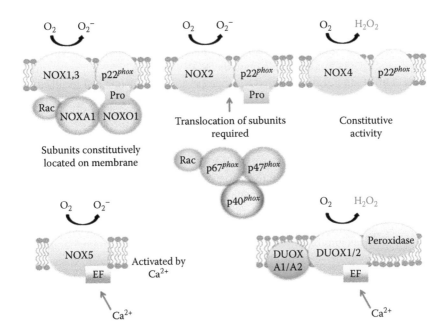

FIGURE 9.2 Basic structures of NOX isoforms. Seven NADPH oxidase isoforms have been described and can be classified by domain structure, regulation, and reactive oxygen product. The phagocytic NADPH oxidase, originally termed gp91^{phox} and now designated NOX2, exists as a membrane complex with p22^{phox}. Activating signals, for example, microbial products and inflammatory mediators, cause assembly of a Rac-containing cytoplasmic complex along with p67^{phox}, p47^{phox}, and p40^{phox}. The nonphagocytic enzymes NOX1 and NOX3 also form a heterodimeric complex with p22^{phox} and assemble with Rac plus the paralogous proteins NOXA1 and NOXO1 to form an active complex. Nox4 also requires p22^{phox}, but not other regulatory subunits, and is constitutively active. Nox5 and the dual oxidases DUOX1 and DUOX2 are activated by Ca^{2+} through a cytoplasmic calcium-binding domain. In addition to a calcium-binding domain, DUOX1 and DUOX2 also possess a separate extracellular peroxidase-homology domain of unknown function. DUOX1 and DUOX2 form an in-membrane complex with DUOXA1 and DUOXA2, respectively; the latter function in membrane trafficking of the DUOX enzymes as well as in modulating their enzymatic activity. The primary product of NOX1, NOX2, NOX3, and NOX5 is superoxide; NOX4, DUOX1, and DUOX2 generate mainly hydrogen peroxide.

systems in nonphagocytic cell types showed superficial similarities with the ROS-generating system in neutrophils, NOX2 was initially not detected, although the NOX2 partner protein p22^{phox} was frequently seen, implying the existence of one or more unknown but related NOX-like partner enzymes. In 1999, NOX1, the first homolog of NOX2, was identified [11]. Over the next several years, additional NOX homologs were identified [12–16], raising the total number of NOX isoforms in mammals to seven. In addition, new regulatory subunits termed NOXO1 (NOX organizer 1) and NOXA1 (NOX activator 1), homologs of p47^{phox} and p67^{phox}, respectively, were identified [17,18].

As indicated in Figure 9.2, the NOX enzymes are categorized into two broad groups: (1) those that require p22^{phox} (NOX1, NOX2, NOX3, and NOX4) and (2) those that are regulated by calcium via an EF hand–containing calcium-binding domain (NOX5, DUOX1, and DUOX2). Depending upon the NOX subunit, p22^{phox} can serve multiple roles: as a maturation factor required for appropriate glycosylation and localization of the NOX subunit, to stabilize the NOX subunit, and to provide a docking site for binding of regulatory subunits. In the latter context, the p22^{phox}-dependent NOX subgroup can be further subdivided: The NOX1, NOX2, and NOX3 isoforms require that their heterodimer partner p22^{phox} possess a proline-rich region that serves as the docking site for either p47^{phox} or NOXO1, which in turn bind respectively to p67^{phox} and NOXA1. NOX4 on the other hand is constitutively active without regulatory subunits, and while requiring p22^{phox} for stability and/or maturation, does not require an intact proline-rich region in p22^{phox} [19]. The calcium-regulated subgroup can be subdivided into NOX5 and the dual oxidase enzymes (DUOX1 and DUOX2), the latter so named because in addition to the calcium-binding EF domain, they contain an N-terminal domain that shows homology to heme-containing peroxidase enzymes [20]. The enzymes can also be categorized according to the species of reactive oxygen that is released. NOX1, NOX2, NOX3, and NOX5 generate primarily superoxide, whereas the major detected product of NOX4, DUOX1, and DUOX2 is hydrogen peroxide [21,22].

NOX Enzyme Expression

NOX enzyme expression is complex and dependent upon factors such as species, tissue, cell type, and physiological or pathological status. Cataloguing NOX expression in all tissues is beyond the scope of this chapter. Rather, we will focus here on the expression of NOX isoforms in lung tissue and changes in expression during viral infections. NOX expression and function in lung are summarized in earlier reviews [23–26].

Levels of NOX mRNA in human tracheal epithelial cells have been examined by qPCR and one study observed the following relative levels: DUOX1, DUOX2 ≫ NOX5 > NOX1, NOX4 > NOX2 [27]. DUOX1 and DUOX2, the most prominently expressed NOX isoforms in this tissue, are localized mainly in the apical cell pole of ciliated epithelia lining the airway. DUOX1 and DUOX2 appear to play a role in host defense against microbial infection by supplying H_2O_2 as a substrate for mucosal lactoperoxidase; the latter generates bactericidal hypothiocyanite (OSCN)– [28] from H_2O_2 and thiocyanate (SCN)-. DUOX1 also functions in wound repair of tracheal epithelium. Silencing of DUOX1 by oropharyngeal administration of DUOX1-targeted siRNA prevented reepithelialization of the trachea in a naphthalene-induced epithelial injury mouse model [29]. The roles of DUOX and the other NOX isoforms in IAV will be discussed in a later section.

ROS AND VIRUS INFECTION

ROS in Viral Infections

Oxidative stress, loosely defined as an abnormal increase in levels of superoxide, hydrogen peroxide, and other oxygen-derived reactive species, is a critical mechanism of pathology in many viral infections. Increased ROS can result from either increased production of radical species (for example by NOX enzymes or mitochondria) or a decrease in their removal by antioxidant systems. Reduced glutathione (GSH) is an abundant small molecule in cells that becomes oxidized to GSSG in the presence of H_2O_2 and other oxidants [30] and lowered GSH levels, with or without a decrease in GSH/GSSG levels, has been used as a marker of oxidative stress, for example, following viral infection [31–35]. Several enzymes also metabolize and thus eliminate ROS: these include superoxide dismutase (SOD), heme oxygenase (HO-1), catalase, glutathione peroxidase, peroxiredoxins, and others. In contrast, eosinophil peroxidase and myeloperoxidase (MPO) metabolize hydrogen peroxide to more reactive species (HOCl, HOSCN) [36], and these can then react with cellular molecules to cause molecular damage.

HIV

Oxidative stress is associated with HIV infection (reviewed in [37–39]). At least five HIV proteins have been shown to cause oxidative stress (see Table 9.2). Of these, Tat [40–45], Nef [46,47], and the envelope glycoprotein gp160 [40,48,49] are linked to increased ROS production via NOX enzymes, particularly NOX2 and NOX4

TABLE 9.2
Effects on ROS by HIV and HCV Proteins

Virus	Protein	Effects	References
HIV	Tat	Increases ROS	[41,42,339–344]
		Decreases GSH and SOD	[339,341–349]
	Nef	Increases ROS	[46,47,225,350]
	gp120	Increases ROS	[40,48,49,342,344,351–356]
		Decreases GSH and GS	[342,344,352,354,355]
		Increases antioxidant proteins	[48]
	Vpr	Increases ROS	[357,358]
	RT	Increases ROS	[359]
		Increases antioxidant proteins	[359]
HCV	Core	Increases ROS	[59,60,62,66,67,69,72,215,360–362]
		Increases antioxidant proteins	[60,65]
		Decreases GSH	[340]
	E1	Increases ROS	[60,215]
		Increases antioxidant proteins	[60]
	E2	Increases ROS	[60,68]
		Increases antioxidant proteins	[60]
	NS3/4A	Increases ROS	[67,70,71,215,362]
	NS4B	Increases ROS	[60,63]
		Increases antioxidant proteins	[60]
	NS5A	Increases ROS	[60,62,64,67,360]
		Increases antioxidant proteins	[60]

[42,43,46,47,49]. On the other hand, in a rat model, expression of transgenic HIV proteins did not lead to increased NOX activity [50]. However, in a study on HIV-1-positive patients, platelet activation was associated with increased platelet NOX2 activation over healthy controls [51]. Thus, the role of NOX isoforms in HIV infection needs further clarification.

Hepatitis Virus

Oxidative stress is also associated with infections by hepatitis C virus (HCV). GSH is decreased in HCV patients' plasma, red blood cells, and peripheral blood mononuclear cells [52,53] indicating a systemic oxidative stress, and increased turnover of glutathione is seen in liver [54]. In hepatocellular carcinoma lines, HCV elicits ROS production by mitochondria [55,56], and by NOX1 and NOX4 [57–59]. HCV-associated proteins also induce the transcription of antioxidant enzymes by activating Nrf2 [60,61]. HCV proteins have been linked to increased ROS production [60,62–67]. In hepatic stellate cells, HCV core proteins cause increased ROS and development of fibrosis [68,69]. In myeloid cells, NOX2 activity was increased by HCV proteins leading indirectly to decreased NK and T cells [70–72], thereby undermining a critical immune response to HCV. ROS elicited by HCV are thought to be a contributing factor to both the liver damage seen with HCV [73] and the progression to hepatocellular carcinoma (reviewed in [74,75]). A variety of other hepatitis viruses, including hepatitis A and hepatitis B [76], implicated in hepatic disease also cause elevated levels of ROS, suggesting that ROS is a common denominator in liver and other tissue damage. HBVx protein increases mitochondrial ROS generation [77]. Hepatitis delta virus (HDV) proteins p27 and p24 both induce ROS production in infected cells. NOX4 expression is increased by p27, which may be connected to a ROS- and p27-independent increase in NF-κB and STAT3 activity [78].

Respiratory Viruses

Rhinovirus, responsible for the common cold, generates oxidative stress in target cells [79], but does not inhibit antioxidant enzyme activity or expression [80]. Rather, this virus activates the ROS-producing enzymes XO [81], NOX1 [82,83], and DUOX2 [84,85] in airway epithelial cells. ROS activate NF-kB, driving transcription of intracellular adhesion molecule-1 (ICAM-1), the receptor for 90% of rhinoviruses [79]. NOX1-derived ROS play a role in disrupting the barrier function of bronchial epithelial cells [83]. Superoxide and/or H_2O_2 signal through NF-κB to increase IL-8 [81,82,86], leading to most of the symptoms of the common cold [87]. This suggests that inhibition of XO, NOX1, and/or DUOX2 may be therapeutic.

Respiratory syncytial virus (RSV) triggers ROS production by airway epithelial cells [88,89], leading to oxidative stress in the airways [35,90]. ROS appear to be mainly derived from mitochondrial superoxide production [88,91]. NOX2 contributes to RSV-induced cytokine signaling downstream of NF-κB [92,93] and possibly STAT activation [94]. ROS are critical for the expression of inflammatory cytokines RANTES, IL-8, and IL-1β, which contribute to lung pathology after RSV infection [82,89,93]. After RSV infection of cells or rodents, antioxidant treatment can decrease the following: ROS, clinical signs of illness, weight loss, neutrophil recruit-ment to the lung, airway hyper-reactivity, viral titers, lipid peroxidation products, DNA damage, cytokines and chemokines, mucin, and the activation of transcription factors NF-κB and IRF3 [88,95–98]. Nrf2-induced genes and peroxiredoxins 1 and 4 protect mice and airway epithelial cells, respectively, from RSV-induced oxidative damage [99,100]. The body of evidence indicates that ROS production by airway epithelial cells contributes to the pathology of RSV, and further studies are war-ranted to determine whether modulation of ROS can be an effective therapeutic goal. Among other respiratory virus infections, influenza virus also elicits ROS in a variety of cell types. The details of the effects of ROS on influenza infection are described in "Nox Enzymes and Influenza" section.

Dengue Virus

Dengue virus (DENV) infection causes oxidative stress proportional to the progres-sion and severity of illness. Both NOX2 and mitochondria contribute to ROS accu-mulation in DENV-infected cells [101]. While oxidative stress is beneficial to DENV infectivity [102,103] and contributes to hemorrhage and encephalitis [104,105], NOX-dependent superoxide leads to activation of antiviral immune response path-ways [101,102,106]. Whether these pathways are ultimately beneficial or detrimental to the human host is unclear [101–103,106,107]. In contrast, ROS facilitate anti-DENV responses downstream of Toll in the mosquito host, *Aedes aegypti* [108]. Further research is necessary to determine the therapeutic potential of NOX inhibi-tors and/or antioxidants in response to dengue.

Herpesviruses

Human cytomegalovirus (HCMV) infection increases ROS production from XO and mitochondria *in vitro* [109–112], which is countered by activation of Nrf2 and induction of HO-1 [111]. HCMV also induces SOD, as well as enzymes important for maintaining GSH levels [31]. This helps maintain cell homeostasis during the relatively long replicative period of HCMV. Hydrogen peroxide also both directly activates the HCMV major intermediate promoter (MIEP) and activates NF-κB, increasing the ability of the intermediate-early protein IE72 to activate the MIEP. *In vitro*, antioxidant treatment of HCMV-infected cells decreases viral titers and cytopathic effects [109,113]. The majority of the ten proteins encoded by CMV [114] induce an increase in ROS production (see Table 9.2) by both NOX-dependent and NOX-independent mechanisms [60,115–117]. Overall, the data suggest that HCMV thrives at a balance point between elevating ROS to propagate its genome and lower-ing ROS to maintain cell survival.

Other herpes viruses also induce ROS after infection of cells. Herpes simplex virus-1 (HSV-1) or bovine herpesvirus-1 infection decreases GSH concentration, and their replication is inhibited by exogenous GSH [34,118,119]. HSV-1 infection increases ROS production downstream of the mitochondria and NOXes, which may implicate an effect on antioxidant mechanisms [120–122]. Peptides derived from the secreted G protein of herpes simplex virus-2 (HSV-2) stimulate (NOX2-dependent) ROS production by human monocytes or neutrophils [123,124]. Two Epstein-Barr virus proteins have been linked to ROS through activation of NOX enzymes: EBNA-1 increases NOX1 and NOX2 expression while LMP1 upregulates p22^{phox} [125]. ROS contribute to DNA damage [126], telomere dysfunction [127], and bone resorption [128] after EBV infection. Kaposi's sarcoma virus (KSHV) induces ROS production in human vascular endothelial cells to increase viral entry by stimulating macropino-cytosis [129]. KSHV-induced ROS also activate Nrf2, which is critical for efficient KSHV replication [130]. In AIDS patients' KSHV tumors, oxidative stress is indi-cated by increased Mn-SOD expression but decreased activity was seen compared with normal tissue [131]. After primary infection, ROS are important for reactivating EBV and KSHV from latency [132–135]. Although HSV-1 is susceptible to oxida-tive degradation, in general, herpes viruses utilize ROS to increase the efficiency of infection and replication.

Other Viruses

Lymphocytic choriomeningitis virus (LCMV) infection leads to increased ROS in fibroblasts, B cells [136,137], granulocytes, and splenic macrophages [138]. The impact of ROS on LCMV infection remains unclear, however. *In vitro*, antioxidant treatment decreases viral titers, correlating with decreased binding of LCMV to the surface of BHK-12 cells [136]. p47^{phox} knockout mice (whose phagocytes fail to gen-erate superoxide and hydrogen peroxide) have improved T cell responses to LCMV [138], while SOD-deficient mice have increased liver inflammation in response to infection [139]. In contrast, *in vivo* antioxidant treatment increases LCMV titers in the spleen, correlating with attenuated T and B cell responses [137,140]. Thus, the effect of systemically administered antioxidants can differ greatly from the effect of selectively targeting a specific ROS-generating enzyme.

Besides the viruses detailed here, ROS and NOX are implicated in the host response to many other viruses. Three porcine viruses—porcine reproductive and respiratory syndrome virus, porcine circovirus type 2, and porcine arterivirus—induce ROS upstream of NF-κB [141–143]. Similarly, NOX- and mitochondrial-derived ROS upstream of NF-κB are critical for efficient replication of enterovirus 71 [144,145]. Human metapneumovirus, rotavirus, and Nipah virus increase oxida-tive stress and/or lipid oxidation and increase most cellular antioxidants [146–148]. Antioxidants are protective in Newcastle disease virus, Nipah virus, and Sendai virus [147,149,150]. Marburg virus protein VP24 indirectly activates Nrf2, protecting cells from ROS and cytotoxicity [151,152]. Parvovirus and coccolithovirus spread are aided by ROS-induced apoptosis [153–155], but chikungunya virus-induced oxida-tive stress forces host cells into autophagy, avoiding apoptosis [156,157]. In addition, ROS are increased after viral infections in fish and plants [158–161]. The evidence suggests that maintaining control of ROS signaling in the host cell is a key aspect of viral parasitism.

GENERAL CONSEQUENCES OF ROS IN VIRAL INFECTIONS

Viral Inactivation

In vitro, hydrogen peroxide and/or myeloperoxidase can inactivate vesicular sto-matitis virus (VSV), HSV-1, polio, and vaccinia, and this might contribute to viral clearance [162–164]. In support of this hypothesis, HSV-1 is partially protected by a viral catalase [165]. The effectiveness of H_2O_2-mediated antiviral activity *in vivo* is unclear [164–166]. Because viruses are not generally restricted to phagosomes, targeting the oxidative burst to the virus is problematic.

Apoptosis

ROS can trigger apoptosis in virus-infected cells [101,105,167–169], but there is a lack of conclusive evidence that apoptotic clearance of infected cells is effective at lowering viral titers. However, the existence of antiapoptotic mechanisms in viruses (e.g., apoptosis-inhibiting proteins in poxviruses [170] and CMV [171,172]) is suggestive of such a role for apoptosis.

Viral Replication

ROS may contribute to antiviral activity by limiting viral replication. ROS can trigger an antiviral phenotype through production of type I IFNs [173,174]. H_2O_2 can inhibit the mammalian target of rapamycin (mTOR), thus inhibiting cell growth, and therefore viral replication [31]. H_2O_2 decreases HBV replication *in vitro* [175]. However, demonstration of ROS-mediated antiviral activity has at times necessitated use of viral or cellular enzyme inhibitors and antiviral agents [31,165,173], suggesting that viruses have developed strategies to avoid ROS-mediated antiviral defenses.

In contrast to an effect of ROS on *lowering* viral titers, a great deal of evidence suggests that ROS contribute to viral replication. Certain antioxidants decrease HIV, HSV-1, RSV, CMV, and enterovirus 71 titers [34,95,100,109,176,177]. The titers of dengue, coronavirus 228E, and enterovirus 71 are increased in patients with glucose-6-phosphate dehydrogenase deficiency, which limits NADPH-dependent antioxidant mechanisms [102,178,179]; the deficiency is associated with lowered levels of intracellular superoxide and nitric oxide, but increased oxidative stress. A prooxidant diet increases the titers and pathology of coxsackie virus [180–186]. Depending on their concentrations, H_2O_2 and superoxide can stimulate host cell activation and proliferation in a wide range of cell types [187–189], supporting virus replication. ROS may also increase viral replication by facilitating virus spread to new cells, for example, via alterations in surface receptors [79,129,136] or by favoring disulfide bonds needed for viral capsid synthesis [34,150]. In addition, antioxidants inhibit the activity of HIV reverse transcriptase [190]. Increasing cellular ROS may be a common strategy for viruses to facilitate replication, but it is important to note that whether ROS is pro- or antiviral very likely will depend upon the localization, amount, and particular molecular species (superoxide, H_2O_2, HOCl, etc.) of ROS.

Effects on Transcription

NF-κB is a ubiquitously expressed transcription factor controlling many genes, including those necessary for early host defense [191,192] and prosurvival phenotype [193–196], both of which make it a target for viral control of host cell fate. In the inactivated state, NF-kB is held in the cytoplasm by complexation with IκB [197]. The signaling pathways both upstream and downstream of NF-kB are redox-regulated (reviewed in [191,198]). By altering the redox status of the cell, viruses may block NF-κB activation to avoid induction of innate immunity [199–202].

Alternatively, viruses may activate NF-κB to stimulate host gene replication machinery [109,197], induce apoptosis [105,203,204], or preserve host cells from death [113,205,206]. Other viruses known to activate NF-κB via increasing ROS include HBV [77,207], HDV [78], HCMV [208], HSV [121], RSV [95,209], DENV [101], and PRRSV [143]. Thus, viruses acquire control of NF-κB, a critical determinant of host cell fate, by altering cellular ROS.

STAT3, another ROS-regulated transcription factor, regulates expression of genes that partially overlap with those controlled by NF-κB [210]. Like NF-κB, STAT3 regulates genes that control cell cycle and proliferation as well as antiapoptotic genes. STATs are activated by their phosphorylation state, which is in turn determined by the ROS-activated Janus kinases (JAKs) [211,212], and ROS-inhibited phosphatases [211–213]. Some viruses activate STAT3 by these mechanisms, leading to host cell survival and proliferation. RSV activates STAT1 and STAT3 in respiratory epithelial cells, causing induction IRF-1 and IRF-7 [94] and contributing to lung inflammation [89]. The HIV gp120 protein induces STAT3 in myeloid-derived suppressor cells (MDSC), which suppress host immunity. In this model, gp120-increased p47^{phox} expression and ROS are involved in the suppressive activity of MDSC [214]. These two examples exemplify the range of ROS-dependent, STAT3-mediated effects that viruses can exert on the host.

Both STAT3 and NF-κB contribute to malignant transformation following virus infection [210]. HCV's core, E1, NS3, and NS5A proteins stimulate ROS and activate STAT3 *in vivo* and *in vitro* [64,69,215–217]. NS5A activates mitochondrial ROS generation indirectly via Ca^{2+} release from the ER [64,218], while core protein stimulates ROS downstream of JAK [69]. The consequences to HCV replication are debated [219,220], but it seems clear that translocation of phosphorylated STAT3 to the nucleus of hepatoma cells contributes to pathological changes associated with hepatic carcinoma. These changes include suppression of matrix metalloproteinase-1, that in turn supports development of liver fibrosis, [69], increased cell survival [219], and toxic iron accumulation [216]. SOD and catalase are generally decreased in tumorous HBV-infected liver tissues compared with surrounding nontumorous tissue [221]. H_2O_2 silences the tumor suppressor gene *RUNX3* in HBV-associated hepatic carcinoma [222]. Furthermore, STAT3 and NF-κB activation due to ROS are reported to be downstream of HBV proteins HBx and MHBs^t [77,207] and HDV protein p27 [78], suggesting a common theme in the development of hepatic carcinoma. ROS activation of STAT3 is also pivotal to transformation of B cells by EBV [223]. These studies demonstrate a link between virus-induced ROS and malignancy.

Effects on Cytokines

Cytokines can both trigger ROS production and can be regulated by ROS. Binding of IL-1, TNF, TGF-β or IFN-γ to their cognate receptors leads to the production of ROS in a cell-specific manner [224], while IL-10 blocks superoxide release by NOX2 [225]. NOX-mediated ROS promotes TNF-α-dependent NF-κB activation in a mouse model [226]. On the other hand, ROS can regulate cytokine production through activation of transcription factors NF-κB, STAT3, and their upstream kinases, for example, JAK. Several viruses have been shown to induce TNF in a ROS-dependent manner [96,188,189]. NF-κB induces cytokines TNF-α, IFN-γ,

IL-1, IL-2, IL-6, IL-17, IL-8, IL-22, IL-23, and IL-27, as well as the chemokines CXCL10 and CXCL12. IL-6, IL-17, IL-22, IL-23, and IL-27 are also inducible by STAT3. ROS are necessary for maximal HIV Tat-induced production of TNF, IL-6, and MCP-1 [44,189]. Importantly, ROS activate transcription factors in a cell- and tissue-specific manner [226–228]. In two different mouse macrophage lines, H_2O_2 treatment activates either NF-κB or AP-1 and TNF-α downstream of p38 and JNK phosphorylation [229,230]. Oxidant levels impact Th1/Th2 polarization of the overall immune response [231]. The positive feedback loops linking cytokines and ROS necessitate robust control mechanisms, such as the range of antioxidants upregulated by NF-κB, to prevent excessive inflammation.

Molecular and Cellular Damage

In addition to affecting cellular homeostasis via signaling pathways [91,232], ROS can directly cause tissue damage. ROS oxidizes biomolecules, leading to reversible or irreversible chemical changes. ROS-mediated oxidation of amino acids [233] was observed in bronchiolar lavage fluid from patients with acute respiratory distress syndrome [234]. Oxidation of proteins often results in loss of their function [233,235], which for surfactant A means an inability to resolve edema [236]. Superoxide effects on mitochondrial uncoupling proteins can also impact mitochondrial respiration [237]. ROS-mediated oxidation of DNA can lead to mutations or lesions contributing to the genetic instability in cancer [235,238,239]. Virus-associated ROS and reactive nitrogen species are associated with DNA double-strand breaks or mutations in HCV, LCMV, RSV, EBV, HTLV-1, and human papilloma virus infections [88,126,139,215,240–242]. For HCV, it is suggested that oxidation of the tyrosine residue in the active site of a critical DNA repair enzyme, NBS1, may contribute to accumulation of DNA damage [243]. Finally, ROS-dependent lipid peroxidation impairs cellular membrane integrity [235,244], as has been observed in HCV infection [215]. The combined impact of these biomolecular modifications leads to cancer in the absence of the protective effects of Nrf1 [245]. Therefore, viral dysregulation of the cellular oxidant balance has the potential to cause cell damage independent of signaling pathways.

Wound Healing

While evolution has led to adaptations to minimize negative impacts of ROS, it has also harnessed ROS as a conserved mechanism of signaling for wound healing [246]. In a tail wound model in zebrafish larvae, H_2O_2 generated at the wound site by DUOX served as a chemoattractant for leukocyte recruitment [247]. Multiple studies have indicated that NOX family enzymes, including DUOX1, NOX2, and NOX4, act both upstream and downstream of growth factor signals to promote wound healing and cell proliferation [248–254]. NOX1, 2, 4, and 5 are expressed in endothelial cells [255,256], where they contribute to angiogenesis [257,258], reviewed in [255,256,259,260]. Wound healing also depends on activation of stem cells to proliferate, and NOX1, 2, and 4 are involved in activation of various stem cells [258,261–265]. HIV Tat, a known angiogenic protein, promotes ROS-dependent cytoskeletal changes in endothelial cells correlating with localization of NOX subunits to the Tat-induced membrane ruffles [266]. Similar angiogenic effects of viral proteins may contribute to

vascularization of virus-associated cancers. Although it has not been formally demonstrated, it seems likely that virus-generated ROS intersect with ROS-driven growth
factor signals to promote cell proliferation and further viral propagation.

NOX ENZYMES AND INFLUENZA

GENERAL CONSIDERATIONS

While diverse enzymatic and other sources of ROS in influenza infection have
been reviewed [267], this section focuses on ROS generated by NOX enzymes.
NOX enzymes are expressed in virtually all cell types of the respiratory system
[26,268–270]. Using both animal models and lung cell lines, basal levels of different NOX isoforms are seen and their levels can change dramatically in response to
various stimuli [271–274]. Influenza A Virus upregulates NOX/DUOX expression
in epithelial cells and increases ROS levels [275,276]. Since ROS influence innate
immune responses to virus infections [277–280], an in-depth understanding of the
expression and activity of different NOX isoforms and their impact on innate and
adaptive immunity to influenza virus infection is needed.

NOX1 and Influenza

NOX1 is expressed in mouse lung parenchyma [281] perhaps suggesting a role in
defense against pathogens. We observed an increase in NOX1 mRNA in response
to influenza A virus in several human lung cell lines derived from epithelial cells
(A549), endothelial cells (HULEC), and monocytes (THP-1) (Gangappa et al.,
unpublished). Lung tissue from influenza (A/PR8 and A/Mexico/4108/2009)-
infected mice also showed an increase in NOX1 mRNA at 24 and 48 h postinfection
(Gangappa et al., unpublished). Based on studies demonstrating NOX1 expression
in epithelial [270] and endothelial cells (Gangappa et al., unpublished), it is possible
that influenza virus entry and initial replication in the upper respiratory tract (epithelium), and the ensuing cytokine/chemokine milieu, may attract NOX1-expressing
inflammatory (monocytic) cells from the circulation into the lung parenchyma.
Consequently, NOX1-induced signaling pathways in such inflammatory cells may
influence the course and outcome of influenza infection. For example, in NOX1-
deficient mice, a decrease in virus-induced inflammation and oxidative stress during
early phase of influenza A virus infection was seen [270]. Our studies [282] using
"knock-in" mice expressing a mutated, catalytically inactive form of NOX1 in place
of wild-type NOX1, showed improved survival after infection, with survival correlating with changes in the phenotype and functionality of virus-specific T cells.
In both studies, NOX1 deficiency did not have any effect on virus replication in the
lung tissue, suggesting that NOX1 deficiency may not have a significant influence on
viral replication or on early antiviral responses, but may suggest a role for NOX1 in
anti-inflammatory responses and adaptive immunity during influenza infection.

NOX2 and Influenza

NOX2 is highly expressed in phagocytic cells as well as in various other cell
types including dendritic cells and B cells [92,283,284]. In addition to its role in

antimicrobial host defense, NOX2 has been implicated in regulation of inflammation, antigen presentation and adaptive immune responses [285]. With regard
to influenza A infection, mice lacking NOX2 showed decreased inflammation in
the airways compared with control mice [286]. Similarly, mice lacking the gene
encoding p47^{phox} [287] showed a decrease in lung pathology caused by influenza
A virus infection [288]. Furthermore, in NOX2-deficinet mice infected with low
pathogenicity influenza A virus (X31), a significant decrease in lung virus titer and
inflammation was seen [289]. Moreover, treatment of influenza virus–infected mice
with apocynin, a NOX inhibitor and antioxidant, led to decreases in lung virus titer,
apoptosis of alveolar epithelial cells, and inflammation. Results from these studies
suggest that NOX2 deficiency benefits the host by augmentation of the early antiviral response and by suppression of virus-induced inflammation. On the other hand,
NOX2-deficient mice showed the same survival as wild-type animals after infection
with influenza A virus [282].

NOX4 and Influenza
Relatively few reports have considered the significance of NOX4 in viral infections [59,61,290]. In the case of influenza, in a cell culture model, virus caused an
increased expression of NOX4, which participated in nuclear transport of viral ribo-
nucleoprotein [275]. Moreover, NOX4 supported virus replication in NCI-H292 and
BALB/c mice-derived primary airway epithelial cells [275]. NOX4 is expressed in
endothelial cells [291,292] where it is implicated in endothelial cell activation as well
as cell dysfunction and injury [293]. Because several influenza virus strains replicate
in endothelial cells [294–296], it is possible that NOX4-dependent ROS generated
in endothelial cells impacts the integrity of the lung endothelial barrier, allowing
extravasation and trafficking of inflammatory cells from the vasculature into the lung
parenchyma.

DUOX and Influenza
DUOX1 and DUOX2 are prominently expressed in the lung epithelium where their
biological roles have been investigated (reviewed in [28]). *In vitro* and *in vivo* studies
examined the role of DUOX1 and 2 on host response to influenza infection [276]. In
NHBE cells, infection caused increased expression of DUOX2 and increased ROS.
Suppression of DUOX1/2 using siRNA led to an increase in lung virus titer [276].
Likewise, siRNA-mediated DUOX1/2 suppression in mice led to increase in lung
virus titer, suggesting a protective function of DUOX during influenza virus infection [276]. These findings underscore the important role of DUOX enzymes in early
antiviral response to influenza infection. A separate *in vitro* study used human nasal
epithelial cells and showed that ROS generated via mitochondria and DUOX2 were
associated with stimulation of the innate immune response as judged by induction of
IFN-lambda [278].

From the studies described here, it is clear that influenza infection alters expression of NOX isoforms and ROS in a variety of lung cell types responsible for entry,
replication, and dissemination of virus in the host tissues [270,276,288]. Several
influenza virus isolates including laboratory strains [270], seasonal influenza clinical isolate (Gangappa et al., unpublished), and avian influenza viruses [288,297]

cause varying degrees of increased NOX expression in both *in vitro* and *in vivo* experimental models. Importantly, mice lacking some NOX isoforms or NOX regulatory protein (e.g., NOX1, NOX2, or p47^{phox}) show decreased oxidative stress and lung pathology following infection [270,288]. Also, NOX1 deficiency alters the kinetics of inflammatory cytokine production [270]. Therefore, changes in virus-induced NOX expression in different cell types certainly alter the cytokine milieu in cell types relevant to the upper and lower respiratory tract and thereby are likely to influence the course of infection. The roles of specific NOX enzymes during influenza infection are likely to depend on such factors as viral strain, immune status of the host, and complicating secondary bacterial infections that often accompany influenza infection [298].

NOX Enzymes and Innate Immunity to Influenza

Epithelial cells lining the upper respiratory mucosal tissue are generally the primary targets of influenza infection [299,300]. Upon infection by influenza A virus, a well-orchestrated host innate immune response is initiated (reviewed in [301]). Like other viral infections, the innate response to influenza is regulated by viral and cellular processes that lead to production of type I IFN and pro-inflammatory cytokines [302]. For influenza and other viruses, ROS from NOX enzymes and other sources regulate the production of antiviral (type I IFN) and pro-inflammatory cytokines [277–280]. Both mitochondrial and DUOX2-generated ROS play a role in inducing type I IFN and in lowering virus titer [278]. DUOX2-derived ROS induced transcription of the pattern recognition receptor RIGI and of MDA5, the two key players involved in recognition of influenza A virus and the early antiviral response [279]. In cells infected with different subtypes of influenza A virus, apocynin led to a decrease in inflammatory mediators via upregulation of suppressors of cytokine signaling (SOCS1 and SOCS3) [280]. Also, in lung tissue of influenza-infected mice, apocynin caused a significant decrease in macrophages and neutrophils and a decrease in viral titer [289]. These studies suggest that suppression of ROS using pharmacological NOX inhibitors will modulate the innate immune response to influenza A virus infection.

It is not yet clear whether NOX-derived ROS participate in the activation of the inflammasome, a cellular protein assembly in myeloid cells that is linked to cellular inflammation [303,304]. The inflammasome can be activated by a variety of microbial and other stimuli including ROS, which acts via the inflammasome protein NLRP3. Several reports have demonstrated that NOX2 is not required for NLRP3 inflammasome activation [305–307]. Specifically, the role of NOX2-derived ROS in inflammasome activation was examined in one study [307]: in NOX2-deficient neutrophils, LPS activated caspase 1 and stimulated secretion of IL-1beta, but no difference in NLRP3 expression was seen. Interestingly, siRNA depletion of NOX2 *in vitro*, as well as the use of NOX2-deficient mice, demonstrated that NOX2 inhibition leads to a decrease in inflammasome activation [304]. Several studies using influenza A virus *in vitro* and *in vivo* have highlighted NLRP3 inflammasome activation by various influenza virus subtypes in lung epithelial cells [308]. Specifically, while recognition of viral RNA by TLRs and RIG-I serves as a primary signal, many other factors including NOX and mitochondria-derived ROS provide a secondary signal for NLRP3 inflammasome activation [308]. Thus, the contribution of NOX isoforms and NOX-derived ROS in various cell types that participate in innate and adaptive immunity to influenza warrants further investigation.

Since different subtypes of influenza virus alter NOX isoform expression and kinetics of induction in different ways [276], these differences are likely to influence early antiviral response, NLRP3 inflammasome activation, and the inflammatory response. Although studies using mice deficient in a specific NOX isoform or regulatory subunit implicate NOX enzymes in immunity to influenza, the results of these studies have yielded a range of results making it difficult to generalize regarding a specific role for NOX isoforms [270,282,286,288,289]. Apparent differences could be due to differences in experimental design, virus strain, virus dose, age, or strain of mice or other factors. Lung epithelial cells derived from conditional NOX isoform knockout mice could provide a tool to better understand the specific roles of NOX isoenzymes. Also, experiments that quantify the kinetics of ROS production *versus* markers of innate immunity following infection are needed, as are experiments using seasonal as well as potentially pandemic avian influenza A virus strains.

NOX Enzymes and Adaptive Immunity to Influenza

Although innate immunity plays an early and a key role in the host response to influenza infection, efficient virus clearance and establishment of long lasting memory to reinfection depends on a well-coordinated adaptive immune response [309]. Several molecular pathways of innate immunity are required to produce an efficient adaptive immune response to influenza infection. For example, recognition of influenza by TLRs, type I IFN induction, and inflammasome recognition of influenza virus play critical roles [310,311]. As described here, these and other facets of the innate immune response to influenza infection are influenced by NOX enzymes and by NOX-derived ROS.

Therefore it is likely that NOX effects on innate immunity will alter the magnitude and/or quality of the subsequent adaptive immune response to influenza. The extent and effects will depend on whether expression of NOX and NOX-induced ROS are modified in cell types relevant to the adaptive immune system. Effector molecules produced by the innate cells as a result of NOX activation might also indirectly regulate the phenotype and functionality of cell types involved in the adaptive immune response.

Adaptive immunity to influenza A virus infection depends on efficient trafficking and uptake of antigens by dendritic cells (DCs) in the upper respiratory tract [312]. Upon antigen uptake, DCs migrate to draining lymph nodes for antigen presentation and activation of naïve T lymphocytes [313]. ROS generation in the DCs influences both antigen degradation and presentation [314]. Since the first demonstration of ROS generation by DCs [315] and subsequent discovery of NOX2 in mouse splenic DCs [316], many studies have supported the notion that NOX2 plays a role in antigen presentation in DC phagosomes (reviewed in [314]).

In addition to antigen presentation, activation of T cells requires signaling via costimulatory molecules [317]. CD40 is a key costimulatory ligand that is expressed on activated DCs and has a role in T cell activation [318]. In recent studies, we found an increase in the percentage of cells expressing CD40 among different DC subsets in the draining lymph nodes of influenza A virus-infected NOX1-deficient mice [282].

Compared to wild-type mice, NOX1-deficient mice showed an increased frequency of CD40 expressing conventional DCs, CD8+DCs, and interstitial DCs. Nevertheless, it was not clear whether the alterations in CD40 expressing DC subsets contributed to the observed improvement in host survival in NOX1-deficient cells, or to observed differences in T cell phenotypes.

While it is known that both transformed [319] and primary human B cells [320] generate ROS from NOX2, little is known about the expression and function of other NOX isoforms in different lymphocyte subtypes. Studies indicate a role for ROS in B cell receptor-mediated signal transduction, TCR-mediated responses, and T cell pro-liferation [321,322]. ROS generation in T cells activated NF-kB, leading to produc-tion of IL-2 [323]. Moreover, ROS generated by accessory immune cells suppressed T cell proliferation [324]. With regard to influenza-specific immunity and NOX1, we observed differences in the proportions and functionality of antigen-specific T cells between NOX1-deficient and control groups of mice. Specifically, influenza virus–infected NOX1-deficient mice showed a decrease in early effector CD8+ T cells and an increased proportion of memory precursor effector cells [282]. Additional investigations addressing the distribution and kinetics of ROS-generating enzymes in lymphocytes and their impact on influenza-specific immune responses are needed. Furthermore, considering that a defective adaptive immune response was observed upon *A. fumigatus* infection of p47^{phox}-deficient mice [325], a detailed evaluation of NOX isoform expression and function in immune cells may have implications for modulation of adaptive immune response to influenza vaccination.

TARGETING NOX-DERIVED ROS AS ADJUNCT THERAPY FOR INFLUENZA INFECTION

Although superoxide and hydrogen peroxide are necessary both as an antimicrobial defense mechanism [326] and for normal cellular homeostasis [327], overproduction can lead to pathological outcomes [328]. In contrast, as demonstrated in chronic granulomatous disease caused by mutations in the NOX enzymes in phagocytes, the absence of oxidants can result in excessive inflammation, even in the absence of infection [329]. Since many NOX enzymes are modulated by pathogens including influenza A viruses, it may be beneficial to limit over production of ROS to control severity of hyperinflammatory conditions (e.g., acute respiratory distress syndrome and others) that relate to morbidity and mortality following infection.

With regard to influenza A–induced disease severity, treatment of mice with a cell-permeant form of superoxide dismutase (SOD) suppressed virus-induced ROS levels and improved survival [330]. Similarly, treatment of influenza-infected mice with a small molecule cell-permeant SOD mimetic decreased virus-induced pathol-ogy in the lung [286]. Ebselen, a glutathione peroxidase mimetic and a NOX2 inhibi-tor [331], decreased levels of inflammatory mediators following influenza infection [332]. Apocynin decreased lung virus titer and inflammatory cells in influenza virus–infected mice [289]. In another study in influenza-infected human and chicken cell lines [280], apocynin decreased cytokine and ROS levels, but did not affect virus titer. In contrast, in a study in NCI-H292 cells, VAS2870, a pan NOX isoform inhibi-tor, produced a significant decrease in virus titer [275]. While the use of pharmaco-logical inhibitors of NOX enzymes in the treatment of viral diseases holds promise, isoform-specific effects appear to be complex, and the development and commercialization of isoform-selective inhibitors remains in its infancy (see [333]).

While several other enzymes [334] and signaling pathways [335] are known to participate in influenza-induced inflammatory diseases, this discussion suggests that the use of NOX isoform-selective inhibitors (along with antiviral compounds) as adjunct therapies may have certain advantages. First, NOX enzymes are expressed in cell types that mediate viral entry, in cells that regulate the trafficking of other cell types responsible for inflammation and cell damage, and in lung cell types where replicating virus leads to significant tissue damage. Second, NOX enzymes are induced within hours after influenza infection [275,276] and play a role in the regulation of antiviral [275,289], anti-inflammatory [270], and adaptive [282] immune responses. Therefore, unlike other host cell–targeted adjunct treatments that promote innate immunity and antiviral cytokines (reviewed in [336]), NOX inhibitors can be used early as well as during the course of an ongoing influenza infection. However, different NOX-selective inhibitors may be more or less effective depending on the stage of infection, the host tissues involved, and the influenza virus strain. For example, while some influenza strains replicate primarily in epithelial cells, a few others also replicate in monocytes and endothelial cells [337,338]. Moreover, while inhibition of NOX-derived ROS has the potential to benefit the infected host by inhibiting virus-induced inflammatory response [289,330,332] and promoting adaptive immunity [282], NOX-derived ROS can also benefit the host during early innate immune responses [277–280]. Thus, the function of NOX enzymes during viral infections may be a double-edged sword, and careful attention to inhibitor properties (isoform selectivity, route of administration, etc.) and dosing schedule may prove important for effective treatment.

REFERENCES

1. Winterbourn, C. C. and M. B. Hampton. 2008. Thiol chemistry and specificity in redox signaling. *Free Radic Biol Med* 45 (5):549–561. doi: 10.1016/j.freeradbiomed.2008.05.004.
2. Parke, D. V. and A. Sapota. 1996. Chemical toxicity and reactive oxygen species. *Int J Occup Med Environ Health* 9 (4):331–340.
3. Monks, T. J. and D. C. Jones. 2002. The metabolism and toxicity of quinones, quinonimines, quinone methides, and quinone-thioethers. *Curr Drug Metab* 3 (4):425–438.
4. Gutierrez, P. L. 2000. The metabolism of quinone-containing alkylating agents: Free radical production and measurement. *Front Biosci* 5:D629–D638.
5. Lambert, A. J. and M. D. Brand. 2009. Reactive oxygen species production by mitochondria. *Methods Mol Biol* 554:165–181. doi: 10.1007/978-1-59745-521-3_11.
6. Quinlan, C. L., I. V. Perevoshchikova, M. Hey-Mogensen, A. L. Orr, and M. D. Brand. 2013. Sites of reactive oxygen species generation by mitochondria oxidizing different substrates. *Redox Biol* 1:304–312. doi: 10.1016/j.redox.2013.04.005.
7. Nohl, H., L. Gille, A. Kozlov, and K. Staniek. 2003. Are mitochondria a spontaneous and permanent source of reactive oxygen species? *Redox Rep* 8 (3):135–141. doi: 10.1179/135100003225001502.
8. Dikalova, A. E., M. C. Gongora, D. G. Harrison, J. D. Lambeth, S. Dikalov, and K. K. Griendling. 2010. Upregulation of Nox1 in vascular smooth muscle leads to impaired endothelium-dependent relaxation via eNOS uncoupling. *Am J Physiol Heart Circ Physiol* 299 (3):H673–H679. doi: 10.1152/ajpheart.00242.2010.

9. Lambeth, J. D. 2004. NOX enzymes and the biology of reactive oxygen. *Nat Rev Immunol* 4 (3):181–189. doi: 10.1038/nri1312.

10. Nauseef, W. M. 2004. Assembly of the phagocyte NADPH oxidase. *Histochem Cell Biol* 122 (4):277–291. doi: 10.1007/s00418-004-0679-8.

11. Suh, Y. A., R. S. Arnold, B. Lassegue, J. Shi, X. Xu, D. Sorescu, A. B. Chung, K. K. Griendling, and J. D. Lambeth. 1999. Cell transformation by the superoxide-generating oxidase Mox1. *Nature* 401 (6748):79–82. doi: 10.1038/43459.

12. Banfi, B., G. Molnar, A. Maturana, K. Steger, B. Hegedus, N. Demaurex, and K. H. Krause. 2001. A Ca(2+)-activated NADPH oxidase in testis, spleen, and lymph nodes. *J Biol Chem* 276 (40):37594–37601. doi: 10.1074/jbc.M103034200.

13. Cheng, G., Z. Cao, X. Xu, E. G. van Meir, and J. D. Lambeth. 2001. Homologs of gp91phox: Cloning and tissue expression of Nox3, Nox4, and Nox5. *Gene* 269 (1–2):131–140. S0378111901004498 [pii].

14. De Deken, X., D. Wang, M. C. Many, S. Costagliola, F. Libert, G. Vassart, J. E. Dumont, and F. Miot. 2000. Cloning of two human thyroid cDNAs encoding new members of the NADPH oxidase family. *J Biol Chem* 275 (30):23227–23233. doi: 10.1074/jbc. M000916200. M000916200 [pii].

15. Geiszt, M., J. B. Kopp, P. Varnai, and T. L. Leto. 2000. Identification of renox, an NAD(P)H oxidase in kidney. *Proc Natl Acad Sci USA* 97 (14):8010–8014. doi: 10.1073/ pnas.130135897. 130135897 [pii].

16. Shiose, A., J. Kuroda, K. Tsuruya, M. Hirai, H. Hirakata, S. Naito, M. Hattori, Y. Sakaki, and H. Sumimoto. 2001. A novel superoxide-producing NAD(P)H oxidase in kidney. *J Biol Chem* 276 (2):1417–1423. doi: 10.1074/jbc.M007597200. M007597200 [pii].

17. Cheng, G. and J. D. Lambeth. 2004. NOXO1, regulation of lipid binding, localization, and activation of Nox1 by the Phox homology (PX) domain. *J Biol Chem* 279 (6):4737–4742. doi: 10.1074/jbc.M305968200.

18. Takeya, R., N. Ueno, K. Kami, M. Taura, M. Kohjima, T. Izaki, H. Nunoi, and H. Sumimoto. 2003. Novel human homologues of p47phox and p67phox participate in activation of superoxide-producing NADPH oxidases. *J Biol Chem* 278 (27):25234–25246. doi: 10.1074/jbc.M212856200.

19. Kawahara, T., D. Ritsick, G. Cheng, and J. D. Lambeth. 2005. Point mutations in the proline-rich region of p22phox are dominant inhibitors of Nox1- and Nox2-dependent reactive oxygen generation. *J Biol Chem* 280 (36):31859–31869. doi: 10.1074/jbc. M501882200.

20. Edens, W. A., L. Sharling, G. Cheng, R. Shapira, J. M. Kinkade, T. Lee, H. A. Edens et al. 2001. Tyrosine cross-linking of extracellular matrix is catalyzed by Duox, a multi-domain oxidase/peroxidase with homology to the phagocyte oxidase subunit gp91phox. *J Cell Biol* 154 (4):879–891. doi: 10.1083/jcb.200103132.

21. Takac, I., K. Schroder, L. Zhang, B. Lardy, N. Anilkumar, J. D. Lambeth, A. M. Shah, F. Morel, and R. P. Brandes. 2011. The E-loop is involved in hydrogen peroxide formation by the NADPH oxidase Nox4. *J Biol Chem* 286 (15):13304–13313. doi: 10.1074/ jbc.M110.192138.

22. Nisimoto, Y., B. A. Diebold, D. Cosentino-Gomes, and J. D. Lambeth. 2014. Nox4: A hydrogen peroxide-generating oxygen sensor. *Biochemistry* 53 (31):5111–5120. doi: 10.1021/bi500331y.

23. Carnesecchi, S., J. C. Pache, and C. Barazzone-Argiroffo. 2012. NOX enzymes: Potential target for the treatment of acute lung injury. *Cell Mol Life Sci* 69 (14):2373–2385. doi: 10.1007/s00018-012-1013-6.

24. Thannickal, V. J. 2012. Mechanisms of pulmonary fibrosis: Role of activated myofibroblasts and NADPH oxidase. *Fibrogenesis Tissue Repair* 5 (Suppl 1):S23. doi: 10.1186/1755-1536-5-S1-S23.

25. Pendyala, S. and V. Natarajan. 2010. Redox regulation of Nox proteins. *Respir Physiol Neurobiol* 174 (3):265–271. doi: 10.1016/j.resp.2010.09.016.

26. Grandvaux, N., M. Mariani, and K. Fink. 2015. Lung epithelial NOX/DUOX and respiratory virus infections. *Clin Sci* 128 (6):337–347.

27. Schwarzer, C., T. E. Machen, B. Illek, and H. Fischer. 2004. NADPH oxidase-dependent acid production in airway epithelial cells. *J Biol Chem* 279 (35):36454–36461. doi: 10.1074/jbc.M404983200.

28. Fischer, H. 2009. Mechanisms and function of DUOX in epithelia of the lung. *Antioxid Redox Signal* 11 (10):2453–2465. doi: 10.1089/ARS.2009.2558.

29. Gorissen, S. H., M. Hristova, A. Habibovic, L. M. Sipsey, P. C. Spiess, Y. M. Janssen-Heininger, and A. van der Vliet. 2013. Dual oxidase-1 is required for airway epithelial cell migration and bronchiolar reepithelialization after injury. *Am J Respir Cell Mol Biol* 48 (3):337–345. doi: 10.1165/rcmb.2012-0393OC.

30. Fraternale, A., M. F. Paoletti, A. Casabianca, L. Nencioni, E. Garaci, A. T. Palamara, and M. Magnani. 2009. GSH and analogs in antiviral therapy. *Mol Aspects Med* 30 (1–2):99–110. doi: 10.1016/j.mam.2008.09.001.

31. Tilton, C., A. J. Clippinger, T. Maguire, and J. C. Alwine. 2011. Human cytomegalovirus induces multiple means to combat reactive oxygen species. *J Virol* 85 (23):12585–12593. doi: 10.1128/JVI.05572-11.

32. Staal, F. J., S. W. Ela, M. Roederer, M. T. Anderson, L. A. Herzenberg, and L. A. Herzenberg. 1992. Glutathione deficiency and human immunodeficiency virus infection. *Lancet* 339 (8798):909–912.

33. Hennet, T., E. Peterhans, and R. Stocker. 1992. Alterations in antioxidant defences in lung and liver of mice infected with influenza A virus. *J Gen Virol* 73 (Pt 1):39–46. doi: 10.1099/0022-1317-73-1-39.

34. Palamara, A. T., C. F. Perno, M. R. Ciriolo, L. Dini, E. Balestra, C. D'Agostini, P. Di Francesco, C. Favalli, G. Rotilio, and E. Garaci. 1995. Evidence for antiviral activity of glutathione: In vitro inhibition of herpes simplex virus type 1 replication. *Antiviral Res* 27 (3):237–253.

35. Hosakote, Y. M., P. D. Jantzi, D. L. Esham, H. Spratt, A. Kurosky, A. Casola, and R. P. Garofalo. 2011. Viral-mediated inhibition of antioxidant enzymes contributes to the pathogenesis of severe respiratory syncytial virus bronchiolitis. *Am J Respir Crit Care Med* 183 (11):1550–1560. doi: 10.1164/rccm.201010-1755OC.

36. van Dalen, C. J., M. W. Whitehouse, C. C. Winterbourn, and A. J. Kettle. 1997. Thiocyanate and chloride as competing substrates for myeloperoxidase. *Biochem J* 327 (Pt 2):487–492.

37. Pace, G. W. and C. D. Leaf. 1995. The role of oxidative stress in HIV disease. *Free Radic Biol Med* 19 (4):523–528.

38. Baruchel, S. and M. A. Wainberg. 1992. The role of oxidative stress in disease progression in individuals infected by the human immunodeficiency virus. *J Leukoc Biol* 52 (1):111–114.

39. Porter, K. M. and R. L. Sutliff. 2012. HIV-1, reactive oxygen species, and vascular complications. *Free Radic Biol Med* 53 (1):143–159. doi: 10.1016/j.freeradbiomed.2012.03.019.

40. Lachgar, A., N. Sojic, S. Arbault, D. Bruce, A. Sarasin, C. Amatore, B. Bizzini, D. Zagury, and M. Vuillaume. 1999. Amplification of the inflammatory cellular redox state by human immunodeficiency virus type 1-immunosuppressive tat and gp160 proteins. *J Virol* 73 (2):1447–1452.

41. Gu, Y., R. F. Wu, Y. C. Xu, S. C. Flores, and L. S. Terada. 2001. HIV Tat activates c-Jun amino-terminal kinase through an oxidant-dependent mechanism. *Virology* 286 (1):62–71. doi: 10.1006/viro.2001.0998.

42. Wu, R. F., Z. Ma, Z. Liu, and L. S. Terada. 2010. Nox4-derived H_2O_2 mediates endoplasmic reticulum signaling through local Ras activation. *Mol Cell Biol* 30 (14):3553–3568. doi: 10.1128/mcb.01445-09.

43. Wu, R. F., Z. Ma, D. P. Myers, and L. S. Terada. 2007. HIV-1 Tat activates dual Nox pathways leading to independent activation of ERK and JNK MAP kinases. *J Biol Chem* 282 (52):37412–37419. doi: 10.1074/jbc.M704481200.

44. Turchan–Cholewo, J., V. M. Dimayuga, S. Gupta, R. M. Charlotte Gorospe, J. N. Keller, and A. J. Bruce–Keller. 2009. NADPH oxidase drives cytokine and neurotoxin release from microglia and macrophages in response to HIV-Tat. *Antioxid Redox Signal* 11 (2):193–204. doi: 10.1089/ARS.2008.2097.

45. Williams, R., H. Yao, F. Peng, Y. Yang, C. Bethel-Brown, and S. Buch. 2010. Cooperative induction of CXCL10 involves NADPH oxidase: Implications for HIV dementia. *Glia* 58 (5):611–621. doi: 10.1002/glia.20949.

46. Vilhardt, F., O. Plastre, M. Sawada, K. Suzuki, M. Wiznerowicz, E. Kiyokawa, D. Trono, and K. H. Krause. 2002. The HIV-1 Nef protein and phagocyte NADPH oxidase activation. *J Biol Chem* 277 (44):42136–42143. doi: 10.1074/jbc.M200862200.

47. Salmen, S., M. Colmenares, D. L. Peterson, E. Reyes, J. D. Rosales, and L. Berrueta. 2010. HIV-1 Nef associates with p22-phox, a component of the NADPH oxidase protein complex. *Cell Immunol* 263 (2):166–171. doi: 10.1016/j.cellimm.2010.03.012.

48. Reddy, P. V., N. Gandhi, T. Samikkannu, Z. Saiyed, M. Agudelo, A. Yndart, P. Khatavkar, and M. P. Nair. 2012. HIV-1 gp120 induces antioxidant response element-mediated expression in primary astrocytes: Role in HIV associated neurocognitive disorder. *Neurochem Int* 61 (5):807–814. doi: 10.1016/j.neuint.2011.06.011.

49. Shah, A., S. Kumar, S. D. Simon, D. P. Singh, and A. Kumar. 2013. HIV gp120- and methamphetamine-mediated oxidative stress induces astrocyte apoptosis via cytochrome P450 2E1. *Cell Death Dis* 4:e850. doi: 10.1038/cddis.2013.374.

50. Kline, E. R., D. J. Kleinhenz, B. Liang, S. Dikalov, D. M. Guidot, C. M. Hart, D. P. Jones, and R. L. Sutliff. 2008. Vascular oxidative stress and nitric oxide depletion in HIV-1 transgenic rats are reversed by glutathione restoration. *Am J Physiol Heart Circ Physiol* 294 (6):H2792–H2804. doi: 10.1152/ajpheart.91447.2007.

51. Pastori, D., A. Esposito, R. Carnevale, S. Bartimoccia, M. Novo, A. Fantauzzi, F. Di Campli, P. Pignatelli, F. Violi, and I. Mezzaroma. 2015. HIV-1 induces in vivo platelet activation by enhancing platelet NOX2 activity. *J Infect* 70 (6):651–658. doi: 10.1016/j.jinf.2015.01.005.

52. Beloqui, O., J. Prieto, M. Suarez, B. Gil, C. H. Qian, N. Garcia, and M. P. Civeira. 1993. N-acetyl cysteine enhances the response to interferon-alpha in chronic hepatitis C: A pilot study. *J Interferon Res* 13 (4):279–282.

53. Levent, G., A. Ali, A. Ahmet, E. C. Polat, C. Aytac, E. Ayse, and S. Ahmet. 2006. Oxidative stress and antioxidant defense in patients with chronic hepatitis C patients before and after pegylated interferon alfa-2b plus ribavirin therapy. *J Transl Med* 4:25. doi: 10.1186/1479-5876-4-25.

54. Farinati, F., R. Cardin, N. De Maria, G. D. Libera, C. Marafin, E. Lecis, P. Burra, A. Floreani, A. Cecchetto, and R. Naccarato. 1995. Iron storage, lipid peroxidation and glutathione turnover in chronic anti-HCV positive hepatitis. *J Hepatol* 22 (4):449–456. doi: 10.1016/0168-8278(95)80108-1.

55. Piccoli, C., R. Scrima, G. Quarato, A. D'Aprile, M. Ripoli, L. Lecce, D. Boffoli, D. Moradpour, and N. Capitanio. 2007. Hepatitis C virus protein expression causes calcium-mediated mitochondrial bioenergetic dysfunction and nitro-oxidative stress. *Hepatology* 46 (1):58–65. doi: 10.1002/hep.21679.

56. Ando, M., M. Korenaga, K. Hino, M. Ikeda, N. Kato, S. Nishina, I. Hidaka, and I. Sakaida. 2008. Mitochondrial electron transport inhibition in full genomic hepatitis C virus replicon cells is restored by reducing viral replication. *Liver Int* 28 (8):1158–1166. doi: 10.1111/j.1478-3231.2008.01720.x.

57. Tariq, M., S. Manzoor, Q. L. Ahmed, M. Khalid, and W. Ashraf. 2013. NOX4 induces oxidative stress and apoptosis through upregulation of caspases 3 and 9 and downregulation of TIGAR in HCV-infected Huh-7 cells. *Future Virol* 8 (7):707–716.

58. de Mochel, N. S., S. Seronello, S. H. Wang, C. Ito, J. X. Zheng, T. J. Liang, J. D. Lambeth, and J. Choi. 2010. Hepatocyte NAD(P)H oxidases as an endogenous source of reactive oxygen species during hepatitis C virus infection. *Hepatology* 52 (1):47–59. doi: 10.1002/hep.23671.

59. Boudreau, H. E., S. U. Emerson, A. Korzeniowska, M. A. Jendrysik, and T. L. Leto. 2009. Hepatitis C virus (HCV) proteins induce NADPH oxidase 4 expression in a transforming growth factor beta-dependent manner: A new contributor to HCV-induced oxidative stress. *J Virol* 83 (24):12934–12946. doi: 10.1128/JVI.01059-09.

60. Ivanov, A. V., O. A. Smirnova, O. N. Ivanova, O. V. Masalova, S. N. Kochetkov, and M. G. Isaguliants. 2011. Hepatitis C virus proteins activate NRF2/ARE pathway by distinct ROS-dependent and independent mechanisms in HUH7 cells. *PLoS ONE* 6 (9):e24957. doi: 10.1371/journal.pone.0024957.

61. Smirnova, O. A., O. N. Ivanova, B. Bartosch, V. T. Valuev-Elliston, F. Mukhtarov, S. N. Kochetkov, and A. V. Ivanov. 2016. Hepatitis C virus NS5A protein triggers oxidative stress by inducing NADPH oxidases 1 and 4 and cytochrome P450 2E1. *Oxid Med Cell Longev* 2016:8341937. doi: 10.1155/2016/8341937.

62. Garcia-Mediavilla, M. V., S. Sanchez-Campos, P. Gonzalez-Perez, M. Gomez-Gonzalo, P. L. Majano, M. Lopez-Cabrera, G. Clemente, C. Garcia-Monzon, and J. Gonzalez-Gallego. 2005. Differential contribution of hepatitis C virus NS5A and core proteins to the induction of oxidative and nitrosative stress in human hepatocyte-derived cells. *J Hepatol* 43 (4):606–613. doi: 10.1016/j.jhep.2005.04.019.

63. Li, S., L. Ye, X. Yu, B. Xu, K. Li, X. Zhu, H. Liu, X. Wu, and L. Kong. 2009. Hepatitis C virus NS4B induces unfolded protein response and endoplasmic reticulum overload response-dependent NF-kappaB activation. *Virology* 391 (2):257–264. doi: 10.1016/j.virol.2009.06.039.

64. Gong, G., G. Waris, R. Tanveer, and A. Siddiqui. 2001. Human hepatitis C virus NS5A protein alters intracellular calcium levels, induces oxidative stress, and activates STAT-3 and NF-kappa B. *Proc Natl Acad Sci USA* 98 (17):9599–9604. doi: 10.1073/pnas.171311298.

65. Okuda, M., K. Li, M. R. Beard, L. A. Showalter, F. Scholle, S. M. Lemon, and S. A. Weinman. 2002. Mitochondrial injury, oxidative stress, and antioxidant gene expression are induced by hepatitis C virus core protein. *Gastroenterology* 122 (2):366–375.

66. Otani, K., M. Korenaga, M. R. Beard, K. Li, T. Qian, L. A. Showalter, A. K. Singh, T. Wang, and S. A. Weinman. 2005. Hepatitis C virus core protein, cytochrome P450 2E1, and alcohol produce combined mitochondrial injury and cytotoxicity in hepatoma cells. *Gastroenterology* 128 (1):96–107.

67. Pal, S., S. J. Polyak, N. Bano, W. C. Qiu, R. L. Carithers, M. Shuhart, D. R. Gretch, and A. Das. 2010. Hepatitis C virus induces oxidative stress, DNA damage and modulates the DNA repair enzyme NEIL1. *J Gastroenterol Hepatol* 25 (3):627–634. doi: 10.1111/j.1440-1746.2009.06128.x.

68. Ming-Ju, H., H. Yih-Shou, C. Tzy-Yen, and C. Hui-Ling. 2011. Hepatitis C virus E2 protein induce reactive oxygen species (ROS)-related fibrogenesis in the HSC-T6 hepatic stellate cell line. *J Cell Biochem* 112 (1):233–243. doi: 10.1002/jcb.22926.

69. Wu, C. F., Y. L. Lin, and Y. T. Huang. 2013. Hepatitis C virus core protein stimulates fibrogenesis in hepatic stellate cells involving the obese receptor. *J Cell Biochem* 114 (3):541–550. doi: 10.1002/jcb.24392.

70. Thoren, F., A. Romero, M. Lindh, C. Dahlgren, and K. Hellstrand. 2004. A hepatitis C virus-encoded, nonstructural protein (NS3) triggers dysfunction and apoptosis in lymphocytes: Role of NADPH oxidase-derived oxygen radicals. *J Leukoc Biol* 76 (6): 1180–1186. doi: 10.1189/jlb.0704387.

71. Bureau, C., J. Bernad, N. Chaouche, C. Orfila, M. Beraud, C. Gonindard, L. Alric, J. P. Vinel, and B. Pipy. 2001. Nonstructural 3 protein of hepatitis C virus triggers an oxidative burst in human monocytes via activation of NADPH oxidase. *J Biol Chem* 276 (25):23077–23083. doi: 10.1074/jbc.M100698200.

72. Tacke, R. S., H. C. Lee, C. Goh, J. Courtney, S. J. Polyak, H. R. Rosen, and Y. S. Hahn. 2012. Myeloid suppressor cells induced by hepatitis C virus suppress T-cell responses through the production of reactive oxygen species. *Hepatology* 55 (2):343–353. doi: 10.1002/hep.24700.

73. Diamond, D. L., A. L. Krasnoselsky, K. E. Burnum, M. E. Monroe, B. J. Webb-Robertson, J. E. McDermott, M. M. Yeh et al. 2012. Proteome and computational analyses reveal new insights into the mechanisms of hepatitis C virus-mediated liver disease posttransplantation. *Hepatology* 56 (1):28–38. doi: 10.1002/hep.25649.

74. Choi, J., N. L. Corder, B. Koduru, and Y. Wang. 2014. Oxidative stress and hepatic Nox proteins in chronic hepatitis C and hepatocellular carcinoma. *Free Radic Biol Med* 72:267–284. doi: 10.1016/j.freeradbiomed.2014.04.020.

75. Cardin, R., M. Piciocchi, M. Bortolami, A. Kotsafti, L. Barzon, E. Lavezzo, A. Sinigaglia, K. I. Rodriguez-Castro, M. Rugge, and F. Farinati. 2014. Oxidative damage in the progression of chronic liver disease to hepatocellular carcinoma: An intricate pathway. *World J Gastroenterol* 20 (12):3078–3086. doi: 10.3748/wjg.v20.i12.3078.

76. Świetek, K. and J. Juszczyk. 1997. Reduced glutathione concentration in erythrocytes of patients with acute and chronic viral hepatitis. *J Viral Hepat* 4 (2):139–141.

77. Waris, G., K. W. Huh, and A. Siddiqui. 2001. Mitochondrially associated hepatitis B virus X protein constitutively activates transcription factors STAT-3 and NF-kappa B via oxidative stress. *Mol Cell Biol* 21 (22):7721–7730. doi: 10.1128/MCB.21.22.7721-7730.2001.

78. Williams, V., S. Brichler, E. Khan, M. Chami, P. Deny, D. Kremsdorf, and E. Gordien. 2012. Large hepatitis delta antigen activates STAT-3 and NF-kappaB via oxidative stress. *J Viral Hepat* 19 (10):744–753. doi: 10.1111/j.1365-2893.2012.01597.x.

79. Papi, A., N. G. Papadopoulos, L. A. Stanciu, C. M. Bellettato, S. Pinamonti, K. Degitz, S. T. Holgate, and S. L. Johnston. 2002. Reducing agents inhibit rhinovirus-induced up-regulation of the rhinovirus receptor intercellular adhesion molecule-1 (ICAM-1) in respiratory epithelial cells. *FASEB J* 16 (14):1934–1936. doi: 10.1096/fj.02-0118fje.

80. Kaul, P., M. C. Biagioli, R. B. Turner, and I. Singh. 1997. Association of alterations of cellular redox pathways with rhinovirus(RV)-induced interleukin-8 (IL-8) elaboration in respiratory epithelial cells. *Pediatr Res* 41 (S4):123.

81. Papi, A., M. Contoli, P. Gasparini, L. Bristot, M. R. Edwards, M. Chicca, M. Leis et al. 2008. Role of xanthine oxidase activation and reduced glutathione depletion in rhinovirus induction of inflammation in respiratory epithelial cells. *J Biol Chem* 283 (42):28595–28606. doi: 10.1074/jbc.M805766200.

82. Kaul, P., M. C. Biagioli, I. Singh, and R. B. Turner. 2000. Rhinovirus-induced oxidative stress and interleukin-8 elaboration involves p47-phox but is independent of attachment to intercellular adhesion molecule-1 and viral replication. *J Infect Dis* 181 (6):1885–1890. doi: 10.1086/315504.

83. Comstock, A. T., S. Ganesan, A. Chattoraj, A. N. Faris, B. L. Margolis, M. B. Hershenson, and U. S. Sajjan. 2011. Rhinovirus-induced barrier dysfunction in polarized airway epithelial cells is mediated by NADPH oxidase 1. *J Virol* 85 (13):6795–6808. doi: 10.1128/jvi.02074-10.

84. Linderholm, A. L., J. Onitsuka, C. Xu, M. Chiu, W. M. Lee, and R. W. Harper. 2010. All-trans retinoic acid mediates DUOX2 expression and function in respiratory tract epithelium. *Am J Physiol Lung Cell Mol Physiol* 299 (2):L215–L221. doi: 10.1152/ajplung.00015.2010.

85. Harper, R. W., C. Xu, J. P. Eiserich, Y. Chen, C. Y. Kao, P. Thai, H. Setiadi, and R. Wu. 2005. Differential regulation of dual NADPH oxidases/peroxidases, Duox1 and Duox2, by Th1 and Th2 cytokines in respiratory tract epithelium. *FEBS Lett* 579 (21):4911–4917. doi: 10.1016/j.febslet.2005.08.002.

86. Biagioli, M. C., P. Kaul, I. Singh, and R. B. Turner. 1999. The role of oxidative stress in rhinovirus induced elaboration of IL-8 by respiratory epithelial cells. *Free Radic Biol Med* 26 (3–4):454–462. doi: 10.1016/S0891-5849(98)00233-0.

87. Douglass, J. A., D. Dhami, C. E. Gurr, M. Bulpitt, J. K. Shute, P. H. Howarth, I. J. Lindley, M. K. Church, and S. T. Holgate. 1994. Influence of interleukin-8 challenge in the nasal mucosa in atopic and nonatopic subjects. *Am J Respir Crit Care Med* 150 (4):1108–1113. doi: 10.1164/ajrccm.150.4.7921444.

88. Martinez, I., V. Garcia-Carpizo, T. Guijarro, A. Garcia-Gomez, D. Navarro, A. Aranda, and A. Zambrano. 2016. Induction of DNA double-strand breaks and cellular senescence by human respiratory syncytial virus. *Virulence* 7 (4):427–442. doi: 10.1080/21505594.2016.1144001.

89. Casola, A., N. Burger, T. Liu, M. Jamaluddin, A. R. Brasier, and R. P. Garofalo. 2001. Oxidant tone regulates RANTES gene expression in airway epithelial cells infected with respiratory syncytial virus. Role in viral-induced interferon regulatory factor activation. *J Biol Chem* 276 (23):19715–19722. doi: 10.1074/jbc.M101526200.

90. Hosakote, Y. M., T. Liu, S. M. Castro, R. P. Garofalo, and A. Casola. 2009. Respiratory syncytial virus induces oxidative stress by modulating antioxidant enzymes. *Am J Respir Cell Mol Biol* 41 (3):348–357. doi: 10.1165/rcmb.2008-0330OC.

91. Ren, K., Y. Lv, Y. Zhuo, C. Chen, H. Shi, L. Guo, G. Yang, Y. Hou, R. X. Tan, and E. Li. 2016. Suppression of IRG-1 reduces inflammatory cell infiltration and lung injury in respiratory syncytial virus infection by reducing production of reactive oxygen species. *J Virol* 90 (16):7313–7322. doi: 10.1128/jvi.00563-16.

92. Fink, K., A. Duval, A. Martel, A. Soucy-Faulkner, and N. Grandvaux. 2008. Dual role of NOX2 in respiratory syncytial virus- and sendai virus-induced activation of NF-kappaB in airway epithelial cells. *J Immunol* 180 (10):6911–6922.

93. Indukuri, H., S. M. Castro, S. M. Liao, L. A. Feeney, M. Dorsch, A. J. Coyle, R. P. Garofalo, A. R. Brasier, and A. Casola. 2006. Ikkepsilon regulates viral-induced interferon regulatory factor-3 activation via a redox-sensitive pathway. *Virology* 353 (1):155–165. doi: 10.1016/j.virol.2006.05.022.

94. Liu, T., S. Castro, A. R. Brasier, M. Jamaluddin, R. P. Garofalo, and A. Casola. 2004. Reactive oxygen species mediate virus-induced STAT activation: Role of tyrosine phosphatases. *J Biol Chem* 279 (4):2461–2469. doi: 10.1074/jbc.M307251200.

95. Hosakote, Y. M., N. Komaravelli, N. Mautemps, T. Liu, R. P. Garofalo, and A. Casola. 2012. Antioxidant mimetics modulate oxidative stress and cellular signaling in airway epithelial cells infected with respiratory syncytial virus. *Am J Physiol Lung Cell Mol Physiol* 303 (11):L991–L1000. doi: 10.1152/ajplung.00192.2012.

96. Mata, M., E. Morcillo, C. Gimeno, and J. Cortijo. 2011. N-acetyl-L-cysteine (NAC) inhibit mucin synthesis and pro-inflammatory mediators in alveolar type II epithelial cells infected with influenza virus A and B and with respiratory syncytial virus (RSV). *Biochem Pharmacol* 82 (5):548–555. doi: 10.1016/j.bcp.2011.05.014.

97. Castro, S. M., A. Guerrero-Plata, G. Suarez-Real, P. A. Adegboyega, G. N. Colasurdo, A. M. Khan, R. P. Garofalo, and A. Casola. 2006. Antioxidant treatment ameliorates respiratory syncytial virus-induced disease and lung inflammation. *Am J Respir Crit Care Med* 174 (12):1361–1369. doi: 10.1164/rccm.200603-319OC.

98. Wyde, P. R., D. K. Moore, D. M. Pimentel, B. E. Gilbert, R. Nimrod, and A. Panet. 1996. Recombinant superoxide dismutase (SOD) administered by aerosol inhibits respiratory syncytial virus infection in cotton rats. *Antiviral Res* 31 (3):173–184.

99. Jamaluddin, M., J. E. Wiktorowicz, K. V. Soman, I. Boldogh, J. D. Forbus, H. Spratt, R. P. Garofalo, and A. R. Brasier. 2010. Role of peroxiredoxin 1 and peroxiredoxin 4 in protection of respiratory syncytial virus-induced cysteinyl oxidation of nuclear cyto-skeletal proteins. *J Virol* 84 (18):9533–9545. doi: 10.1128/jvi.01005-10.

100. Cho, H. Y., F. Imani, L. Miller-DeGraff, D. Walters, G. A. Melendi, M. Yamamoto, F. P. Polack, and S. R. Kleeberger. 2009. Antiviral activity of Nrf2 in a murine model of respiratory syncytial virus disease. *Am J Respir Crit Care Med* 179 (2):138–150. doi: 10.1164/rccm.200804-535OC.

101. Olagnier, D., S. Peri, C. Steel, N. van Montfoort, C. Chiang, V. Beljanski, M. Slifker et al. 2014. Cellular oxidative stress response controls the antiviral and apoptotic pro-grams in dengue virus-infected dendritic cells. *PLoS Pathog* 10 (12):e1004566. doi: 10.1371/journal.ppat.1004566.

102. Al-Alimi, A. A., S. A. Ali, F. M. Al-Hassan, F. M. Idris, S. Y. Teow, and N. Mohd Yusoff. 2014. Dengue virus type 2 (DENV2)-induced oxidative responses in monocytes from glucose-6-phosphate dehydrogenase (G6PD)-deficient and G6PD normal subjects. *PLoS Negl Trop Dis* 8 (3):e2711. doi: 10.1371/journal.pntd.0002711.

103. Wang, J., Y. Chen, N. Gao, Y. Wang, Y. Tian, J. Wu, J. Zhang, J. Zhu, D. Fan, and J. An. 2013. Inhibitory effect of glutathione on oxidative liver injury induced by den-gue virus serotype 2 infections in mice. *PLoS One* 8 (1):e55407. doi: 10.1371/journal. pone.0055407.

104. Yen, Y. T., H. C. Chen, Y. D. Lin, C. C. Shieh, and B. A. Wu-Hsieh. 2008. Enhancement by tumor necrosis factor alpha of Dengue virus-induced endothelial cell production of reactive nitrogen and oxygen species is key to hemorrhage development. *J Virol* 82 (24):12312–12324. doi: 10.1128/jvi.00968-08.

105. Jan, J. T., B. H. Chen, S. H. Ma, C. I. Liu, H. P. Tsai, H. C. Wu, S. Y. Jiang, K. D. Yang, and M. F. Shaio. 2000. Potential dengue virus-triggered apoptotic pathway in human neuroblastoma cells: Arachidonic acid, superoxide anion, and NF-kappaB are sequen-tially involved. *J Virol* 74 (18):8680–8691.

106. Immenschuh, S., P. Rahayu, B. Bayat, H. Saragih, A. Rachman, and S. Santoso. 2013. Antibodies against dengue virus nonstructural protein-1 induce heme oxygenase-1 via a redox-dependent pathway in human endothelial cells. *Free Radic Biol Med* 54:85–92. doi: 10.1016/j.freeradbiomed.2012.10.551.

107. Seet, R. C., C. Y. Lee, E. C. Lim, A. M. Quek, L. L. Yeo, S. H. Huang, and B. Halliwell. 2009. Oxidative damage in dengue fever. *Free Radic Biol Med* 47 (4):375–380. doi: 10.1016/j.freeradbiomed.2009.04.035.

108. Pan, X., G. Zhou, J. Wu, G. Bian, P. Lu, A. S. Raikhel, and Z. Xi. 2012. *Wolbachia* induces reactive oxygen species (ROS)-dependent activation of the Toll pathway to control dengue virus in the mosquito *Aedes aegypti*. *Proc Natl Acad Sci USA* 109 (1): E23–E31. doi: 10.1073/pnas.1116932108.

109. Speir, E., T. Shibutani, Z. X. Yu, V. Ferrans, and S. E. Epstein. 1996. Role of reactive oxygen intermediates in cytomegalovirus gene expression and in the response of human smooth muscle cells to viral infection. *Circ Res* 79 (6):1143–1152.

110. Kaarbo, M., E. Ager-Wick, P. O. Osenbroch, A. Kilander, R. Skinnes, F. Muller, and L. Eide. 2011. Human cytomegalovirus infection increases mitochondrial biogenesis. *Mitochondrion* 11 (6):935–945. doi: 10.1016/j.mito.2011.08.008.

111. Lee, J., K. Koh, Y. E. Kim, J. H. Ahn, and S. Kim. 2013. Upregulation of Nrf2 expres-sion by human cytomegalovirus infection protects host cells from oxidative stress. *J Gen Virol* 94 (Pt 7):1658–1668. doi: 10.1099/vir.0.052142-0.

112. Dhaunsi, G. S., J. Kaur, and R. B. Turner. 2003. Role of NADPH oxidase in cytomega-lovirus-induced proliferation of human coronary artery smooth muscle cells. *J Biomed Sci* 10 (5):505–509. doi: 72377.

113. McCormick, A. L., L. Roback, G. Wynn, and E. S. Mocarski. 2013. Multiplicity-dependent activation of a serine protease-dependent cytomegalovirus-associated programmed cell death pathway. *Virology* 435 (2):250–257. doi: 10.1016/j.virol.2012.08.042.

114. Fields, B. N., D. M. Knipe, and P. M. Howley. 2007. *Fields Virology*, 5th ed., vol. 2. Philadelphia, PA: Wolters Kluwer Health/Lippincott Williams & Wilkins.

115. Ivanov, A. V., B. Bartosch, O. A. Smirnova, M. G. Isaguliants, and S. N. Kochetkov. 2013. HCV and oxidative stress in the liver. *Viruses* 5 (2):439–469. doi: 10.3390/v5020439.

116. Paik, Y. H., J. Kim, T. Aoyama, S. De Minicis, R. Bataller, and D. A. Brenner. 2014. Role of NADPH oxidases in liver fibrosis. *Antioxid Redox Signal* 20 (17):2854–2872. doi: 10.1089/ars.2013.5619.

117. Paracha, U. Z., K. Fatima, M. Alqahtani, A. Chaudhary, A. Abuzenadah, G. Damanhouri, and I. Qadri. 2013. Oxidative stress and hepatitis C virus. *Virol J* 10:251. doi: 10.1186/1743-422X-10-251.

118. Yuan, C., X. Fu, L. Huang, Y. Ma, X. Ding, L. Zhu, and G. Zhu. 2016. The synergistic antiviral effects of GSH in combination with acyclovir against BoHV-1 infection in vitro. *Acta Virol* 60 (3):328–332.

119. Nucci, C., A. T. Palamara, M. R. Ciriolo, L. Nencioni, P. Savini, C. D'Agostini, G. Rotilio, L. Cerulli, and E. Garaci. 2000. Imbalance in corneal redox state during herpes simplex virus 1-induced keratitis in rabbits. Effectiveness of exogenous glutathione supply. *Exp Eye Res* 70 (2):215–220. doi: 10.1006/exer.1999.0782.

120. Hu, S., W. S. Sheng, S. J. Schachtele, and J. R. Lokensgard. 2011. Reactive oxygen species drive herpes simplex virus (HSV)-1-induced proinflammatory cytokine production by murine microglia. *J Neuroinflammation* 8 (1):1–9. doi: 10.1186/1742-2094-8-123.

121. Mogensen, T. H., J. Melchjorsen, P. Hollsberg, and S. R. Paludan. 2003. Activation of NF-kappa B in virus-infected macrophages is dependent on mitochondrial oxidative stress and intracellular calcium: Downstream involvement of the kinases TGF-beta-activated kinase 1, mitogen-activated kinase/extracellular signal-regulated kinase kinase 1, and I kappa B kinase. *J Immunol* 170 (12):6224–6233.

122. Gonzalez-Dosal, R., K. A. Horan, and S. R. Paludan. 2012. Mitochondria-derived reactive oxygen species negatively regulates immune innate signaling pathways triggered by a DNA virus, but not by an RNA virus. *Biochem Biophys Res Commun* 418 (4):806–810. doi: 10.1016/j.bbrc.2012.01.108.

123. Bellner, L., F. Thorén, E. Nygren, J. Å. Liljeqvist, A. Karlsson, and K. Eriksson. 2005. A proinflammatory peptide from herpes simplex virus type 2 glycoprotein G affects neutrophil, monocyte, and NK cell functions. *J Immunol* 174 (4):2235–2241. doi: 10.4049/jimmunol.174.4.2235.

124. Bellner, L., J. Karlsson, F. Huamei, F. Boulay, C. Dahlgren, K. Eriksson, and A. Karlsson. 2007. A monocyte-specific peptide from herpes simplex virus type 2 glycoprotein G activates the NADPH-oxidase but not chemotaxis through a G-protein-coupled receptor distinct from the members of the formyl peptide receptor family. *J Immunol* 179 (9):6080–6087. doi: 10.4049/jimmunol.179.9.6080.

125. Sun, J., C. Hu, Y. Zhu, R. Sun, Y. Fang, Y. Fan, and F. Xu. 2015. LMP1 increases expression of NADPH oxidase (NOX) and its regulatory subunit p22 in NP69 nasopharyngeal cells and makes them sensitive to a treatment by a NOX inhibitor. *PLoS One* 10 (8):e0134896. doi: 10.1371/journal.pone.0134896.

126. Gruhne, B., R. Sompallae, D. Marescotti, S. A. Kamranvar, S. Gastaldello, and M. G. Masucci. 2009. The Epstein-Barr virus nuclear antigen-1 promotes genomic instability via induction of reactive oxygen species. *Proc Natl Acad Sci USA* 106 (7):2313–2318. doi: 10.1073/pnas.0810619106.

127. Kamranvar, S. A. and M. G. Masucci. 2011. The Epstein-Barr virus nuclear antigen-1 promotes telomere dysfunction via induction of oxidative stress. *Leukemia* 25 (6):1017–1025. doi: 10.1038/leu.2011.35.

128. Jakovljevic, A., M. Andric, M. Miletic, K. Beljic-Ivanovic, A. Knezevic, S. Mojsilovic, and J. Milasin. 2016. Epstein-Barr virus infection induces bone resorption in apical periodontitis via increased production of reactive oxygen species. *Med Hypotheses* 94:40–42. doi: 10.1016/j.mehy.2016.06.020.

129. Bottero, V., S. Chakraborty, and B. Chandran. 2013. Reactive oxygen species are induced by Kaposi's sarcoma-associated herpesvirus early during primary infection of endothelial cells to promote virus entry. *J Virol* 87 (3):1733–1749. doi: 10.1128/jvi.02958-12.

130. Gjyshi, O., V. Bottero, M. V. Veettil, S. Dutta, V. V. Singh, L. Chikoti, and B. Chandran. 2014. Kaposi's sarcoma-associated herpesvirus induces Nrf2 during de novo infection of endothelial cells to create a microenvironment conducive to infection. *PLoS Pathog* 10 (10):e1004460. doi: 10.1371/journal.ppat.1004460.

131. Mallery, S. R., P. Pei, D. J. Landwehr, C. M. Clark, J. E. Bradburn, G. M. Ness, and F. M. Robertson. 2004. Implications for oxidative and nitrative stress in the pathogenesis of AIDS-related Kaposi's sarcoma. *Carcinogenesis* 25 (4):597–603. doi: 10.1093/carcin/bgh042.

132. Huang, S. Y., C. Y. Fang, C. C. Wu, C. H. Tsai, S. F. Lin, and J. Y. Chen. 2013. Reactive oxygen species mediate Epstein-Barr virus reactivation by N-methyl-N'-nitro-N-nitrosoguanidine. *PLoS One* 8 (12):e84919. doi: 10.1371/journal.pone.0084919.

133. Ye, F. and S. J. Gao. 2011. A novel role of hydrogen peroxide in Kaposi sarcoma-associated herpesvirus reactivation. *Cell Cycle* 10 (19):3237–3238. doi: 10.4161/cc.10.19.17299.

134. Ye, F., F. Zhou, R. G. Bedolla, T. Jones, X. Lei, T. Kang, M. Guadalupe, and S. J. Gao. 2011. Reactive oxygen species hydrogen peroxide mediates Kaposi's sarcoma-associated herpesvirus reactivation from latency. *PLoS Pathog* 7 (5):e1002054. doi: 10.1371/journal.ppat.1002054.

135. Li, X., J. Feng, and R. Sun. 2011. Oxidative stress induces reactivation of Kaposi's sarcoma-associated herpesvirus and death of primary effusion lymphoma cells. *J Virol* 85 (2):715–724. doi: 10.1128/JVI.01742-10.

136. Michalek, R. D., S. T. Pellom, B. C. Holbrook, and J. M. Grayson. 2008. The requirement of reactive oxygen intermediates for lymphocytic choriomeningitis virus binding and growth. *Virology* 379 (2):205–212. doi: 10.1016/j.virol.2008.07.004.

137. Crump, K. E., P. K. Langston, S. Rajkarnikar, and J. M. Grayson. 2013. Antioxidant treatment regulates the humoral immune response during acute viral infection. *J Virol* 87 (5):2577–2586. doi: 10.1128/jvi.02714-12.

138. Lang, P. A., H. C. Xu, M. Grusdat, D. R. McIlwain, A. A. Pandyra, I. S. Harris, N. Shaabani et al. 2013. Reactive oxygen species delay control of lymphocytic choriomeningitis virus. *Cell Death Differ* 20 (4):649–658. doi: 10.1038/cdd.2012.167.

139. Bhattacharya, A., A. N. Hegazy, N. Deigendesch, L. Kosack, J. Cupovic, R. K. Kandasamy, A. Hildebrandt et al. 2015. Superoxide dismutase 1 protects hepatocytes from type I interferon-driven oxidative damage. *Immunity* 43 (5):974–986. doi: 10.1016/j.immuni.2015.10.013.

140. Laniewski, N. G. and J. M. Grayson. 2004. Antioxidant treatment reduces expansion and contraction of antigen-specific CD[8+] T cells during primary but not secondary viral infection. *J Virol* 78 (20):11246–11257. doi: 10.1128/JVI.78.20.11246-11257.2004.

141. Yan, Y., A. Xin, Q. Liu, H. Huang, Z. Shao, Y. Zang, L. Chen, Y. Sun, and H. Gao. 2015. Induction of ROS generation and NF-kappaB activation in MARC-145 cells by a novel porcine reproductive and respiratory syndrome virus in Southwest of China isolate. *BMC Vet Res* 11:232. doi: 10.1186/s12917-015-0480-z.

142. Chen, X., F. Ren, J. Hesketh, X. Shi, J. Li, F. Gan, and K. Huang. 2012. Reactive oxygen species regulate the replication of porcine circovirus type 2 via NF-kappaB pathway. *Virology* 426 (1):66–72. doi: 10.1016/j.virol.2012.01.023.
143. Lee, S. M. and S. B. Kleiboeker. 2005. Porcine arterivirus activates the NF-kappaB pathway through IkappaB degradation. *Virology* 342 (1):47–59. doi: 10.1016/j.virol.2005.07.034.
144. Tung, W. H., H. L. Hsieh, I. T. Lee, and C. M. Yang. 2011. Enterovirus 71 induces integrin beta1/EGFR-Rac1-dependent oxidative stress in SK-N-SH cells: Role of HO-1/CO in viral replication. *J Cell Physiol* 226 (12):3316–3329. doi: 10.1002/jcp.22677.
145. Cheng, M. L., S. F. Weng, C. H. Kuo, and H. Y. Ho. 2014. Enterovirus 71 induces mitochondrial reactive oxygen species generation that is required for efficient replication. *PLoS One* 9 (11):e113234. doi: 10.1371/journal.pone.0113234.
146. Bao, X., M. Sinha, T. Liu, C. Hong, B. A. Luxon, R. P. Garofalo, and A. Casola. 2008. Identification of human metapneumovirus-induced gene networks in airway epithelial cells by microarray analysis. *Virology* 374 (1):114–127. doi: 10.1016/j.virol.2007.12.024.
147. Escaffre, O., H. Halliday, V. Borisevich, A. Casola, and B. Rockx. 2015. Oxidative stress in Nipah virus-infected human small airway epithelial cells. *J Gen Virol* 96 (10):2961–2970. doi: 10.1099/jgv.0.000243.
148. Guerrero, C. A. and O. Acosta. 2016. Inflammatory and oxidative stress in rotavirus infection. *World J Virol* 5 (2):38–62. doi: 10.5501/wjv.v5.i2.38.
149. Brugh, M. 1977. Butylated hydroxytoluene protects chickens exposed to Newcastle disease virus. *Science* 197 (4310):1291–1292.
150. Garaci, E., A. T. Palamara, P. Di Francesco, C. Favalli, M. R. Ciriolo, and G. Rotilio. 1992. Glutathione inhibits replication and expression of viral proteins in cultured cells infected with Sendai virus. *Biochem Biophys Res Commun* 188 (3):1090–1096.
151. Page, A., V. A. Volchkova, S. P. Reid, M. Mateo, A. Bagnaud-Baule, K. Nemirov, A. C. Shurtleff et al. 2014. Marburgvirus hijacks nrf2-dependent pathway by targeting nrf2-negative regulator keap1. *Cell Rep* 6 (6):1026–1036. doi: 10.1016/j.celrep.2014.02.027.
152. Edwards, M. R., B. Johnson, C. E. Mire, W. Xu, R. S. Shabman, L. N. Speller, D. W. Leung, T. W. Geisbert, G. K. Amarasinghe, and C. F. Basler. 2014. The Marburg virus VP24 protein interacts with Keap1 to activate the cytoprotective antioxidant response pathway. *Cell Rep* 6 (6):1017–1025. doi: 10.1016/j.celrep.2014.01.043.
153. Hristov, G., M. Kramer, J. Li, N. El-Andaloussi, R. Mora, L. Daeffler, H. Zentgraf, J. Rommelaere, and A. Marchini. 2010. Through its nonstructural protein NS1, parvovirus H-1 induces apoptosis via accumulation of reactive oxygen species. *J Virol* 84 (12):5909–5922. doi: 10.1128/jvi.01797-09.
154. Zhao, X., H. Xiang, X. Bai, N. Fei, Y. Huang, X. Song, H. Zhang, L. Zhang, and D. Tong. 2016. Porcine parvovirus infection activates mitochondria-mediated apoptotic signaling pathway by inducing ROS accumulation. *Virol J* 13:26. doi: 10.1186/s12985-016-0480-z.
155. Vardi, A., L. Haramaty, B. A. Van Mooy, H. F. Fredricks, S. A. Kimmance, A. Larsen, and K. D. Bidle. 2012. Host-virus dynamics and subcellular controls of cell fate in a natural coccolithophore population. *Proc Natl Acad Sci USA* 109 (47):19327–19332. doi: 10.1073/pnas.1208895109.
156. Joubert, P. E., S. Werneke, C. de la Calle, F. Guivel-Benhassine, A. Giodini, L. Peduto, B. Levine, O. Schwartz, D. Lenschow, and M. L. Albert. 2012. Chikungunya-induced cell death is limited by ER and oxidative stress-induced autophagy. *Autophagy* 8 (8):1261–1263. doi: 10.4161/auto.20751.
157. Joubert, P. E., S. W. Werneke, C. de la Calle, F. Guivel-Benhassine, A. Giodini, L. Peduto, B. Levine, O. Schwartz, D. J. Lenschow, and M. L. Albert. 2012. Chikungunya virus-induced autophagy delays caspase-dependent cell death. *J Exp Med* 209 (5):1029–1047. doi: 10.1084/jem.20110996.

158. Wang, X., Y. Liu, L. Chen, D. Zhao, X. Wang, and Z. Zhang. 2013. Wheat resistome in response to barley yellow dwarf virus infection. *Funct Integr Genomics* 13(2):155–165. doi: 10.1007/s10142-013-0309-4.

159. Shao, J., J. Huang, Y. Guo, L. Li, X. Liu, X. Chen, and J. Yuan. 2016. Up-regulation of nuclear factor E2-related factor 2 (Nrf2) represses the replication of SVCV. *Fish Shellfish Immunol* 58:474–482. doi: 10.1016/j.fsi.2016.09.012.

160. Olavarria, V. H., S. Valdivia, B. Salas, M. Villalba, R. Sandoval, H. Oliva, S. Valdebenito, and A. Yanez. 2015. ISA virus regulates the generation of reactive oxygen species and p47phox expression in a p38 MAPK-dependent manner in *Salmo salar*. *Mol Immunol* 63 (2):227–234. doi: 10.1016/j.molimm.2014.07.016.

161. Manacorda, C. A., C. Mansilla, H. J. Debat, D. Zavallo, F. Sanchez, F. Ponz, and S. Asurmendi. 2013. Salicylic acid determines differential senescence produced by two Turnip mosaic virus strains involving reactive oxygen species and early transcriptomic changes. *Mol Plant Microbe Interact* 26 (12):1486–1498. doi: 10.1094/MPMI-07-13-0190-R.

162. Belding, M. E., S. J. Klebanoff, and C. G. Ray. 1970. Peroxidase-mediated virucidal systems. *Science* 167 (3915):195–196.

163. Hayashi, K., L. C. Hooper, T. Okuno, Y. Takada, and J. J. Hooks. 2012. Inhibition of HSV-1 by chemoattracted neutrophils: Supernatants of corneal epithelial cells (HCE) and macrophages (THP-1) treated with virus components chemoattract neutrophils (PMN), and supernatants of PMN treated with these conditioned media inhibit viral growth. *Arch Virol* 157 (7):1377–1381. doi: 10.1007/s00705-012-1306-y.

164. Rager-Zisman, B., M. Kunkel, Y. Tanaka, and B. R. Bloom. 1982. Role of macrophage oxidative metabolism in resistance to vesicular stomatitis virus infection. *Infect Immun* 36 (3):1229–1237.

165. Newcomb, W. W. and J. C. Brown. 2012. Internal catalase protects herpes simplex virus from inactivation by hydrogen peroxide. *J Virol* 86 (21):11931–11934. doi: 10.1128/JVI.01349-12.

166. Schachtele, S. J., S. Hu, and J. R. Lokensgard. 2012. Modulation of experimental herpes encephalitis-associated neurotoxicity through sulforaphane treatment. *PLoS One* 7 (4):e36216. doi: 10.1371/journal.pone.0036216.

167. Chang, C. W., Y. C. Su, G. M. Her, C. F. Ken, and J. R. Hong. 2011. Betanodavirus induces oxidative stress-mediated cell death that prevented by anti-oxidants and zfcatalase in fish cells. *PLoS One* 6 (10):e25853. doi: 10.1371/journal.pone.0025853.

168. Riva, D. A., M. C. de Molina, I. Rocchetta, E. Gerhardt, F. C. Coulombie, and S. E. Mersich. 2006. Oxidative stress in vero cells infected with vesicular stomatitis virus. *Intervirology* 49 (5):294–298. doi: 10.1159/000094245.

169. Rozzi, S. J., G. Borelli, K. Ryan, J. P. Steiner, D. Reglodi, I. Mocchetti, and V. Avdoshina. 2014. PACAP27 is protective against tat-induced neurotoxicity. *J Mol Neurosci* 54 (3):485–493. doi: 10.1007/s12031-014-0273-z.

170. Dobo, J., R. Swanson, G. S. Salvesen, S. T. Olson, and P. G. Gettins. 2006. Cytokine response modifier a inhibition of initiator caspases results in covalent complex formation and dissociation of the caspase tetramer. *J Biol Chem* 281 (50):38781–38790. doi: 10.1074/jbc.M605151200.

171. Upton, J. W., W. J. Kaiser, and E. S. Mocarski. 2010. Virus inhibition of RIP3-dependent necrosis. *Cell Host Microbe* 7 (4):302–313. doi: 10.1016/j.chom.2010.03.006.

172. Upton, J. W., W. J. Kaiser, and E. S. Mocarski. 2008. Cytomegalovirus M45 cell death suppression requires receptor-interacting protein (RIP) homotypic interaction motif (RHIM)-dependent interaction with RIP1. *J Biol Chem* 283 (25):16966–16970. doi: 10.1074/jbc.C800051200.

173. Tal, M. C., M. Sasai, H. K. Lee, B. Yordy, G. S. Shadel, and A. Iwasaki. 2009. Absence of autophagy results in reactive oxygen species-dependent amplification of RLR signaling. *Proc Natl Acad Sci USA* 106 (8):2770–2775. doi: 10.1073/pnas.0807694106.

174. Huang, T. T., E. J. Carlson, L. B. Epstein, and C. J. Epstein. 1992. The role of superoxide anions in the establishment of an interferon-alpha-mediated antiviral state. *Free Radic Res Commun* 17 (1):59–72.

175. Zheng, Y. W. and T. S. Yen. 1994. Negative regulation of hepatitis B virus gene expression and replication by oxidative stress. *J Biol Chem* 269 (12):8857–8862.

176. Ho, H. Y., M. L. Cheng, S. F. Weng, Y. L. Leu, and D. T. Chiu. 2009. Antiviral effect of epigallocatechin gallate on enterovirus 71. *J Agric Food Chem* 57 (14):6140–6147. doi: 10.1021/jf901128u.

177. Kalebic, T., A. Kinter, G. Poli, M. E. Anderson, A. Meister, and A. S. Fauci. 1991. Suppression of human immunodeficiency virus expression in chronically infected monocytic cells by glutathione, glutathione ester, and N-acetylcysteine. *Proc Natl Acad Sci USA* 88 (3):986–990.

178. Wu, Y. H., C. P. Tseng, M. L. Cheng, H. Y. Ho, S. R. Shih, and D. T. Chiu. 2008. Glucose-6-phosphate dehydrogenase deficiency enhances human coronavirus 229E infection. *J Infect Dis* 197 (6):812–816. doi: 10.1086/528377.

179. Ho, H. Y., M. L. Cheng, S. F. Weng, L. Chang, T. T. Yeh, S. R. Shih, and D. T. Chiu. 2008. Glucose-6-phosphate dehydrogenase deficiency enhances enterovirus 71 infection. *J Gen Virol* 89 (Pt 9):2080–2089. doi: 10.1099/vir.0.2008/001404-0.

180. Beck, M. A. 1997. Increased virulence of coxsackievirus B3 in mice due to vitamin E or selenium deficiency. *J Nutr* 127 (Suppl 5):966s–970s.

181. Beck, M. A. 1998. The influence of antioxidant nutrients on viral infection. *Nutr Rev* 56 (1 Pt 2):S140–S146.

182. Beck, M. A., J. Handy, and O. A. Levander. 2000. The role of oxidative stress in viral infections. *Ann N Y Acad Sci* 917:906–912.

183. Beck, M. A., P. C. Kolbeck, L. H. Rohr, Q. Shi, V. C. Morris, and O. A. Levander. 1994. Benign human enterovirus becomes virulent in selenium-deficient mice. *J Med Virol* 43 (2):166–170.

184. Beck, M. A., O. A. Levander, and J. Handy. 2003. Selenium deficiency and viral infection. *J Nutr* 133 (5 Suppl 1):1463s–1467s.

185. Beck, M. A., Q. Shi, V. C. Morris, and O. A. Levander. 2005. Benign coxsackievirus damages heart muscle in iron-loaded vitamin E-deficient mice. *Free Radic Biol Med* 38 (1):112–116. doi: 10.1016/j.freeradbiomed.2004.10.007.

186. Beck, M. A., D. Williams-Toone, and O. A. Levander. 2003. Coxsackievirus B3-resistant mice become susceptible in Se/vitamin E deficiency. *Free Radic Biol Med* 34 (10):1263–1270.

187. Burdon, R. H. 1995. Superoxide and hydrogen peroxide in relation to mammalian cell proliferation. *Free Radic Biol Med* 18 (4):775–794.

188. Schreck, R. and P. A. Baeuerle. 1991. A role for oxygen radicals as second messengers. *Trends Cell Biol* 1 (2–3):39–42.

189. Israel, N., M. A. Gougerot-Pocidalo, F. Aillet, and J. L. Virelizier. 1992. Redox status of cells influences constitutive or induced NF-kappa B translocation and HIV long terminal repeat activity in human T and monocytic cell lines. *J Immunol* 149 (10):3386–3393.

190. Kameoka, M., Y. Okada, M. Tobiume, T. Kimura, and K. Ikuta. 1996. Intracellular glutathione as a possible direct blocker of HIV type 1 reverse transcription. *AIDS Res Hum Retroviruses* 12 (17):1635–1638. doi: 10.1089/aid.1996.12.1635.

191. Pahl, H. L. and P. A. Baeuerle. 1994. Oxygen and the control of gene expression. *Bioessays* 16 (7):497–502. doi: 10.1002/bies.950160709.

192. Schreck, R., K. Albermann, and P. A. Baeuerle. 1992. Nuclear factor kappa B: An oxidative stress-responsive transcription factor of eukaryotic cells (a review). *Free Radic Res Commun* 17 (4):221–237.

193. Teshima, S., H. Kutsumi, T. Kawahara, K. Kishi, and K. Rokutan. 2000. Regulation of growth and apoptosis of cultured guinea pig gastric mucosal cells by mitogenic oxidase 1. *Am J Physiol Gastrointest Liver Physiol* 279 (6):G1169–G1176.

194. Giri, D. K. and B. B. Aggarwal. 1998. Constitutive activation of NF-κB causes resistance to apoptosis in human cutaneous T cell lymphoma HuT-78 cells: Autocrine role of tumor necrosis factor and reactive oxygen intermediates. *J Biol Chem* 273 (22):14008–14014. doi: 10.1074/jbc.273.22.14008.

195. von Knethen, A., D. Callsen, and B. Brüne. 1999. Superoxide attenuates macrophage apoptosis by NF-κB and AP-1 activation that promotes cyclooxygenase-2 expression. *J Immunol* 163 (5):2858–2866.

196. Karin, M. and A. Lin. 2002. NF-kappaB at the crossroads of life and death. *Nat Immunol* 3 (3):221–227. doi: 10.1038/ni0302-221.

197. Schreck, R., P. Rieber, and P. A. Baeuerle. 1991. Reactive oxygen intermediates as apparently widely used messengers in the activation of the NF-kappa B transcription factor and HIV-1. *EMBO J* 10 (8):2247–2258.

198. Morgan, M. J. and Z. G. Liu. 2011. Crosstalk of reactive oxygen species and NF-κB signaling. *Cell Res* 21 (1):103–115. doi: 10.1038/cr.2010.178.

199. Powell, P. P., L. K. Dixon, and R. M. Parkhouse. 1996. An IkappaB homolog encoded by African swine fever virus provides a novel mechanism for downregulation of proinflammatory cytokine responses in host macrophages. *J Virol* 70 (12):8527–8533.

200. Wang, X., M. Li, H. Zheng, T. Muster, P. Palese, A. A. Beg, and A. Garcia-Sastre. 2000. Influenza A virus NS1 protein prevents activation of NF-kappaB and induction of alpha/beta interferon. *J Virol* 74 (24):11566–11573.

201. Shisler, J. L. and X. L. Jin. 2004. The vaccinia virus K1L gene product inhibits host NF-kappaB activation by preventing IkappaBalpha degradation. *J Virol* 78 (7):3553–3560.

202. Zoll, J., W. J. Melchers, J. M. Galama, and F. J. van Kuppeveld. 2002. The mengovirus leader protein suppresses alpha/beta interferon production by inhibition of the iron/ferritin-mediated activation of NF-kappa B. *J Virol* 76 (19):9664–9672.

203. Connolly, J. L., S. E. Rodgers, P. Clarke, D. W. Ballard, L. D. Kerr, K. L. Tyler, and T. S. Dermody. 2000. Reovirus-induced apoptosis requires activation of transcription factor NF-kappaB. *J Virol* 74 (7):2981–2989.

204. Lin, R. J., C. L. Liao, and Y. L. Lin. 2004. Replication-incompetent virions of Japanese encephalitis virus trigger neuronal cell death by oxidative stress in a culture system. *J Gen Virol* 85 (Pt 2):521–533. doi: 10.1099/vir.0.19496-0.

205. Goodkin, M. L., A. T. Ting, and J. A. Blaho. 2003. NF-kappaB is required for apoptosis prevention during herpes simplex virus type 1 infection. *J Virol* 77 (13):7261–7280.

206. Waris, G., A. Livolsi, V. Imbert, J. F. Peyron, and A. Siddiqui. 2003. Hepatitis C virus NS5A and subgenomic replicon activate NF-kappaB via tyrosine phosphorylation of IkappaBalpha and its degradation by calpain protease. *J Biol Chem* 278 (42):40778–40787. doi: 10.1074/jbc.M303248200.

207. Meyer, M., W. H. Caselmann, V. Schluter, R. Schreck, P. H. Hofschneider, and P. A. Baeuerle. 1992. Hepatitis B virus transactivator MHBst: Activation of NF-kappa B, selective inhibition by antioxidants and integral membrane localization. *EMBO J* 11 (8):2991–3001.

208. Speir, E. 2000. Cytomegalovirus gene regulation by reactive oxygen species. Agents in atherosclerosis. *Ann N Y Acad Sci* 899:363–374.

209. Jamaluddin, M., B. Tian, I. Boldogh, R. P. Garofalo, and A. R. Brasier. 2009. Respiratory syncytial virus infection induces a reactive oxygen species-MSK1-phospho-Ser-276 RelA pathway required for cytokine expression. *J Virol* 83 (20):10605–10615. doi: 10.1128/JVI.01090-09.

210. Grivennikov, S. and M. Karin. 2010. Dangerous liaisons: STAT3 and NF-κB collaboration and crosstalk in cancer. *Cytokine Growth Factor Rev* 21 (1):11–19. doi: 10.1016/j.cytogfr.2009.11.005.

211. Duhe, R. J. 2013. Redox regulation of Janus kinase: The elephant in the room. *JAKSTAT* 2 (4):e26141. doi: 10.4161/jkst.26141.

212. Smith, J. K., C. N. Patil, S. Patlolla, B. W. Gunter, G. W. Booz, and R. J. Duhé. 2012. Identification of a redox-sensitive switch within the JAK2 catalytic domain. *Free Radic Biol Med* 52 (6):1101–1110. doi: 10.1016/j.freeradbiomed.2011.12.025.

213. Carballo, M., M. Conde, R. El Bekay, J. Martin-Nieto, M. J. Camacho, J. Monteseirin, J. Conde, F. J. Bedoya, and F. Sobrino. 1999. Oxidative stress triggers STAT3 tyrosine phosphorylation and nuclear translocation in human lymphocytes. *J Biol Chem* 274 (25):17580–17586.

214. Garg, A. and S. A. Spector. 2014. HIV type 1 gp120-induced expansion of myeloid derived suppressor cells is dependent on interleukin 6 and suppresses immunity. *J Infect Dis* 209 (3):441–451. doi: 10.1093/infdis/jit469.

215. Machida, K., K. T. H. Cheng, C. K. Lai, K. S. Jeng, V. M. H. Sung, and M. M. C. Lai. 2006. Hepatitis C virus triggers mitochondrial permeability transition with production of reactive oxygen species, leading to DNA damage and STAT3 activation. *J Virol* 80 (14):7199–7207.

216. Miura, K., K. Taura, Y. Kodama, B. Schnabl, and D. A. Brenner. 2008. Hepatitis C virus-induced oxidative stress suppresses hepcidin expression through increased histone deacetylase activity. *Hepatology* 48 (5):1420–1429. doi: 10.1002/hep.22486.

217. Wang, A. G., D. S. Lee, H. B. Moon, J. M. Kim, K. H. Cho, S. H. Choi, H. L. Ha et al. 2009. Non-structural 5A protein of hepatitis C virus induces a range of liver pathology in transgenic mice. *J Pathol* 219 (2):253–262. doi: 10.1002/path.2592.

218. Zhang, K. and R. J. Kaufman. 2008. From endoplasmic-reticulum stress to the inflammatory response. *Nature* 454 (7203):455–462. doi: 10.1038/nature07203.

219. Waris, G., J. Turkson, T. Hassanein, and A. Siddiqui. 2005. Hepatitis C virus (HCV) constitutively activates STAT-3 via oxidative stress: Role of STAT-3 in HCV replication. *J Virol* 79 (3):1569–1580. doi: 10.1128/JVI.79.3.1569-1580.2005.

220. Choi, J., K. J. Lee, Y. Zheng, A. K. Yamaga, M. M. Lai, and J. H. Ou. 2004. Reactive oxygen species suppress hepatitis C virus RNA replication in human hepatoma cells. *Hepatology* 39 (1):81–89. doi: 10.1002/hep.20001.

221. Kim, W., S. Oe Lim, J. S. Kim, Y. H. Ryu, J. Y. Byeon, H. J. Kim, Y. I. Kim, J. S. Heo, Y. M. Park, and G. Jung. 2003. Comparison of proteome between hepatitis B virus- and hepatitis C virus-associated hepatocellular carcinoma. *Clin Cancer Res* 9 (15): 5493–5500.

222. Poungpairoj, P., P. Whongsiri, S. Suwannasin, A. Khlaiphuengsin, P. Tangkijvanich, and C. Boonla. 2015. Increased oxidative stress and RUNX3 hypermethylation in patients with hepatitis B virus-associated hepatocellular carcinoma (HCC) and induction of RUNX3 hypermethylation by reactive oxygen species in HCC cells. *Asian Pac J Cancer Prev* 16 (13):5343–5348.

223. Chen, X., S. A. Kamranvar, and M. G. Masucci. 2016. Oxidative stress enables Epstein-Barr virus-induced B-cell transformation by posttranscriptional regulation of viral and cellular growth-promoting factors. *Oncogene* 35 (29):3807–3816. doi: 10.1038/onc.2015.450.

224. Bonizzi, G., J. Piette, S. Schoonbroodt, R. Greimers, L. Havard, M. P. Merville, and V. Bours. 1999. Reactive oxygen intermediate-dependent NF-kappaB activation by interleukin-1beta requires 5-lipoxygenase or NADPH oxidase activity. *Mol Cell Biol* 19 (3):1950–1960.

225. Olivetta, E., D. Pietraforte, I. Schiavoni, M. Minetti, M. Federico, and M. Sanchez. 2005. HIV-1 Nef regulates the release of superoxide anions from human macrophages. *Biochem J* 390 (Pt 2):591–602. doi: 10.1042/BJ20042139.

226. Zhang, W. J., H. Wei, Y. T. Tien, and B. Frei. 2011. Genetic ablation of phagocytic NADPH oxidase in mice limits TNFalpha-induced inflammation in the lungs but not other tissues. *Free Radic Biol Med* 50 (11):1517–1525. doi: 10.1016/j.freeradbiomed.2011.02.027.

227. Zmijewski, J. W., X. Zhao, Z. Xu, and E. Abraham. 2007. Exposure to hydrogen peroxide diminishes NF-kappaB activation, IkappaB-alpha degradation, and protea-some activity in neutrophils. *Am J Physiol Cell Physiol* 293 (1):C255. doi: 10.1152/ajpcell.00618.2006.

228. Droge, W. 2002. Free radicals in the physiological control of cell function. *Physiol Rev* 82 (1):47–95. doi: 10.1152/physrev.00018.2001.

229. Nakao, N., T. Kurokawa, T. Nonami, G. Tumurkhuu, N. Koide, and T. Yokochi. 2008. Hydrogen peroxide induces the production of tumor necrosis factor-alpha in RAW 264.7 macrophage cells via activation of p38 and stress-activated protein kinase. *Innate Immun* 14 (3):190–196. doi: 10.1177/1753425908093932.

230. Kaul, N. and H. J. Forman. 1996. Activation of NF kappa B by the respiratory burst of macrophages. *Free Radic Biol Med* 21 (3):401–405.

231. Murata, Y., M. Amao, J. Yoneda, and J. Hamuro. 2002. Intracellular thiol redox status of macrophages directs the Th1 skewing in thioredoxin transgenic mice during aging. *Mol Immunol* 38 (10):747–757.

232. Mittal, M., M. R. Siddiqui, K. Tran, S. P. Reddy, and A. B. Malik. 2013. Reactive oxy-gen species in inflammation and tissue injury. *Antioxid Redox Signal* 20 (7):1126–1167. doi: 10.1089/ars.2012.5149.

233. Dickinson, B. C. and C. J. Chang. 2011. Chemistry and biology of reactive oxygen species in signaling or stress responses. *Nat Chem Biol* 7 (8):504–511. doi: 10.1038/nchembio.607.

234. Lamb, N. J., J. M. Gutteridge, C. Baker, T. W. Evans, and G. J. Quinlan. 1999. Oxidative damage to proteins of bronchoalveolar lavage fluid in patients with acute respiratory distress syndrome: Evidence for neutrophil-mediated hydroxylation, nitration, and chlo-rination. *Crit Care Med* 27 (9):1738–1744.

235. Chabot, F., J. A. Mitchell, J. M. Gutteridge, and T. W. Evans. 1998. Reactive oxygen species in acute lung injury. *Eur Respir J* 11 (3):745–757.

236. Matthay, M. A., T. Geiser, S. Matalon, and H. Ischiropoulos. 1999. Oxidant-mediated lung injury in the acute respiratory distress syndrome. *Crit Care Med* 27 (9):2028–2030.

237. Echtay, K. S., D. Roussel, J. St-Pierre, M. B. Jekabsons, S. Cadenas, J. A. Stuart, J. A. Harper et al. 2002. Superoxide activates mitochondrial uncoupling proteins. *Nature* 415 (6867):96–99. doi: 10.1038/415096a.

238. Russo, M. T., M. F. Blasi, F. Chiera, P. Fortini, P. Degan, P. Macpherson, M. Furuichi et al. 2004. The oxidized deoxynucleoside triphosphate pool is a significant contributor to genetic instability in mismatch repair-deficient cells. *Mol Cell Biol* 24 (1):465–474.

239. Maeda, H. and T. Akaike. 1998. Nitric oxide and oxygen radicals in infection, inflamma-tion, and cancer. *Biochemistry (Mosc)* 63 (7):854–865.

240. Kinjo, T., J. Ham-Terhune, J. M. Peloponese, Jr., and K. T. Jeang. 2010. Induction of reactive oxygen species by human T-cell leukemia virus type 1 tax correlates with DNA damage and expression of cellular senescence marker. *J Virol* 84 (10):5431–5437. doi: 10.1128/jvi.02460-09.

241. Marullo, R., E. Werner, H. Zhang, G. Z. Chen, D. M. Shin, and P. W. Doetsch. 2015. HPV16 E6 and E7 proteins induce a chronic oxidative stress response via NOX2 that causes genomic instability and increased susceptibility to DNA damage in head and neck cancer cells. *Carcinogenesis* 36 (11):1397–1406. doi: 10.1093/carcin/bgv126.

242. Gil, L., G. Martinez, I. Gonzalez, A. Tarinas, A. Alvarez, A. Giuliani, R. Molina, R. Tapanes, J. Perez, and O. S. Leon. 2003. Contribution to characterization of oxidative stress in HIV/AIDS patients. *Pharmacol Res* 47 (3):217–224.

243. Machida, K., G. McNamara, K. T. Cheng, J. Huang, C. H. Wang, L. Comai, J. H. Ou, and M. M. Lai. 2010. Hepatitis C virus inhibits DNA damage repair through reactive oxygen and nitrogen species and by interfering with the ATM-NBS1/Mre11/Rad50 DNA repair pathway in monocytes and hepatocytes. *J Immunol* 185 (11):6985–6998. doi: 10.4049/jimmunol.1000618.

244. Pamplona, R. 2008. Membrane phospholipids, lipoxidative damage and molecular integrity: A causal role in aging and longevity. *Biochim Biophys Acta* 1777 (10):1249–1262. doi: 10.1016/j.bbabio.2008.07.003.

245. Xu, Z., L. Chen, L. Leung, T. S. Yen, C. Lee, and J. Y. Chan. 2005. Liver-specific inactivation of the Nrf1 gene in adult mouse leads to nonalcoholic steatohepatitis and hepatic neoplasia. *Proc Natl Acad Sci USA* 102 (11):4120–4125. doi: 10.1073/pnas.0500660102.

246. Sen, C. K. and S. Roy. 2008. Redox signals in wound healing. *Biochim Biophys Acta* 1780 (11):1348–1361. doi: 10.1016/j.bbagen.2008.01.006.

247. Niethammer, P., C. Grabher, A. T. Look, and T. J. Mitchison. 2009. A tissue-scale gradient of hydrogen peroxide mediates rapid wound detection in zebrafish. *Nature* 459 (7249):996–999. doi: 10.1038/nature08119.

248. Sen, C. K., S. Khanna, B. M. Babior, T. K. Hunt, E. C. Ellison, and S. Roy. 2002. Oxidant-induced vascular endothelial growth factor expression in human keratinocytes and cutaneous wound healing. *J Biol Chem* 277 (36):33284–33290. doi: 10.1074/jbc. M203391200.

249. Nakai, K., K. Yoneda, J. Igarashi, T. Moriue, H. Kosaka, and Y. Kubota. 2008. Angiotensin II enhances EGF receptor expression levels via ROS formation in HaCaT cells. *J Dermatol Sci* 51 (3):181–189. doi: 10.1016/j.jdermsci.2008.03.004.

250. Peshavariya, H., G. J. Dusting, F. Jiang, L. R. Halmos, C. G. Sobey, G. R. Drummond, and S. Selemidis. 2009. NADPH oxidase isoform selective regulation of endothelial cell proliferation and survival. *Naunyn Schmiedebergs Arch Pharmacol* 380 (2):193–204. doi: 10.1007/s00210-009-0413-0.

251. Qian, Y., K. J. Liu, Y. Chen, D. C. Flynn, V. Castranova, and X. Shi. 2005. Cdc42 regulates arsenic-induced NADPH oxidase activation and cell migration through actin filament reorganization. *J Biol Chem* 280 (5):3875–3884. doi: 10.1074/jbc.M403788200.

252. Datla, S. R., H. Peshavariya, G. J. Dusting, K. Mahadev, B. J. Goldstein, and F. Jiang. 2007. Important role of Nox4 type NADPH oxidase in angiogenic responses in human microvascular endothelial cells in vitro. *Arterioscler Thromb Vasc Biol* 27 (11):2319–2324. doi: 10.1161/ATVBAHA.107.149450.

253. Koff, J. L., M. X. Shao, S. Kim, I. F. Ueki, and J. A. Nadel. 2006. Pseudomonas lipopolysaccharide accelerates wound repair via activation of a novel epithelial cell signaling cascade. *J Immunol* 177 (12):8693–8700.

254. Petry, A., T. Djordjevic, M. Weitnauer, T. Kietzmann, J. Hess, and A. Gorlach. 2006. NOX2 and NOX4 mediate proliferative response in endothelial cells. *Antioxid Redox Signal* 8 (9–10):1473–1484. doi: 10.1089/ars.2006.8.1473.

255. Pendyala, S., I. A. Gorshkova, P. V. Usatyuk, D. He, A. Pennathur, J. D. Lambeth, V. J. Thannickal, and V. Natarajan. 2009. Role of Nox4 and Nox2 in hyperoxia-induced reactive oxygen species generation and migration of human lung endothelial cells. *Antioxid Redox Signal* 11 (4):747–764. doi: 10.1089/ARS.2008.2203.

256. Ushio-Fukai, M. 2009. Compartmentalization of redox signaling through NADPH oxidase-derived ROS. *Antioxid Redox Signal* 11 (6):1289–1299. doi: 10.1089/ars.2008.2333.

257. Tojo, T., M. Ushio-Fukai, M. Yamaoka-Tojo, S. Ikeda, N. Patrushev, and R. W. Alexander. 2005. Role of gp91phox (Nox2)-containing NAD(P)H oxidase in angiogenesis in response to hindlimb ischemia. *Circulation* 111 (18):2347–2355. doi: 10.1161/01. CIR.0000164261.62586.14.

258. Lange, S., J. Heger, G. Euler, M. Wartenberg, H. M. Piper, and H. Sauer. 2009. Platelet-derived growth factor BB stimulates vasculogenesis of embryonic stem cell-derived endothelial cells by calcium-mediated generation of reactive oxygen species. *Cardiovasc Res* 81 (1):159–168. doi: 10.1093/cvr/cvn258.

259. Ushio-Fukai, M. 2006. Redox signaling in angiogenesis: Role of NADPH oxidase. *Cardiovasc Res* 71 (2):226–235. doi: 10.1016/j.cardiores.2006.04.015.

260. Jiang, F., Y. Zhang, and G. J. Dusting. 2011. NADPH oxidase-mediated redox signaling: Roles in cellular stress response, stress tolerance, and tissue repair. *Pharmacol Rev* 63 (1):218–242. doi: 10.1124/pr.110.002980.

261. Schmelter, M., B. Ateghang, S. Helmig, M. Wartenberg, and H. Sauer. 2006. Embryonic stem cells utilize reactive oxygen species as transducers of mechanical strain-induced cardiovascular differentiation. *FASEB J* 20 (8):1182–1184. doi: 10.1096/fj.05-4723fje.

262. Sauer, H., G. Rahimi, J. Hescheler, and M. Wartenberg. 2000. Role of reactive oxygen species and phosphatidylinositol 3-kinase in cardiomyocyte differentiation of embryonic stem cells. *FEBS Lett* 476 (3):218–223.

263. Wo, Y. B., D. Y. Zhu, Y. Hu, Z. Q. Wang, J. Liu, and Y. J. Lou. 2008. Reactive oxygen species involved in prenylflavonoids, icariin and icaritin, initiating cardiac differentiation of mouse embryonic stem cells. *J Cell Biochem* 103 (5):1536–1550. doi: 10.1002/jcb.21541.

264. Xiao, Q., Z. Luo, A. E. Pepe, A. Margariti, L. Zeng, and Q. Xu. 2009. Embryonic stem cell differentiation into smooth muscle cells is mediated by Nox4-produced H_2O_2. *Am J Physiol Cell Physiol* 296 (4):C711–C723. doi: 10.1152/ajpcell.00442.2008.

265. Mofarrahi, M., R. P. Brandes, A. Gorlach, J. Hanze, L. S. Terada, M. T. Quinn, D. Mayaki, B. Petrof, and S. N. Hussain. 2008. Regulation of proliferation of skeletal muscle precursor cells by NADPH oxidase. *Antioxid Redox Signal* 10 (3):559–574. doi: 10.1089/ars.2007.1792.

266. Wu, R. F., Y. Gu, Y. C. Xu, S. Mitola, F. Bussolino, and L. S. Terada. 2004. Human immunodeficiency virus type 1 Tat regulates endothelial cell actin cytoskeletal dynamics through PAK1 activation and oxidant production. *J Virol* 78 (2):779–789.

267. Nathan, C. and A. Cunningham-Bussel. 2013. Beyond oxidative stress: An immunologist's guide to reactive oxygen species. *Nat Rev Immunol* 13 (5):349–361. doi: 10.1038/nri3423.

268. Drummond, G. R. and C. G. Sobey. 2014. Endothelial NADPH oxidases: Which NOX to target in vascular disease? *Trends Endocrinol Metab* 25 (9):452–463. doi: 10.1016/j.tem.2014.06.012.

269. Takac, I., K. Schroder, and R. P. Brandes. 2012. The Nox family of NADPH oxidases: Friend or foe of the vascular system? *Curr Hypertens Rep* 14 (1):70–78. doi: 10.1007/s11906-011-0238-3.

270. Selemidis, S., H. J. Seow, B. R. Broughton, A. Vinh, S. Bozinovski, C. G. Sobey, G. R. Drummond, and R. Vlahos. 2013. Nox1 oxidase suppresses influenza a virus-induced lung inflammation and oxidative stress. *PLoS One* 8 (4):e60792. doi: 10.1371/journal.pone.0060792.

271. Kolarova, H., L. Bino, K. Pejchalova, and L. Kubala. 2010. The expression of NADPH oxidases and production of reactive oxygen species by human lung adenocarcinoma epithelial cell line A549. *Folia Biol (Praha)* 56 (5):211–217.

272. Hsu, H. T., Y. T. Tseng, Y. Y. Hsu, K. I. Cheng, S. H. Chou, and Y. C. Lo. 2015. Propofol attenuates lipopolysaccharide-induced reactive oxygen species production through activation of Nrf2/GSH and suppression of NADPH oxidase in human alveolar epithelial cells. *Inflammation* 38 (1):415–423. doi: 10.1007/s10753-014-0046-4.

273. Zhang, C., T. Lan, J. Hou, J. Li, R. Fang, Z. Yang, M. Zhang, J. Liu, and B. Liu. 2014. NOX4 promotes non-small cell lung cancer cell proliferation and metastasis through positive feedback regulation of PI3K/Akt signaling. *Oncotarget* 5 (12):4392–4405. doi: 10.18632/oncotarget.2025.

274. Li, W., F. Yan, H. Zhou, X. Lin, Y. Wu, C. Chen, N. Zhou, Z. Chen, J. D. Li, and H. Shen. 2013. *P. aeruginosa* lipopolysaccharide-induced MUC5AC and CLCA3 expression is partly through Duox1 in vitro and in vivo. *PLoS One* 8 (5):e63945. doi: 10.1371/journal. pone.0063945.

275. Amatore, D., R. Sgarbanti, K. Aquilano, S. Baldelli, D. Limongi, L. Civitelli, L. Nencioni, E. Garaci, M. R. Ciriolo, and A. T. Palamara. 2015. Influenza virus replication in lung epithelial cells depends on redox-sensitive pathways activated by NOX4-derived ROS. *Cell Microbiol* 17 (1):131–145. doi: 10.1111/cmi.12343.

276. Strengert, M., R. Jennings, S. Davanture, P. Hayes, G. Gabriel, and U. G. Knaus. 2014. Mucosal reactive oxygen species are required for antiviral response: Role of Duox in influenza a virus infection. *Antioxid Redox Signal* 20 (17):2695–2709. doi: 10.1089/ars.2013.5353.

277. Soucy-Faulkner, A., E. Mukawera, K. Fink, A. Martel, L. Jouan, Y. Nzengue, D. Lamarre, C. Vande Velde, and N. Grandvaux. 2010. Requirement of NOX2 and reactive oxygen species for efficient RIG-I-mediated antiviral response through regulation of MAVS expression. *PLoS Pathog* 6 (6):e1000930. doi: 10.1371/journal.ppat.1000930.

278. Kim, H. J., C. H. Kim, J. H. Ryu, M. J. Kim, C. Y. Park, J. M. Lee, M. J. Holtzman, and J. H. Yoon. 2013. Reactive oxygen species induce antiviral innate immune response through IFN-lambda regulation in human nasal epithelial cells. *Am J Respir Cell Mol Biol* 49 (5):855–865. doi: 10.1165/rcmb.2013-0003OC.

279. Kim, H. J., C. H. Kim, M. J. Kim, J. H. Ryu, S. Y. Seong, S. Kim, S. J. Lim, M. J. Holtzman, and J. H. Yoon. 2015. The induction of pattern-recognition receptor expression against influenza A virus through Duox2-derived reactive oxygen species in nasal mucosa. *Am J Respir Cell Mol Biol* 53 (4):525–535. doi: 10.1165/rcmb.2014-0334OC.

280. Ye, S., S. Lowther, and J. Stambas. 2015. Inhibition of reactive oxygen species production ameliorates inflammation induced by influenza A viruses via upregulation of SOCS1 and SOCS3. *J Virol* 89 (5):2672–2683. doi: 10.1128/JVI.03529-14.

281. Trocme, C., C. Deffert, J. Cachat, Y. Donati, C. Tissot, S. Papacatzis, V. Braunersreuther et al. 2015. Macrophage-specific NOX2 contributes to the development of lung emphysema through modulation of SIRT1/MMP-9 pathways. *J Pathol* 235 (1):65–78. doi: 10.1002/path.4423.

282. Hofstetter, A. R., J. A. De La Cruz, W. Cao, J. Patel, J. A. Belser, J. McCoy, J. S. Liepkalns et al. 2016. NADPH oxidase 1 is associated with altered host survival and T cell phenotypes after influenza A virus infection in mice. *PLoS One* 11 (2):e0149864. doi: 10.1371/journal.pone.0149864.

283. Ano, Y., A. Sakudo, T. Kimata, R. Uraki, K. Sugiura, and T. Onodera. 2010. Oxidative damage to neurons caused by the induction of microglial NADPH oxidase in encephalomyocarditis virus infection. *Neurosci Lett* 469 (1):39–43. doi: 10.1016/j.neulet.2009.11.040.

284. Cachat, J., C. Deffert, S. Hugues, and K. H. Krause. 2015. Phagocyte NADPH oxidase and specific immunity. *Clin Sci (Lond)* 128 (10):635–648. doi: 10.1042/CS20140635.

285. Singel, K. L. and B. H. Segal. 2016. NOX2-dependent regulation of inflammation. *Clin Sci (Lond)* 130 (7):479–490. doi: 10.1042/CS20150660.

286. Snelgrove, R. J., L. Edwards, A. J. Rae, and T. Hussell. 2006. An absence of reactive oxygen species improves the resolution of lung influenza infection. *Eur J Immunol* 36 (6):1364–1373. doi: 10.1002/eji.200635977.

287. Hultqvist, M., P. Olofsson, J. Holmberg, B. T. Backstrom, J. Tordsson, and R. Holmdahl. 2004. Enhanced autoimmunity, arthritis, and encephalomyelitis in mice with a reduced oxidative burst due to a mutation in the Ncf1 gene. *Proc Natl Acad Sci USA* 101 (34):12646–12651. doi: 10.1073/pnas.0403831101.

288. Imai, Y., K. Kuba, G. G. Neely, R. Yaghubian-Malhami, T. Perkmann, G. van Loo, M. Ermolaeva et al. 2008. Identification of oxidative stress and Toll-like receptor 4 signaling as a key pathway of acute lung injury. *Cell* 133 (2):235–249. doi: 10.1016/j. cell.2008.02.043.

289. Vlahos, R., J. Stambas, S. Bozinovski, B. R. Broughton, G. R. Drummond, and S. Selemidis. 2011. Inhibition of Nox2 oxidase activity ameliorates influenza A virus-induced lung inflammation. *PLoS Pathog* 7 (2):e1001271. doi: 10.1371/journal.ppat.1001271.
290. Jadhav, V. S., K. H. Krause, and S. K. Singh. 2014. HIV-1 Tat C modulates NOX2 and NOX4 expressions through miR-17 in a human microglial cell line. *J Neurochem* 131 (6):803–815. doi: 10.1111/jnc.12933.
291. Kim, J., M. Seo, S. K. Kim, and Y. S. Bae. 2016. Flagellin-induced NADPH oxidase 4 activation is involved in atherosclerosis. *Sci Rep* 6:25437. doi: 10.1038/srep25437.
292. Mistry, R. K., T. V. Murray, O. Prysyazhna, D. Martin, J. R. Burgoyne, C. Santos, P. Eaton, A. M. Shah, and A. C. Brewer. 2016. Transcriptional regulation of cystathionine-gamma-lyase in endothelial cells by NADPH oxidase 4-dependent signaling. *J Biol Chem* 291 (4):1774–1788. doi: 10.1074/jbc.M115.685578.
293. Guo, S. and X. Chen. 2015. The human Nox4: Gene, structure, physiological function and pathological significance. *J Drug Target* 23 (10):888–896. doi: 10.3109/1061186X.2015.1036276.
294. Sun, X., H. Zeng, A. Kumar, J. A. Belser, T. R. Maines, and T. M. Tumpey. 2016. Constitutively expressed IFITM3 protein in human endothelial cells poses an early infection block to human influenza viruses. *J Virol* 90 (24):11157–11167. doi: 10.1128/JVI.01254-16.
295. Wang, W., X. Mu, L. Zhao, J. Wang, Y. Chu, X. Feng, B. Feng, X. Wang, J. Zhang, and J. Qiao. 2015. Transcriptional response of human umbilical vein endothelial cell to H9N2 influenza virus infection. *Virology* 482:117–127. doi: 10.1016/j.virol.2015.03.037.
296. Short, K. R., E. J. Veldhuis Kroeze, L. A. Reperant, M. Richard, and T. Kuiken. 2014. Influenza virus and endothelial cells: A species specific relationship. *Front Microbiol* 5:653. doi: 10.3389/fmicb.2014.00653.
297. Lin, X., R. Wang, W. Zou, X. Sun, X. Liu, L. Zhao, S. Wang, and M. Jin. 2016. The influenza virus H5N1 infection can induce ROS production for viral replication and host cell death in A549 cells modulated by human Cu/Zn superoxide dismutase (SOD1) overexpression. *Viruses* 8 (1):1–16. doi: 10.3390/v8010013.
298. Rynda-Apple, A., K. M. Robinson, and J. F. Alcorn. 2015. Influenza and bacterial superinfection: Illuminating the immunologic mechanisms of disease. *Infect Immun* 83 (10):3764–3770. doi: 10.1128/IAI.00298-15.
299. Bhatia, A. and R. E. Kast. 2007. How influenza's neuraminidase promotes virulence and creates localized lung mucosa immunodeficiency. *Cell Mol Biol Lett* 12 (1):111–119. doi: 10.2478/s11658-006-0055-x.
300. Hillaire, M. L., H. P. Haagsman, A. D. Osterhaus, G. F. Rimmelzwaan, and M. van Eijk. 2013. Pulmonary surfactant protein D in first-line innate defence against influenza A virus infections. *J Innate Immun* 5 (3):197–208. doi: 10.1159/000346374.
301. Goraya, M. U., S. Wang, M. Munir, and J. L. Chen. 2015. Induction of innate immunity and its perturbation by influenza viruses. *Protein Cell* 6 (10):712–721. doi: 10.1007/s13238-015-0191-z.
302. Pizzolla, A., J. M. Smith, A. G. Brooks, and P. C. Reading. 2016. Pattern recognition receptor immunomodulation of innate immunity as a strategy to limit the impact of influenza virus. *J Leukoc Biol* 101(4):851–861. doi: 10.1189/jlb.4MR0716-290R.
303. Zheng, Q., Y. Ren, P. S. Reinach, Y. She, B. Xiao, S. Hua, J. Qu, and W. Chen. 2014. Reactive oxygen species activated NLRP3 inflammasomes prime environment-induced murine dry eye. *Exp Eye Res* 125:1–8. doi: 10.1016/j.exer.2014.05.001.
304. Abais, J. M., M. Xia, G. Li, T. W. Gehr, K. M. Boini, and P. L. Li. 2014. Contribution of endogenously produced reactive oxygen species to the activation of podocyte NLRP3 inflammasomes in hyperhomocysteinemia. *Free Radic Biol Med* 67:211–220. doi: 10.1016/j.freeradbiomed.2013.10.009.

305. van Bruggen, R., M. Y. Koker, M. Jansen, M. van Houdt, D. Roos, T. W. Kuijpers, and T. K. van den Berg. 2010. Human NLRP3 inflammasome activation is Nox1-4 independent. *Blood* 115 (26):5398–5400. doi: 10.1182/blood-2009-10-250803.
306. van de Veerdonk, F. L., S. P. Smeekens, L. A. Joosten, B. J. Kullberg, C. A. Dinarello, J. W. van der Meer, and M. G. Netea. 2010. Reactive oxygen species-independent activation of the IL-1beta inflammasome in cells from patients with chronic granulomatous disease. *Proc Natl Acad Sci USA* 107 (7):3030–3033. doi: 10.1073/pnas.0914795107.
307. Gabelloni, M. L., F. Sabbione, C. Jancic, J. Fuxman Bass, I. Keitelman, L. Iula, M. Oleastro, J. R. Geffner, and A. S. Trevani. 2013. NADPH oxidase derived reactive oxygen species are involved in human neutrophil IL-1beta secretion but not in inflammasome activation. *Eur J Immunol* 43 (12):3324–3335. doi: 10.1002/eji.201243089.
308. Ong, J. D., A. Mansell, and M. D. Tate. 2016. Hero turned villain: NLRP3 inflammasome-induced inflammation during influenza A virus infection. *J Leukoc Biol* 101(4):863–874. doi: 10.1189/jlb.4MR0616-288R.
309. Kohlmeier, J. E. and D. L. Woodland. 2009. Immunity to respiratory viruses. *Annu Rev Immunol* 27:61–82. doi: 10.1146/annurev.immunol.021908.132625.
310. Ichinohe, T. 2010. Respective roles of TLR, RIG-I and NLRP3 in influenza virus infection and immunity: Impact on vaccine design. *Expert Rev Vaccines* 9 (11):1315–1324. doi: 10.1586/erv.10.118.
311. Ichinohe, T., H. K. Lee, Y. Ogura, R. Flavell, and A. Iwasaki. 2009. Inflammasome recognition of influenza virus is essential for adaptive immune responses. *J Exp Med* 206 (1):79–87. doi: 10.1084/jem.20081667.
312. Kim, T. H. and H. K. Lee. 2014. Differential roles of lung dendritic cell subsets against respiratory virus infection. *Immune Netw* 14 (3):128–137. doi: 10.4110/in.2014.14.3.128.
313. Worbs, T., S. I. Hammerschmidt, and R. Forster. 2016. Dendritic cell migration in health and disease. *Nat Rev Immunol* 17(1):30–48. doi: 10.1038/nri.2016.116.
314. Kotsias, F., E. Hoffmann, S. Amigorena, and A. Savina. 2013. Reactive oxygen species production in the phagosome: Impact on antigen presentation in dendritic cells. *Antioxid Redox Signal* 18 (6):714–729. doi: 10.1089/ars.2012.4557.
315. Matsue, H., D. Edelbaum, D. Shalhevet, N. Mizumoto, C. Yang, M. E. Mummert, J. Oeda, H. Masayasu, and A. Takashima. 2003. Generation and function of reactive oxygen species in dendritic cells during antigen presentation. *J Immunol* 171 (6):3010–3018.
316. Elsen, S., J. Doussiere, C. L. Villiers, M. Faure, R. Berthier, A. Papaioannou, N. Grandvaux, P. N. Marche, and P. V. Vignais. 2004. Cryptic O_2-generating NADPH oxidase in dendritic cells. *J Cell Sci* 117 (Pt 11):2215–2226. doi: 10.1242/jcs.01085.
317. Crawford, A. and E. J. Wherry. 2009. The diversity of costimulatory and inhibitory receptor pathways and the regulation of antiviral T cell responses. *Curr Opin Immunol* 21 (2):179–186. doi: 10.1016/j.coi.2009.01.010.
318. Grewal, I. S. and R. A. Flavell. 1996. The role of CD40 ligand in costimulation and T-cell activation. *Immunol Rev* 153:85–106.
319. Maly, F. E., A. R. Cross, O. T. Jones, G. Wolf-Vorbeck, C. Walker, C. A. Dahinden, and A. L. De Weck. 1988. The superoxide generating system of B cell lines. Structural homology with the phagocytic oxidase and triggering via surface Ig. *J Immunol* 140 (7):2334–2339.
320. Maly, F. E., M. Nakamura, J. F. Gauchat, A. Urwyler, C. Walker, C. A. Dahinden, A. R. Cross, O. T. Jones, and A. L. de Weck. 1989. Superoxide-dependent nitroblue tetrazolium reduction and expression of cytochrome b-245 components by human tonsillar B lymphocytes and B cell lines. *J Immunol* 142 (4):1260–1267.
321. Singh, D. K., D. Kumar, Z. Siddiqui, S. K. Basu, V. Kumar, and K. V. Rao. 2005. The strength of receptor signaling is centrally controlled through a cooperative loop between Ca^{2+} and an oxidant signal. *Cell* 121 (2):281–293. doi: 10.1016/j.cell.2005.02.036.

322. Devadas, S., L. Zaritskaya, S. G. Rhee, L. Oberley, and M. S. Williams. 2002. Discrete generation of superoxide and hydrogen peroxide by T cell receptor stimulation: Selective regulation of mitogen-activated protein kinase activation and Fas ligand expression. *J Exp Med* 195 (1):59–70.

323. Los, M., H. Schenk, K. Hexel, P. A. Baeuerle, W. Droge, and K. Schulze-Osthoff. 1995. IL-2 gene expression and NF-kappa B activation through CD28 requires reactive oxygen production by 5-lipoxygenase. *EMBO J* 14 (15):3731–3740.

324. Fisher, R. I. and F. Bostick-Bruton. 1982. Depressed T cell proliferative responses in Hodgkin's disease: Role of monocyte-mediated suppression via prostaglandins and hydrogen peroxide. *J Immunol* 129 (4):1770–1774.

325. De Luca, A., R. G. Iannitti, S. Bozza, R. Beau, A. Casagrande, C. D'Angelo, S. Moretti et al. 2012. CD4(+) T cell vaccination overcomes defective cross-presentation of fungal antigens in a mouse model of chronic granulomatous disease. *J Clin Invest* 122 (5):1816–1831. doi: 10.1172/JCI60862.

326. Fang, F. C. 2011. Antimicrobial actions of reactive oxygen species. *MBio* 2 (5):1–6. doi: 10.1128/mBio.00141-11.

327. Lobo, V., A. Patil, A. Phatak, and N. Chandra. 2010. Free radicals, antioxidants and functional foods: Impact on human health. *Pharmacogn Rev* 4 (8):118–126. doi: 10.4103/0973-7847.70902.

328. Silva, J. P. and O. P. Coutinho. 2010. Free radicals in the regulation of damage and cell death—Basic mechanisms and prevention. *Drug Discov Ther* 4 (3):144–167.

329. Kuijpers, T. and R. Lutter. 2012. Inflammation and repeated infections in CGD: Two sides of a coin. *Cell Mol Life Sci* 69 (1):7–15. doi: 10.1007/s00018-011-0834-z.

330. Oda, T., T. Akaike, T. Hamamoto, F. Suzuki, T. Hirano, and H. Maeda. 1989. Oxygen radicals in influenza-induced pathogenesis and treatment with pyran polymer-conjugated SOD. *Science* 244 (4907):974–976. doi: 10.1126/science.2543070.

331. Smith, S. M., J. Min, T. Ganesh, B. Diebold, T. Kawahara, Y. Zhu, J. McCoy et al. 2012. Ebselen and congeners inhibit NADPH oxidase 2-dependent superoxide generation by interrupting the binding of regulatory subunits. *Chem Biol* 19 (6):752–763. doi: 10.1016/j.chembiol.2012.04.015.

332. Yatmaz, S., H. J. Seow, R. C. Gualano, Z. X. Wong, J. Stambas, S. Selemidis, P. J. Crack, S. Bozinovski, G. P. Anderson, and R. Vlahos. 2012. Glutathione peroxidase-1 (GPx-1) reduces influenza A virus-induced lung inflammation. *Am J Respir Cell Mol Biol* 48 (1):17–26. doi: 10.1165/rcmb.2011-0345OC.

333. Jaquet, V., L. Scapozza, R. A. Clark, K. H. Krause, and J. D. Lambeth. 2009. Small-molecule NOX inhibitors: ROS-generating NADPH oxidases as therapeutic targets. *Antioxid Redox Signal* 11 (10):2535–2552. doi: 10.1089/ARS.2009.2585.

334. Zheng, B. J., K. W. Chan, Y. P. Lin, G. Y. Zhao, C. Chan, H. J. Zhang, H. L. Chen et al. 2008. Delayed antiviral plus immunomodulator treatment still reduces mortality in mice infected by high inoculum of influenza A/H5N1 virus. *Proc Natl Acad Sci USA* 105 (23):8091–8096. doi: 10.1073/pnas.0711942105.

335. Walsh, K. B., J. R. Teijaro, P. R. Wilker, A. Jatzek, D. M. Fremgen, S. C. Das, T. Watanabe et al. 2011. Suppression of cytokine storm with a sphingosine analog provides protection against pathogenic influenza virus. *Proc Natl Acad Sci USA* 108 (29):12018–12023. doi: 10.1073/pnas.1107024108.

336. Unterholzner, L. and A. G. Bowie. 2008. The interplay between viruses and innate immune signaling: Recent insights and therapeutic opportunities. *Biochem Pharmacol* 75 (3):589–602. doi: 10.1016/j.bcp.2007.07.043.

337. Cao, W., A. K. Taylor, R. E. Biber, W. G. Davis, J. H. Kim, A. J. Reber, T. Chirkova et al. 2012. Rapid differentiation of monocytes into type I IFN-producing myeloid dendritic cells as an antiviral strategy against influenza virus infection. *J Immunol* 189 (5):2257–2265. doi: 10.4049/jimmunol.1200168.

338. Zeng, H., C. Pappas, J. A. Belser, K. V. Houser, W. Zhong, D. A. Wadford, T. Stevens, R. Balczon, J. M. Katz, and T. M. Tumpey. 2012. Human pulmonary microvascular endothelial cells support productive replication of highly pathogenic avian influenza viruses: Possible involvement in the pathogenesis of human H5N1 virus infection. *J Virol* 86 (2):667–678. doi: 10.1128/JVI.06348-11.

339. Toborek, M., Y. W. Lee, H. Pu, A. Malecki, G. Flora, R. Garrido, B. Hennig, H. C. Bauer, and A. Nath. 2003. HIV-Tat protein induces oxidative and inflammatory pathways in brain endothelium. *J Neurochem* 84 (1):169–179.

340. Korenaga, M., T. Wang, Y. Li, L. A. Showalter, T. Chan, J. Sun, and S. A. Weinman. 2005. Hepatitis C virus core protein inhibits mitochondrial electron transport and increases reactive oxygen species (ROS) production. *J Biol Chem* 280 (45):37481–37488. doi: 10.1074/jbc.M506412200.

341. Zhang, H. S., H. Y. Li, Y. Zhou, M. R. Wu, and H. S. Zhou. 2009. Nrf2 is involved in inhibiting Tat-induced HIV-1 long terminal repeat transactivation. *Free Radic Biol Med* 47 (3):261–268. doi: 10.1016/j.freeradbiomed.2009.04.028.

342. Banerjee, A., X. Zhang, K. R. Manda, W. A. Banks, and N. Ercal. 2010. HIV proteins (gp120 and Tat) and methamphetamine in oxidative stress-induced damage in the brain: Potential role of the thiol antioxidant N-acetylcysteine amide. *Free Radic Biol Med* 48 (10):1388–1398. doi: 10.1016/j.freeradbiomed.2010.02.023.

343. Richard, M. J., P. Guiraud, C. Didier, M. Seve, S. C. Flores, and A. Favier. 2001. Human immunodeficiency virus type 1 Tat protein impairs selenoglutathione peroxidase expression and activity by a mechanism independent of cellular selenium uptake: Consequences on cellular resistance to UV-A radiation. *Arch Biochem Biophys* 386 (2):213–220. doi: 10.1006/abbi.2000.2197.

344. Price, T. O., N. Ercal, R. Nakaoke, and W. A. Banks. 2005. HIV-1 viral proteins gp120 and Tat induce oxidative stress in brain endothelial cells. *Brain Res* 1045 (1–2):57–63. doi: 10.1016/j.brainres.2005.03.031.

345. Flores, S. C., J. C. Marecki, K. P. Harper, S. K. Bose, S. K. Nelson, and J. M. McCord. 1993. Tat protein of human immunodeficiency virus type 1 represses expression of manganese superoxide dismutase in HeLa cells. *Proc Natl Acad Sci USA* 90 (16):7632–7636.

346. Westendorp, M., V. A. Shatrov, K. Schulze-Osthoff, R. Frank, M. Kraft, M. Los, P. H. Krammer, W. Dröge, and V. Lehmann. 1995. HIV-1 Tat potentiates TNF-induced NF-kappa B activation and cytotoxicity by altering the cellular redox state. *EMBO J* 14 (3):546.

347. Choi, J., R. M. Liu, R. K. Kundu, F. Sangiorgi, W. Wu, R. Maxson, and H. J. Forman. 2000. Molecular mechanism of decreased glutathione content in human immunodeficiency virus type 1 Tat-transgenic mice. *J Biol Chem* 275 (5):3693–3698.

348. Opalenik, S. R., Q. Ding, S. R. Mallery, and J. A. Thompson. 1998. Glutathione depletion associated with the HIV-1 TAT protein mediates the extracellular appearance of acidic fibroblast growth factor. *Arch Biochem Biophys* 351 (1):17–26. doi: 10.1006/abbi.1997.0566.

349. Marecki, J. C., A. Cota-Gomez, G. M. Vaitaitis, J. R. Honda, S. Porntadavity, D. K. St Clair, and S. C. Flores. 2004. HIV-1 Tat regulates the SOD2 basal promoter by altering Sp1/Sp3 binding activity. *Free Radic Biol Med* 37 (6):869–880.

350. Duffy, P., X. Wang, P. H. Lin, Q. Yao, and C. Chen. 2009. HIV Nef protein causes endothelial dysfunction in porcine pulmonary arteries and human pulmonary artery endothelial cells. *J Surg Res* 156 (2):257–264. doi: 10.1016/j.jss.2009.02.005.

351. Louboutin, J. P., L. Agrawal, B. A. Reyes, E. J. Van Bockstaele, and D. S. Strayer. 2010. HIV-1 gp120-induced injury to the blood-brain barrier: Role of metalloproteinases 2 and 9 and relationship to oxidative stress. *J Neuropathol Exp Neurol* 69 (8):801–816. doi: 10.1097/NEN.0b013e3181e8c96f.

352. Shatrov, V. A., F. Ratter, A. Gruber, W. Droge, and V. Lehmann. 1996. HIV type 1 glycoprotein 120 amplifies tumor necrosis factor-induced NF-kappa B activation in Jurkat cells. *AIDS Res Hum Retroviruses* 12 (13):1209–1216. doi: 10.1089/aid.1996.12.1209.

353. Foga, I. O., A. Nath, B. B. Hasinoff, and J. D. Geiger. 1997. Antioxidants and dipyridamole inhibit HIV-1 gp120-induced free radical-based oxidative damage to human monocytoid cells. *J Acquir Immune Defic Syndr Hum Retrovirol* 16 (4):223–229.

354. Visalli, V., C. Muscoli, I. Sacco, F. Sculco, E. Palma, N. Costa, C. Colica, D. Rotiroti, and V. Mollace. 2007. N-acetylcysteine prevents HIV gp 120-related damage of human cultured astrocytes: Correlation with glutamine synthase dysfunction. *BMC Neurosci* 8:106. doi: 10.1186/1471-2202-8-106.

355. Ronaldson, P. T. and R. Bendayan. 2008. HIV-1 viral envelope glycoprotein gp120 produces oxidative stress and regulates the functional expression of multidrug resistance protein-1 (Mrp1) in glial cells. *J Neurochem* 106 (3):1298–1313. doi: 10.1111/j.1471-4159.2008.05479.x.

356. Reddy, P. V., M. Agudelo, V. S. Atluri, and M. P. Nair. 2012. Inhibition of nuclear factor erythroid 2-related factor 2 exacerbates HIV-1 gp120-induced oxidative and inflammatory response: Role in HIV associated neurocognitive disorder. *Neurochem Res* 37 (8):1697–1706. doi: 10.1007/s11064-012-0779-0.

357. Deshmane, S. L., R. Mukerjee, S. Fan, L. Del Valle, C. Michiels, T. Sweet, I. Rom, K. Khalili, J. Rappaport, S. Amini, and B. E. Sawaya. 2009. Activation of the oxidative stress pathway by HIV-1 Vpr leads to induction of hypoxia-inducible factor 1alpha expression. *J Biol Chem* 284 (17):11364–11373. doi: 10.1074/jbc.M809266200.

358. Hoshino, S., M. Konishi, M. Mori, M. Shimura, C. Nishitani, Y. Kuroki, Y. Koyanagi, S. Kano, H. Itabe, and Y. Ishizaka. 2010. HIV-1 Vpr induces TLR4/MyD88-mediated IL-6 production and reactivates viral production from latency. *J Leukoc Biol* 87 (6):1133–1143. doi: 10.1189/jlb.0809547.

359. Isaguliants, M., O. Smirnova, A. V. Ivanov, A. Kilpelainen, Y. Kuzmenko, S. Petkov, A. Latanova et al. 2013. Oxidative stress induced by HIV-1 reverse transcriptase modulates the enzyme's performance in gene immunization. *Hum Vaccin Immunother* 9 (10):2111–2119. doi: 10.4161/hv.25813.

360. Dionisio, N., M. V. Garcia-Mediavilla, S. Sanchez-Campos, P. L. Majano, I. Benedicto, J. A. Rosado, G. M. Salido, and J. Gonzalez-Gallego. 2009. Hepatitis C virus NS5A and core proteins induce oxidative stress-mediated calcium signalling alterations in hepatocytes. *J Hepatol* 50 (5):872–882. doi: 10.1016/j.jhep.2008.12.026.

361. Li, Y., D. F. Boehning, T. Qian, V. L. Popov, and S. A. Weinman. 2007. Hepatitis C virus core protein increases mitochondrial ROS production by stimulation of Ca^{2+} uniporter activity. *FASEB J* 21 (10):2474–2485. doi: 10.1096/fj.06-7345com.

362. Bataller, R., Y. H. Paik, J. N. Lindquist, J. J. Lemasters, and D. A. Brenner. 2004. Hepatitis C virus core and nonstructural proteins induce fibrogenic effects in hepatic stellate cells. *Gastroenterology* 126 (2):529–540.

10 The Neutrophil NADPH Oxidase*

William M. Nauseef

CONTENTS

INTRODUCTION

Insights in biomedical science highlight the informative interface between clinical practice and basic science research and illustrate that explication of a given disease frequently reveals fundamental aspects of normal biology not previously recognized. Few disorders demonstrate this principle more than does the relationship between recognition of the clinical disorder of chronic granulomatous disease (CGD) and elucidation of the structure and function of the neutrophil NADPH oxidase. At the annual meeting of the American Pediatric Society in 1954, Janeway et al. described five boys with hepatosplenomegaly, elevated γ-globulins, generalized lymphadenopathy, and severe recurrent infections [1]. A subsequent report

---

* The author has no conflicts of interest.

by Bridges et al. described a cadre of similarly affected boys and speculated that their phenotype "could represent an alteration in immunological response turned against the reticuloendothelial system" [2]. A decade later, the Good laboratory at the University of Minnesota linked the clinical syndrome with neutrophil dysfunction [3], and thus began decades of clinical observations and laboratory experimentation that *together* identified the composition and activity of the NADPH oxidase in normal neutrophils and in parallel elucidated the molecular defects that underlie the clinical disorder of CGD.

Three specific examples demonstrate how clinical observations of patients with CGD provided the foundation for fundamental insights into the neutrophil NADPH oxidase. The Orkin lab exploited the coincidence of CGD, retinitis pigmentosa, and Duchenne's muscular dystrophy in two boys with a known interstitial deletion of Xp21 to identify the "X-linked CGD protein" [4], now known as gp91^{phox} or NOX2, the membrane protein that serves as the electron transferase of the NADPH oxidase. Recognition that one cytosol among a group of samples obtained from patients with autosomal complemented the activity of the other cytosols in a broken cell assay of NADPH oxidase activity made interpretation of immunoblots possible and thus catalyzed the identification of two of the cytoplasmic components of the NADPH oxidase [5,6] that are essential for assembly and activity of the functioning enzyme. More recently, identification of the genetic basis for CGD in a patient with an atypical clinical presentation provided critical insight into the importance of p40^{phox} for targeting the NADPH oxidase selectively to phagosomal membranes [7]. These three examples demonstrate how many investigators operating in diverse disciplines have contributed to our understanding of the neutrophil NADPH oxidase and, by extension, of the NADPH oxidase (NOX) protein family. The goal of this review is to provide an overview of the neutrophil NADPH oxidase, including its composition, activation, regulation, and function. Excluded from the presentation is the discussion of NADPH oxidase activity in other phagocytes and other NOX proteins, both of which have been reviewed recently [8–13].

Known components of the neutrophil NADPH oxidase include membrane proteins in distinct cellular compartments and soluble proteins in cytoplasm. In unstimulated neutrophils, individual components reside in separate places, leaving the oxidase unassembled and inactive, thereby utilizing spatial segregation as a regulatory feature of the phagocyte oxidase.

MEMBRANE COMPONENTS

The redox center of the NADPH oxidase resides in a heterodimeric transmembrane heme protein, flavocytochrome b (aka flavocytochrome b$_{558}$ or cytochrome b$_{-245}$, referring to its peak in reduced-minus-oxidized spectra or its low midpoint potential, respectively), comprised of gp91^{phox} and p22^{phox}. Although Japanese investigators had noted the presence of an unusual b-type cytochrome in neutrophils from rabbits and horses and speculated that it may contribute to NADPH oxidase activity [14–16], Segal et al. observed its presence in normal but absence from membranes from patients with X-linked CGD [17] and in over the subsequent decade demonstrated convincingly that it serves as the electron transferase in the neutrophil

NADPH oxidase [18–29]. The Orkin lab subsequently cloned and identified the protein missing from patients with X-linked CGD as a 91 kDa glycoprotein, soon after named gp91^{phox} [4,30]. Its partner in flavocytochrome b, p22^{phox}, was subsequently purified and its gene cloned and sequenced by the Orkin lab [31].

Mature neutrophils express flavocytochrome b in plasma membrane as well as in the membranes of two intracellular compartments, secretory vesicles and the peroxidase-negative granules (i.e., specific and gelatinase granules [32–34]). The plasma membrane of unstimulated neutrophils contains only 15% of the total cellular flavocytochrome b, the remainder housed intracellularly and serving as a reservoir readily recruited when neutrophils are primed or stimulated.

gp91^{phox}

Human gp91^{phox} is an extensively glycosylated protein of 570 amino acids encoded by *CYBB* [4]. Of note, the *CYBB* gene product in mice lacks the carbohydrate sidechains and exists in murine phagocytes as a ~55 kDa protein, still associated in a heterodimer with p22^{phox} [35,36]. During biosynthesis in the endoplasmic reticulum (ER), nascent gp91^{phox} undergoes cotranslational glycosylation to generate the 65 kDa intermediate product that associates with p22^{phox} [37–41]. Heterodimer formation and heme insertion occur in the ER and are required for successful export of gp65-p22 to the Golgi, where modifications of the three N-linked carbohydrate sidechains on gp65 result in the mature gp91^{phox}-p22^{phox}. When heme synthesis in cultured human promyelocytes is inhibited, heterodimer formation fails and the free subunits undergo degradation in the proteasome. Biochemical features of neutrophil oxidase activity in patients with CGD confirm the importance of heterodimer formation for the fate of the individual components of the heterodimer: phagocytes from patients with mutations in *CYBB* lack both gp91^{phox} and p22^{phox}, and those with a genetic absence of p22^{phox} do not express gp91^{phox} [42–44].

The structural organization of gp91^{phox}, intuited from its sequence and other features that will be discussed later, resembles that of ferric reductase proteins widely expressed in microbes and eukaryotes. One homologue of gp91^{phox}, the ferric reductase of *Saccharomyces cerevisiae* (FRE1)[45], contains six transmembrane helices with carboxy-terminal NADPH- and flavin adenine dinucleotide (FAD)-binding domains in a cytoplasmic region and two heme groups, which are stacked between two helices in the membrane and serve to transport electrons from cytoplasm [46]. FRE1 and gp91^{phox} both exhibit similar spectral properties and very low redox potentials (e.g., −250 mV for FRE1) [47]. Both operate as electron transferases, although they utilize different substrates. FRE1 reduces Fe^{+3} in the extracellular space to enable its transport into yeast to support vital cellular functions [45]. FRE1 can reduce oxygen, but very inefficiently [20–100 pmol min^{-1} (mg wet weight)$^{-1}$] [48]. Gp91^{phox} does not exhibit ferric reductase activity [49] and utilizes only oxygen as substrate. The phylogeny of the ferric reductase domain proteins is reviewed in [50].

In the absence of crystal structure information, mutagenesis (both experimentally targeted and naturally occurring), physicochemical analytical techniques and well-characterized monoclonal antibodies provide indirect characterization of the organization of flavocytochrome b [40,51–62]. Expressed as an integral membrane protein,

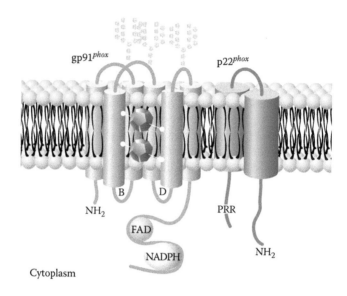

FIGURE 10.1 Flavocytochrome b. Composed of gp91*phox* and p22*phox*, flavocytochrome b is a transmembrane protein that serves as the redox center of the neutrophil NADPH oxidase. The current model for human gp91*phox* includes six transmembrane helices and amino (NH$_2$) and carboxy-termini both on the cytoplasmic face of the membrane. Between helices III and V are two inequivalent heme groups (purple polygons) that transport electrons across the membrane. There are three N-linked carbohydrate chains on the second and third extracellular loops, and two cytoplasmic loops (B and D). The carboxy terminal region includes both FAD and NADPH binding sites. Models for human p22*phox* likewise have both amino and carboxy termini on the cytoplasmic side of the membrane but the extracellular loops are not glycosylated. The carboxy region of p22*phox* contains a proline-rich region (PRR) that associates with p47*phox* during oxidase assembly (see text).

gp91*phox* includes six membrane-spanning helices, with both N- and C-termini expressed intracellularly (Figure 10.1). The hemes in the bis-heme cytochrome b subunit of the mitochondrial cytochrome bc1 complex are bis-histidine linked to the same pair of parallel transmembrane helices and the ligating histidines in each helix are separated by 13 amino acids [63]. Although gp91*phox* possesses seven histidine residues within proposed transmembrane domains, only the pairs His[101]-His[115] and His[209]-His[222] are 13 and 12 residues apart and thus mirror the arrangement in the mitochondrial b-cytochrome. Replacement by site-directed mutagenesis of any of these four but not the other histidines in gp91*phox* results in a loss of the characteristic spectrum of flavocytochrome b, failure of surface expression of the heterodimer, and the absence of NADPH oxidase activity [40]. Based on these data, His[101] and His[115] in helix III and His[209] and His[222] in helix V likely coordinate the heme groups in gp91*phox*, which are likely stacked one over the other, as suggested in the model of yeast iron reductase FRE1 [46].

The midpoint redox potentials of the hemes in gp91*phox* differ (E_{m7} = −225 and −265 mV [64]) and thereby mediate the sequential transfer of electrons across the membrane, from the cytosolic face into the phagosomal lumen or into the extracellular space.

Cytoplasmic NADPH serves as the source of electrons for the NADPH oxidase, which has a Km for NADPH of 33 μM (vs 930 μM for NADH) [65,66]. In the first step in electron transfer, flavin adenine dinucleotide (FAD) associated with the FAD-binding domain in the cytosolic tail of gp91^{phox} accepts a pair of electrons from NADPH, also associated with C-terminus of gp91^{phox} [67]. The flavin group delivers each electron in the pair sequentially to the innermost heme of gp91^{phox}. Each electron subsequently shifts to the outermost heme and then to molecular oxygen, culminating in the generation of superoxide anion, the proximal product of the neutrophil NADPH oxidase. The overall reaction occurs with the following stoichiometry:

$$NADPH + 2O_2 \longrightarrow NADP^+ + H^+ + 2O_2^{\cdot -}.$$

p22^{phox}

Purification of flavocytochrome b recovers both gp91^{phox} and p22^{phox}, a nonglycosylated protein of 195 amino acids encoded by *CYBA* and located on chromosome 16 [68,69]. Like its partner gp91^{phox}, p22^{phox} has eluded crystallization, leaving *in silico* modeling, physicochemical analyses such as nuclear magnetic resonance, and epitope mapping with monoclonal antibodies as the sources for structural information [51,70–72]. For the most part, data suggest that both the amino- and carboxy-termini reside intracellularly and that p22^{phox} contains two transmembrane helices [54,56,58,62,73,74] (Figure 10.1). In contrast to the transferase activity of gp91^{phox}, p22^{phox} lacks intrinsic enzymatic activity and instead provides structural support in two important ways; it contributes to the integrity of the heterodimer and provides a critical docking site for cytosolic oxidase components during assembly of the NADPH oxidase complex.

As noted earlier, heterodimer formation in the ER is prerequisite for its maturation, and the individual subunits of flavocytochrome b undergo proteasomal degradation in the absence of association with its partner. Patients with CGD secondary to mutations in *CYBA* lack *both* p22^{phox} and gp91^{phox} from their phagocytes [4,26,75]. The structural basis for the formation of the gp91^{phox}-p22^{phox} heterodimer remains incompletely defined, although mutational analysis of transfectants has suggested that the terminal eleven amino acids and residues 65–90, a region in the putative first transmembrane domain of p22^{phox}, interact with gp91^{phox} [62]. A similar experimental approach, but using mutations that naturally occur in patients with X$^-$ CGD that have missense mutations that compromise but do not eliminate flavocytochrome b synthesis [76], provides insight into important sites in gp91^{phox} that contribute to heterodimer formation. Data suggest that residues both in the first transmembrane helix and in several regions near the cytosolic dehydrogenase domain and the FAD binding site of gp91^{phox} contribute to the stability of flavocytochrome b [77].

CYTOPLASMIC COMPONENTS

Recognition that components of the phagocyte NADPH oxidase reside in cytoplasm rests largely on the development of cell-free assays for oxidase activity. In a seminal paper that reported the role unsaturated fatty acids in the stimulation of NADPH oxidase activity of guinea pig macrophages [78], Bromberg and Pick described an *in vitro* cell-free oxidase system that utilizes membranes and cytosol from unstimulated phagocytes to generate superoxide anion in an FAD- and NADPH-dependent fashion. Confirmed independently soon after by other laboratories [79–81], the cell-free system established the existence of one or more cytosolic components by demonstrating that the cytosol of neutrophils from patients with the autosomal form of CGD and a normal flavocytochrome b was inactive in the cell-free system [6,82,83]. The cell-free system proved to be the analytical tool essential to the discovery of the cytosolic components of the NADPH oxidase [5,6].

The soluble components of the neutrophil NADPH oxidase exist in the cytosol of unstimulated cells in two complexes, one containing p47^{phox}, p67^{phox}, and p40^{phox} and the with Rac and RhoGDI, and each complex translocates independently to target membranes during cell activation [84] (*vide infra*).

p47^{phox}

A protein of 390 amino acids, p47^{phox} possesses structural domains that enable it to serve as an organizing adaptor protein in assembly of a functioning oxidase (Figure 10.2). The amino-terminal end of p47^{phox} contains a PX domain with its phosphoinositide-recognition motif, originally identified during a search of proteins containing C2 domains, a structural feature associated with membrane targeting [85,86]. Searches of protein databases revealed the presence of PX domains in both p47^{phox} and p40^{phox}, as well as the yeast protein Bem1p and a large number of proteins homologous with human sorting nexins [85]. The PX domains

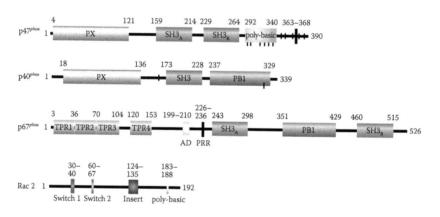

FIGURE 10.2 Cytosolic components of the neutrophil NADPH oxidase. The four currently identified cytosolic components of the neutrophil NADPH oxidase are p47^{phox}, p40^{phox}, p67^{phox}, and Rac2. Each has structural domains that mediate interactions with membrane phospholipids (PX) and with other oxidase components both in the resting and active state of the oxidase (see text for details). The residue number of amino acids delineating motifs and protein domains are indicated. (Modified from Groemping, Y. and Rittinger, K., *Biochem. J.*, 386, 401, 2005.)

in both p47^{phox} and p40^{phox} promote association with phosphoinositides in target membranes, although each recognizes different phospholipid structures [87,88]. Sequencing of p47^{phox} and p67^{phox} demonstrated the presence of two Src homology 3 (SH3) domains [89–91]. In a wide variety of proteins, SH3 domains mediate interactions between proteins and thereby direct their cellular localization (reviewed in [92]).

Adjacent to the carboxy-terminal SH3 domain sits the polybasic autoinhibitory domain, which contains several serine residues clustered among basic amino acids. Phosphorylation of serines in the polybasic region during oxidase activation alters the conformation of p47^{phox} and makes its potentially interactive motifs accessible for binding to targets (discussed in "Activation" section). At the carboxy terminus of p47^{phox} sits a proline-rich region (PRR) that contains the PxxP motif, which in a variety of proteins supports binding to SH3 domains [93,94]. As discussed in more detail later, these structural motifs support both intra- and intermolecular interactions critical to the regulation of NADPH oxidase assembly and activity.

p40^{phox}

p40^{phox}, a 339 amino acid protein encoded by NCF4 on chromosome 22q13.1 [95] and expressed in most nonerythroid hematopoietic cells, was first identified as a protein that copurified with the cytosolic oxidase components from unstimulated neutrophils [96]. Three structural domains have been identified in p40^{phox} (Figure 10.2). Like p47^{phox}, p40^{phox} contains a PX motif near its amino-terminus. However, the PX domains of p40^{phox} and p47^{phox} associate with different phospholipids and thus provide a mechanism for selectively directing oxidase components to specific membranes. p40^{phox} possesses a single SH3 domain and, near its carboxy terminus, a PB1 domain [97,98]. First identified in **p67**^{phox} and the yeast protein **Bem1** [97], PB1 has been identified as a protein interaction motif conserved throughout the plant and animal kingdoms (reviewed in [99]). All three structural domains—PX, SH3, and PB1—participate cooperatively in the interaction of p40^{phox} with phospholipids in target membranes and with its cytosolic oxidase partners.

p67^{phox}

Human p67^{phox} contains 526 amino acids and is encoded by NCF2 located on chromosome 1q25.3 [91]. Whereas p47^{phox} and p40^{phox} operate primarily as adaptor proteins that organize the structure and assembly of the NADPH oxidase, p67^{phox} serves as an essential catalytic agent in the assembled enzyme complex. Structural elements in the terminal 300 amino acids of p67^{phox} include a PRR, two SH3 domains, and, as mentioned here, a PB1 domain (Figure 10.2). Starting at the amino terminus are four tetratricopeptide repeat (TPR) motifs and an activation domain (AA 199–210). The TPR domains collectively support binding of Rac-GTP [51,52,100,101] and the activation domain associates with gp91^{phox} to catalyze electron transfer [102,103].

RacGTP

In addition to the complex of cytosolic phox proteins, GDP-bound Rac2 associated with RhoGDI exists in the cytoplasm of unstimulated human neutrophils. The important role of small molecular weight GTPases to NADPH oxidase activity predated the discovery of the phox proteins, but the identity of the specific protein eluded investigators, a saga chronicled in an informative and clearly written review by Pick [104].

Both Rac2, in the case of neutrophils, and Rac1 from macrophages were purified from cytosol in a prenylated state and complexed to RhoGDI, and the corresponding GTP-loaded Rac support optimal activity of cell-free oxidase systems [105–111]. Structural features of Rac2, the form expressed in human neutrophils, whereas Rac1 is ubiquitously expressed [112,113], include two regions near the amino-terminus termed Switch 1 (AA 30–40) and Switch 2 (AA 60–67) that figure prominently in conformational changes that regulate transition from inactive to active states [51,114,115] (Figure 10.2). Of special importance to the NADPH oxidase is the insert helix (amino acids 123–135), which, despite some controversy with respect to parts of the story (discussed in depth in [51]), promotes interactions with p67^{phox} that are essential for electron transfer through flavocytochrome b.

ASSOCIATED PROTEINS

In addition to the NADPH oxidase components already presented, proteins that copurify or appear functionally linked to the active oxidase have been reported. Included among such proteins is Rap1A, a small molecular weight GTPase that copurified with flavocytochrome b [116]. Subsequent studies have demonstrated that flavocytochrome b free of associated Rap1A supports complete oxidase activity in a cell-free system [117,118], indicating that Rap1A is not required for oxidase activity under these experimental conditions. Studies using murine neutrophils suggest that Rap1A may participate in upstream events important for oxidase activity in intact cells [119] and thus leave unresolved the contribution of Rap1a to NADPH oxidase activity.

Several lines of evidence suggest that a phospholipase in the cytoplasm, cytosolic phospholipase 2Aα (cPLA$_2$), contributes to the activity of the assembled oxidase. First isolated from the macrophage-like cell line U937 [120], cPLA$_2$ acts on fatty acids situated at the sn-2 position and prefers hydrolyzing arachidonic acid, thus providing a mechanism for modulating local eicosanoid production. cPLA$_2$ translocates to the assembled oxidase, with the PX domain of p47^{phox} serving as the docking site, and reduction in cPLA$_2$ correlates with decreases in oxidase activity in neutrophils, monocytes, and cultured human promyelocytes [121–123]. However, the precise role of cPLA$_2$ in the operation of the oxidase remains unclear, since neutrophils from a murine cPLA$_2$ knockout exhibit normal oxidase activity [124].

The calcium binding proteins S100A8 and S100A9, aka myeloid-related protein (MRP) 8 and 16, respectively [125], have been implicated in promoting oxidase activation [126,127]. In a calcium-dependent fashion, S100A8 drives the delivery of arachidonic acid in plasma membranes to flavocytochrome b already associated with both p67^{phox} and Rac2 [128,129]. Collectively, these events increase the affinity of p67^{phox} for flavocytochrome b and concomitantly alter its conformation in the membrane [130,131].

The activated NADPH oxidase complex from neutrophils also contains 6-phosphofructo-2-kinase/fructose-2,6-bisphosphatase (PFK-2/FBase-2) [130], a bifunctional enzyme with distinct kinase and phosphatase domains and a major role in glycolysis [132]. Pharmacologic inhibition or targeted depletion with siRNA reduces agonist-triggered NADPH oxidase regulation [133]. The implications of these observations await additional study, but the data suggest an important functional link between the NADPH oxidase and glycolysis.

ACTIVATION

The neutrophil NADPH oxidase in unstimulated cells exists unassembled and inactive, its components spatially segregated as membrane and soluble proteins (reviewed in [51,52,134–136]). Agonists engage signaling pathways that initiate a cascade of biochemical events that causes posttranslational modifications of the components, with phosphorylation as the predominant change. The concatenation of biochemical reactions culminates in conformational changes that promote the assembly of the oxidase enzyme complex and the relay of electrons from cytosolic NADPH to FAD, across the two inequivalent hemes stacked in flavocytochrome b, and eventually to molecular oxygen [137]. As described here, the oxidase components contain several structural motifs that support binding with proteins and lipid targets. Multiple interactions between these binding domains support intra- and intermolecular links that maintain the trimeric phox protein complex in a 1:1:1 stoichiometry in resting neutrophils [138–140], although there is evidence that p47^{phox} is free in the cytosol of resting neutrophils and associates with the complex during the activation cascade [141].

Intramolecular interactions within p47^{phox} depends on its SH3 domains to maintain its conformation in the cytosol of resting neutrophils. Although models early on suggested that intramolecular associations between the SH3 domains and the PRR maintained p47^{phox} in its resting conformation [142,143], the current consensus is that the region carboxy terminal to the second SH3 and rich in basic amino acids (polybasic region) interacts with *both* SH3 domains [71,144–148] (Figure 10.3a).

At least three important intermolecular interactions among oxidase components operate in the cytosol of resting neutrophils, with p67^{phox} envisioned as a bridge joining the two adaptor proteins p47^{phox} and p40^{phox} in the phox cytosolic protein complex (Figure 10.3b). First, the PB1 domain in p67^{phox} supports a constitutive and firm association with the PB1 region in p40^{phox} [97,139,149–151] and the crystal structure of the PB1 p67^{phox}-p40^{phox} dimer has been solved [139,152]. The association between p67^{phox} and p40^{phox} seems to benefit the integrity of each protein. Recombinant p67^{phox} is less stable in the absence of p40^{phox}, and neutrophils from p67^{phox}-deficient CGD patients contain less than normal amounts of p40^{phox} [91,150,151,153–155]. Second, the carboxy terminal SH3 domain of p67^{phox} binds the PRR of p47^{phox} with an affinity atypically strong for SH3-PRR associations [139,156], likely reflecting input from additional interactions outside of the PRR [150,155,157–159]. An interaction between the PRR of p47^{phox} and the SH3 in

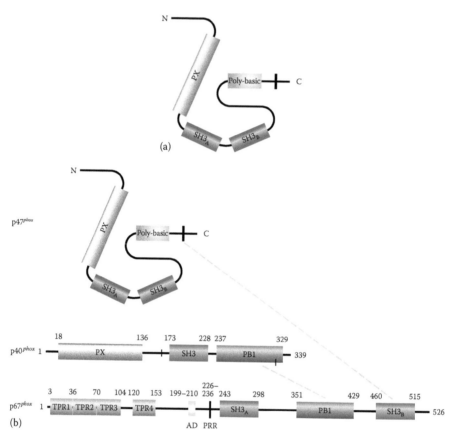

FIGURE 10.3 Organization of cytosolic phox proteins in unstimulated neutrophils. (a) Conformation of p47^{phox} in the cytosol of unstimulated neutrophils. In the cytosol of unstimulated neutrophils, p47^{phox} adopts a conformation wherein its PX and SH3 domains, which mediate association with specific phospholipids in membrane and the PRR of p22^{phox} in the assembled oxidase, respectively, are cryptic and unavailable for docking with their targets. The intramolecular interaction between the polybasic region and the two SH3 domains retains p47^{phox} in its resting configuration until phosphorylation of several serines in the polybasic region occurs as part of oxidase assembly. (b) The cytosolic phox proteins exist in resting neutrophils in a trimeric complex with 1:1:1 stoichiometry. Critical intermolecular interactions between the polybasic region of p47^{phox} and the carboxy terminal SH3 domain of p67^{phox} (SH$_B$) and between the PB1 domains of p40^{phox} and p67^{phox} support the trimeric complex.

p40^{phox} has been proposed [150,160,161] but likely has lower affinity than that between p47^{phox} PRR and p67^{phox} SH3 [139,156] and not likely to occur for a variety of reasons, as lucidly presented by Groemping and Rittinger [51]. Lastly, RhoGDI is bound to the switch II region of Rac [162–164].

The multiple inter- and intramolecular interactions between and within cytosolic oxidase components render unavailable or hidden protein domains that otherwise would associate with complementary sites in flavocytochrome b or target membranes. The docking of p47^{phox} at the target plasma membrane depends on at least two interactions, namely binding of its SH3 to the PRR in p22^{phox} and association of its PX domain with phosphoinositide 3,4 phosphate [PI(3,4)P$_2$] and, to a lesser extent, other phosphoinositides, phosphatidylserine, and phosphoinositol. Missense or deletion mutations in the PRR compromise oxidase activity, thereby underscoring the contribution of the PRR in p22^{phox} to oxidase assembly, [56,62,142,165–170].

Despite their presence in both p47^{phox} and p40^{phox}, the individual PX domains exhibit important differences with respect to the phosphoinositides that they recognize and the membranes that they target. Whereas the p47^{phox} PX binds best to PI(3,4)P$_2$ and to plasma membranes [171], p40^{phox} PX associates with PI(3)P on phagosomal membranes [172–176]. In one patient in whom the p40^{phox} PX had a missense mutation in the region critical for phospholipid binding, the NADPH oxidase fails to assemble on phagosomes, and particulate stimuli do not trigger oxidant production, despite normal responses to soluble agonists phorbol myristate acetate and fMLF [7]. This very selective defect in oxidase assembly demonstrates the specificity of targeting determinants on cytosolic components during oxidase activation.

Similar to the arrangement in the trimeric phox protein complex in cytosol, Rac in its resting state bound to RhoGDI maintains its potentially interactive domains concealed. Upon stimulation, RhoGDI dissociates from Rac, and GTP replaces GDP, allowing translocation of RacGTP to target membranes independent of the phox proteins [84,177,178]. Once exposed in RacGTP, the carboxy terminal prenyl group and the polybasic region together support membrane association (reviewed in [51,179]).

The conformational changes in the cytosolic phox protein complex that are prerequisite for NADPH oxidase assembly and activation depend in large part on phosphorylation of serine residues in the polybasic autoinhibitory domain of p47^{phox} [180–184]. Nearly a dozen serines in this arginine-rich region have been mapped as targets for phosphorylation by a variety of kinases, including protein kinase C isoforms β, γ, and ζ, protein kinase A, p38-activated protein kinase, p21-activated kinase, casein kinase 2, and PKB/Akt [183–193]. The sequential phosphorylation of multiple serines in p47^{phox} relieves the conformational constraints that conceal its PX and SH3 domains and thereby allows translocation to the target PI(3,4)P$_2$ and PRR in the plasma membrane and p22^{phox} (Figure 10.4), respectively [142,146,166,168,170,194–196]. Because of interactions with the B loop of flavocytochrome b [197,198], translocated p47^{phox} brings associated p67^{phox} near both the redox center and membrane-bound prenylated Rac2. After a Rac2-triggered conformational change, p67^{phox} prompts electron transferase activity by gp91^{phox} [178,199]. Although it is generally accepted that the activation domain of p67^{phox} regulates electron flow [102,103,200], the precise action of the insertion domain of p67^{phox} during oxidase activity remains controversial (reviewed in [104]).

Compelling data from many labs implicate the lipid environment in target membranes as important modulators of oxidase activity. In part, and as discussed earlier, changes in the membrane phosphoinositide composition by PI-3-kinases in neutrophils [176,201–203] alter the recognition sites for differential targeting of PX domains in p47^{phox} and p40^{phox} [204–209]. In addition, alterations in the lipid environment near the assembled oxidase trigger functionally significant conformational changes in flavocytochrome b [134,210,211]. Changes in lipid metabolism (reviewed in [210]) and organization of flavocytochrome b into lipid rafts within the membrane

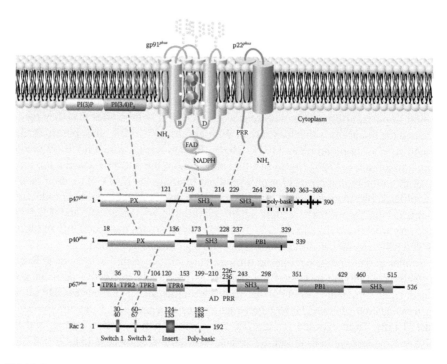

FIGURE 10.4 Oxidase assembly. Assembly of the NADPH oxidase in stimulated neutrophils depends on multiple intermolecular interactions among the individual components and between the PX domains of p47^{phox} and p40^{phox} with PI(3,4)P$_2$ in plasma membrane and PI(3)P in phagosomal membrane, respectively. Shown are those interactions that mediate assembly. For ease of illustration, interactions present in resting neutrophils are omitted.

[212,213] and in association with S100A8 [126] serve to promote electron transfer. More refined definition of the specific changes and the particular lipid changes responsible await further study.

PRIMING

Presenting activation of the neutrophil NADPH oxidase as a binary state—that is, either on or off—serves a practical goal, as it facilitates the discussion of components and processes integral to the structure and activation of the NADPH oxidase. However, such a depiction misrepresents the cell biology. Many bioactive agents modulate the structural and functional state of the oxidase, rendering it more susceptible to subsequent activation without driving it to a fully activated state. Known as "priming," the phenomenon of heightened susceptibility to activation was first observed after neutrophils were treated with lipopolysaccharide prior to exposure to a known agonist [214,215]. Priming of the NADPH oxidase can be promoted also by treatment with other Toll-like receptor agonists, chemoattractants, and cytokines. Elegant studies from the El Benna lab, recently reviewed in [216], provide detailed description of the molecular changes in cytosolic oxidase components in primed human neutrophils.

Partial phosphorylation of p47^{phox} and the resultant conformational changes lie at the heart of priming of the neutrophil NADPH oxidase. In the case of priming by TNFα or GM-CSF, phosphorylation of Ser^{345} in p47^{phox} mediated by p38 MAPK or ERK1/2 provides the essential first step in transformation of the oxidase from its resting state [217]. Simultaneous with prompting the phosphorylation of Ser^{345}, priming agents activate the proline isomerase Pin 1. In general, activated Pin 1 binds phosphorylated serine or threonine residues that are adjacent to a proline and triggers changes in the conformation of the proline [218–221]. In p47^{phox}, the resultant conformational shift from resting to primed state exposes sites for phosphorylation by protein kinase C, subsequent "hyperactivation" of the oxidase and exposure of protein domains that mediate oxidase assembly [216] (*vide supra*). Other oxidase components undergo phosphorylation during TNFα priming or activation, including p67^{phox} [222], gp91^{phox} [223], p40^{phox} [224], and p22^{phox} [225], although the functional correlates are incompletely defined.

REGULATION

In large part, spatial segregation regulates the activation state of the neutrophil NADPH oxidase. The partitioning of factors in cytosol apart from the redox center in the membranes renders the oxidase unassembled and inactive in resting neutrophils. To the extent that phosphorylation promotes the conformational changes in p47^{phox} that allow oxidase assembly, kinases can be considered as regulatory agents.

Once activated, the neutrophil oxidase depends on a variety of biochemical inputs to sustain its action. Of course, oxidant generation requires a supply of molecular oxygen as substrate, and the active oxidase has a Km for oxygen of ~5 to 10 μM [67], although estimates vary. Neutrophils fed opsonized zymosan at varied concentrations of oxygen decrease oxidase activity gradually down to 1% oxygen levels, after which oxidant production decreases precipitously and becomes almost nil at 0.25% oxygen [226]. However, it is unlikely that tissue hypoxia limits oxidase function in clinical settings, as enzyme activity remains approximately half-maximal at as low as 0.35% oxygen, a level well below the ~3% (~25 mm pO$_2$) present in abscesses [227].

As the source of electrons essential for activity, NADPH levels in neutrophil cytosol dictate the enzymatic capacity of the oxidase. Fully activated, the neutrophil oxidase has a turnover rate of 300–330 electrons per second [28,228–230] and would thus consume the 50–100 μM NADPH in neutrophil cytosol within ~100 ms [231]. Activation of the hexose monophosphate shunt precludes the consumption of cytosolic NADPH and premature termination of oxidase activity by providing NADPH to supply electrons for the oxidase in parallel with enzyme activation [232–235]. Underscoring the requirement for a steady supply of NADPH from a functional hexose monophosphate shunt are reports of defective antimicrobial activity by neutrophils with low levels of glucose-6-phosphatase, an enzyme essential for shunt activity [236–238]. Levels of glucose-6-phosphatase in neutrophils below 5% of normal compromise oxidase activity [239], and affected patients may exhibit the clinical phenotype of CGD, with recurrent and severe infection [239–241]. Individuals with an oxidase with low affinity for NADPH likewise can present with CGD [242].

The electron transfer from cytoplasm across plasma or phagosomal membranes poses a challenge to neutrophils: how to compensate for the charge inequality and subsequent depolarization of the membrane. As discussed by DeCoursey, the failure to compensate the charge from electron transfer would cause NADPH oxidase activity to stop prematurely [231,243,244] because of membrane depolarization caused by oxidase activity [245–247]. The voltage-gated proton channel Hv1, first identified in snail neurons [248–250], supports proton export from cytosol and provides the bulk of the charge compensation required by oxidase activity, as reviewed in detail recently by DeCoursey [251]. Hv1 is expressed in the membrane of neutrophil phagosomes [252], and one Hv1 can compensate for the electrons transferred by 100 molecules of the NAPDH oxidase. There are estimated to be ~20 molecules of NADPH oxidase per Hv1 [253]. Neutrophils that lack Hv1 initiate oxidase activity normally but do not sustain superoxide production over time, illustrating the impact of membrane depolarization on electron transfer through the oxidase. Fluxes of other ions also contribute to charge compensation [254–256], but Hv1 shoulders the bulk of the load and thus serves as a key cofactor for sustained NADPH oxidase activity. In addition to its support of the oxidase, charge compensation by Hv1 sustains calcium influx and calcium-dependent neutrophil activities, including adhesion and chemotaxis [257]. Hv1 serves other purposes as well (reviewed in [258]) and appears to be coupled to the release of azurophilic granules, although the mechanisms and functional implications remain to be more fully elucidated [259,260].

The activity of the NADPH oxidase in neutrophils in suspension begins rapidly and typically persists for 0.1–100 min, depending on the agonist used (reviewed in [231]). The lag time between agonist treatment and response, and the rate, magnitude, and duration of oxidant production by neutrophils adherent to surfaces, particularly those that engage integrins, differ markedly from the properties of suspended neutrophils [261]. In all cases, however, activity is finite. Under physiologic conditions, neither exhausted NADPH nor limited oxygen dictates termination of oxidase activity [262,263].

Continued oxidase activity requires the multicomponent complex to remain intact at the target membrane [264]. Several lines of evidence suggest that turnover of the oxidase complexes sustains oxidant production, whereby spent complexes are replaced by new, active complexes [265–271]. Although data supporting this notion are many and sound, the precise events in cycling oxidase complexes remains to be defined. DeCoursey and Ligeti suggest two very different possible mechanisms to explain the observations: (1) dissociation of spent complexes from target membranes with immediate replacement by newly recruited, active complexes, or (2) modulation of the activity of assembled complexes by a cytosolic factor that serves to convert the enzyme between "on" and "off" states [231]. Better understanding of the mechanisms underlying cytosolic oxidase cycling requires additional exploration.

Irrespective of oxidase complex recycling, additional biochemical events figure in the overall regulation of enzyme function. Oxidant production depends on repetitive cycles of phosphorylation-dephosphorylation of p47^{phox} [272], catalyzed by balanced activities of several kinases and phosphatases [266,273,274]. These balanced activities create a tonic state of p47^{phox} phosphorylation that becomes manifest when inhibitors of one of the classes of enzymes are applied. A protein kinase C inhibitor such as staurosporine terminates oxidase action [180,272], whereas those that block phosphotyrosine phosphatase activity [275,276], but not serine-threonine phosphatases [277,278], activate the neutrophil oxidase. Such data underscore the tight regulation of oxidase activity imposed by the state of p47^{phox} phosphorylation at any given moment.

In addition to p47^{phox}, Rac2 provides a means to terminate oxidase action. Because only the GTP-bound form of Rac2 supports oxidase activity [279,280], hydrolysis of GTP by GTPases shuts down oxidant production. Conversely, activity in the cell-free system is sustained with the addition of the nonhydrolyzable GTP analogue GTP-γ-S or GTPase inhibitors such as fluoride [281,282]. Readers interested in learning more about GTPase regulation of the oxidase are encouraged to access the thoughtful and detail discussion by DeCoursey and Ligeti [231].

Lastly, data implicate MPO-dependent chemistry as participating at some level in the cessation of oxidase activity. Inhibition of MPO augments superoxide production in a cell-free oxidase system [262] and neutrophils that lack MPO or in whom MPO is inhibited exhibit more sustained oxidant generation [283–292]. The extent to which the prolonged and more productive oxidase activity compensates for the absence of MPO with respect to antimicrobial activity of MPO-deficient neutrophils is not known.

ANTIMICROBIAL ACTIVITY

Human neutrophils work collaboratively with their phagocytic compatriots, macrophages [293], in a wide range of inflammatory and immune responses [294], although antimicrobial activity during infection likely has been the most extensively studied. As expected from their phagocytic nature, the bulk of antimicrobial attack occurs in phagosomes, the specialized membrane-bound compartments absent from unstimulated neutrophils but created by virtue of ingestion of a target (reviewed in [295]). Once confined to the phagosome, the ingested microbe encounters assaults by multiple toxic agents generated by the NADPH oxidase (reviewed in [296,297]) and released from recruited granules. Although categorization of antimicrobial as oxygen "dependent" or "independent" achieves a useful pedagogical end, as it facilitates appreciation of the diverse mechanisms by which microbes can be killed, it understates the complexity of biochemical reactions and interactions that culminate in the death and destruction of a given organism. Several fundamental aspects of these events merit recognition, including the variety of potential substrates for attack and the plethora of reactive species generated by the NAPDH oxidase and its products.

Starting with the premise that damage mediated in neutrophil phagosomes is the result of biochemical attack, it follows that variability of substrate composition in the targets would result in differential susceptibility to modification. For example, Gram positive bacteria such as *Staphylococcus aureus* provide an array of proteins, carbohydrates, and lipids for attack that differ in part from those presented by a Gram negative bacterium such as *Escherichia coli* or the mycolic acid-rich *Mycobacterium tuberculosis*. Not only do the potential substrates for biochemical attack differ among species of microbes but also between organisms of the same strain that are in different phases of growth. Surface-expressed and secreted proteins of *S. aureus* in stationary phase of growth differ markedly from those in log phase [298], thus providing different substrates to react with active agents in the phagosome. In the case of secreted proteins, some will act as competitive substrates, thereby consuming toxic molecules that otherwise might engage vulnerable targets on the organism and cause damage. Lastly, it is important to recognize that unlike opsonized inert particles, such as latex beads or zymosan fragments, ingested microbes are alive and respond instantly to their immediate environment. *E.coli* ingested by normal neutrophils rapidly sense and respond to oxidants in the phagosome, a response not seen when ingested by CGD neutrophils [299]. Likewise, *S. aureus* possesses complex oxygen-sensing machinery that prompts cellular responses that cause structural and functional changes in the bacterium [300–302]. In many ways, the exchange in phagosomes between host and microbe is a conversation, not a monologue. Taken together, these issues underscore the complexity posed by the broad spectrum of microbial targets presented to toxic agents generated by the neutrophil [303–305]. However, this complexity is magnified by the plethora and range of toxic species generated in neutrophil.

As comprehensively discussed by Winterbourn and colleagues [306] and by Hurst [303], complex oxidant biochemistry in neutrophil phagosomes yields predominantly HOCl, the reaction product of H_2O_2 from the NADPH oxidase and the granule protein myeloperoxidase (MPO) [307]. HOCl reacts with a wide variety of biomolecules, but there is a hierarchy among the potential substrates. For example, HOCl reacts with methionine residues far more readily than with tyrosine residues [306,308,309]. Furthermore, protein chloramines, aldehydes, and low molecular weight chloramines, all downstream products of the reaction of HOCl with its substrates, provide numerous long-lived oxidants with their own spectrum of reactivity [310–316]. However, the biochemical reactions are indiscriminant with respect to the origin of substrates. Methionine residues in bacterial or mammalian proteins are equally susceptible to attack, a fact particularly germane to consideration of events in the neutrophil phagosome, where host proteins predominate. Concurrent with NADPH oxidase activation, granules fuse with nascent phagosomes [317], thereby adding their contents to the mix of oxidants produced in the phagosomal lumen [296,297]. The granule proteins include those with direct antimicrobial action, such as bactericidal permeability increasing protein [318], as well as those with enzymatic activity, such as the four neutrophil serine proteases [319]. Synergistic interactions predominate within the phagosome, including collaborations between oxidants and granule proteins and among the individual granule proteins (reviewed in [320]). Interactions can be productive, as when MPO catalyzes the generation of the potent microbicide HOCl, or counterproductive, as when MPO-generated HOCl attacks granule proteins, thereby eliminating them from the assembly of antimicrobial agents. MPO-generated HOCl inactivates many granule enzymes, including matrix metalloproteinases 7, 8 and 9, cathepsin G, and elastase [321–326]. In fact, MPO-deficient neutrophils release more enzymatically active elastase than do normal neutrophils [326]. HOCl-mediated inactivation extends beyond elastase and applies to all vulnerable granule proteins. The majority of proteins halogenated by activated neutrophils, the biochemical fingerprint of MPO activity, are of host, not microbial origin [327,328], a finding that simply reflects the predominance of host-derived proteins in the phagosomal lumen. However, chlorination and methionine oxidation of microbial proteins correlate with loss of viability of ingested bacteria [327,329–331], findings that are consistent with the MPO-H_2O_2-chloride system as the predominant microbicidal system in normal human neutrophils [332].

PHAGOSOME AS AN EMERGENT SYSTEM

The interconnected and interdependent relationships among phagosomal contents support potent activity against a remarkably diverse range of microbes that might threaten a host. By the same token, the same complexity makes it difficult to assign to any one component a specific, immutable role in the antimicrobial system of neutrophils. The combination of complex biochemistry, reactive interactions among individual components in the phagosomal milieu, and dynamic responses to the assault by captive microbes leave reductionist approaches to the study of antimicrobial mechanisms in neutrophils as crude approximations of reality. In many ways, the inherent complexity of antimicrobial events within phagosomes displays some features of an emergent system, whose properties have long been recognized in science and philosophy [333] and currently provide a rationale for the analytical approaches in "network biology" [334]. In short, the Aristotelean notion posits that output of a complete system *differs from* the sum of activities of its individual components. Depending on the specific synergies among components that may exist at any given time, the overall impact of the complete system may be greater or lesser than the sum of the activities of individual elements [335]. Casadevall and colleagues reject a reductionist view of microbial virulence and instead argue convincingly that it should be seen as an emerging property, a reflection of the net outcome of inputs from both host and microbe [336]. Although this perspective fails to provide a unifying theory of neutrophil antimicrobial action, the recognition and acceptance of the emergent properties of the neutrophil phagosome brings some solace when trying to understand the clinical consequences of specific neutrophil deficiencies.

Given that HOCl-mediated modifications mediate killing and inactivate some antimicrobial granule proteins, it is clear that the NADPH oxidase profoundly influences the nature of neutrophil antimicrobial function beyond simply serving as a source of oxidants to act directly on ingested organisms. In many ways, oxidants alter the overall antimicrobial tone in the phagosome and, consequently, their absence eliminates the contributions of many microbicides *in situ*. In contrast to the pervasive influence that the lack of oxidants exerts on neutrophil killing, the absence of a single granule protein has a much lesser impact. The contrasting clinical phenotypes of CGD and MPO deficiency illustrate this point. CGD neutrophils *in vitro* fail to kill a variety of bacteria, including *S. aureus*, although killing of *E. coli* is normal. Affected patients suffer frequent and severe infection from *S. aureus*, *Serratia*, *Burkholderia*, *Nocardia*, *Aspergillus*, and other microbes (reviewed in [337]). Thus, at the level of both the isolated neutrophil and the patient, the absence of oxidase activity in neutrophils profoundly compromises host defense.

In contrast, MPO-deficient individuals do not incur infection more frequently, and *in vitro* their neutrophils eventually kill most bacteria, albeit at a much slower rate than do normal neutrophils, as initially described by Klebanoff [338].

However, MPO-deficient neutrophils are unable *in vitro* to kill strains of *Candida*, including *C. albicans* [283,339–342], and there are several case reports of systemic candidal infections occurring in MPO-deficient subjects with concurrent diabetes mellitus (reviewed in [343]). Taken together, the contrasting consequences between the absence of oxidants or MPO illustrate two principles underlying the complex antimicrobial activity of human neutrophils. First, oxidants support optimal antimicrobial action, characterized by killing the broadest spectrum of microbes (but not all) with the greatest efficiency. In the absence of oxidants, both the range of susceptible organisms and the rapidity of killing decrease. Second, limitations in antimicrobial action imposed by the absence of single granule protein (e.g., MPO) or class of granule proteins (e.g., serine proteases [344,345]) are minimized by overlapping and redundant actions of the remaining granule proteins.

FUNCTIONS UNRELATED TO ANTIMICROBIAL ACTION

Neutrophil-derived oxidants participate in biological events in addition to those germane to antimicrobial defense. Since the initial recognition of CGD, its clinical picture includes a predisposition for granuloma formation, which can cause outlet obstruction of the genitourinary or gastrointestinal tracts. Studies from the Dinauer laboratory elegantly demonstrated that granuloma formation in CGD reflects dysregulated inflammation and is not an exclusive manifestation of their antimicrobial defect. Airway challenge of gp91*phox*-deficient mice with killed *Aspergillus* elicits robust granuloma formation that exceeds in quantity and duration that seen in normal mice identically challenged [346]. Investigators using different experimental models of noninfectious inflammation have confirmed and extended findings that link the absence of phagocyte oxidase activity to a defect in the capacity to resolve inflammation normally [347–356]. Of note, data from the Krause lab suggest that lack of oxidase activity in macrophages and dendritic cells, not neutrophils, may be the cause of excess granuloma formation [357].

Extracellular oxidants derived from the neutrophil NADPH oxidase can modify susceptible targets and thus operate as either pro- or anti-inflammatory agents, depending on the specific target and context (e.g., [358–360]). Intracellularly, oxidants contribute to selective signaling pathways, including those that modulate cell death, as summarized in an excellent review highlighting key questions in the field and identifying specific methodologic needs [136].

SUMMARY

The neutrophil NADPH oxidase constitutes a robust, multicomponent enzyme complex that shuttles electrons from cytoplasmic NADPH across plasma or phagosomal membranes to molecular oxygen and thereby generates superoxide anion. Essential components of the phagocyte NADPH oxidase that have been identified to date include the membrane heterodimeric protein flavocytochrome b, composed of gp91*phox* and p22*phox*, and two cytoplasmic protein complexes of p47*phox*-p67*phox*-p40*phox* and Rac2-RhoGDI. Agonist-induced changes in conformation of cytosolic components allow for their translocation to the membrane, assembly of a functioning oxidase, and generation of oxidants. In concert with granule proteins and circulating plasma proteins (e.g., Group IIA phospholipase A2 [361]), products of the NADPH oxidase contribute in great measure to acute host defense against microbial infection. In addition, the products of the neutrophil oxidase alter host constituents that more broadly modulate the inflammatory response and, in some situations, extend tissue damage. Elucidation of the composition and regulation of the neutrophil oxidase has set the stage for exploration of the biology of other members of the NADPH oxidase protein family (NOX family), which has members expressed throughout nearly all cellular systems in both the plant and animal kingdoms (reviewed in [8–13]).

ACKNOWLEDGMENTS

Although I worked hard to include work from as many of the important contributors who have advanced our understanding of the neutrophil NADPH oxidase, I recognized that I likely fell short of perfection and omitted some excellent scientists and their seminal contributions. I apologize for any oversights and emphasize that they were unintentional.

Figures were created by Tom Nelson, Medical Media Chief at the Iowa City VA Health Care System. Work in the Nauseef lab is supported by funding from National Institute of Health grant AI116546 (WMN) and by a Merit Review award and use of facilities at the Iowa City Department of Veterans Affairs Medical Center, Iowa City, IA.

REFERENCES

1. Janeway, C. A., J. Craig, M. Davidson, W. Downey, D. Gitlin, and J. C. Sullivan. 1954. Hypergammaglobulinemia associated with severe recurrent and chronic non-specific infection. *American Journal of Diseases of Children* 88:388–392.
2. Bridges, R. A., H. Berendes, and R. A. Good. 1959. A fatal granulomatous disease of childhood. *American Journal of Diseases of Children* 97:387–408.
3. Holmes, B., P. G. Quie, D. B. Windhorst, and R. A. Good. 1966. Fatal granulomatous disease of childhood. *Lancet* 1:1225–1228.
4. Dinauer, M. C., S. H. Orkin, R. Brown, A. J. Jesaitis, and C. A. Parkos. 1987. The glycoprotein encoded by the X-linked chronic granulomatous disease locus is a component of the neutrophil cytochrome b complex. *Nature* 327:717–720.
5. Volpp, B. D., W. M. Nauseef, and R. A. Clark. 1988. Two cytosolic neutrophil NADPH oxidase components absent in autosomal chronic granulomatous disease. *Science* 242:1295–1298.
6. Nunoi, H., D. Rotrosen, J. I. Gallin, and H. L. Malech. 1988. Two forms of autosomal chronic granulomatous disease lack distinct neutrophil cytosol factors. *Science* 242:1298–1301.
7. Matute, J. D., A. A. Arias, N. A. M. Wright, I. Wrobel, C. C. M. Waterhouse, X. J. Li, C. C. Archal et al. 2009. A new genetic subgroup of chronic granulomatous disease with autosomal recessive mutations in p40 phox and selective defects in neutrophil NADPH oxidase activity. *Blood* 114 (15):3309–3315.
8. Sirokmany, G., A. Donko, and M. Geiszt. 2016. Nox/Duox family of NADPH Oxidases: Lessons from knockout mouse models. *Trends in Pharmacological Sciences* 37 (4):318–327. doi: 10.1016/j.tips.2016.01.006.

9. Lambeth, J. D. and A. S. Neish. 2014. Nox enzymes and new thinking on reactive oxygen: A double-edged Sword revisited. *Annual Review of Pathology* 9:119–145.

10. Brandes, R. P., N. Weissmann, and K. Schroder. 2014. Nox family NADPH oxidases: Molecular mechanisms of activation. *Free Radical Biology and Medicine* 76c:208–226. doi: 10.1016/j.freeradbiomed.2014.07.046.

11. Wang, Y. J., X. Y. Wei, X. Q. Jing, Y. L. Chang, C. H. Hu, X. Wang, and K. M. Chen. 2016. The fundamental role of NOX family proteins in plant immunity and their regulation. *International Journal of Molecular Science* 17:pii:E805.

12. Aguirre, J. and J. D. Lambeth. 2010. Nox enzymes from fungus to fly to fish and what they tell us about nox function in mammals. *Free Radical Biology and Medicine* 49:1342–1353.

13. Bedard, K. and K. H. Krause. 2007. The NOX family of ROS-generating NADPH oxidases: Physiology and pathophysiology. *Physiological Review* 87:245–313.

14. Hattori, H. 1961. Studies on the labile, stable Nadi oxidase and peroxidase staining reactions in the isolated particles of horse granulocyte. *Nagoya Journal of Medical Science* 23:362–378.

15. Shinagawa, Y., C. Tanaka, A. Teroaka, and Y. Shinagawa. 1966. A new cytochrome in neutrophilic granules of rabbit leucocyte. *Journal of Biochemistry* 59:622–624.

16. Shinagawa, Y., C. Tanaka, and A. Teroaka. 1966. Electron microscopic and biochemical study of the neutrophilic granules from leucocytes. *Journal of Electron Microscope (Tokyo)* 15:81–85.

17. Segal, A. W., O. T. Jones, D. Webster, and A. C. Allison. 1978. Absence of a newly described cytochrome b from neutrophils of patients with chronic granulomatous disease. *Lancet* 2:446–449.

18. Segal, A. W. and O. T. G. Jones. 1978. Novel cytochrome b system in phagocytic vacuoles of human granulocytes. *Nature* 276:515–517.

19. Segal, A. W., I. West, F. Wientjes, J. H. A. Nugent, A. J. Chavan, B. Haley, R. C. Garcia, H. Rosen, and G. Scrace. 1992. Cytochrome b-245 is a flavocytochrome containing FAD and the NADPH-binding site of the microbicidal oxidase of phagocytes. *Biochemical Journal* 284:781–788.

20. Segal, A. W. 1987. Absence of both cytochrome b-245 subunits from neutrophils in X-linked chronic granulomatous disease. *Nature* 326:88–91.

21. Segal, A. W. 1979. The subcellular distribution and some properties of the cytochrome b componenet of the microbicidal oxidase system of human neutrophils. *Biochemical Journal* 182:181–188.

22. Harper, A. M., M. F. Chaplin, and A. W. Segal. 1985. Cytochrome b-245 from human neutrophils is a glycoprotein. *Biochemical Journal* 227:783–788.

23. Cross, A. R., O. T. G. Jones, R. Garcia, and A. W. Segal. 1982. The association of FAD with the cytochrome b-245 of human neutrophils. *Biochemical Journal* 208:759–763.

24. Segal, A. W., A. R. Cross, R. C. Garcia, N. Borregaard, N. H. Valerius, J. F. Soothill, and O. T. G. Jones. 1983. Absence of cytochrome b-245 in chronic granulomatous disease. A multicenter European evaluation of its incidence and relevance. *New England Journal of Medicine* 308:245–251.

25. Harper, A. M., M. J. Dunne, and A. W. Segal. 1984. Purification of cytochrome b-245 from human neutrophils. *Biochemical Journal* 219:519–527.

26. Teahan, C., P. Rowe, P. Parker, N. Totty, and A. W. Segal. 1987. The X-linked chronic granulomatous disease gene codes for the b-chain of cytochrome b-245. *Nature* 327:720–721.

27. Nugent, J. H. A., W. Gratzer, and A. W. Segal. 1989. Identification of the haem-binding subunit of cytochrome-245. *Biochemical Journal* 264:921–924.

28. Cross, A. R., O. T. G. Jones, A. M. Harper, and A. W. Segal. 1981. Oxidation-reduction properties of the cytochrome b found in the plasma-membrane fraction of human neutrophils. A possible oxidase in the respiratory burst. *Biochemical Journal* 194:599–606.

29. Cross, A. R., F. K. Higson, and O. T. G. Jones. 1982. The enzymic reduction and kinetics of oxidation of cytochrome b of neutrophils. *Biochemical Journal* 204:479–485.

30. Royer-Pokora, B., L. M. Kunkel, A. P. Monaco, S. C. Goff, P. E. Newburger, R. L. Baehner, F. S. Cole, J. T. Curnutte, and S. H. Orkin. 1986. Cloning the gene for an inherited human disorder—Chronic granulomatous disease—On the basis of its chromosomal location. *Nature* 322:32–38.

31. Dinauer, M. C., E. A. Pierce, G. A. P. Bruns, J. T. Curnutte, and S. H. Orkin. 1990. Human neutrophil cytochrome b light chain (p 22-phox). Gene structure, chromosomal location, and mutations in cytochrome-negative autosomal recessive chronic granulomatous disease. *Journal of Clinical Investigation* 86:1729–1737.

32. Rorvig, S., O. Ostergaard, N. H. Heegaard, and N. Borregaard. 2013. Proteome profiling of human neutrophil granule subsets, secretory vesicles, and cell membrane: Correlation with transcriptome profiling of neutrophil precursors. *Journal of Leukocyte Biology* 94 (4):711–721. doi: 10.1189/jlb.1212619.

33. Lominadze, G., D. W. Powell, G. C. Luerman, A. J. Link, R. A. Ward, and K. R. McLeish. 2005. Proteomic analysis of human neutrophil granules. *Molecular and Cellular Proteomics* 4:1503–1521.

34. Cowland, J. B. and N. Borregaard. 2016. Granulopoiesis and granules of human neutrophils. *Immunological Review* 273 (1):11–28. doi: 10.1111/imr.12440.

35. Pollock, J. D., D. A. Williams, M. A. C. Gifford, L. L. Li, X. Du, J. Fisherman, S. H. Orkin, C. M. Doerschuk, and M. C. Dinauer. 1995. Mouse model of X-linked chronic granulomatous disease, an inherited defect in phagocyte superoxide production. *Nature Genetics* 9:202–209.

36. Björgvinsdóttir, H., L. Zhen, and M. C. Dinauer. 1996. Cloning of murine gp91 phox cDNA and functional expression in a human X-linked chronic granulomatous disease cell line. *Blood* 87:2005–2010.

37. Yu, L. and M. C. Dinauer. 1997. Biosynthesis of the phagocyte NADPH oxidase cytochrome b 558. *Journal of Biological Chemistry* 272:27288–27294.

38. Yu, L., F. R. DeLeo, K. J. Biberstine-Kinkade, W. Renee, W. M. Nauseef, and M. C. Dinauer. 1999. Biosynthesis of flavocytochrome b558. *Journal of Biological Chemistry* 274:4364–4369.

39. DeLeo, F. R., J. B. Burritt, L. Yu, A. J. Jesaitis, M. C. Dinauer, and W. M. Nauseef. 2000. Processing and maturation of flavocytochrome b558 includes incorporation of heme as a prerequisite for heterodimer assembly. *Journal of Biological Chemistry* 275:13986–13993.

40. Biberstine-Kinkade, K. J., F. R. DeLeo, R. I. Epstein, B. A. LeRoy, W. M. Nauseef, and M. C. Dinauer. 2001. Heme-ligating histidines in flavocytochrome b 558. *Journal of Biological Chemistry* 276:31105–31112.

41. Biberstine-Kinkade, K. J., L. Yu, N. Stull, B. LeRoy, S. Bennett, A. Cross, and M. C. Dinauer. 2002. Mutagenesis of p22 phox histidine 94. *Journal of Biological Chemistry* 277 (33):30368–30374.

42. Heyworth, P. G., A. R. Cross, and J. T. Curnutte. 2003. Chronic granulomatous disease. *Current Opinion in Immunology* 15:578–584.

43. Roos, D., D. B. Kuhns, A. Maddalena, J. Roesler, J. A. Lopez, T. Ariga, T. Avcin et al. 2010. Hematologically important mutations: X-linked chronic granulomatous disease (third update). *Blood Cells, Molecules, and Diseases* 45 (3):246–265.

44. Roos, D. and M. de Boer. 2014. Molecular diagnosis of chronic granulomatous disease. *Clinical and Experimental Immunology* 175 (2):139–149. doi: 10.1111/cei.12202.

45. Dancis, A., D. G. Roman, G. J. Anderson, A. G. Hinnebusch, and R. D. Klausner. 1992. Ferric reductase of *Saccharomyces cerevisiae*: Molecular characterization, role in iron uptake, and transcriptional control by iron. *Proceedings of the National Academy of Sciences of the United States of America* 89:3869–3873.

46. Finegold, A. A., K. P. Shatwell, A. W. Segal, R. D. Klausner, and A. Dancis. 1996. Intramembrane bis-heme motif for transmembrane electron transport conserved in a yeast iron reductase and the human NADPH oxidase. *Journal of Biological Chemistry* 271 (49):31021–31024.

47. Shatwell, K. P., A. Dancis, A. R. Cross, R. D. Klausner, and A. W. Segal. 1996. The FRE1 ferric reductase of *Saccharomyces cerevisiae* is a cytochrome b similar to that of NADPH oxidase. *Journal of Biological Chemistry* 271:14240–14244.

48. Lesuisse, E., M. Casteras-Simon, and P. Labbe. 1996. Evidence for the *Saccharomyces cerevisiae* ferrireductase system being a multicomponent electron transport chain. *Journal of Biological Chemistry* 271:13578–13583.

49. DeLeo, F. R., O. Olakanmi, G. T. Rasmussen, S. J. McCormick, W. M. Nauseef, and B. E. Britigan. 1999. Despite structural similarities between gp91 phox and FRE1, flavocytochrome b 558 does not mediate iron uptake by myeloid cells. *Journal of Laboratory and Clinical Medicine* 134:275–282.

50. Zhang, X., K. H. Krause, I. Xenarios, T. Soldati, and B. Boeckmann. 2013. Evolution of the ferric reductase domain (FRD) superfamily: Modularity, functional diversification, and signature motifs. *PLoS One* 8 (3):e58126. doi: 10.1371/journal.pone.0058126.

51. Groemping, Y. and K. Rittinger. 2005. Activation and assembly of the NADPH oxidase: A structural perspective. *Biochemical Journal* 386:401–416.

52. Sumimoto, H. 2008. Structure, regulation, and evolution of Nox-family NADPH oxidases that produce reactive oxygen species. *FEBS Journal* 275:3249–3277.

53. Burritt, J. B., M. T. Quinn, M. A. Jutila, C. W. Bond, and A. J. Jesaitis. 1995. Topological mapping of neutrophil cytochrome b epitopes with phage-display libraries. *Journal of Biological Chemistry* 270 (28):16974–16980.

54. Burritt, J. B., S. C. Busse, D. Gizachew, D. W. Siemsen, M. T. Quinn, C. W. Bond, E. A. Dratz, and A. J. Jesaitis. 1998. Antibody imprint of a membrane protein surface. *Journal of Biological Chemistry* 273 (38):24847–24852.

55. Burritt, J. B., F. R. DeLeo, C. L. McDonald, J. R. Prigge, M. C. Dinauer, M. Nakamura, W. M. Nauseef, and A. J. Jesaitis. 2001. Phage display epitope mapping of human neutrophil flavocytochrome b 558. Identification of two juxtaposed extracellular domains. *Journal of Biological Chemistry* 276:2053–2061.

56. Taylor, R. M., J. B. Burritt, D. Baniulis, T. R. Foubert, C. I. Lord, M. C. Dinauer, C. A. Parkos, and A. J. Jesaitis. 2004. Site-specific inhibitors of NADPH oxidase activity and structural probes of flavocytochrome b: Characterization of six monoclonal antibodies to the p22 phox subunit. *Journal of Immunology* 173 7349–7357.

57. Taylor, R. M., C. I. Lord, M. H. Riesselman, J. M. Gripentrog, T. L. Leto, L. C. McPhail, Y. Berdichevsky, E. Pick, and A. J. Jesaitis. 2007. Characterization of surface structure and p47 phox. SH3 domain-mediated conformational changes for human neutrophil flavocytochrone b. *Biochemistry* 46 (49):14291–14304.

58. Taylor, R. M., D. Baniulis, J. B. Burritt, J. M. Gripentrog, C. I. Lord, M. H. Riesselman, W. Maaty et al. 2006. Analysis of human phagocyte flavocytochrome B by mass spectrometry. *Journal of Biological Chemistry* 281 (48):37045–37056.

59. Taylor, R. M., E. A. Dratz, and A. J. Jesaitis. 2011. Invariant local conformation in p22 phox p.Y72H polymorphisms suggested by mass spectral analysis of crosslinked human neutrophil flavocytochrome b. *Biochimie* 93:1502–1509.

60. Campion, Y., A. J. Jesaitis, M. V. C. Nguyen, A. Grichine, Y. Herenger, A. Baillet, S. Berthier, F. Morel, and M. H. Paclet. 2009. New p22-phox monoclonal antibodies: Identification of a conformational probe for cytochrome b 558. *Journal of Innate Immunity* 1:556–569.

61. Campion, Y., M. H. Paclet, A. J. Jesaitis, B. Marques, A. Grichine, S. Berthier, J. L. Lenormand, B. Lardy, M. J. Stasia, and F. Morel. 2007. New insights into the membrane topology of the phagocyte NADPH oxidase: Characterization of an anti-gp91-phox conformational antibody. *Biochemie* 89:1145–1158.
62. Zhu, Y., C. Marchal, A. J. Casbon, N. Stull, A. J. Jesaitis, S. McCormick, W. M. Nauseef, and M. C. Dinauer. 2006. Deletion mutagenesis of p22 phox subunit of flavocytochrome b558: Identification of regions critical for gp91 phox maturation and NADPH oxidase activity. *Journal of Biological Chemistry* 281 (41):30336–30346.
63. Robertson, D., R. Farid, C. Moser, J. Urbauer, S. Mulholland, R. Pidikiti, J. Lear, A. Wand, W. DeGrado, and P. Dutton. 1994. Design and synthesis of multi-haem proteins. *Nature* 368:425–432.
64. Cross, A. R., J. Rae, and J. T. Curnutte. 1995. Cytochrome b-245 of the neutrophil superoxide-generating system contains two nonidentical hemes. Potentiometric studies of a mutant form of gp91*phox*. *Journal of Biological Chemistry* 270:17075–17077.
65. Gabig, T. G. and B. Babior. 1979. The O_2^--forming oxidase responsible for the respiratory burst in human neutrophils. *Journal of Biological Chemistry* 254 (18):9070–9074.
66. Rossi, F. 1986. The O_2^--forming NADPH oxidase of the phagocytes: Nature, mechanisms of activation and function. *Biochimica et Biophysica Acta* 853 (1):65–89.
67. Cross, A. R. and A. W. Segal. 2004. The NADPH oxidase of professional phagocytes-prototype of the NOX electron transport chain systems. *Biochimica et Biophysica Acta* 1657:1–22.
68. Parkos, C. A., M. C. Dinauer, L. E. Walker, R. A. Allen, A. J. Jesaitis, and S. H. Orkin. 1988. Primary structure and unique expression of the 22-kilodalton light chain of human neutrophil cytochrome b. *Proceedings of the National Academy of Sciences of the United States of America* 85:3319–3323.
69. Bu-Ghanim, H. N., C. M. Casimir, S. Povey, and A. W. Segal. 1990. The alpha subunit of cytochrome b-245 mapped to chromosome 16. *Genomics* 8 (3):568–570.
70. Meijles, D. N., B. J. Howlin, and J. M. Li. 2012. Consensus in silico computational modelling of the p22*phox* subunit of the NADPH oxidase. *Computational Biology and Chemistry* 39:6–13. doi: 10.1016/j.compbiolchem.2012.05.001.
71. Groemping, Y., K. Lapouge, S. J. Smerdon, and K. Rittinger. 2003. Molecular basis of phosphorylation-induced activation of the NADPH oxidase. *Cell* 113:343–355.
72. Ogura, K., I. Nobuhisa, S. Yuzawa, R. Takeya, S. Torikai, K. Saikawa, H. Sumimoto, and F. Inagaki. 2006. NMR solution structure of the tandem Src homology 3 domains of p47 phox complexed with a p22 phox-derived proline-rich peptide. *Journal of Biological Chemistry* 281 (6):3660–3668.
73. Imajoh-Ohmi, S., K. Tokita, H. Ochiai, M. Nakamura, and S. Kanegasaki. 1992. Topology of cytochrome b 558 in neutrophil membrane analyzed by anti-peptide antibodies and proteolysis. *Journal of Biological Chemistry* 267:180–184.
74. Dahan, I., I. Issaeva, Y. Gorzalczany, N. Sigal, M. Hirshberg, and E. Pick. 2002. Mapping of functional domains in the p22(phox) subunit of flavocytochrome b(559) participating in the assembly of the NADPH oxidase complex by "peptide walking". *Journal of Biological Chemistry* 277:8421–8432.
75. Parkos, C. A., M. C. Dinauer, A. J. Jesaitis, S. H. Orkin, and J. T. Curnutte. 1989. Absence of both the 91kD and 22kD subunits of human neutrophil cytochrome b in two genetic forms of chronic granulomatous disease. *Blood* 73:1416–1420.
76. Roos, D., D. B. Kuhns, A. Maddalena, J. Bustamante, C. Kannengiesser, M. De Boer, K. van Leeuwen et al. 2010. Hematologically important mutations: The autosomal recessive forms of chronic granulomatous disease (second update). *Blood Cells, Molecules, and Diseases* 44 (4):291–299.
77. Beaumel, S., D. Grunwald, F. Fieschi, and M. J. Stasia. 2014. Identification of NOX2 regions for normal biosynthesis of cytochrome b558 in phagocytes highlighting essential residues for p22phox binding. *Biochemical Journal* 464 (3):425–437. doi: 10.1042/bj20140555.

78. Bromberg, Y. and E. Pick. 1984. Unsaturated fatty acids stimulate NADPH-dependent superoxide production by cell-free system derived from macrophages. *Cellular Immunology* 88 (1):213–221.

79. Heynemann, R. A. and R. E. Vercauteren. 1984. Activation of a NADPH-dependent oxidase from horse polymorphonuclear leukocytes in a cell-free system. *Journal of Leukocyte Biology* 36:751–759.

80. McPhail, L. C., P. S. Shirley, C. C. Clayton, and R. Snyderman. 1985. Activation of the respiratory burst enzyme from human neutrophils in a cell-free system: Evidence for a soluble cofactor. *Journal of Clinical Investigation* 75:1735–1739.

81. Curnutte, J. T. 1985. Activation of human neutrophil nicotinamide adenine dinucleotide phosphate, reduced (triphosphopyridine nucleotide, reduced) oxidase by arachidonic acid in a cell-free system. *Journal of Clinical Investigation* 75:1740–1743.

82. Curnutte, J. T., R. L. Berkow, R. L. Roberts, S. B. Shurin, and P. J. Scott. 1988. Chronic granulomatous disease due to a defect in the cytosolic factor required for nicotinamide adenine dinucleotide phosphate oxidase activation. *Journal of Clinical Investigation* 81:606–610.

83. Caldwell, S. E., C. E. McCall, C. L. Hendricks, P. A. Leone, D. A. Bass, and L. C. McPhail. 1988. Coregulation of NADPH oxidase activation and phosphorylation of a 48-kD protein(s) by a cytosolic factor defective in autosomal recessive chronic granulomatous disease. *Journal of Clinical Investigation* 81:1485–1496.

84. Heyworth, P. G., B. P. Bohl, G. M. Bokoch, and J. T. Curnutte. 1994. Rac translocates independently of the neutrophil NADPH oxidase components p47 phox and p67 phox. Evidence for its interaction with flavocytochrome b 558. *Journal of Biological Chemistry* 269:30749–30752.

85. Ponting, C. P. 1996. Novel domains in NADPH oxidase subunits, sorting nexins, and PtdIns 3-kinases: Binding partners of SH3 domains. *Protein Science* 5:2353–2357.

86. Ellson, C. D., S. Andrews, L. R. Stephens, and P. T. Hawkins. 2002. The PX domain: A new phosphoinositide-binding module. *Journal of Cell Science* 115 (6):1099–1105.

87. Yaffe, M. B. 2002. The p47phox PX domain: Two heads are better than one! *Structure* 10:1288–1290.

88. Kanai, F., H. Liu, S. Field, H. Akbary, T. Matsuo, G. Brown, L. Cantley, and M. Yaffe. 2001. The PX domains of p47phox and p40phox bind to lipid products of PI(3)K. *Natural Cell Biology* 3:675–678.

89. Volpp, B. D., W. M. Nauseef, J. E. Donelson, D. R. Moser, and R. A. Clark. 1989. Cloning of the cDNA and functional expression of the 47 kilodalton cytosolic component of the human neutrophil respiratory burst oxidase. *Proceedings of the National Academy of Sciences of the United States of America* 86:7195–7199.

90. Lomax, K. J., T. L. Leto, H. Nunoi, J. I. Gallin, and H. L. Malech. 1989. Recombinant 47-kilodalton cytosol factor restores NADPH oxidase in chronic granulomatous disease. *Science* 245:409–412.

91. Leto, T. L., K. J. Lomax, B. D. Volpp, H. Nunoi, J. M. G. Sechler, W. M. Nauseef, R. A. Clark, J. I. Gallin, and H. L. Malech. 1990. Cloning of a 67 K neutrophil cytosolic factor and its similarity to a noncatalytic region of p60c-src. *Science* 248:727–730.

92. Kaneko, T., L. Li, and S. S. Li. 2008. The SH3 domain—A family of versatile peptide- and protein-recognition module. *Frontiers in Bioscience* 13:4938–4952.

93. Ren, R., B. J. Mayer, P. Cicchetti, and D. Baltimore. 1993. Identification of a ten-amino acid proline-rich SH3 binding site. *Science* 259:1157–1161.

94. Yu, H., J. K. Chen, S. Feng, D. C. Dalgarno, A. W. Brauer, and S. L. Schreiber. 1994. Structural basis for the binding of proline-rich peptides to SH3 domains. *Cell* 76:933–945.

95. Zhan, S., N. Vazquez, S. Zhan, F. B. Wientjes, M. L. Budarf, E. Schrock, T. Ried, E. D. Green, and S. J. Chanock. 1996. Genomic structure, chromosomal localization, start of transcription, and tissue expression of the human p40-phox, a new component of the nicotinamide adenine dinucleotide phosphate-oxidase complex. *Blood* 88 (7):2714–2721.

96. Wientjes, F. B., J. J. Hsuan, N. F. Totty, and A. W. Segal. 1993. p40phox, a third cyto-solic component of the activation complex of the NADPH oxidase to contain src homol-ogy 3 domains. *Biochemical Journal* 296:557–561.
97. Ito, T., Y. Matsui, T. Ago, K. Ota, and H. Sumimoto. 2001. Novel modular domain PB1 recognizes PC motif to mediate functional protein-protein interactions. *The EMBO Journal* 20 (15):1–9.
98. Honbou, K., R. Minakami, S. Yuzawa, R. Takeya, N. N. Suzuki, S. Kamakura, H. Sumimoto, and F. Inagaki. 2007. Full-length p40 phox structure suggests a basis for regulation mechanism of its membrane binding. *The EMBO Journal* 26 (4):1176–1186.
99. Sumimoto, H., S. Kamakura, and T. Ito. 2007. Structure and function of the PB1 domain, a protein interaction module conserved in animals, fungi, amoebas, and plants. *Science STKE* 2007 (401):re6. doi: 10.1126/stke.4012007re6.
100. Lapouge, K., S. J. M. Smith, P. A. Walker, S. J. Gamblin, S. J. Smerdon, and K. Rittinger. 2000. Structure of the TPR domain of p67 phox in complex with Rac-GTP. *Molecular Cell* 6:899–907.
101. Koga, H., H. Terasawa, H. Nunoi, K. Takeshige, F. Inagaki, and H. Sumimoto. 1999. Tetratricopeptide repeat (TPR) motifs of p67 phox participate in interaction with the small GTPase Rac and activation of the phagocyte NADPH oxidase. *Journal of Biological Chemistry* 274:25051–25060.
102. Nisimoto, Y., S. Motalebi, C. H. Han, and J. D. Lambeth. 1999. The p67 phox activa-tion domain regulates electron flow from NADPH to flavin in flavocytochrome b 558. *Journal of Biological Chemistry* 274 (33):22999–23005.
103. Han, C. H., J. L. R. Freeman, T. Lee, S. A. Motalebi, and J. D. Lambeth. 1998. Regulation of the neutrophil respiratory oxidase. *Journal of Biological Chemistry* 273:16663–16668.
104. Pick, E. 2014. Role of the Rho GTPase Rac in the activation of the phagocyte NADPH oxidase: Outsourcing a key task. *Small GTPases* 5:e27952. doi: 10.4161/sgtp.27952.
105. Knaus, U. G., P. G. Heyworth, T. Evans, J. T. Curnutte, and G. M. Bokoch. 1991. Regulation of phagocyte oxygen radical production by the GTP-binding protein Rac 2. *Science* 254:1512–1515.
106. Knaus, U. G., P. G. Heyworth, B. T. Kinsella, J. T. Curnutte, and G. M. Bokoch. 1992. Purification and characterization of Rac 2. *Journal of Biological Chemistry* 267:23575–23582.
107. Abo, A. and E. Pick. 1991. Purification and characterization of a third cytosolic com-ponent of the superoxide-generating NADPH oxidase of macrophages. *Journal of Biological Chemistry* 266:23577–23585.
108. Abo, A., E. Pick, A. Hall, N. Totty, C. G. Teahan, and A. W. Segal. 1991. Activation of the NADPH oxidase involves the small GTP-binding protein p21 rac1. *Nature* 353:668–670.
109. Pick, E., Y. Gorzalczany, and S. Engel. 1993. Role of the rac1 p21-GDP-dissociation inhibitor for rho heterodimer in the activation of the superoxide-forming NADPH oxi-dase of macrophages. *European Journal of Biochemistry* 217 (1):441–455.
110. Mizuno, T., K. Kaibuchi, S. Ando, T. Musha, K. Hiraoka, K. Takaishi, M. Asada, H. Nunoi, I. Matsuda, and Y. Takai. 1992. Regulation of the superoxide-generating NADPH oxidase by a small GTP-binding protein and its stimulatory and inhibitory GDP/GTP exchange proteins. *Journal of Biological Chemistry* 267:10215–10218.
111. Kwong, C. H., H. L. Malech, D. Rotrosen, and T. L. Leto. 1993. Regulation of the human neutrophil NADPH oxidase by rho—Related G-proteins. *Biochemistry* 32:5711–5717.
112. Roberts, A. W., C. Kim, L. Zhen, J. B. Lowe, R. Kapur, B. Petryniak, A. Spaetti et al. 1999. Deficiency of the hematopoietic cell-specific Rho family GTPase Rac2 is characterized by abnormalities in neutrophil function and host defense. *Immunity* 10 (2):183–196.

113. Gu, Y., B. Jia, F. C. Yang, M. D'Souza, C. E. Harris, C. W. Derrow, Y. Zheng, and D. A. Williams. 2001. Biochemical and biological characterization of a human Rac2 GTPase mutant associated with phagocytic immunodeficiency. *Journal of Biological Chemistry* 276 (19):15929–15938. doi: 10.1074/jbc.M010445200.
114. Sprang, S. R. 1997. G protein mechanisms: Insights from structural analysis. *Annual Review of Biochemistry* 66:639–678. doi: 10.1146/annurev.biochem.66.1.639.
115. Vetter, I. R. and A. Wittinghofer. 2001. The guanine nucleotide-binding switch in three dimensions. *Science* 294 (5545):1299–1304. doi: 10.1126/science.1062023.
116. Quinn, M. T., C. A. Parkos, L. Walker, S. H. Orkin, M. C. Dinauer, and A. J. Jesaitis. 1989. Association of a ras-related protein with cytochrome b of human neutrophils. *Nature* 342:198–200.
117. Abo, A., A. Boyhan, I. West, A. J. Thrasher, and A. W. Segal. 1992. Reconstitution of neutrophil NADPH oxidase activity in the cell-free system by four components: p67-phox, p47-phox, p21 rac 1, and cytochrome b-245. *Journal of Biological Chemistry* 267:16767–16770.
118. Knoller, S., S. Shpungin, and E. Pick. 1991. The membrane-associated component of the amphiphile-activated, cytosol-dependent superoxide-forming NADPH oxidase of macrophages is identical to cytochrome b 559. *Journal of Biological Chemistry* 266:2795–2804.
119. Li, Y., J. Yan, P. De, H. C. Chang, A. Yamauchi, K. W. Christopherson, 2nd, N. C. Paranavitana et al. 2007. Rap1a null mice have altered myeloid cell functions suggesting distinct roles for the closely related Rap1a and 1b proteins. *Journal of Immunology* 179 (12):8322–8331.
120. Clark, J. D., N. Milona, and J. L. Knopf. 1990. Purification of a 110-kilodalton cytosolic phospholipase A2 from the human monocytic cell line U937. *Proceedings of the National Academy Sciences of the United States of America* 87 (19):7708–7712.
121. Dana, R., T. L. Leto, H. L. Malech, and R. Levy. 1998. Essential requirement of cytosolic phospholipase A2 for activation of the phagocyte NADPH oxidase. *Journal of Biological Chemistry* 273 (1):441–445.
122. Li, Q. and M. K. Cathcart. 1997. Selective inhibition of cytosolic phospholipase A 2 in activated human mononcytes. *Journal of Biological Chemistry* 272 (4):2404–2411.
123. Bae, Y., Y. Kim, J. Kim, T. Lee, Y. Kim, P. Suh, and S. Ryu. 2000. Independent functioning of cytosolic phospholipase A 2 and phospholipase D 1 in Trp-Lys-Tyr-Met-Val-D-Met induced superoxide generationin human monocytes. *Journal of Immunology* 164:4089–4096.
124. Rubin, B. B., G. P. Downey, A. Koh, N. Degousee, F. Ghomashchi, L. Nallan, E. Stefanski et al. 2005. Cytosolic phospholipase A 2-α is necessary for platelet-activating factor biosynthesis, efficient neutrophil-mediated bacteial killing, and the innate immune response to pulmonary infect. cPLA2-α does not regulate neutrophil nadph oxidase activity. *Journal of Biological Chemistry* 280 (9):7519–7529.
125. Roth, J., T. Vogl, C. Sorg, and C. Sunderkötter. 2003. Phagocyte-specific S100 proteins: A novel group of proinflammatory molecules. *Trends in Immunology* 24 (4):155–158.
126. Berthier, S., M. H. Paclet, S. Lerouge, F. Roux, S. Vergnaud, A. W. Coleman, and F. Morel. 2003. Changing the conformation state of cytochrome b558 initiates NADPH oxidase activation. *Journal of Biological Chemistry* 278 (28):25499–25508.
127. Berthier, S., M. V. C. Nguyen, A. Baillert, M. A. Hograindleur, M. H. Paclet, B. Polack, and F. Morel. 2012. Molecular interface of S100A8 with cytochrome b 558 and NADPH oxidase activation. *PLoS ONE* 7 (7):e40277.
128. Doussiere, J., F. Bouzidi, and P. V. Vignais. 2002. The S100A8/A9 protein as a partner for the cytosolic factors of NADPH oxidase activation in neutrophils. *European Journal of Biochemistry* 269:3246–3255.

129. Kerkhoff, C., W. Nacken, M. Benedyk, M. C. Dagher, C. Sopalla, and J. Doussiere. 2005. The arachidonic acid-binding protein S100A8/A9 promotes NADPH oxidase activation by interaction with p67phox and Rac-2. *FASEB Journal* 19 (3):467–469. doi: 10.1096/fj.04-2377fje.

130. Paclet, M. H., S. Berthier, L. Kuhn, J. Garin, and F. Morel. 2007. Regulation of phagocyte NADPH oxidase activity: Identification of two cytochrome b 558 activation states *FASEB Journal* 21:1244–1255.

131. Taylor, R. M., M. H. Riesselman, C. I. Lord, J. M. Gripentrog, and A. J. Jesaitis. 2012. Anionic lipid-induced conformational changes in human phagocyte flavocytochrome b precede assembly and activation of the NADPH oxidase complex. *Archives of Biochemical and Biophysics* 52 (1–2):24–31.

132. Rider, M. H., L. Bertrand, D. Vertommen, P. A. Michels, G. G. Rousseau, and L. Hue. 2004. 6-Phosphofructo-2-kinase/fructose-2,6-bisphosphatase: Head-to-head with a bifunctional enzyme that controls glycolysis. *Biochemical Journal* 381 (Pt 3):561–579. doi: 10.1042/bj20040752.

133. Baillet, A., M. A. Hograindleur, J. El Benna, A. Grichine, S. Berthier, F. Morel, and M. H. Paclet. 2016. Unexpected function of the phagocyte NADPH oxidase in supporting hyperglycolysis in stimulated neutrophils: Key role of 6-phosphofructo-2-kinase. *FASEB Journal* 31:663–673.

134. Nauseef, W. M. 2004. Assembly of the phagocyte NADPH oxidase. *Histochemistry and Cell Biology* 122 (4):277–291.

135. Karimi, G., C. Houee Levin, M. C. Dagher, L. Baciou, and T. Bizouarn. 2014. Assembly of phagocyte NADPH oxidase: A concerted binding process? *Biochimica et Biophysica Acta* 1840 (11):3277–3283. doi: 10.1016/j.bbagen.2014.07.022.

136. Dupre-Crochet, S., M. Erard, and O. Nüβe. 2013. ROS production in phagocytes: Why, when, and where? *Journal of Leukocyte Biology* 94 (4):657–670. doi: 10.1189/jlb.1012544.

137. Pickard, R. T., X. G. Chiou, B. A. Strifler, M. R. DeFelippis, P. A. Hyslop, A. L. Tebbe, Y. K. Yee et al. 1996. Identification of essential residues for the catalytic funtion of 85-kDa cytosolic phospholipase A 2. Probing the role of histidine, aspartic acid, cysteine, and arginine. *Journal of Biological Chemistry* 271:19225–19231.

138. Iyer, S. S., D. W. Pearson, W. M. Nauseef, and R. A. Clark. 1994. Evidence for a readily dissociable complex of p47phox and p67phox in cytosol of unstimulated human neutrophils. *Journal of Biological Chemistry* 269:22405–22411.

139. Lapouge, K., S. J. M. Smith, Y. Groemping, and K. Rittinger. 2002. Architecture of the p40-p47-p67 phox complex in the resting state of the NADPH oxidase. *Journal of Biological Chemistry* 277 (12):10121–10128.

140. Park, J. W., J. El Benna, K. E. Scott, B. L. Christensen, S. J. Chanock, and B. M. Babior. 1994. Isolation of a complex of respiratory burst oxidase components from resting neutrophil cytosol. *Biochemistry* 33:2907–2911.

141. Brown, G. E., M. Q. Stewart, H. Liu, V. L. Ha, and M. B. Yaffe. 2003. A novel assay system implicates PtdIns(3,4)P 2, PtdIns(3)P, and PKCδ in intracellular production of reactive oxygen species by the NADPH oxidase. *Molecular Cell* 11 (1):35–47.

142. Leto, T. L., A. G. Adams, and I. de Mendez. 1994. Assembly of the phagocyte NADPH oxidase: Binding of Src homology 3 domains to proline-rich targets. *Proceedings of the National Academy of Sciences of the United States of America* 91:10650–10654.

143. Finan, P., Y. Shimizu, I. Gout, J. Hsuan, O. Truong, C. Butcher, P. Bennett, M. D. Waterfield, and S. Kellie. 1994. An SH3 domain and proline-rich sequence mediate an interaction between two components of the phagocyte NADPH oxidase complex. *Journal of Biological Chemistry* 269:13752–13755.

144. Ago, T., H. Nunoi, T. Ito, and H. Sumimoto. 1999. Mechanism for phosphorylation-induced activation of the phagocyte NADPH oxidase protein p47 phox. *Journal of Biological Chemistry* 274:33644–33653.

145. Huang, J. and M. E. Kleinberg. 1999. Activation of the phagocyte NADPH oxidase protein p47 phox: Phosphorylation controls SH3 domain-dependent binding to p22 phox *Journal of Biological Chemistry* 274:19731–19737.

146. de Mendez, I., N. Homayounpour, and T. L. Leto. 1997. Specificity of p47 phox SH3 domain interactions in NADPH oxidase assembly and activation. *Molecular and Cellular Biology* 17 (4):2177–2184.

147. Yuzawa, S., N. N. Suzuki, Y. Fujioka, K. Ogura, H. Sumimoto, and F. Inagaki. 2004. A molecular mechanism for autoinhibition of the tandem SH3 domains of p47phox, the regulatory subunit of the phagocyte NADPH oxidase. *Genes Cells* 9 (5):443–456. doi: 10.1111/j.1356-9597.2004.00733.x.

148. Yuzawa, S., K. Ogura, M. Horiuchi, N. N. Suzuki, Y. Fujioka, M. Kataoka, H. Sumimoto, and F. Inagaki. 2004. Solution structure of the tandem Src homology 3 domains of p47phox in an autoinhibited form. *Journal of Biological Chemistry* 279 (28):29752–29760. doi: 10.1074/jbc.M401457200.

149. Tsunawaki, S., S. Kagara, K. Yoshikawa, L. S. Yoshida, T. Kuratsuji, and H. Namiki. 1996. Involvement of p47phox in activation of phagocyte NADPH oxidase through association of its carboxyl-terminal, but not its amino-terminal, with p67 phox. *Journal of Experimental Medicine* 184:893–902.

150. Fuchs, A., M. C. Dagher, J. Fauré, and P. V. Vignais. 1996. Topological organization of the cytosolic activating complex of the superoxide-generating NADPH-oxidase. Pinpointing the sites of interaction between p47phox, p67phox and p40phox using the two-hybrid system. *Biochimica et Biophysica Acta: Molecular Cell Research* 1312:39–47.

151. Fuchs, A., M. C. Dagher, and P. V. Vignais. 1995. Mapping the domains of interactions of p40phox with both p47phox and p67phox of the neutrophil oxidase complex using the two-hybrid system. *Journal of Biological Chemistry* 270:5695–5697.

152. Wilson, M. I., D. J. Gill, O. Perisic, M. T. Quinn, and R. L. Williams. 2003. PB1 domain-mediated heterodimerization in NADPH oxidase and signaling complexes of atypical protein kinase C with Par6 and p62. *Molecular Cell* 12 (1):39–50.

153. Dusi, S., M. Donini, and F. Rossi. 1996. Mechanisms of NADPH oxidase activation: Translocation of p40 phox, Rac1 and Rac2 from the cytosol to the membranes in human neutrophils lacking p47 phox or p67 phox. *Biochemical Journal* 314:409–412.

154. Tsunawaki, S., H. Mizunari, M. Nagata, O. Tatsuzawa, and T. Kuratsuji. 1994. A novel cytosolic component, p40 phox, of respiratory burst oxidase associates with p67 phox and is absent in patients with chronic granulomatous disease who lack p67 phox. *Biochemical and Biophysical Research Communications* 199:1378–1387.

155. Sathyamoorthy, M., I. de Mendez, A. G. Adams, and T. L. Leto. 1997. p40 phox down-regulates NADPH oxidase activity through interactions with its SH3 domain. *Journal of Biological Chemistry* 272:9141–9146.

156. Massenet, C., S. Chenavas, C. Cohen-Addad, M. C. Dagher, G. Brandolin, E. Pebay-Peyroula, and F. Fieschi. 2005. Effects of p47 phox C terminus phosphorylations on binding interactions with p40 phox and p67 phox. Structural and functional comparison of p40 phox and p67 phox SH3 domains. *Journal of Biological Chemistry* 280 (14):13752–13761.

157. Kami, K., R. Takeya, H. Sumimoto, and D. Kohda. 2002. Diverse recognition of non-PxxP peptide ligands by the SH3 domains from p67 phox, Grb2 and Pex13p. *EMBO Journal* 21 (16):4268–4276.

158. Grizot, S., N. Grandvaux, F. Fieschi, J. Fauré, C. Massenet, J. P. Andrieu, A. Fuchs et al. 2001. Small angle neutron scattering and gel filtration analyses of neutrophil NADPH oxidase cytosolic factors highlight the role of the C-terminal end of p47 phox in the association with p40 phox. *Biochemistry* 40 (10):3127–3133.

159. Wilson, L., C. Butcher, P. Finan, and S. Kellie. 1997. SH3 domain-mediated interactions involving the phox components of the NADPH oxidase. *Inflammation Research* 46:265–271.
160. Ito, T., R. Nakamura, H. Sumimoto, K. Takeshige, and Y. Sakaki. 1996. An SH3 domain-mediated interaction between the phagocyte NADPH oxidase factors p40phox and p47phox. *FEBS Letters* 385 (3):229–232.
161. Wientjes, F. B., G. Panayotou, E. Reeves, and A. W. Segal. 1996. Interactions between cytosolic components of the NADPH oxidase: p40 phox interacts with both p67 phox and p47 phox. *Biochemical Journal* 317:919–924.
162. Regazzi, R., A. Kikuchi, Y. Takai, and C. B. Wollheim. 1992. The small GTP-binding proteins in the cytosol of insulin-secreting cells are complexed to GDP dissociation inhibitor proteins. *Journal of Biological Chemistry* 267:17512–17519.
163. Scheffzek, K., I. Stephan, O. N. Jensen, D. Illenberger, and P. Gierschik. 2000. The Rac-RhoGDI complex and the structural basis for the regulation of Rho proteins by RhoGDI. *Nature Structural and Biology* 7 (2):122–126. doi: 10.1038/72392.
164. Hoffmann, G. R., N. Nassar, and R. A. Cerione. 2000. Structure of the Rho family GTP-binding protein Cdc42 in complex with the multifunctional regulator RhoGDI. *Cell* 100:345–356.
165. von Lohneysen, K., D. Noack, M. R. Wood, J. S. Friedman, and U. G. Knaus. 2010. Structural insights into Nox4 and Nox2: Motifs involved in function and cellular localization. *Molecular Cell and Biology* 30 (4):961–975. doi: 10.1128/mcb.01393-09.
166. Shiose, A. and H. Sumimoto. 2000. Arachidonic acid and phosphorylation synergistically induce a conformational change of p47*phox* to activate the phagocyte NADPH oxidase. *Journal of Biological Chemistry* 275 (18):13793–13801.
167. Kawahara, T., D. Ritsick, G. Cheng, and J. D. Lambeth. 2005. Point mutations in the proline-rich region of p22 phox are dominant inhibitors of NOX1- and NOX2-dependent reactive oxygen generation. *Journal of Biological Chemistry* 280 31859–31869.
168. Sumimoto, H., Y. Kage, H. Nunoi, H. Sasaki, T. Nose, Y. Fukumaki, M. Ohno, S. Minakami, and K. Takeshige. 1994. Role of Src homology 3 domains in assembly and activation of the phagocyte NADPH oxidase. *Proceedings of the National Academy of Sciences of the United States of America* 91:5345–5349.
169. Rae, J., D. Noack, P. Heyworth, B. Ellis, J. Curnutte, and A. Cross. 2000. Molecular analysis of 9 new families with chronic granulomatous disease caused by mutations in CYBA, the gene encoding p22 phox *Blood* 96 (3):1106–1111.
170. Sumimoto, H., K. Hata, K. Mizuki, T. Ito, Y. Kage, Y. Sakaki, Y. Fukumaki, M. Nakamura, and K. Takeshige. 1996. Assembly and activation of the phagocyte NADPH oxidase. *Journal of Biological Chemistry* 271 (36):22152–22158.
171. Li, X. J., C. C. Marchal, N. D. Stull, R. V. Stahelin, and M. C. Dinauer. 2010. p47phox Phox homology domain regulates plasma membrane but not phagosome neutrophil NADPH oxidase activation. *Journal of Biological Chemistry* 285 (45):35169–35179. doi: 10.1074/jbc.M110.164475.
172. Ellson, C., S. Gobert-Gosse, K. Anderson, K. Davidson, H. Erdjument-Bromage, P. Tempst, J. Thuring et al. 2001. PtdIns(3)P regulates the neutrophil oxidase complex by binding to the PX domain of p40 phox. *Nature and Cell Biology* 3:679–682.
173. Ellson, C., K. Davidson, K. Anderson, L. R. Stephens, and P. T. Hawkins. 2006. PtdIns3P binding to the PX domain of p40 phox is a physiological signal in NADPH oxidase activation. *The EMBO Journal* 25 (19):4468–4478.
174. Ellson, C. D., K. Davidson, G. J. Ferguson, R. O'Connor, L. R. Stephens, and P. T. Hawkins. 2006. Neutrophils from p40 phox-1-mice exhibit severe defects in NADPH oxidase regulation and oxidant-dependent bacterial killing. *Journal of Experimental Medicine* 203 (8):1927–1937.

175. Suh, C. I., N. D. Stull, X. J. Li, W. Tian, M. O. Price, S. Grinstein, M. B. Yaffe, S. Atkinson, and M. C. Dinauer. 2006. The phosphoinositide-binding protein p40 phox activates the NADPH oxidase during FcgIIA receptor-induced phagocytosis. *Journal of Experimental Medicine* 203 (8):1915–1925.

176. Tian, W., X. J. Li, N. D. Stull, W. Ming, C. I. Suh, S. A. Bissonnette, M. B. Yaffe, S. Grinstein, S. J. Atkinson, and M. C. Dinauer. 2008. Fc g R-stimulated activation of the NADPH oxidase: Phosphoinositide-binding protein p40 phox regulates NADPH oxidase activity after enzyme assembly on the phagosome. *Blood* 112 (9):3867–3877.

177. Dorseuil, O., M. T. Quinn, and G. M. Bokoch. 1995. Dissociation of Rac translocation from p47 phox/p67 phox movements in human neutrophils by tyrosine kinase inhibitors. *Journal of Leukocyte Biology* 58:108–113.

178. Gorzalczany, Y., N. Sigal, M. Itan, O. Lotan, and E. Pick. 2000. Targeting of rac1 to the phagocyte membrane is sufficient for the induction of NADPH oxidase assembly. *Journal of Biological Chemistry* 275 (51):40073–40081.

179. Lu, K. P. and X. Z. Zhou. 2007. The prolyl isomerase PIN1: A pivotal new twist in phosphorylation signalling and disease. *Nature Reviews Molecular Cell Biology* 8:904–916.

180. Nauseef, W. M., S. McCormick, J. Renee, K. G. Leidal, and R. A. Clark. 1993. Functional domain in an arginine-rich carboxy terminal region of p47 phox. *Journal of Biological Chemistry* 268:23646–23651.

181. El Benna, J., L. R. P. Faust, and B. Babior. 1996. The phosphorylation of the respiratory burst oxidase component p47 phox during neutrophil activation. *Journal of Biological Chemistry* 269:23431–23436.

182. El Benna, J., P. M. Dang, M. A. Gougerot-Pocidalo, J. C. Marie, and F. Braut-Boucher. 2009. p47phox, the phagocyte NADPH oxidase/NOX2 organizer: Structure, phosphorylation and implication in diseases. *Experimental Molecular Medicine* 41 (4):217–225. doi: 10.3858/emm.2009.41.4.058.

183. El Benna, J., L. P. Faust, J. L. Johnson, and B. M. Babior. 1996. Phosphorylation of the respiratory burst oxidase subunit p47*phox* as determined by two-dimensional phosphopeptide mapping. *Journal of Biological Chemistry* 271 (11):6374–6378.

184. Faust, L. P., J. El Benna, B. M. Babior, and S. J. Chanock. 1995. The phosphorylation targets of p47*phox*, a subunit of the respiratory burst oxidase. Functions of the individual target serines as evaluated by site-directed mutagenesis. *Journal of Clinical Investigation* 96:1499–1505.

185. el Benna, J., L. P. Faust, and B. M. Babior. 1994. The phosphorylation of the respiratory burst oxidase component p47phox during neutrophil activation. Phosphorylation of sites recognized by protein kinase C and by proline-directed kinases. *Journal of Biological Chemistry* 269 (38):23431–23436.

186. El Benna, J., J. H. Han, J. W. Park, E. Schmid, R. J. Ulevitch, and B. M. Babior. 1996. Activation of p38 in stimulated human neutrophils: Phosphorylation of the oxidase component p47 phox by p38 and ERK but not by JNK. *Archives of Biochemistry and Biophysics* 334:395–400.

187. Fontayne, A., P. M. C. Dang, M. A. Gougerot-Pocidalo, and J. El Benna. 2002. Phosphorylation of p47 phox sites by PKC α, ßII, δ, and zeta: Effect on binding to p22 phox and on NADPH oxidase activation. *Biochemistry* 41 (24):7743–7750.

188. Inanami, O., J. L. Johnson, J. K. McAdara, J. El Benna, L. R. P. Faust, P. E. Newburger, and B. M. Babior. 1998. Activation of the leukocyte NADPH oxidase by phorbol ester requires the phosphorylation of p47 PHOX on serine 303 or 304. *Journal of Biological Chemistry* 273:9539–9543.

189. Dang, P. M., A. Fontayne, J. Hakim, J. El Benna, and A. Perianin. 2001. Protein kinase C zeta phosphorylates a subset of selective sites of the NADPH oxidase component p47phox and participates in formyl peptide-mediated neutrophil respiratory burst. *Journal of Immunology* 166 (2):1206–1213.

190. Chen, Q., D. W. Powell, M. J. Rane, S. Singh, W. Butt, J. B. Klein, and K. R. McLeish. 2003. Akt phosphorylates p47phox and mediates respiratory burst activity in human neutrophils. *Journal of Immunology* 170 (10):5302–5308.

191. Hoyal, C. R., A. Gutierrez, B. M. Young, S. D. Catz, J. H. Lin, P. N. Tsichlis, and B. M. Babior. 2003. Modulation of p47 phox activity by site-specific phosphorylation: Akt-dependent activation of the NADPH oxidase. *Proceedings of the National Academy of Sciences of the United States of America* 100 (9):5130–5135.

192. Park, H. S., S. M. Lee, J. H. Lee, Y. S. Kim, Y. S. Bae, and J. W. Park. 2001. Phosphorylation of the leucocyte NADPH oxidase subunit p47 phox by casein kinase 2: Conformation-dependent phosphorylation and modulation of oxidase activity. *Biochemical Journal* 358 (3):783–790.

193. Knaus, U. G., S. Morris, H. J. Dong, J. Chernoff, and G. M. Bokoch. 1995. Regulation of human leukocyte p21-activated kinases through G protein-coupled receptors. *Science* 269:221–223.

194. DeLeo, F. R., W. M. Nauseef, A. J. Jesaitis, J. B. Burritt, R. A. Clark, and M. T. Quinn. 1995. A domain of p47 phox that interacts with human neutrophil flavocytochrome b 558 *Journal of Biological Chemistry* 270 (44):26246–26251.

195. Nauseef, W. M., B. D. Volpp, S. McCormick, K. G. Leidal, and R. A. Clark. 1991. Assembly of the neutrophil respiratory burst oxidase. Protein kinase C promotes cytoskeletal and membrane association of cytosolic oxidase components. *Journal of Biological Chemistry* 266:5911–5917.

196. Heyworth, P. G., J. T. Curnutte, W. M. Nauseef, B. D. Volpp, D. W. Pearson, H. Rosen, and R. A. Clark. 1991. Neutrophil NADPH oxidase assembly. Membrane transloca-tion of p47-phox and p67-phox requires interaction between p47-phox and cytochrome b558. *Journal of Clinical Investigation* 87:352–356.

197. DeLeo, F. R., L. Yu, J. B. Burritt, L. R. Loetterle, C. W. Bond, A. J. Jesaitis, and M. T. Quinn. 1995. Mapping sites of interaction of p47-phox and flavocytochrome b with random-sequence peptide phage display libraries. *Proceedings of the National Academy of Sciences of the United States of America* 92:7110–7114.

198. Biberstine-Kinkade, K. J., L. Yu, and M. C. Dinauer. 1999. Mutagenesis of an arginine- and lysine-rich domain in the gp91 phox subunit of the phagocyte NADPH-oxidase flavocytochrome b 558. *Journal of Biological Chemistry* 274 (15):10451–10457.

199. Ahmed, S., E. Prigmore, S. Govind, C. Veryard, R. Kozma, F. B. Wientjes, A. W. Segal, and L. Lim. 1998. Cryptic Rac-binding and p21 Cdc42HS/Rac-activated kinse phosphorylation sites of NADPH oxidase component p67 phox. *Journal of Biological Chemistry* 273:15693–15701.

200. Sarfstein, R., Y. Gorzalczany, A. Mizrahi, Y. Berdichevsky, S. Molshanski-Mor, C. Weinbaum, M. Hirshberg, M. C. Dagher, and E. Pick. 2004. Dual role of rac in the assembly of NADPH oxidase: Tethering to the membrane and activation of p67 phox. *Journal of Biological Chemistry* 279:16007–16016.

201. Vieira, O. V., R. J. Botelho, L. Rameh, S. M. Brachmann, T. Matsuo, H. W. Davidson, A. Schreiber, J. M. Backer, L. C. Cantley, and S. Grinstein. 2001. Distinct roles of class I and class III phosphatidylinositol 3-kinases in phagosome formation and maturation. *Journal of Cellular Biology* 155 (1):19–25. doi: 10.1083/jcb.200107069.

202. Ellson, C. D., K. E. Anderson, G. Morgan, E. R. Chilvers, P. Lipp, L. R. Stephens, and P. T. Hawkins. 2001. Phosphatidylinositol 3-phosphate is generated in phagosomal membranes. *Current Biology* 11:1631–1635.

203. Minakami, R., Y. Maehara, S. Kamakura, O. Kumano, K. Miyano, and H. Sumimoto. 2010. Membrane phospholipid metabolism during phagocytosis in human neutrophils. *Genes Cells* 15 (5):409–424. doi: 10.1111/j.1365-2443.2010.01393.x.

204. Hawkins, P. T., L. R. Stephens, S. Suire, and M. Wilson. 2010. PI3K signaling in neutrophils. *Current Topics in Microbiological Immunology* 346:183–202. doi: 10.1007/82_2010_40.
205. Norton, L., Y. Lindsay, A. Deladeriere, T. Chessa, H. Guillou, S. Suire, J. Lucocq et al. 2016. Localizing the lipid products of PI3Kgamma in neutrophils. *Advances in Biological Regulation* 60:36–45. doi: 10.1016/j.jbior.2015.10.005.
206. Raiborg, C., K. O. Schink, and H. Stenmark. 2013. Class III phosphatidylinositol 3-kinase and its catalytic product PtdIns3P in regulation of endocytic membrane traffic. *FEBS Journal* 280 (12):2730–2742. doi: 10.1111/febs.12116.
207. Perisic, O., M. I. Wilson, D. Karathanassis, J. Bravo, M. E. Pacold, C. D. Ellson, P. T. Hawkins, L. Stephens, and R. L. Williams. 2004. The role of phosphoinositides and phosphorylation in regulation of NADPH oxidase. *Advances in Enzyme Regulation* 44:279–298. doi: 10.1016/j.advenzreg.2003.11.003.
208. Hawkins, P. T., K. E. Anderson, K. Davidson, and L. R. Stephens. 2006. Signalling through Class I PI3Ks in mammalian cells. *Biochemical Society Transactions* 34 (Pt 5):647–662. doi: 10.1042/bst0340647.
209. Vanhaesebroeck, B., J. Guillermet-Guibert, M. Graupera, and B. Bilanges. 2010. The emerging mechanisms of isoform-specific PI3K signalling. *Nature Reviews Molecular Cell Biology* 11 (5):329–341. doi: 10.1038/nrm2882.
210. Brechard, S., S. Plancon, and E. J. Tschirhart. 2013. New insights into the regulation of neutrophil NADPH oxidase activity in the phagosome: A focus on the role of lipid and Ca(2+) signaling. *Antioxidants Redox Signaling* 18 (6):661–676. doi: 10.1089/ars.2012.4773.
211. Foubert, T. R., J. B. Burritt, R. M. Taylor, and A. J. Jesaitis. 2002. Structural changes are induced in human neutrophil cytochrome b by NADPH oxidase activators, LDS, SDS, and arachidonate: Intermolecular resonance energy transfer between trisulfopyrenyl-wheat germ agglutinnin and cytochrome b 558. *Biochimica et Biophysica Acta* 78380:1–11.
212. Shao, D. M., A. W. Segal, and L. V. Dekker. 2003. Lipid rafts determine efficiency of NADPH oxidase activation in neutrophils. *FEBS Letters* 550 (1–3):101–106.
213. Jin, S., F. Zhou, F. Katirai, and P. L. Li. 2011. Lipid raft redox signaling: Molecular mechanisms in health and disease. *Antioxidant Redox Signaling* 15 (4):1043–1083. doi: 10.1089/ars.2010.3619.
214. Guthrie, L. A., L. C. McPhail, P. M. Henson, and R. B. Johnston, Jr. 1984. Priming of neutrophils for enhanced release of oxygen metabolites by bacterial lipopolysaccharide. *Journal of Experimental Medicine* 160:1656–1671.
215. Worthen, G. S., J. F. Seccombe, K. L. Clay, L. A. Guthrie, and R. B. Johnston, Jr. 1988. The priming of neutrophils by lipopolysaccharide for production of intracellular platelet-activating factor. *Journal of Immunology* 140:3553–3559.
216. El-Benna, J., M. Hurtado-Nedelec, V. Marzaioli, J. C. Marie, M. A. Gougerot-Pocidalo, and P. M. Dang. 2016. Priming of the neutrophil respiratory burst: Role in host defense and inflammation. *Immunological Review* 273 (1):180–193. doi: 10.1111/imr.12447.
217. Dang, P. M. C., A. Stensballe, T. Boussetta, H. Raad, C. Dewas, Y. Kroviarski, G. Hayem, O. N. Jensen, M. A. Gougerot-Pocidalo, and J. El-Benna. 2006. A specific p47 phox -serine phosphorylated by convergent MAPKs mediates neutrophil NADPH oxidase priming at inflammatory sites. *Journal of Clinical Investigation* 116 (7):2033–2043.
218. Makni-Maalej, K., T. Boussetta, M. Hurtado-Nedelec, S. A. Belambri, M. A. Gougerot-Pocidalo, and J. El-Benna. 2012. The TLR7/8 agonist CL097 primes N-formyl-methionyl-leucyl-phenylalanine-stimulated NADPH oxidase activation in human neutrophils: Critical role of p47phox phosphorylation and the proline isomerase Pin1. *Journal of Immunology* 189 (9):4657–4665. doi:10.4049/jimmunol.1201007.

219. Boussetta, T., M. A. Gougerot-Pocidalo, G. Hayem, S. Ciappelloni, H. Raad, R. A. Derkawi, O. Bournier et al. 2010. The prolyl isomerase Pin1 acts as a novel molecular switch for TNF-α-induced priming of the NADPH oxidase in human neutrophils. *Blood* 116 (26):5795–5802.
220. Yaffe, M. B., M. Schutkowski, M. Shen, X. Z. Zhou, P. T. Stukenberg, J. U. Rahfeld, J. Xu et al. 1997. Sequence-specific and phosphorylation-dependent proline isomerization: A potential mitotic regulatory mechanism. *Science* 278:1957–1960.
221. Liou, Y. C., X. Z. Zhou, and K. P. Lu. 2011. Prolyl isomerase Pin1 as a molecular switch to determine the fate of phosphoproteins. *Trends in Biochemical Science* 36 (10):501–514. doi: 10.1016/j.tibs.2011.07.001.
222. Brown, G. E., M. Q. Stewart, S. A. Bissonnette, A. E. H. Elia, E. Wilker, and M. B. Yaffe. 2004. Distinct ligand-dependent roles for p38 MAPK in priming and activation of the neutrophil NADPH oxidase. *Journal of Biological Chemistry* 279 (26):27059–27068.
223. Raad, H., M. H. Paclet, T. Boussetta, Y. Kroviarski, F. Morel, M. T. Quinn, M. A. Gougerot-Pocidalo, P. M. C. Dang, and J. El-Benna. 2009. Regulation of the phagocyte NADPH oxidase activity: Phosphorylation of gp91 phox/NOX2 by protein kinase C enhances its diaphorase activity and binding to Rac2, p67 phox, and p47 phox. *The FASEB Journal* 23:1011–1022.
224. Chessa, T. A. M., K. E. Anderson, Y. Hu, Q. Xu, O. Rausch, L. R. Stephens, and P. T. Hawkins. 2010. Phosphorylation of threonine 154 in p40 phox is an important physiological signal for activation of the neutrophil NADPH oxidase. *Blood* 116 (26):6027–6036.
225. Lewis, E. M., S. Sergeant, B. Ledford, N. Stull, M. C. Dinauer, and L. C. McPhail. 2010. Phosphorylation of p22 phox on theronine 147 enhances NADPH oxidase activity by promoting p47 phox binding. *Journal of Biological Chemistry* 285:2959–2967.
226. Gabig, T. G., S. I. Bearman, and B. M. Babior. 1979. Effects of oxygen tension and pH on the respiratory burst of human neutrophils. *Blood* 53 (6):1133–1139.
227. Hays, R. C. and G. L. Mandell. 1974. pO2, pH, and redox potential of experimental abscesses. *Proceedings of the Society for Experimental Biology and Medicine* 147:29–30.
228. Cross, A. R., R. W. Erickson, and J. T. Curnutte. 1999. The mechanism of activation of NADPH oxidase in the cell-free system: The activation process is primarily catalytic and not through the formation of a stoichiometric complex. *Biochemical Journal* 341 (2):251–255.
229. Koshkin, V., O. Lotan, and E. Pick. 1997. Electron transfer in the superoxide-generating NADPH oxidase complex reconstituted in vitro. *Biochimica et Biophysica Acta* 1319:139–146.
230. Cross, A. R., J. F. Parkinson, and O. T. G. Jones. 1985. Mechanism of the superoxide-producing oxidase of neutrophils. O_2 is necessary for the fast reduction of cytochrome b-245 by NADPH. *Biochemical Journal* 226:881–884.
231. DeCoursey, T. E. and E. Ligeti. 2005. Regulation and termination of NADPH oxidase activity. *Cellular and Molecular Life Sciences* 62:2173–2193.
232. Borregaard, N., J. H. Schwartz, and A. I. Tauber. 1984. Proton secretion by stimulated neutrophils. Significance of hexose monophosphate shunt activity as source of electrons and protons for the respiratory burst. *Journal of Clinical Investigation* 74 (2):455–459. doi:10.1172/jci111442.
233. Iyer, G. Y. N., D. M. F. Islam, and J. H. Quastel. 1961. Biochemical aspects of phagocytosis. *Nature* 192:535–541.
234. Zatti, M. and F. Rossi. 1965. Early changes of hexose monophosphate pathway activity and of NADPH oxidation in phagocytizing leucocytes. *Biochimica et Biophysica Acta* 99 (3):557–561.

235. Repine, J. E., J. G. White, C. C. Clawson, and B. M. Holmes. 1974. Effects of phorbol myristate acetate on the metabolism and ultrastructure of neutrophils in chronic granulomatous disease. *Journal of Clinical Investigation* 54 (1):83–90. doi: 10.1172/jci107752.

236. Cooper, M. R., L. R. DeChatelet, C. E. McCall, M. F. LaVia, C. L. Spurr, and R. L. Baehner. 1972. Complete deficiency of leukocyte glucose-6-phosphate dehydrogenase with defective bactericidal activity. *Journal of Clinical Investigation* 51 (4):769–778. doi: 10.1172/jci106871.

237. Vives Corrons, J. L., E. Feliu, M. A. Pujades, F. Cardellach, C. Rozman, A. Carreras, J. M. Jou, M. T. Vallespi, and F. J. Zuazu. 1982. Severe-glucose-6-phosphate dehydrogenase (G6PD) deficiency associated with chronic hemolytic anemia, granulocyte dysfunction, and increased susceptibility to infections: Description of a new molecular variant (G6PD Barcelona). *Blood* 59 (2):428–434.

238. Mamlok, R. J., V. Mamlok, G. C. Mills, C. W. Daeschner, 3rd, F. C. Schmalstieg, and D. C. Anderson. 1987. Glucose-6-phosphate dehydrogenase deficiency, neutrophil dysfunction and *Chromobacterium violaceum* sepsis. *Journal of Pediatrics* 111 (6 Pt 1):852–854.

239. Baehner, R. L., R. B. Johnston, Jr., and D. G. Nathan. 1972. Comparative study of the metabolic and bactericidal characteristics of severely glucose-6-phosphate dehydrogenase-deficient polymorphonuclear leukocytes and leukocytes from children with chronic granulomatous disease. *Journal of Reticuloendothelial Society* 12 (2):150–169.

240. Roos, D., R. van Zwieten, J. T. Wijnen, F. Gomez-Gallego, M. de Boer, D. Stevens, C. J. Pronk-Admiraal et al. 1999. Molecular basis and enzymatic properties of glucose 6-phosphate dehydrogenase volendam, leading to chronic nonspherocytic anemia, granulocyte dysfunction, and increased susceptibility to infections. *Blood* 94 (9):2955–2962.

241. Gray, G. R., G. Stamatoyannopoulos, S. C. Naiman, M. R. Kliman, S. J. Klebanoff, T. Austin, A. Yoshida, and G. C. Robinson. 1973. Neutrophil dysfunction, chronic granulomatous disease, and non-spherocytic haemolytic anaemia caused by complete deficiency of glucose-6-phosphate dehydrogenase. *Lancet* 2 (7828):530–534.

242. Lew, P. D., F. S. Southwick, T. P. Stossel, J. C. Whitin, E. Simons, and H. J. Cohen. 1981. A variant of chronic granulomatous disease: Deficient oxidative metabolism due to a low-affinity NADPH oxidase. *New England Journal of Medicine* 305:1329–1333.

243. DeCoursey, T. E., D. Morgan, and V. V. Cherny. 2003. The voltage dependence of NADPH oxidase reveals why phagocytes need proton channels. *Nature* 422:531–534.

244. DeCoursey, T. E. 2010. Voltage-gated proton channels find their dream job managing the respiratory burst in phagocytes. *Physiology* 25:27–40.

245. Henderson, L. M., J. B. Chappell, and O. T. G. Jones. 1987. The superoxide-generating NADPH oxidase of human neutrophils is electrogenic and associated with an H + channel. *Biochemical Journal* 246:325–329.

246. Henderson, L. M., J. B. Chappell, and O. T. Jones. 1988. Superoxide generation by the electrogenic NADPH oxidase of human neutrophils is limited by the movement of a compensating charge. *Biochemical Journal* 255 (1):285–290.

247. Henderson, L. M., J. B. Chappell, and O. T. Jones. 1988. Internal pH changes associated with the activity of NADPH oxidase of human neutrophils. Further evidence for the presence of an H+ conducting channel. *Biochemical Journal* 251 (2):563–567.

248. Meech, R. 2012. A contribution to the history of the proton channel. *Wiley Interdisciplinary Reviews: Membrane Transport and Signaling* 1 (5):533–557. doi: 10.1002/wmts.59.

249. Thomas, R. C. and R. W. Meech. 1982. Hydrogen ion currents and intracellular pH in depolarized voltage-clamped snail neurones. *Nature* 299 (5886):826–828.

250. Byerly, L., R. Meech, and W. Moody, Jr. 1984. Rapidly activating hydrogen ion currents in perfused neurones of the snail, *Lymnaea stagnalis*. *Journal of Physiology* 351:199–216.

251. DeCoursey, T. E. 2016. The intimate and controversial relationship between voltage-gated proton channels and the phagocyte NADPH oxidase. *Immunological Reviews* 273 (1):194–218. doi: 10.1111/imr.12437.

252. Okochi, Y., M. Saski, H. Iwasaki, and Y. Okamura. 2009. Voltage-gated proton channel is expressed on phagosomes. *Biochemical and Biophysical Research Communications* 382 (2):274–279.

253. DeCoursey, T. E. 2003. Voltage-gated proton channels and other proton transfer pathways. *Physiological Reviews* 83 (2):475–579.

254. Nunes, P., N. Demaurex, and M. C. Dinauer. 2013. Regulation of the NADPH oxidase and associated ion fluxes during phagocytosis. *Traffic* 14 (11):1118–1131. doi: 10.1111/tra.12115.

255. Aiken, M. L., R. G. Painter, Y. Zhou, and G. Wang. 2012. Chloride transport in functionally active phagosomes isolated from Human neutrophils. *Free Radical and Biological Medicine* 53 (12):2308–2317. doi: 10.1016/j.freeradbiomed.2012.10.542.

256. Painter, R. G., R. W. Bonvillain, V. G. Valentine, G. A. Laombard, S. G. Laplace, W. M. Nauseef, and G. Wang. 2008. The role of chloride anion and CFTR in killing of *Pseudomonas aeruginosa* by normal and CF neutrophils. *Journal of Leukocyte Biology* 83:1345–1353.

257. El Chemaly, A., Y. Okochi, M. Sasaki, S. Arnaudeau, Y. Okamura, and N. Demaurex. 2010. VSOP/Hv1 proton channels sustain calcium entry, neutrophil migration, and superoxide production by limiting cell depolarization and acidification. *Journal of Experimental Medicine* 207:129–139.

258. Demaurex, N. and A. El Chemaly. 2010. Physiological roles of voltage-gated proton channels in leukocytes. *Journal of Physiology* 588 (Pt 23):4659–4665. doi: 10.1113/jphysiol.2010.194225.

259. Okochi, Y., Y. Aratani, H. A. Adissu, N. Miyawaki, M. Sasaki, K. Suzuki, and Y. Okamura. 2016. The voltage-gated proton channel Hv1/VSOP inhibits neutrophil granule release. *Journal of Leukocyte Biology* 99 (1):7–19. doi: 10.1189/jlb.3HI0814-393R.

260. Clark, R. A. 2016. Editorial: Proton pathway paradox: Hv1 H+ channel sustains neutrophil Nox2 activity, yet suppresses HOCl formation. *Journal of Leukocyte Biology* 99 (1):1–4. doi: 10.1189/jlb.4CE0515-188RR.

261. Nathan, C. 1987. Neutrophil activation on biological surfaces: Massive secretion of hydrogen peroxide in response to products of macrophages and lymphocytes. *Journal of Clinical Investigation* 80:1550–1560.

262. Jandl, R. C., J. Andre-Schwartz, L. Borges-DuBois, R. S. Kipnes, B. J. McMurrich, and B. M. Babior. 1978. Termination of the respiratory burst in human neutrophils. *Journal of Clinical Investigation* 61:1176–1185.

263. Kimura, S. and M. Ikeda-Saito. 1988. Human myeloperoxidase and thyroid peroxidase, two enzymes with separate and distinct physiological functions, are evolutionarily related members of the same gene family. *Proteins: Structure, Function, and Genetics* 3:113–120.

264. Tamura, M., M. Takeshita, J. T. Curnutte, D. J. Uhlinger, and J. D. Lambeth. 1992. Stabilization of human neutrophil NADPH oxidase activated in a cell-free system by cytosolic proteins and by 1-ethyl-3-(3-dimethylaminopropyl) carbodiimide. *Journal of Biological Chemistry* 267:7529–7538.

265. Akard, L. P., D. English, and T. G. Gabig. 1988. Rapid deactivation of NADPH oxidase in neutrophils: Continuous replacement by newly activated enzyme sustains the respiratory burst. *Blood* 72:322–327.

266. Gillibert, M., Z. Dehry, M. Terrier, J. El Benna, and F. Lederer. 2005. Another biological effect of tosylphenylalanylchloromethane (TPCK): It prevents p47phox phosphorylation and translocation upon neutrophil stimulation. *Biochemical Journal* 386 (Pt 3):549–556. doi: 10.1042/bj20041475.

267. Chollet-Przednowed, E. and F. Lederer. 1993. Aminoacyl chloromethanes as tools to study the requirements of NADPH oxidase activation in human neutrophils. *European Journal of Biochemistry* 218:83–93.

268. van Bruggen, R., E. Anthony, M. Fernandez-Borja, and D. Roos. 2004. Continuous translocation of Rac2 and the NADPH oxidase component p67phox during phagocytosis. *Journal of Biological Chemistry* 279:9097–9102.

269. Yang, S., A. Panoskaltsis-Mortari, M. Shukla, B. R. Blazar, and I. Y. Haddad. 2002. Exuberant inflammation in nicotinamide adenine dinucleotide phosphate-oxidase-deficient mice after allogeneic marrow transplantation. *Journal of Immunity* 168:5840–5847.

270. Morgan, D., V. V. Cherny, R. Murphy, W. Xu, L. L. Thomas, and T. E. DeCoursey. 2003. Temperature dependence of NADPH oxidase in human eosinophils. *Journal of Physiology* 550 (2):447–458.

271. Dusi, S., V. Della Bianca, M. Grzeskowiak, and F. Rossi. 1993. Relationship between phosphorylation and translocation to the plasma membrane of p47phox and p67phox and activation of the NADPH oxidase in normal and Ca^{2+}-depleted human neutrophils. *Biochemical Journal* 290:173–178.

272. Heyworth, P. G. and J. A. Badwey. 1990. Continuous phosphorylation of both the 47 and the 49 kDa proteins occurs during superoxide production by neutrophils. *Biochimica et Biophysica Acta: Molecular Cell Research* 1052:299–305.

273. Ding, J. and J. A. Badwey. 1992. Effects of antagonists of protein phosphatases on superoxide release by neutrophils. *Journal of Biological Chemistry* 267:6442–6448.

274. Curnutte, J. T., R. W. Erickson, J. Ding, and J. A. Badwey. 1994. Reciprocal interactions between protein kinase C and components of the NADPH oxidase complex may regulate superoxide production by neutrophils stimulated with a phorbol ester. *Journal of Biological Chemistry* 269:10813–10819.

275. Grinstein, S., W. Furuya, D. J. Lu, and G. B. Mills. 1990. Vanadate stimulates oxygen consumption and tyrosine phosphorylation in electropermeabilized human neutrophils. *Journal of Biological Chemistry* 265 (1):318–327.

276. Bennett, P. A., P. M. Finan, R. J. Dixon, and S. Kellie. 1995. Tyrosine phosphatase antagonist-induced activation of the neutrophil NADPH oxidase: A possible role for protein kinase C. *Immunology* 85 (2):304–310.

277. Lu, D. J., A. Takai, T. L. Leto, and S. Grinstein. 1992. Modulation of neutrophil activation by okadaic acid, a protein phosphatase inhibitor. *American Journal of Physiology: Cell Physiology* 262:C39–C49.

278. Gay, J. C., K. Raddassi, A. P. Truett, 3rd, and J. J. Murray. 1997. Phosphatase activity regulates superoxide anion generation and intracellular signaling in human neutrophils. *Biochimica et Biophysica Acta* 1336 (2):243–253.

279. Bokoch, G. M. and B. A. Diebold. 2002. Current molecular models for NADPH oxidase regulation by Rac GTPase. *Blood* 100 (8):2692–2696.

280. Diebold, B. and G. Bokoch. 2001. Molecular basis for Rac2 regulation of phagocytic NADPH oxidase. *Nature Immunology* 2 (3):211–215.

281. Wölfl, J., M. C. Dagher, A. Fuchs, M. Geiszt, and E. Ligeti. 1996. In vitro activation of the NADPH oxidase by fluoride. Possible involvement of a factor activating GTP hydrolysis on Rac (Rac-GAP). *European Journal of Biochemistry* 239:369–375.

282. Molnar, G., M. C. Dagher, M. Geiszt, J. Settleman, and E. Ligeti. 2001. Role of prenylation in the interaction of Rho-family small GTPases with GTPase activating proteins. *Biochemistry* 40 (35):10542–10549.

283. Cech, P., A. Papathanassiou, G. Boreux, P. Roth, and P. A. Miescher. 1979. Hereditary myeloperoxidase deficiency. *Blood* 53:403–411.

284. Cramer, R., M. R. Soranzo, P. Dri, G. D. Rottini, M. Bramezza, S. Cirielli, and P. Patriarca. 1982. Incidence of myeloperoxidase deficiency in an area of northern Italy: Histochemical, biochemical, and functional studies. *British Journal of Haematology* 51:81–87.

285. Kitahara, M., H. J. Eyre, Y. Simonian, C. L. Atkin, and S. J. Hasstedt. 1981. Hereditary myeloperoxidase deficiency. *Blood* 57:888–893.
286. Klebanoff, S. J. and C. B. Hamon. 1972. Role of myeloperoxidase-mediated anti-microbial systems in intact leukocytes. *Journal of the Reticuloendothelial Society* 12:170–196.
287. Klebanoff, S. J. and S. H. Pincus. 1971. Hydrogen peroxide utilization in myeloperoxidase-deficient leukocytes: A possible microbicidal control mechanism. *Journal of Clinical Investigation* 50:2226–2229.
288. Nauseef, W. M., J. A. Metcalf, and R. K. Root. 1983. Role of myeloperoxidase in the respiratory burst of human neutrophils. *Blood* 61:483–491.
289. Patriarca, P., P. Dri, K. Kakinuma, F. Tedesco, and F. Rossi. 1975. Studies on the mechanism of metabolic stimulation in polymorphonuclear leucocytes during phagocytosis. I. Evidence for superoxide anion involvement in the oxidation of NADPH$_2$. *Biochimica et Biophysica Acta* 385 (2):380–386.
290. Robertson, C. F., Y. H. Thong, G. L. Hodge, and K. Cheney. 1979. Primary myeloperoxidase deficiency associated with impaired neutrophil margination and chemotaxis. *Acta Paediatrica Scandinavica* 68:915–919.
291. Rosen, H. and S. J. Klebanoff. 1976. Chemiluminescence and superoxide production by myeloperoxidase-deficient leukocytes. *Journal of Clinical Investigation* 58:50–60.
292. Stendahl, O., B. I. Coble, C. Dahlgren, J. Hed, and L. Molin. 1984. Myeloperoxidase modulates the phagocytic activity of polymorphonuclear leukocytes. Studies with cells from a myeloperoxidase-deficient patient. *Journal of Clinical Investigation* 73:366–373.
293. Silva, M. T. 2010. When two is better than one: Macrophages and neutrophils work in concert in innate immunity as complementary and cooperative partners of a myeloid phagocyte system. *Journal of Leukocyte Biology* 87:93–106.
294. Nauseef, W. M. and N. Borregaard. 2014. Neutrophils at work. *Natural Immunology* 15 (7):602–611. doi: 10.1038/ni.2921.
295. Levin, R., S. Grinstein, and J. Canton. 2016. The life cycle of phagosomes: Formation, maturation, and resolution. *Immunological Reviews* 273 (1):156–179. doi: 10.1111/imr.12439.
296. Winterbourn, C. C., A. J. Kettle, and M. B. Hampton. 2016. Reactive oxygen species and neutrophil function. *Annual Review in Biochemistry* 85:765–792.
297. Winterbourn, C. C. and A. J. Kettle. 2013. Redox reactions and microbial killing in the neutrophil phagosome. *Antioxidants Redox Signaling* 18 (6):642–660. doi: 10.1089/ars.2012.4827.
298. Lowy, F. D. 2011. How *Staphylococcus aureus* adapts to its host. *New England Journal of Medicine* 364 (21):1987–1990.
299. Staudinger, B. J., M. A. Oberdoerster, P. J. Lewis, and H. Rosen. 2002. mRNA expression profiles for *Escherichia coli* ingested by normal and phagocyte oxidase-deficient human neutrophils. *Journal of Clinical Investigation* 110 (8):1151–1163.
300. Rigby, K. M. and F. R. DeLeo. 2012. Neutrophils in innate host defense against *Staphylococcus aureus* infections. *Seminars in Immunopathology* 34:237–259.
301. Palazzolo-Ballance, A. M., M. L. Reniere, K. R. Braughton, D. E. Sturdevant, B. N. Kreiswirth, E. P. Skaar, and F. R. DeLeo. 2008. Neutrophil microbicides induce a pathogen survival response in community-associated methicillin resistant *Staphylococcus aureus* (CA-MRSA). *Journal of Immunology* 180 (1):500–509.
302. Pang, Y. Y., J. Schwartz, M. Thoendal, L. W. Ackerman, A. R. Horswill, and W. M. Nauseef. 2010. agr-Dependent interactions of *Staphylococcus aureus* USA300 with human polymorphonuclear neutrophils. *Journal of Innate Immunity* 2 (6):546–549.
303. Hurst, J. K. 2012. What really happens in the neutrophil phagosome? *Free Radical Biology and Medicine* 53 (3):508–520. doi: 10.1016/j.freeradbiomed.2012.05.008.

304. Suquet, C., J. J. Warren, N. Seth, and J. K. Hurst. 2010. Comparative study of HOCl-inflicted damage to bacterial DNA ex vivo and within cells. *Archives in Biochemistry and Biophysics* 493:135–142.
305. King, D. A., M. W. Sheafor, and J. K. Hurst. 2006. Comparative toxicities of putative phagocyte-generated oxidizing radicals toward a bacterium (*Escherichia coli*) and a yeast (*Saccharomyces cerevisiae*). *Free Radical Biology and Medicine* 41:765–774.
306. Winterbourn, C. C., M. B. Hampton, J. H. Livesey, and A. J. Kettle. 2006. Modeling the reactions of superoxide and myeloperoxidase in the neutrophil phagosome. *Journal of Biological Chemistry* 281 (52):39860–39869.
307. Nauseef, W. M. 2014. Myeloperoxidase in human neutrophil host defence. *Cell Microbiology* 16 (8):1146–1155. doi: 10.1111/cmi.12312.
308. Pattison, D. I. and M. J. Davies. 2006. Reactions of myeloperoxidase-derived oxidants with biological substrates: Gaining chemical insight into human inflammatory diseases. *Current Medicinal Chemistry* 13:3271–3290.
309. Winterbourn, C. C. and A. J. Kettle. 2000. Biomarkers of myeloperoxidase-derived hypochlorous acid. *Free Radical Biology and Medicine* 29 (5):403–409.
310. Thomas, E. L. 1979. Myeloperoxidase-hydrogen peroxide-chloride antimicrobial system: Effect of exogenous amines on antibacterial action against *Escherichia coli*. *Infection and Immunity* 25:110–116.
311. Grisham, M. B., M. M. Jefferson, D. F. Melton, and E. L. Thomas. 1984. Chlorination of endogenous amines by isolated neutrophils. Ammonia-dependent bactericidal, cytotoxic, and cytolytic activities of the chloramines. *Journal of Biological Chemistry* 259 (16):10404–10413.
312. Zgliczynski, J. M., T. Stelmaszynska, J. Domanski, and W. Ostrowski. 1971. Chloramines as intermediates of oxidative reaction of amino acids by myeloperoxidase. *Biochemica et Biophysica Acta* 235:419–424.
313. Hazen, S. L., A. d'Avignon, M. M. Anderson, F. F. Hsu, and J. W. Heinecke. 1998. Human neutrophils employ the myeloperoxidase-hydrogen peroxide-chloride system to oxidize α-amino acids to a family of reactive aldehydes. *Journal of Biological Chemistry* 273 (9):4997–5005.
314. Hazen, S. L., F. F. Hsu, A. d'Avignon, and J. W. Heinecke. 1998. Human neutrophils employ myeloperoxidase to convert α-amino acids to a battery of reactive aldehydes: A pathway for aldehyde generation at sites of inflammation. *Biochemistry* 37:6864–6873.
315. Weiss, S. J., M. B. Lampert, and S. T. Test. 1983. Long-lived oxidants generated by human neutrophils: Characterization and bioactivity. *Science* 222:625–628.
316. Coker, M. S. A., W. P. Hu, S. T. Senthilmohan, and A. Kettle. 2008. Pathways for the decay of organic dichloramines and liberation of antimicrobial chloramine gases. *Chemical Research Toxicology* 21 (12):2334–2343.
317. Ramadass, M. and S. D. Catz. 2016. Molecular mechanisms regulating secretory organelles and endosomes in neutrophils and their implications for inflammation. *Immunological Review* 273 (1):249–265. doi: 10.1111/imr.12452.
318. Weiss, J. 2003. Bactericidal/permeability-increasing protein (BPI) and lipopolysaccharide-binding protein (LBP): Structure, function and regulation in host defense against gram-negative bacteria. *Biochemical Society* 31:785–790.
319. Kettritz, R. 2016. Neutral serine proteases of neutrophils. *Immunological Review* 273 (1):232–248. doi: 10.1111/imr.12441.
320. Nauseef, W. M. 2007. How human neutrophils kill and degrade microbes: An integrated view. *Immunological Reviews* 219:88–102.
321. Voetman, A. A., R. S. Weening, M. N. Hamers, L. J. Meerhof, A. A. Bot, and D. Roos. 1981. Phagocytosing human neutrophils inactivate their own granular enzymes. *Journal of Clinical Investigation* 67 (5):1541–1549.

322. Clark, R. A., and N. Borregaard. 1985. Neutrophils autoinactivate secretory products by myeloperoxidase-catalyzed oxidation. *Blood* 65:375–381.
323. Fu, X., S. Y. Kassim, W. C. Parks, and J. W. Heinecke. 2001. Hypochlorous acid oxygenates the cysteine switch domain of pro-matrilysin (MMP-7). *Journal of Biological Chemistry* 276 (44):41279–41287.
324. Fu, X., S. Y. Kassim, W. C. Parks, and J. W. Heinecke. 2003. Hypochlorous acid generated by myeloperoxidase modifies adjacent tryptophan and glycine residues in the catalytic domain of matrix metalloproteinase-7 (matrilysin). *Journal of Biological Chemistry* 278 (31):28403–28409.
325. Hawkins, C. L. and M. J. Davies. 2005. Inactivation of protease inhibitors and lysozyme by hypochlorous acid: Role of side-chain oxidation and protein unfolding in loss of biological function. *Chemical Research Toxicology* 18:1600–1610.
326. Hirche, T. O., J. P. Gaut, J. W. Heinecke, and A. Belaaouaj. 2005. Myeloperoxidase plays critical roles in killing *Klebsiella pneumoniae* and inactivating neutrophil elastase: Effects on host defense. *Journal of Immunology* 174:1557–1565.
327. Chapman, A. L. P., M. B. Hampton, R. Senthilmohan, C. C. Winterbourn, and A. J. Kettle. 2002. Chlorination of bacterial and neutrophil proteins during phagocytosis and killing of *Staphylococcus aureus*. *Journal of Biological Chemistry* 277 (12):9757–9762.
328. Segal, A. W., R. C. Garcia, and A. M. Harper. 1983. Iodination by stimulated human neutrophils. *Biochemical Journal* 210:215–225.
329. Green, J. N., A. J. Kettle, and C. C. Winterbourn. 2014. Protein chlorination in neutrophil phagosomes and correlation with bacterial killing. *Free Radical in Biological Medicine* 77:49–56.
330. Rosen, H., S. J. Klebanoff, Y. Wang, N. Brot, J. W. Heinecke, and X. Fu. 2009. Methionine oxidation contributes to bacterial killing by the myeloperoxidase system of neutrophils. *Proceedings of the National Academic Sciences of the United States of America* 106 (44):18686–18691.
331. Rosen, H., J. R. Crowley, and J. W. Heinecke. 2002. Human neutrophils use the myeloperoxidase-hydrogen peroxide-chloride system to chlorinate but not nitrate bacterial proteins during phagocytosis. *Journal of Biological Chemistry* 277 (34):30463–30468.
332. Klebanoff, S. J., A. J. Kettle, H. Rosen, C. C. Winterbourn, and W. M. Nauseef. 2013. Myeloperoxidase: A front-line defender against phagocytosed microorganisms. *Journal of Leukocyte Biology* 93:185–198.
333. Novikoff, A. B. 1945. The concept of integrative levels and biology. *Science* 101 (2618):209–215. doi: 10.1126/science.101.2618.209.
334. Mazzocchi, F. 2008. Complexity in biology. Exceeding the limits of reductionism and determinism using complexity theory. *EMBO Reports* 9 (1):10–14. doi: 10.1038/sj.embor.7401147.
335. Corning, P. A. 2002. The re-emergence of "Emergence": A venerable concept in search of a theory. *Complexity* 7 (6):18–30.
336. Casadevall, A., F. C. Fang, and L. A. Pirofski. 2011. Microbial virulence as an emergent property: Consequences and opportunities. *PLoS Pathogens* 7 (7):e1002136. doi: 10.1371/journal.ppat.1002136.
337. Marciano, B. E., C. Spalding, A. Fitzgerald, D. Mann, T. Brown, S. Osgood, L. Yockey et al. 2015. Common severe infections in chronic granulomatous disease. *Clinical Infectious Diseases* 60 (8):1176–1183. doi: 10.1093/cid/ciu1154.
338. Klebanoff, S. 1982. The Iron-H_2O_2-iodide cytotoxic system. *Journal of Experimental Medicine* 156:1262–1267.
339. Larrocha, C., M. F. de Castro, G. Fontan, A. Vitoria, J. L. Fernandez-Chacon, and C. Jimenez. 1982. Hereditary myeloperoxidase deficiency: Study of 12 cases. *Scandinavian Journal of Haematology* 29:389–397.

340. Lehrer, R. I. and M. J. Cline. 1969. Leukocyte myeloperoxidase deficiency and disseminated candidiasis: The role of myeloperoxidase in resistance to *Candida* infection. *Journal of Clinical Investigation* 48:1478–1488.

341. Moosmann, K. and A. Bojanowsky. 1975. Rezidivierende candidiosis bei myeloperoxydasemangel. *Monatsschrift Kinderheilkde* 123:407–409.

342. Parry, M. F., R. K. Root, J. A. Metcalf, K. K. Delaney, L. S. Kaplow, and N. J. Richar. 1981. Myeloperoxidase deficiency: Prevalence and clinical significance. *Annals of Internal Medicine* 95:293–301.

343. Nauseef, W. M. 1988. Myeloperoxidase deficiency. In *Hematology/Oncology Clinics of North America*, ed. J. T. Curnutte, pp. 135–158. Philadelphia, PA: W.B. Sanders.

344. Roberts, H., P. White, I. Dias, S. McKaig, R. Veeramachaneni, N. Thakker, M. Grant, and I. Chapple. 2016. Characterization of neutrophil function in Papillon-Lefevre syndrome. *Journal of Leukocyte Biology* 100 (2):433–444. doi: 10.1189/jlb.5A1015-489R.

345. Sorensen, O. E., S. N. Clemmensen, S. L. Dahl, O. Ostergaard, N. H. Heegaard, A. Glenthoj, F. C. Nielsen, and N. Borregaard. 2014. Papillon-Lefevre syndrome patient reveals species-dependent requirements for neutrophil defenses. *Journal of Clinical Investigation* 124 (10):4539–4548. doi: 10.1172/jci76009.

346. Morgenstern, D. E., M. A. C. Gifford, L. L. Li, C. M. Doerschuk, and M. C. Dinauer. 1997. Absence of respiratory burst in X-linked chronic granulomatous disease mice leads to abnormalities in both host defense and inflammatory response to *Aspergillus fumigatus*. *Journal of Experimental Medicine* 185 (2):207–218.

347. Kuijpers, T. and R. Lutter. 2012. Inflammation and repeated infections in CGD: Two sides of a coin. *Cellular and Molecular Life Sciences* 69:7–15.

348. Schäppi, M. G., V. Jaquet, D. C. Belli, and K. H. Krause. 2008. Hyperinflammation in chronic granulomatous disease and anti-inflammatory role of the phagocyte NADPH oxidase. *Seminars in Immunopathology* 30:255–271.

349. Whitmore, L. C., K. L. Goss, E. A. Newell, B. M. Hilkin, J. S. Hook, and J. G. Moreland. 2014. NOX2 protects against progressive lung injury and multiple organ dysfunction syndrome. *American Journal of Physiology Lung Cellular and Molecular Physiology* 307 (1):L71–L82. doi: 10.1152/ajplung.00054.2014.

350. Zhang, W. J., H. Wei, and B. Frei. 2009. Genetic deficiency of NADPH oxidase does not diminish, but rather enhances, LPS-induced acute inflammatory responses in vivo. *Free Radical Biology and Medicine* 46 (6):791–798. doi: 10.1016/j.freeradbiomed.2008.12.003.

351. Segal, B. H., M. J. Grimm, A. N. H. Khan, W. Han, and T. S. Blackwell. 2012. Regulation of innate immunity by NADPH oxidase *Free Radical Biology and Medicine* 53 (1):72–80.

352. Zeng, M. Y., D. Pham, J. Bagaitkar, J. Liu, K. Otero, M. Shan, T. A. Wynn et al. 2013. An efferocytosis-induced, IL-4-dependent macrophage-iNKT cell circuit suppresses sterile inflammation and is defective in murine CGD. *Blood* 121 (17):3473–3483. doi: 10.1182/blood-2012-10-461913.

353. Endo, D., K. Fujimoto, R. Hirose, H. Yamanaka, M. Homme, K. I. Ishibashi, N. Miura, N. Ohno, and Y. Aratani. 2016. Genetic phagocyte NADPH oxidase deficiency enhances nonviable *Candida albicans*-induced inflammation in mouse lungs. *Inflammation* 40:123–135.

354. Segal, B. H., W. Han, J. J. Bushey, M. Joo, Z. Bhatti, J. Feminella, C. G. Dennis et al. 2010. NADPH oxidase limits innate immune responses in the lungs in mice. *PLoS ONE* 5 (3):e9631.

355. Petersen, J. E., T. S. Hiran, W. S. Goebel, C. Johnson, R. C. Murphy, F. H. Azmi, A. F. Hood, J. B. Travers, and M. C. Dinauer. 2002. Enhanced cutaneous inflammatory reactions to *Aspergillus fumigatus* in a murine model of chronic granulomatous disease. *Journal of Investigative Dermatology* 118 (3):424–429.

356. Rajakariar, R., J. Newson, E. Jackson, P. Sawmynaden, A. Smith, F. Rahman, M. M. Yaqoob, and D. W. Gilroy. 2009. Nonrevolving inflammation in gp91 phox-I-mice, a model of human chronic granulomatous disease, has lower adenosine and cyclic adenosine 5'-monophosphate. *Journal of Immunology* 182 (5):3262–3269.

357. Deffert, C., S. Carnesecchi, H. Yuan, A. L. Rougemont, T. Kelkka, R. Holmdahl, K. H. Krause, and M. Schäppi. 2012. Hyperinflammation of chronic granulomatous disease is abolished by NOX2 reconstitution in macrophages and dendritic cells. *Journal of Pathology* 228:341–350.

358. Wang, Y., H. Rosen, D. K. Madtes, B. Shao, T. R. Martin, J. W. Heinecke, and X. Fu. 2007. Myeloperoxidase inactivates TIMP-1 by oxidizing its N-terminal cysteine residue: An oxidative mechanism for regulating proteolysis during inflammation. *Journal of Biological Chemistry* 282 (44):31826–31834.

359. Sareila, O., T. Kelkka, A. Pizzolla, M. Hultqvist, and R. Holmdahl. 2011. NOX2 complex-derived ROS as immune regulators. *Antioxidants Redox Signaling* 15 (8):2197–2208. doi: 10.1089/ars.2010.3635.

360. Jaeschke, H. 2011. Reactive oxygen and mechanisms of inflammatory liver injury: Present concepts. *Journal of Gastroenterological Hepatology* 26 (Suppl 1):173–179. doi: 10.1111/j.1440-1746.2010.06592.x.

361. Femling, J. K., W. M. Nauseef, and J. P. Weiss. 2005. Synergy between extracellular Group IIA phospholipase A2 and phagocyte NADPH oxidase in digestion of phospholipids of *Staphylococcus aureus* ingested by human neutrophils. *Journal of Immunology* 175 (7):4653–4661.

Section III

Myeloperoxidase and Derived Oxidants

11 Myeloperoxidase
Unleashing the Power of Hydrogen Peroxide

Louisa V. Forbes and Anthony J. Kettle

CONTENTS

INTRODUCTION

Christine Winterbourn's visit to London in 1984 sparked a new research interest for her that eventually became a major focus of her work on free radicals in biology and medicine. At that time, Christine was on sabbatical leave in Tony Segal's laboratory at University College London. She was intrigued by how free radicals may contribute to host defense. Tony suggested she investigate how myeloperoxidase—an abundant antimicrobial enzyme of neutrophils—transforms hydrogen peroxide and superoxide into reactive species that kill microorganisms. Myeloperoxidase was then virtually neglected as a major player in oxidant production by neutrophils. Most researchers in the free radical field were engrossed in the reactions of superoxide—recently discovered by Bernard Babior and his team to be produced by the NADPH oxidase of neutrophils [1]. Seymour Klebanoff was one of the few scientists extolling the capabilities of myeloperoxidase and how it uses hydrogen peroxide to oxidize ubiquitous chloride to hypochlorous acid—a lethal oxidant that kills all pathogens [2]. Christine initially investigated the redox transformations of myeloperoxidase that occur when it reacts with hydrogen peroxide and superoxide [3]. She found that, depending on the concentration of hydrogen peroxide, the enzyme could act either as a producer of hypochlorous acid or a catalase. Since that time, Christine has published over a hundred papers involving myeloperoxidase. With her team, she has demonstrated that when neutrophils phagocytose bacteria, myeloperoxidase consumes the majority of superoxide and hydrogen they generate to produce hypochlorous acid, which kills ingested bacteria [4]. She has also advanced our understanding of how myeloperoxidase uses hydrogen peroxide to promote inflammatory tissue damage.

Neutrophils, the most abundant white blood cell in humans, must generate hydrogen peroxide to kill many of the different pathogens they ingest during infections [5]. However, hydrogen peroxide itself is a weak antimicrobial agent, even though it is a thermodynamically strong oxidant. It is an ineffectual oxidant of biological molecules because of the high activation energies for most of its reactions [6]. Myeloperoxidase lowers these barriers by transforming hydrogen peroxide into several kinetically reactive species. It is a classical heme peroxidase that catalyzes the oxidation of numerous organic substrates to free radicals [7]. During oxidation of these substrates, the enzyme cycles through three redox intermediates—the native ferric heme iron species, and compound I and compound II, which are high-valent iron species [8]. However, what sets myeloperoxidase apart from other mammalian enzymes is the exceptional ability of compound I to oxidize chloride to chlorine bleach or hypochlorous acid—a strong oxidant that kills all bacteria and is toxic to human cells [9]. In this chapter, we will focus on how myeloperoxidase unleashes the oxidation potential of hydrogen peroxide and exploits it to produce oxidants that are highly toxic to microorganisms and human cells.

STRUCTURE OF MYELOPEROXIDASE

Myeloperoxidase is a strikingly beautiful dark green heme protein. At 146 kDa in size, it is composed of two identical heterodimers containing heavy (58.5 kDa) and light (14.5 kDa) subunits. Cysteine residues on the heavy subunits link the heterodimers. The three dimensional structure of myeloperoxidase has been determined at high resolution and provides useful insights into how it exploits the oxidation potential of hydrogen peroxide (Figure 11.1). The secondary structure is predominantly α-helical [10]. Each identical half consists of a core of five helices—four from the larger subunit plus one from the small—and a covalently bound heme in the heavy subunit. The cavities containing the heme prosthetic groups are on the same side of the protein and about 40 Å apart. There are five N-glycans on asparagine residues at positions 323, 355, 391, 483, and 729 on each heavy subunit [11]. The majority of the glycosylated residues are on the side opposite to the heme cavities and located in the interface between the dimers. Deglycosylation, especially of Asn355, lowers enzyme activity. Each heterodimer contains a tightly bound calcium ion that is ligated to the heavy and light subunits [12]. It maintains the stability of the heterodimer but may also affect redox properties of the enzyme because it is coordinated to Asp96, which is close to His95—the distal histidine that activates hydrogen peroxide for reaction with the heme (see Figure 11.4).

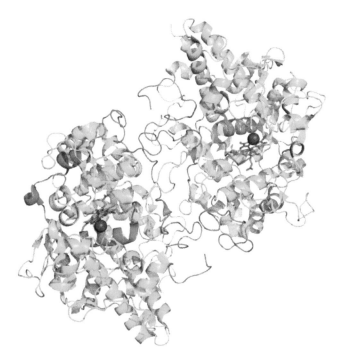

FIGURE 11.1   The structure of myeloperoxidase. The crystal structure of myeloperoxidase showing the heavy (green) and light (blue) subunits plus the heme prosthetic groups (orange) with a bromide (red) bound in the distal pockets. The structure was drawn with Pymol using data from 1CXP in the Protein Data Bank.

The two cavities containing the heme prosthetic groups have identical structures. There is a large opening at the entrance of each cavity that is partially covered on one side by the C-terminal residue (Ala104) on the small subunits and a loop containing His216 on the heavy subunits (Figure 11.2a) [10]. These residues would restrict direct entry of substrates into the active site. However, there is relatively unrestricted access on the opposite side. From here, the cavity narrows down to a channel that extends into the core of the protein. There is a large hydrophobic region just beyond this narrowing. The heme prosthetic group is located within the channel with its D pyrrole ring and propionate group poking out of the entrance (Figure 11.2d). The B pyrrole ring is at the back of the channel. The architecture of the substrate channel of myeloperoxidase allows numerous potential substrates to interact with the catalytic heme. The active site is more restricted than that of horseradish peroxidase (Figure 11.2b), but much more accessible than that of catalase (Figure 11.2c) [13,14]. There is no evidence to date that the two active sites interact during catalysis.

The structure of the heme groups and their interactions with both proximal and distal residues gives myeloperoxidase its unique redox properties [15]. Each heme is an uncommon derivative of protoporphyrin IX where the methyl groups on the A and C pyrrole rings are covalently linked to the protein (Figure 11.2d). The A pyrrole ring is attached to the heavy subunit via an ester bond with Glu242 and a positively

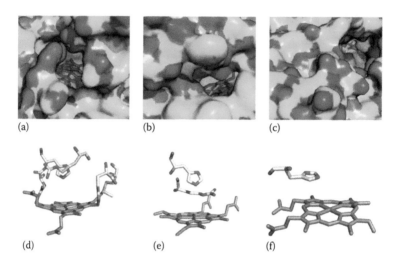

FIGURE 11.2 Residues in the distal site for myeloperoxidase (a, d) and horseradish peroxidase (b, e) are histidine and arginine, while just histidine is shown for catalase (c, f). Heme prosthetic groups are shown in orange. The structure were drawn with Pymol using data from 1CXP, 1HCH, and 2IQF in the Protein data bank.

charged vinyl-sulfonium bond to Met243. The C pyrrole ring is linked to the light subunit via a second ester bond with Asp94. These bonds disrupt the planarity of the heme. When viewed across the distal half of the heme pocket, the heme is noticeably bent with the edges of pyrrole rings A and C bowed upward (Figure 11.2d). The heme iron sits slightly below the bottom of the heme curvature and is coordinated to His336 on its proximal side [10]. His336 forms a triad with Asn421 and Arg333 that link the heme iron with the propionate group of the D pyrrole ring. Disruption of this proximal triad lowers enzyme activity and alters the spectral properties of the protein [16,17]. Important residues on the distal side of the heme that are most likely to be involved in catalysis include His95, Arg239, and Gln91. These residues, along with the propionate group on pyrrole ring C, are hydrogen bonded to four of the five water molecules that are located in the distal pocket [10]. The heme group (Figure 11.2d) is considerably different from those of horseradish peroxidase (Figure 11.2e) and catalase (Figure 11.2f), but the perpendicular orientation of histidine and location of the arginine in the distal space resemble that of the plant peroxidase [14].

MYELOPEROXIDASE LIBERATES THE OXIDATION POTENTIAL OF HYDROGEN PEROXIDE

Myeloperoxidase is not a typical enzyme. In contrast to textbook examples, it has multiple substrates and several activities. However, the common feature to all the reactions myeloperoxidase catalyzes is that hydrogen peroxide provides the thermodynamic driving force to generate kinetically reactive oxygen and halogen species. Hydrogen peroxide is either used directly by the enzyme or via the ability of

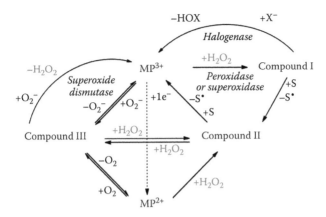

FIGURE 11.3 The established reactions catalyzed by myeloperoxidase. X$^-$ is chloride, bromide, iodide, or thiocyanate, which are oxidized to hypohalous acids (HOX) when the enzyme acts as a *halogenase*. S is for organic substrates, such as serotonin, urate, tyrosine, and ascorbate, when myeloperoxidase acts as a classical *peroxidase* to produce free radical S$^•$, or S is superoxide when myeloperoxidase acts as a *superoxidase*. S can also be hydrogen peroxide when it reacts with compound I to form compound II and superoxide.

myeloperoxidase to dismutate superoxide. The chief physiological activity is oxidation of halides, predominantly chloride and the pseudohalide thiocyanate, to hypohalous acids (Figure 11.3) [18,19]. It is also an efficient classical peroxidase that generates free radicals from myriad substrates (Figure 11.3). In fact, it is a much more powerful classical peroxidase than plant and other mammalian peroxidases [20]. We have shown that myeloperoxidase also uses superoxide to hydroxylate aromatic compounds in a reaction similar to that of cytochrome P450 [21]. Superoxide should indeed be viewed as a major physiological substrate for myeloperoxidase. The propinquity of superoxide and myeloperoxidase in time and space, plus their rapid reactions (Figure 11.3), ensures an intimate interaction that neutrophils must exploit to defend us against pathogens. Finally, myeloperoxidase also has catalase activity [22,23]. That is, it degrades hydrogen peroxide to oxygen and water without the need for other substrates. The mechanism is contentious and has yet to be satisfactorily clarified. We believe the details of this mechanism need to be defined because they may alter our perspective on the physiological functions of myeloperoxidase.

ACTIVATION OF HYDROGEN PEROXIDE VIA COMPOUND I

The oxidation potential of hydrogen peroxide is harnessed by myeloperoxidase when it reacts with hydrogen peroxide to form compound I. This reaction is fast ($k = 2 \times 10^7$ M^{-1} s^{-1}). In the process, two electrons are removed from the heme prosthetic group and an oxygen atom is added to its iron to form compound I (Figure 11.4) [24,25]. Water is released as a by-product. The oxygen bound to the heme iron in compound I contains only six valence electrons, making it a potent oxidant. Compound I is then poised to unleash the oxidation potential of hydrogen

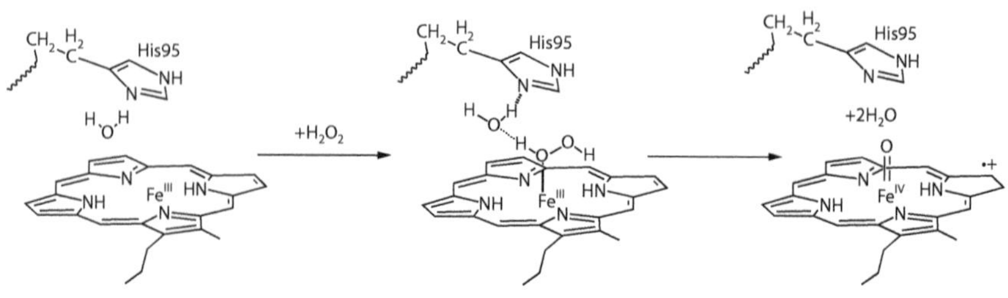

FIGURE 11.4 Proposed formation of myeloperoxidase compound I. Based on the analogous reaction with horseradish peroxidase, hydrogen peroxide becomes associated with the heme iron through a hydrogen bond relay with a water molecule and histidine 95 and rapidly rearranges to compound I with an iron (IV) heme and a π-cation radical delocalized over the porphyrin.

peroxide by channeling the enzyme into either the halogenation or peroxidation cycles (Figure 11.3). As with other peroxidases, it is likely that the reaction of hydrogen peroxide with ferric myeloperoxidase is facilitated by His95 and Arg239 in the distal heme cavity (Figure 11.4). The distal histidine, via hydrogen bonding with a nearby water, could act as an acid–base catalyst that promotes heterolytic cleavage of hydrogen peroxide [10]. At first, the proton would be removed from the oxygen atom that becomes bound to the ferric iron, then transfer the proton to the other oxygen. By swapping the proton from one oxygen atom to the other, it would facilitate the heterolytic cleavage of the O–O bond. The distal arginine (see Figure 11.2d) could decrease the pK_a of His95 and also stabilize the developing negative charge on the leaving hydroxide anion during bond rupture. It would play an ancillary role by stabilizing the resultant oxyferryl center of compound I. In myeloperoxidase, the distal histidine is also hydrogen bonded to a buried water molecule, which is linked by His250 to a series of four other water molecules that form a chain of hydrogen bonds extending to the surface of the protein. This chain may function as a conduit for transfer of protons away from the distal histidine and enable it to accept a proton from hydrogen peroxide.

Similar to other peroxidases [14], formation of compound I may be preceded by an initial binding of hydrogen peroxide onto the ferric enzyme to form compound 0 (see Figure 11.4). Whether myeloperoxidase compound 0 exists as a fleeting redox intermediate has yet to be investigated. Compound I is often described as a π-cation radical in which heme iron is present as ferryl iron (IV) and a cation radical is delocalized on the heme (Figure 11.4). However, this electron configuration is based on that for horseradish peroxidase and no study has verified the presence of a π-cation radical in myeloperoxidase compound I [26]. Nevertheless, compound I has high reduction potentials and oxidizes numerous substrates by removing either one ($E^{0'}$ vs NHE pH 7.0 Compound I/Compound II = 1.36 V) or two electrons ($E^{0'}$ vs NHE pH 7.0 Compound I/Compound II = 1.16 V) [27]. In the absence of reducing substrates, compound I rapidly and irreversibly inactivates the enzyme [28], presumably via transfer of the π-cation radical on the heme group to a nearby amino acid residue on the protein.

HALOGENATION ACTIVITY OF MYELOPEROXIDASE

Compound I removes two electrons from chloride, bromide, iodide, and thiocyanate to produce their respective hypohalous acids and regenerate the ferric enzyme (Figure 11.3) [19,29,30]. Crystallographic studies support a model in which chloride, bromide, and thiocyanate occupy the distal heme cavity binding site [10,31]. Compound I is highly unstable and cannot be studied by x-ray crystallography. However, the cyanide complex with ferric myeloperoxidase has been proposed to be a useful model for compound I because its electron paramagnetic resonance spectrum is affected by halides [32]. Crystal structures show that cyanide interferes with the binding of bromide but has little effect on the location of thiocyanate [10,31]. From these structures, it is apparent that the halides and thiocyanate are favorably positioned to enable them to transfer electrons to the heme and accept oxygen from compound I to yield hypohalous acids as products. The halide binding site is close to the bridge between the A and D pyrrole rings, which may be electron deficient due to the adjacent and positively charged sulfonium ion that links Met243 to the heme. Thus, the halides may transfer two electrons to an electron-deficient heme π-cation radical of compound I, followed by the incorporation of oxyferryl oxygen into the hypohalous acid product.

The rate constant for oxidation of halides is inversely related to their two-electron reduction potential [29,30]. The similar rate constants for thiocyanate and iodide, despite the lower reduction potential for iodide, suggest that iodide is too big to fit favorably into the halide binding site. At physiological concentrations, chloride and thiocyanate are the preferred substrates for compound I because rates of oxidation are determined by both the rate constant and substrate concentration [29,30]. In physiologically relevant mixtures of halides, bromide is also oxidized to a limited degree by the purified enzyme or isolated neutrophils [33–35]. The concentration of iodide is too low in biological fluids for it to be a substrate.

It has been proposed that chloride reacts with compound I to form an enzyme-bound chlorinating intermediate that oxidizes taurine to taurine chloramine [24]. This proposal is supported by kinetic analysis of the reaction of chloride with compound I over a range of chloride concentrations [36]. In addition, myeloperoxidase selectively chlorinated the N-terminal amine of apolipoprotein B-100, implying that this residue must enter the catalytic site and react with the putative chlorinating intermediate [37]. The existence of a chlorinating intermediate, however, was dismissed when the myeloperoxidase-generated oxidant was found to have similar reactivity to free hypochlorous acid [38]. More recent studies suggest that depending on accessibility of the substrate to the active site, myeloperoxidase may either release free hypochlorous acid or chlorinate it directly [39]. Further investigation is required to establish the functional relevance of a chlorinating intermediate in the catalytic reactions of myeloperoxidase.

Production of hypochlorous acid by compound I is a reversible reaction because ferric myeloperoxidase is rapidly oxidized by hypochlorous acid to produce compound I and chloride ($k = 2 \times 10^8$ M^{-1} s^{-1}) [40]. Reaction of hypochlorous acid with the enzyme's heme group is accompanied by chlorination of protein residues and irreversible inactivation [41]. Oxidation of myeloperoxidase by hypochlorous acid is likely to be a minor reaction because hypochlorous acid released from the enzyme will be rapidly scavenged by multiple other reactive targets. However, it may be relevant to inactivation of the enzyme when local reducing species have been depleted, and provide a mechanism for curtailing unwanted oxidative activity by myeloperoxidase. Peroxynitrous acid (HOONO) also oxidizes ferric myeloperoxidase with a rate comparable to that of hydrogen peroxide. In this case, however, the end product is compound II without observable formation of compound I [42]. Under some conditions this reaction can be expected to divert the enzyme from its normal activity and inhibit halogenation.

Chloride binds reversibly to ferric myeloperoxidase at physiologically relevant concentrations having a dissociation constant of 0.5 M at neutral pH [43]. Binding is stronger at acidic pH where chloride competes with hydrogen peroxide and inhibits production of hypochlorous acid [43,44]. Chloride most likely inhibits reactions of substrates with ferric myeloperoxidase when it binds to the substrate pocket within the distal heme cavity and thereby prevents other substrates from reacting with the heme iron [10]. Nitric oxide also binds reversibly to ferric myeloperoxidase (k_{on} = 1.1 × 10^6 M^{-1} s^{-1}; k_{off} 10.8 s^{-1}) and modulates chlorination activity [45].

PEROXIDATION ACTIVITY OF MYELOPEROXIDASE

Myeloperoxidase functions as a classical peroxidase when compound I oxidizes substrates by removing a single electron to produce compound II and a substrate free radical (Figure 11.3). Compound II—an oxoferryl complex—reacts with a second substrate molecule to produce another radical and regenerate the ferric enzyme. Physiological substrates that are readily oxidized by the classical peroxidase cycle include superoxide [46], serotonin [47], urate [48], tyrosine [49], ascorbate [50], and hydrogen sulfide [51]. Although nitrite is readily oxidized by compound I [52], it is a relatively poor substrate for compound II and requires the presence of cosubstrates to promote its oxidation via turnover of compound II [53]. Numerous xenobiotics are also oxidized via the classical peroxidation cycle including anti-inflammatory and anticancer drugs as well as dietary polyphenols [7].

The high one-electron reduction potential of compound I allows it to oxidize a wide variety of substrates. In fact, the magnitude of its reduction potential is so high that the rate constants for its reduction by substrates are largely independent of the substrate reduction potential. However, when a series of benzoic acid hydrazides was investigated as substrates for compound I and compound II, it was found that their rate constants for reduction of the redox intermediates were dependent on the electron-donating capacity of substituents on the aromatic ring [54]. This suggests that the reduction potential of the substrate influences the ease of their oxidation by compound I and compound II. The effect was greater for compound II than compound I. With compound I, rates of reaction were influenced more by the structure and charge of the substrate. For example, the aromatic nature of tryptophan, tyrosine, and serotonin is more important for oxidation by compound I than the reducing capabilities of urate and ascorbate. Hence, interaction of the aromatic rings with the heme must favor oxidation. Negatively charged substrates hinder oxidation by both compound I and compound II. For example, comparison of rate constants for aliphatic thiols indicates that the absence of a carboxylic acid group increases the ease of oxidation by several orders of magnitude [55]. It is likely that negatively charged substrates are repelled by the propionate group on the D pyrrole ring of the heme prosthetic group.

It has been proposed that substrates for horseradish peroxidase transfer a single electron to the heme periphery [56]. An analogous mechanism is likely to operate with for myeloperoxidase. Based on the crystal structure of hydroxamates bound in the entrance of the heme cavity [57], substrates are ideally positioned above and parallel to the plane of the D pyrrole ring to transfer electrons to the electron-deficient locus of the heme group. This proposal is supported by the covalent binding of 2-thioxanthines to the methyl group of the D pyrrole ring, which protrudes out of the heme cavity. 2-Thioxanthines are suicide substrates that are oxidized to radicals that subsequently oxidize the heme and become covalently attached [58]. Electron transfer most likely occurs at the heme edge, where the substrate initially binds, and before the product radical can leave the active site.

In contrast to compound I, compound II has a more restricted substrate preference due to its lower reduction potential and possibly a narrower substrate channel [59,60]. As a consequence, most substrates react up to several orders of magnitude slower with compound II than with compound I. Therefore, reduction of compound II is the rate determining step and controls the rate at which the enzyme turns over in the classical peroxidation cycle. A substrate's reduction potential, structure, and charge all influence the rate at which it is oxidized by compound II. Serotonin and superoxide are the best physiological substrates for compound II.

Superoxide is as a physiological substrate for compound I and compound II; reacting rapidly with these redox intermediates (Figure 11.3) [61,62]. Thus, myeloperoxidase can be viewed as a superoxidase that uses hydrogen peroxide to oxidize superoxide. The product is presumed to be molecular oxygen but formation of singlet oxygen is thermodynamically feasible [63]. Oxidation of superoxide is inhibited by chloride, which at physiological concentrations outcompetes superoxide for reaction with compound I [23]. The superoxidase activity could be relevant during bacterial killing and may account for singlet oxygen detected within neutrophil phagosomes [64].

Superoxide Dismutase Activity of Myeloperoxidase

Even though the reaction of hydrogen peroxide with the enzyme is highly favorable, superoxide should also be regarded as a principal physiological substrate for native ferric myeloperoxidase [46,62]. Superoxide is the primary species formed by the NADPH oxidase in neutrophil phagosomes and should react with ferric myeloperoxidase before any hydrogen peroxide is produced [4]. Superoxide reacts rapidly with ferric myeloperoxidase to form oxymyeloperoxidase or compound III ($k = 2.0 \times 10^6$ M^{-1} s^{-1}) [65], and diverts the enzyme from its main oxidative cycles (Figure 11.3). Thus, superoxide and hydrogen peroxide will compete for reaction with ferric myeloperoxidase and, as a result, formation of compound III will inhibit both the chlorination and peroxidation activity of the enzyme [46].

Compound III can be represented as the electronically equivalent structures of superoxide bound to ferric iron or molecular oxygen bound to ferrous iron. Its main physiological substrate appears to be superoxide which reduces it in a favorable reaction ($k = 1.3 \times 10^5$ M^{-1} s^{-1}) to either compound I or ferric enzyme plus hydrogen peroxide [62]. Thus, superoxide reacts with both ferric myeloperoxidase and compound III. The products are hydrogen peroxide and molecular oxygen (Figure 11.3). Consequently, myeloperoxidase has superoxide dismutase activity [62,66]. Experimental evidence for the cycle comes from the finding that superoxide generating systems are able to convert myeloperoxidase to a maximum of 90% compound III at neutral pH, indicating that superoxide must react with compound III to maintain its turnover [62,67]. Myeloperoxidase is much less efficient than superoxide dismutases at dismutating superoxide. However, when present at high concentrations, such as in neutrophil phagosomes, it could dismutate superoxide to hydrogen peroxide.

Compound III is also reduced by ascorbate ($k = 4 \times 10^2$ M^{-1} s^{-1}) [67] and serotonin [68,69], and is involved in the hydroxylation of salicylate [21] and oxidation of melatonin [70]. Benzoquinone reacts readily with compound III to form ferric myeloperoxidase and presumably benzosemiquinone and oxygen [71]. In the absence of reducing agents, compound III decays slowly to ferric myeloperoxidase [3].

CATALASE ACTIVITY OF MYELOPEROXIDASE

Christine, in her early collaboration with Tony Segal, was the first to show that myeloperoxidase could degrade hydrogen peroxide to oxygen and water, and consequently suggested that myeloperoxidase could function as a catalase [3]. Soon after, Hiroyuki Iwamoto and colleagues at Kyoto University investigated the kinetics of hydrogen peroxide loss and oxygen liberation by myeloperoxidase. They proposed that the enzyme had true catalase activity [22,23]. That is, hydrogen peroxide reacts with both the ferric enzyme and compound I in a ping-pong mechanism to produce oxygen and water. The thermodynamics for two-electron reduction of compound I ($E^{0'}$ Compound I/MP^{3+} = 1.16 V) [27] by hydrogen peroxide are highly favorable ($E^{0'}$ H_2O_2/O_2 = 0.28 V) [63]. Indeed, if compound I can pull two electrons from chloride, oxidation of hydrogen peroxide to oxygen should occur with relative ease. Subsequently, Christine's team investigated the reaction of hydrogen peroxide with myeloperoxidase in more detail and agreed with the proposal of the Japanese researchers [23]. They concluded that myeloperoxidase has true catalase activity because hydrogen peroxide consumption and production of molecular oxygen mirrored each other and had a distinct burst-phase that slowed to a steady-state rate as the enzyme was diverted to compound II [23]. Furthermore, substrates that prevented accumulation of compound II, such as tyrosine and superoxide, enhanced the conversion of hydrogen peroxide to molecular oxygen.

The proposal that compound I reacts with hydrogen peroxide as a true catalase to produce oxygen met with considerable skepticism [8,72]. When Marquez and Dunford investigated the pre-steady-state kinetics for reaction of hydrogen peroxide with ferric myeloperoxidase, they found that at least a 10-fold excess of hydrogen peroxide over the enzyme was required to convert the enzyme predominantly, but not completely, to compound I [25]. In contrast, a one to one ratio is sufficient to form compound I of HRP or catalase [14]. The large excess of hydrogen peroxide suggests that formation of compound I is reversible. Indeed, plots of the concentration of hydrogen peroxide against the observed rate constant for the reaction had a positive intercept, which is indicative of a reversible reaction. They found hydrogen peroxide reduced compound I to compound II in a relatively slow reaction ($k = 8 \times 10^4$ M^{-1} s^{-1}) [25]. There was no spectral evidence that hydrogen peroxide reduced compound I to the ferric enzyme. These findings were confirmed by Obinger's group [30]. Marquez and Dunford proposed that compound I ($E^{0'}$ compound I/MP^{3+} = 1.16 V) [27] is reduced back to ferric myeloperoxidase by water ($E^{0'}$ H_2O/H_2O_2; 1.35 V) [25,63]. But the unfavorable thermodynamics suggests that this reaction is unlikely to explain the reversibility of compound I formation.

The skepticism leveled at true catalase activity for hydrogen peroxide breakdown is understandable because the mechanism fails to take into account the relative difficulty of forming compound I [72,73]. However, the detractors of the true catalase activity have failed to acknowledge that myeloperoxidase does indeed degrade hydrogen peroxide to oxygen and water in a rapid reaction that precedes formation of compound II. Oxygen production cannot be explained by other much slower reactions of hydrogen peroxide with compound II, compound III and the ferrous enzyme (Figure 11.3). Consequently, the apparently simple reaction of hydrogen peroxide with myeloperoxidase is at a mechanistic impasse and needs resolution.

INFLUENCE OF pH

In vivo, myeloperoxidase may operate anywhere from pH 6–9. It is important, therefore, to appreciate how pH affects the different reactions of the enzyme. pH has a major effect on myeloperoxidase activity by altering the charge on His95 in the distal heme cavity (see Figure 11.4). Formation of compound I is unaffected in the range of pH 5–9 but is inhibited below pH 5, suggesting that His95 has a pK_a of approx. 4.5–4.9 [24,25,73]. As described, when His95 is uncharged, it can form a hydrogen bond relay that helps to orientate hydrogen peroxide correctly and optimize its reaction with the ferric enzyme and form compound I. In contrast, reaction of several substrates with compound I, including the halides, requires that His953 is protonated [30,52,74]. Optimal oxidation of these substrates occurs below pH 5 and decreases with increasing pH. Presumably the negatively charged substrates need to interact with the positively charged His95 to favor their oxidation. Interestingly, reduction of compound I by hydrogen peroxide to form compound II has a similar pH profile to the formation of compound I [25,75]. That is, the reaction is independent of pH above pH 5 but inhibited below this pH.

Formation of compound I and its subsequent reactions with halides and hydrogen peroxide determines the pH optimum for hypohalous acid production. For example, when the steady-state concentration of hydrogen peroxide is low and chloride is present at physiological concentrations of 100–140 mM, then the formation of compound I will be rate determining and the enzyme will function optimally anywhere from pH 5–9 or above [44,76]. Conversely, at high concentrations of hydrogen peroxide, reactions of compound I will be rate determining. These reactions include reduction of compound I by halides and hydrogen peroxide. Increasing acidity will favor reaction with chloride and inhibit formation of compound II. Hence, under these conditions, the optimal pH will be around pH 5.0. These contrasting pH optima were observed when enzyme activity was monitored with a hydrogen peroxide electrode at varying initial concentrations of hydrogen peroxide [76].

Introduction of peroxidase substrates adds complexity to the observed pH optimum because enzyme turnover becomes dependent on the rate of reduction of compound II. For most substrates, such as ascorbate, nitrite, and urate, plus nonphysiological substrates including 3,3′,5,5′-tetramethylbenzidine and monochlorodimedon, the reaction is optimal at low pH [48,52,77–79]. As with the halides, it is likely that His95 needs to be protonated to favor their reaction with the heme center. There are, however, exceptions to this rule. The rates of oxidation of phenols increase with increasing pH [47,49,80]. This is because their phenolate forms (pKa > 10) reduce compound II. Based on these considerations, it is expected that under physiological concentrations of low hydrogen peroxide but high chloride, urate, and ascorbate, turnover of the enzyme will be dependent on the rate of reduction of compound II, and the enzyme will have an acidic pH optimum.

PHYSIOLOGICAL ENVIRONMENTS OF MPO AND H_2O_2 ACTIVATION

Humans make about 0.5 g of myeloperoxidase per day. This impressive amount is derived from 5 μg of myeloperoxidase per million neutrophils, and a normal daily production of 10^{11} neutrophils [81]. In some pathologies when neutrophil and myeloperoxidase production goes into overdrive, body fluids become visibly green, for example pus and phlegm. As the predominant neutrophil protein at up to 5% of dry cell weight, and neutrophils the most abundant white blood cell in normal circulation [82], it is no wonder myeloperoxidase is scrutinized for its impact on health and disease. Synthesis of myeloperoxidase occurs in the bone marrow at an early promyelocytic stage of neutrophil development, where it is packaged into granules together with other antimicrobial agents [83]. Neutrophils circulate in the bloodstream as a monitor for host distress, and when infection or inflammation is detected they enter a complex activation cascade, ultimately resulting in degranulation. Some stimuli cause the neutrophil to undergo intracellular release of myeloperoxidase, and others lead to the release of myeloperoxidase into the extracellular milieu. These are very different environments for the enzyme, including with respect to its exposure to hydrogen peroxide.

MYELOPEROXIDASE IN NEUTROPHIL PHAGOSOMES

Neutrophils are professional phagocytic cells of the innate immune system. They trap bacteria within intracellular vacuoles called phagosomes, into which they discharge their granule contents and generate superoxide [84] (Figure 11.5). The superoxide is produced via the NADPH oxidase (NOX2) on the phagosomal membrane, which shuttles electrons from NADPH in the cytosol to dioxygen in the phagosome [85]. The intraphagosomal space around an ingested bacterium is tight, estimated to be only ~1 × 10^{-15} l in volume. Within this compartment, the consumption rate of

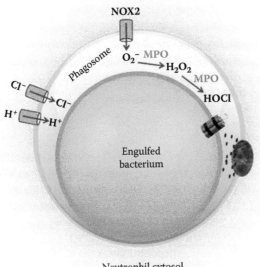

NOX2

O_2^- MPO $\rightarrow H_2O_2$

Phagosome

MPO

HOCl

Cl^- $\rightarrow Cl^-$

H^+ $\rightarrow H^+$

Engulfed
bacterium

Neutrophil cytosol

FIGURE 11.5 Myeloperoxidase activity in the neutrophil phagosome. A bacterium (yellow) is depicted inside a neutrophil phagosome (green) into which a granule (dark green) is releasing its contents, including myeloperoxidase. Superoxide is formed inside the phagosome by the NADPH oxidase (NOX2). It is dismutated by ferric myeloperoxidase and compound III to form hydrogen peroxide, which converts the enzyme to compound I. Compound I then oxidizes chloride to hypochlorous acid. Protons and chloride are pumped into the phagosomes through specific channels (gray). Through a combination of subsequent oxidative and proteolytic assaults, the ingested microorganism is killed in the phagosome.

dioxygen is calculated to be approx. 2.5 mM s^{-1} [4]. This enormous respiratory burst creates a superoxide flux of about 5 mM s^{-1} in the phagosome. Concomitant degranulation occuring in the tiny phagocytic space subjects the ingested bacterium to a strong cocktail of antimicrobial peptides and proteins. Phagosomal myeloperoxidase is estimated to approach a concentration of 1 mM [4]. Traditionally it was assumed that superoxide would spontaneously dismutate and deliver hydrogen peroxide to myeloperoxidase, but the kinetics for this are too slow. Rather, with myeloperoxidase present at such a high concentration, superoxide will preferentially react with it, and dismutate to hydrogen peroxide via the cycling of ferric enzyme and compound III (Figure 11.3) [4].

Following the initial respiratory burst, a steady state is established in the sealed phagosome. The rapid myeloperoxidase-dependent dismutation of superoxide initially causes an influx of protons and an increase in pH to 7.8–8.0 [86,87]. As hydrogen peroxide accumulates toward a steady-state concentration of 1 µM, it will react with the ferric enzyme to form compound I, which in turn oxidizes chloride to hypochlorous acid. There are no other substrates present at sufficiently high concentrations to have a substantive effect on oxidant production. Some one-electron reductants may reduce compound I to compound II, but superoxide would readily reduce compound II back to the active enzyme and prevent it accumulating. Computer modeling suggests that, provided chloride is not limiting, myeloperoxidase will convert the majority of superoxide to hypochlorous acid [4]. The initial chloride concentration in the phagosome has been measured at roughly 70 mM [88,89]. There are multiple pumps to replenish chloride, including the cystic fibrosis transmembrane conductance regulator (CFTR) [88,89].

Most of the hypochlorous acid will react with neutrophil proteins inside the phagosome, this has been substantiated experimentally [90]. There could be specific protein targets that once oxidized, promote death pathways. Enough hypochlorous acid might react directly on the bacterium to kill it, or protein chloramines could breakdown to ammonia chloramine (NH_2Cl) and ammonia dichloramine ($NHCl_2$), which are both bactericidal [91]. Protein chloramines may also act via histidine residues to chlorinate bacterial proteins [92]. Given its estimated high steady-state concentration of 20 μM [4], superoxide could have a bactericidal effect independent of myeloperoxidase or react with protein chloramines to generate radical species [93]. Thus, within neutrophil phagosomes, superoxide rather than hydrogen peroxide is the dominant form of reactive oxygen, and myeloperoxidase will act mainly as a superoxide dismutase and a chlorinating enzyme.

EXTRACELLULAR ACTIVITY OF MYELOPEROXIDASE

In addition to myeloperoxidase exerting its powers within the confines of the phagosome, it is active extracellularly. There are various mechanisms by which it is released outside neutrophils, and some remain poorly understood. First, not all neutrophil stimuli lead to phagocytosis, the oxidative burst can be activated with the assembly of NOX2 on the external plasma membrane and release of myeloperoxidase to the outside. Second, the term frustrated phagocytosis has long been used to describe degranulation leakage from an incompletely sealed phagosome, but it may be that this form of extracellular release is more actively controlled than simply "toxic spill." Over the last decade much has been learnt about the exportation of granule contents outside the cell.

Phagocytic and inflammatory stimuli induce neutrophils to cast out web-like structures called neutrophil extracellular traps (NETs) that consist of DNA, histones, and an array of other proteins, including myeloperoxidase. NETs trap bacteria and other microorganisms [94] and, when supplied with hydrogen peroxide, kill bacteria entangled within them [95]. The myeloperoxidase tethered to the NETS is active [96]. Also, the repackaging of neutrophil contents into extracellular vesicles is an active process not just associated with apoptotic cell death. Vesicles produced by neutrophils that have been stimulated by opsonized particles contain myeloperoxidase and can exert antibacterial activity [97]. It is acknowledged that myeloperoxidase may be released from neutrophils for different purposes. Aside from microbicidal activity, other functions of extracellular myeloperoxidase have been identified such as mediation of neutrophil recruitment [98], as well as a role in the resolution of inflammation [99] or suppression of the immune response [100].

Elevated extracellular myeloperoxidase is a consequence of the proinflammatory response in numerous acute and chronic diseases. The neutrophil is not only a frontline defender against microbial infection, but it is also involved in wide-ranging noninfectious inflammatory etiologies. The neutrophil population soars in the circulation leading to an invasion of afflicted tissues. Myeloperoxidase activity is always central to the neutrophil's impact and is implicated in the oxidative damage at sites of inflammation. Myeloperoxidase generates reactive oxidants in septic blood [101], in joints of patients with rheumatoid arthritis [102], in the lungs of children with cystic fibrosis [103], and in the brains of the elderly with Alzheimer's disease [104]. It is also active in atherosclerotic plaques [105] and solid tumors [106]. At the onset of a heart attack or stroke, neutrophils release large amounts of myeloperoxidase into circulation where it sticks to the endothelium and impacts the pathology of these often fatal clinical events [107–109]. More recently, it has been found that when neutrophils infiltrate adipose tissue, myeloperoxidase contributes to the development of obesity and obesity-associated insulin resistance [110].

Enzymatic cycling of myeloperoxidase in the extracellular milieu will vary depending on the available substrates. Unlike the situation in the phagosome, exposure to reactive oxygen species is likely to be fleeting. Hydrogen peroxide could be supplied by stimulated neutrophils (NOX2), other NADPH oxidases, particularly NOX4 on the endothelium [111], or xanthine oxidase [112,113]. The presence of ceruloplasmin, the major endogenous inhibitor currently known [114–116], will also determine the extent of myeloperoxidase activity. This copper containing protein is abundant in plasma, binds to myeloperoxidase, reduces it to compound II, and prevents turnover of the enzyme [117]. Ceruloplasmin inhibits myeloperoxidase in plasma but whether it is effective *in vivo* has yet to be established [117]. Nitric oxide may also modulate the activity of myeloperoxidase and lower its production of oxidants by binding to the ferric and ferrous redox intermediates [118,119].

Hypochlorous acid and hypothiocyanite will be the main oxidants generated in the extracellular milieu because chloride and thiocyanate will be the preferred substrates for compound I [29,30]. The concentration of myeloperoxidase will be too low for it to act as an effective superoxide dismutase. Within the oral cavity, where thiocyanate is present at millimolar concentrations, hypothiocyanite will be the major oxidant produced [120]. The numerous one-electron substrates of compound I will compete with chloride and thiocyanate, and continually reduce a small proportion of the enzyme to compound II. Urate will be the major one-electron substrate for compound I because of its high concentration in extracellular fluids compared to other substrates, and its favorable rate constant [48]. The enzyme will not accumulate as compound II because its facile reactions with urate, ascorbate, and superoxide should ensure continued turnover (Figure 11.3). Consequently, under normal physiological conditions, myeloperoxidase is expected to produce predominantly urate free radicals via its classical peroxidation cycle. The proportion of urate radicals produced by myeloperoxidase will be enhanced in individuals with hyperuricemia, where urate concentrations can be as high as 900 μM [48,121]. In summary, in the extracellular environment, myeloperoxidase will act mainly as a halogenating enzyme producing both hypochlorous acid and hypothiocyanite. The classical peroxidase activity of myeloperoxidase will also operate to a small degree and generate free radicals, especially the urate radical.

SUMMARY

Since the time Christine ventured to London and started studying the enzymology of myeloperoxidase, much has been learnt about how this fascinating green protein contributes to host defense and inflammation. We now have a better appreciation of its complex behavior and multiple potential activities. Within neutrophil phagosomes, it will act both as a superoxide dismutase and a halogenase, producing hypochlorous acid to kill bacteria. Hypochlorous acid, hypothiocyanite, and urate free radicals will be its major extracellular products and contribute to oxidative stress during inflammation. However, there are still several aspects of its enzymology that should be teased out. Most importantly, the precise mechanistic features of the formation and reactivity of compound I need to be probed with more informative techniques than have been used to date. The details that emerge will be essential to fully grasp how this enzyme captures, then unleashes, the oxidation potential of hydrogen peroxide and drives oxidative stress exerted by neutrophils. Perhaps, as an additional bonus, a mechanism will come to light for Christine's first discovery that myeloperoxidase is also a catalase, and whether this activity is relevant to the biology of neutrophils.

REFERENCES

1. Babior, B. M., R. S. Kipnes, and J. T. Curnutte. 1973. Biological defense mechanisms: The production by leukocytes of superoxide, a potential bactericidal agent. *The Journal of Clinical Investigation* 52:741–744.
2. Klebanoff, S. J. and C. B. Hamon. 1972. Role of myeloperoxidase mediated antimicrobial systems in intact leukocytes. *Journal of the Reticuloendothelial Society* 12:170–196.
3. Winterbourn, C. C., R. C. Garcia, and A. W. Segal. 1985. Production of the superoxide adduct of myeloperoxidase (compound III) by stimulated human neutrophils and its reactivity with hydrogen peroxide and chloride. *Biochemical Journal* 228 (3):583–592.
4. Winterbourn, C. C., M. B. Hampton, J. H. Livesey, and A. J. Kettle. 2006. Modeling the reactions of superoxide and myeloperoxidase in the neutrophil phagosome: Implications for microbial killing. *Journal of Biological Chemistry* 281 (52):39860–39869.
5. Klebanoff, S. J., A. J. Kettle, H. Rosen, C. C. Winterbourn, and W. M. Nauseef. 2013. Myeloperoxidase: A front-line defender against phagocytosed microorganisms. *Journal of Leukocyte Biology* 93 (2):185–198.
6. Winterbourn, C. C. 2008. Reconciling the chemistry and biology of reactive oxygen species. *Nature Chemical Biology* 4 (5):278–286.
7. Kettle, A. J. and C. C. Winterbourn. 2016. Myeloperoxidase: Structure and function of the green heme enzyme of neutrophils. In *Heme Peroxidases*, eds. E. L. Raven and H. B. Dunford, pp. 272–308. Cambridge, U.K.: Royal Society of Chemistry.
8. Furtmuller, P. G., M. Zederbauer, W. Jantschko, J. Helm, M. Bogner, C. Jakopitsch, and C. Obinger. 2006. Active site structure and catalytic mechanisms of human peroxidases. *Archives of Biochemistry and Biophysics* 445 (2):199–213.
9. Kettle, A. J., A. M. Albrett, A. L. Chapman, N. Dickerhof, L. V. Forbes, I. Khalilova, and R. Turner. 2014. Measuring chlorine bleach in biology and medicine. *Biochimica et Biophysica Acta* 1840 (2):781–793.
10. Fiedler, T. J., C. A. Davey, and R. E. Fenna. 2000. X-ray crystal structure and characterization of halide-binding sites of human myeloperoxidase at 1.8 A resolution. *Journal of Biological Chemistry* 275 (16):11964–11971.

11. Van Antwerpen, P., M. C. Slomianny, K. Z. Boudjeltia, C. Delporte, V. Faid, D. Calay, A. Rousseau et al. 2010. Glycosylation pattern of mature dimeric leukocyte and recombinant monomeric myeloperoxidase: Glycosylation is required for optimal enzymatic activity. *The Journal of Biological Chemistry* 285 (21):16351–16359.
12. Zeng, J. and R. E. Fenna. 1992. X-ray crystal structure of canine myeloperoxidase at 3 A resolution. *Journal of Molecular Biology* 226:185–207.
13. Campomanes, P., U. Rothlisberger, M. Alfonso-Prieto, and C. Rovira. 2015. The molecular mechanism of the catalase-like activity in horseradish peroxidase. *Journal of the American Chemical Society* 137 (34):11170–11178.
14. Poulos, T. L. 2014. Heme enzyme structure and function. *Chemical Reviews* 114 (7):3919–3962.
15. Devarajan, A., A. V. Gaenko, and U. Ryde. 2008. Effect of covalent links on the structure, spectra, and redox properties of myeloperoxidase—A density functional study. *Journal of Inorganic Biochemistry* 102 (8):1549–1557.
16. Carpena, X., P. Vidossich, K. Schroettner, B. M. Calisto, S. Banerjee, J. Stampler, M. Soudi et al. 2009. Essential role of proximal histidine-asparagine interaction in mammalian peroxidases. *The Journal of Biological Chemistry* 284 (38):25929–25937.
17. Stampler, J., M. Bellei, M. Soudi, C. Gruber, G. Battistuzzi, P. G. Furtmuller, and C. Obinger. 2011. Manipulating the proximal triad His-Asn-Arg in human myeloperoxidase. *Archives of Biochemistry and Biophysics* 516 (1):21–28.
18. Klebanoff, S. J. 1968. Myeloperoxidase-halide-hydrogen peroxide antibacterial system. *Journal of Bacteriology* 95:2131–2138.
19. Harrison, J. E. and J. Shultz. 1976. Studies on the chlorinating activity of myeloperoxidase. *Journal of Biological Chemistry* 251:1371–1374.
20. Jantschko, W., P. G. Furtmuller, M. Allegra, M. A. Livrea, C. Jakopitsch, G. Regelsberger, and C. Obinger. 2002. Redox intermediates of plant and mammalian peroxidases: A comparative transient-kinetic study of their reactivity toward indole derivatives. *Archives of Biochemistry and Biophysics* 398 (1):12–22.
21. Kettle, A. J. and C. C. Winterbourn. 1994. Superoxide-dependent hydroxylation by myeloperoxidase. *Journal of Biological Chemistry* 269 (25):17146–17151.
22. Iwamoto, H., T. Kobayashi, E. Hasegawa, and Y. Morita. 1987. Reaction of human myeloperoxidase with hydrogen peroxide and its true catalase activity. *Journal of Biochemistry* 101:1407–1412.
23. Kettle, A. J. and C. C. Winterbourn. 2001. A kinetic analysis of the catalase activity of myeloperoxidase. *Biochemistry* 40 (34):10204–10212.
24. Marquez, L. A. and H. B. Dunford. 1994. Chlorination of taurine by myeloperoxidase. *Journal of Biological Chemistry* 269 (11):7950–7956.
25. Marquez, L. A., J. T. Huang, and H. B. Dunford. 1994. Spectral and kinetic studies on the formation of myeloperoxidase compounds I and II: Roles of hydrogen peroxide and superoxide. *Biochemistry* 33:1447–1454.
26. Jones, P. and H. B. Dunford. 2005. The mechanism of compound I formation revisited. *Journal of Inorganic Biochemistry* 99 (12):2292–2298.
27. Arnhold, J., P. G. Furtmuller, G. Regelsberger, and C. Obinger. 2001. Redox properties of the couple compound I/native enzyme of myeloperoxidase and eosinophil peroxidase. *European Journal of Biochemistry* 268 (19):5142–5148.
28. Paumann-Page, M., P. G. Furtmuller, S. Hofbauer, L. N. Paton, C. Obinger, and A. J. Kettle. 2013. Inactivation of human myeloperoxidase by hydrogen peroxide. *Archives of Biochemistry and Biophysics* 539 (1):51–62.
29. van Dalen, C. J., M. W. Whitehouse, C. C. Winterbourn, and A. J. Kettle. 1997. Thiocyanate and chloride as competing substrates for myeloperoxidase. *Biochemical Journal* 327:487–492.

30. Fürtmuller, P. G., U. Burner, and C. Obinger. 1998. Reaction of myeloperoxidase compound I with chloride, bromide, iodide, and thiocyanate. *Biochemistry* 37:17923–17930.
31. Blair-Johnson, M., T. Fiedler, and R. Fenna. 2001. Human myeloperoxidase: Structure of a cyanide complex and its interaction with bromide and thiocyanate substrates at 1.9 A resolution. *Biochemistry* 40 (46):13990–13997.
32. Lee, H. C., K. S. Booth, W. S. Caughey, and M. Ikeda-Saito. 1991. Interaction of halides with the cyanide complex of myeloperoxidase: A model for substrate binding to compound I. *Biochimica et Biophysica Acta* 1076 (2):317–320.
33. Senthilmohan, R. and A. J. Kettle. 2006. Bromination and chlorination reactions of myeloperoxidase at physiological concentrations of bromide and chloride. *Archives of Biochemical and Biophysics* 445 (2):235–244.
34. Chapman, A. L., O. Skaff, R. Senthilmohan, A. J. Kettle, and M. J. Davies. 2009. Hypobromous acid and bromamine production by neutrophils and modulation by superoxide. *Biochemical Journal* 417 (3):773–781.
35. Gaut, J. P., G. C. Yeh, H. D. Tran, J. Byun, J. P. Henderson, G. M. Richter, M. L. Brennan et al. 2001. Neutrophils employ the myeloperoxidase system to generate antimicrobial brominating and chlorinating oxidants during sepsis. *Proceedings of the National Academy of Science USA* 98 (21):11961–11966.
36. Furtmuller, P. G., C. Obinger, Y. Hsuanyu, and H. B. Dunford. 2000. Mechanism of reaction of myeloperoxidase with hydrogen peroxide and chloride ion. *European Journal of Biochemistry* 267 (19):5858–5864.
37. Yang, C., J. Wang, A. N. Krutchinsky, B. T. Chait, J. D. Morrisett, and C. V. Smith. 2001. Selective oxidation in vitro by myeloperoxidase of the N-terminal amine in apolipoprotein B-100. *Journal of Lipid Research* 42 (11):1891–1896.
38. Winterbourn, C. C. 1985. Comparative reactivities of various biological compounds with myeloperoxidase-hydrogen peroxide-chloride, and similarity of the oxidant to hypochlorite. *Biochimica et Biophysica Acta* 840 (2):204–210.
39. Ramos, D. R., M. V. Garcia, L. M. Canle, J. A. Santaballa, P. G. Furtmuller, and C. Obinger. 2008. Myeloperoxidase-catalyzed chlorination: The quest for the active species. *Journal of Inorganic Biochemistry* 102 (5–6):1300–1311.
40. Furtmuller, P. G., U. Burner, W. Jantschko, G. Regelsberger, and C. Obinger. 2000. The reactivity of myeloperoxidase compound I formed with hypochlorous acid. *Redox Report* 5 (4):173–178.
41. Abu-Soud, H. M., D. Maitra, F. Shaeib, S. N. Khan, J. Byun, I. Abdulhamid, Z. Yang et al. 2014. Disruption of heme-peptide covalent cross-linking in mammalian peroxidases by hypochlorous acid. *Journal of Inorganic Biochemistry* 140:245–254.
42. Floris, R., S. R. Piersma, G. Yang, P. Jones, and R. Wever. 1993. Interaction of myeloperoxidase with peroxynitrite: Comparison with lactoperoxidase, horseradish peroxidase and catalase. *European Journal of Biochemistry* 215:767–775.
43. Bakkenist, A. R. J., J. E. D. De Boer, H. Plat, and R. Wever. 1980. The halide complexes of myeloperoxidase and the mechanism of the halogenation reactions. *Biochimca et Biophysica Acta* 613:349–358.
44. Andrews, P. C. and N. I. Krinsky. 1982. A kinetic analysis of the interaction of human myeloperoxidase with hydrogen peroxide, chloride ions, and protons. *The Journal of Biological Chemistry* 257 (22):13240–13245.
45. Abu-Soud, H. M. and S. L. Hazen. 2000. Nitric oxide modulates the catalytic activity of myeloperoxidase. *The Journal of Biological Chemistry* 275 (8):5425–5430.
46. Kettle, A. J. and C. C. Winterbourn. 1988. Superoxide modulates the activity of myeloperoxidase and optimizes the production of hypochlorous acid. *Biochemical Journal* 252:529–536.

47. Dunford, H. B. and Y. Hsuanyu. 1999. Kinetics of oxidation of serotonin by myeloperoxidase compounds I and II. *Biochemistry and Cell Biology* 77 (5):449–457.
48. Meotti, F. C., G. N. Jameson, R. Turner, D. T. Harwood, S. Stockwell, M. D. Rees, S. R. Thomas, and A. J. Kettle. 2011. Urate as a physiological substrate for myeloperoxidase: Implications for hyperuricemia and inflammation. *Journal of Biological Chemistry* 286 (15):12901–12911.
49. Heinecke, J. W., W. Li, H. L. Daehnke, and J. A. Goldsteiin. 1993. Dityrosine, a specific marker of oxidation, is synthesized by the myeloperoxidase-hydrogen peroxide system of human neutrophils and macrophages. *Journal of Biological Chemistry* 268 (6):4069–4077.
50. Bolscher, B. G. J. M., G. R. Zoutberg, R. A. Cuperus, and R. Wever. 1984. Vitamin C stimulates the chlorinating activity of human myeloperoxidase. *Biochimica et Biophysica Acta* 784:189–191.
51. Palinkas, Z., P. G. Furtmuller, A. Nagy, C. Jakopitsch, K. F. Pirker, M. Magierowski, K. Jasnos, J. L. Wallace, C. Obinger, and P. Nagy. 2014. Interactions of hydrogen sulfide with myeloperoxidase. *British Journal of Pharmacology* 172:1516–1532.
52. Burner, U., P. G. Furtmuller, A. J. Kettle, W. H. Koppenol, and C. Obinger. 2000. Mechanism of reaction of myeloperoxidase with nitrite. *Journal of Biological Chemistry* 275 (27):20597–20601.
53. van Dalen, C. J., C. C. Winterbourn, R. Senthilmohan, and A. J. Kettle. 2000. Nitrite as a substrate and inhibitor of myeloperoxidase. Implications for nitration and hypochlorous acid production at sites of inflammation. *Journal of Biological Chemistry* 275 (16):11638–11644.
54. Burner, U., C. Obinger, M. Paumann, P. G. Furtmuller, and A. J. Kettle. 1999. Transient and steady-state kinetics of the oxidation of substituted benzoic acid hydrazides by myeloperoxidase. *Journal of Biological Chemistry* 274 (14):9494–9502.
55. Burner, U., W. Jantschko, and C. Obinger. 1999. Kinetics of oxidation of aliphatic and aromatic thiols by myeloperoxidase compounds I and II. *FEBS Letters* 443:290–296.
56. Ator, M. A. and P. R. Ortiz de Montellano. 1987. Protein control of prosthetic heme reactivity: Reaction of substrates with the heme edge of horseradish peroxidase. *Journal of Biological Chemistry* 262 (4):1542–1551.
57. Forbes, L. V., T. Sjogren, F. Auchere, D. W. Jenkins, B. Thong, D. Laughton, P. Hemsley et al. 2013. Potent reversible inhibition of myeloperoxidase by aromatic hydroxamates. *Journal of Biological Chemistry* 288 (51):36636–36647.
58. Tiden, A. K., T. Sjogren, M. Svesson, A. Bernlind, R. Senthilmohan, F. Auchere, H. Norman et al. 2011. 2-Thioxanthines are suicide inhibitors of myeloperoxidase that block oxidative stress during inflammation. *Journal of Biological Chemistry* 286 (43):37578–37589.
59. Abu-Soud, H. M. and S. L. Hazen. 2001. Interrogation of heme pocket environment of mammalian peroxidases with diatomic ligands. *Biochemistry* 40 (36):10747–10755.
60. Furtmuller, P. G., J. Arnhold, W. Jantschko, H. Pichler, and C. Obinger. 2003. Redox properties of the couples compound I/compound II and compound II/native enzyme of human myeloperoxidase. *Biochemical and Biophysical Research Communications* 301 (2):551–557.
61. Kettle, A. J., A. Maroz, G. Woodroffe, C. C. Winterbourn, and R. F. Anderson. 2011. Spectral and kinetic evidence for reaction of superoxide with compound I of myeloperoxidase. *Free Radical Biology and Medicine* 51:2190–2194.
62. Kettle, A. J., R. F. Anderson, M. B. Hampton, and C. C. Winterbourn. 2007. Reactions of superoxide with myeloperoxidase. *Biochemistry* 46 (16):4888–4897.
63. Koppenol, W. H., D. M. Stanbury, and P. L. Bounds. 2010. Electrode potentials of partially reduced oxygen species, from dioxygen to water. *Free Radical Biology and Medicine* 49 (3):317–322.

64. Steinbeck, M. J., A. U. Khan, and M. J. Karnovsky. 1992. Intracellular singlet oxygen generation by phagocytosing neutrophils in response to particles coated with a chemical trap. *Journal of Biological Chemistry* 267 (19):13425–13433.

65. Kettle, A. J., D. F. Sangster, J. M. Gebicki, and C. C. Winterbourn. 1988. A pulse radiolysis investigation of the reactions of myeloperoxidase with superoxide and hydrogen peroxide. *Biochimica et Biophysica Acta* 956:58–62.

66. Cuperus, R. A., A. O. Muijsers, and R. Wever. 1986. The superoxidase activity of myeloperoxidase: Formation of compound III. *Biochimica et Biophysica Acta* 871:78–84.

67. Marquez, L. A. and H. B. Dunford. 1990. Reaction of compound III of myeloperoxidase with ascorbic acid. *Journal of Biological Chemistry* 265:6074–6078.

68. Hsuanyu, Y. and H. B. Dunford. 1999. Oxidation of clozapine and ascorbate by myeloperoxidase. *Archives of Biochemical and Biophysics* 368 (2):413–420.

69. Ximenes, V. F., G. J. Maghzal, R. Turner, Y. Kato, C. C. Winterbourn, and A. J. Kettle. 2010. Serotonin as a physiological substrate for myeloperoxidase and its superoxide-dependent oxidation to cytotoxic tryptamine-4,5-dione. *Biochemical Journal* 425 (1):285–293.

70. Ximenes, V. F., S. O. Silva, M. R. Rodrigues, L. H. Catalani, G. J. Maghzal, A. J. Kettle, and A. Campa. 2005. Superoxide-dependent oxidation of melatonin by myeloperoxidase. *Journal of Biological Chemistry* 280 (46):38160–38169.

71. Kettle, A. J. and C. C. Winterbourn. 1992. Oxidation of hydroquinone by myeloperoxidase: Mechanism of stimulation by benzoquinone. *Journal of Biological Chemistry* 267:8319–8324.

72. Dunford, H. B. 2010. *Peroxidases and Catalases*, 2nd ed. New York: Wiley-VCH.

73. Bolscher, B. G. J. M. and R. Wever. 1984. A kinetic study of the reaction between human myeloperoxidase, hydroperoxides and cyanide. Inhibition by chloride and thiocyanate. *Biochimica et Biophysica Acta* 788:1–10.

74. Allegra, M., P. G. Furtmuller, G. Regelsberger, M. L. Turco-Liveri, L. Tesoriere, M. Perretti, M. A. Livrea, and C. Obinger. 2001. Mechanism of reaction of melatonin with human myeloperoxidase. *Biochemistry and Biophysics Research Community* 282 (2):380–386.

75. Hoogland, H., H. L. Dekker, C. van Riel, A. van Kuilenburg, A. O. Muijsers, and R. Wever. 1988. A steady state study on the formation of compounds II and III of myeloperoxidase. *Biochimica et Biophysica Acta* 955:337–345.

76. Kettle, A. J. and C. C. Winterbourn. 1989. Influence of superoxide on myeloperoxidase kinetics measured with a hydrogen peroxide electrode. *Biochemical Journal* 263 (3):823–828.

77. Marquez, L. A., H. B. Dunford, and H. Van Wart. 1990. Kinetic studies on the reaction of compound II of myeloperoxidase with ascorbic acid. *Journal of Biological Chemistry* 265 (10):5666–5670.

78. Marquez, L. A. and H. B. Dunford. 1997. Mechanism of the oxidation of 3,5,3′,5′-tetramethylbenzidine by myeloperoxidase determined by transient- and steady-state kinetics. *Biochemistry* 36 (31):9349–9355.

79. Kettle, A. J. and C. C. Winterbourn. 1988. The mechanism of myeloperoxidase-dependent chlorination of monochlorodimedon. *Biochimica et Biophysica Acta* 957:185–191.

80. Marquez, L. A. and H. B. Dunford. 1996. Kinetics of oxidation of tyrosine and dityrosine by myeloperoxidase compounds I and II. *Journal of Biological Chemistry* 270:30434–30440.

81. Bakkenist, A. R. J., R. Wever, T. Vulsma, H. Plat, and B. F. van Gelder. 1978. Isolation procedure and some properties of myeloperoxidase from human leucocytes. *Biochimica et Biophysica Acta* 524:45–54.

82. Klebanoff, S. J. 1988. Phagocytic cells: Products of oxygen metabolism. In *Inflammation: Basic Principles and Clinical Correlates*, eds. J. I. Gallin, I. M. Goldstein, and R. Snyderman. New York: Raven Press Ltd.

83. Johnson, K. R. and W. M. Nauseef. 1991. Molecular biology of MPO. In *Peroxidases in Chemistry and Biology*, eds. J. Everse, K. E. Everse, and M. B. Grisham, pp. 63–81. Boca Raton, FL: CRC Press.
84. Winterbourn, C. C. and A. J. Kettle. 2013. Redox reactions and microbial killing in the neutrophil phagosome. *Antioxidants & Redox Signaling* 18 (6):642–660.
85. Nauseef, W. M. 2007. How human neutrophils kill and degrade microbes: An integrated view. *Immunological Review* 219 (1):88–102.
86. Segal, A. W., M. Geisow, R. Garcia, A. Harper, and R. Miller. 1981. The respiratory burst of phagocytic cells is associated with a rise in vacuolar pH. *Nature* 290:406–409.
87. Cech, P. and R. I. Lehrer. 1984. Phagolysosomal pH of human neutrophils. *Blood* 63:88–95.
88. Painter, R. G. and G. Wang. 2006. Direct measurement of free chloride concentrations in the phagolysosomes of human neutrophils. *Analytical Chemistry* 78 (9):3133–3137.
89. Aiken, M. L., R. G. Painter, Y. Zhou, and G. Wang. 2012. Chloride transport in functionally active phagosomes isolated from human neutrophils. *Free Radical Biology and Medicine* 53 (12):2308–2317.
90. Green, J. N., A. J. Kettle, and C. C. Winterbourn. 2014. Protein chlorination in neutrophil phagosomes and correlation with bacterial killing. *Free Radical Biology and Medicine* 77:49–56.
91. Coker, M. S. A., W. Hu, S. T. Senthilmohan, and A. J. Kettle. 2008. Pathways for the decay of organic dichloramines and liberation of antimicrobial chloramine gases. *Chemical Research Toxicology* 21 (12):2334–2343.
92. Roemeling, M. D., J. Williams, J. S. Beckman, and J. K. Hurst. 2015. Imidazole catalyzes chlorination by unreactive primary chloramines. *Free Radical Biology and Medicine* 82:167–178.
93. Hawkins, C. L., M. D. Rees, and M. J. Davies. 2002. Superoxide radicals can act synergistically with hypochlorite to induce damage to proteins. *FEBS Letters* 510 (1–2):41–44.
94. Brinkmann, V. and A. Zychlinsky. 2012. Neutrophil extracellular traps: Is immunity the second function of chromatin? *The Journal of Cell Biology* 198 (5):773–783.
95. Parker, H., M. Dragunow, M. B. Hampton, A. J. Kettle, and C. C. Winterbourn. 2012. Requirements for NADPH oxidase and myeloperoxidase in neutrophil extracellular trap formation differ depending on the stimulus. *Journal of Leukocyte Biology* 92 (4):841–849.
96. Parker, H., A. M. Albrett, A. J. Kettle, and C. C. Winterbourn. 2012. Myeloperoxidase associated with neutrophil extracellular traps is active and mediates bacterial killing in the presence of hydrogen peroxide. *Journal of Leukocyte Biology* 91:369–376.
97. Timar, C. I., A. M. Lorincz, R. Csepanyi-Komi, A. Valyi-Nagy, G. Nagy, E. I. Buzas, Z. Ivanyi et al. 2013. Antibacterial effect of microvesicles released from human neutrophilic granulocytes. *Blood* 121 (3):510–518.
98. Klinke, A., C. Nussbaum, L. Kubala, K. Friedrichs, T. K. Rudolph, V. Rudolph, H. J. Paust et al. 2011. Myeloperoxidase attracts neutrophils by physical forces. *Blood* 117 (4):1350–1358.
99. van der Veen, B. S., M. P. de Winther, and P. Heeringa. 2009. Myeloperoxidase: Molecular mechanisms of action and their relevance to human health and disease. *Antioxidants & Redox Signaling* 11 (11):2899–2937.
100. Odobasic, D., R. C. Muljadi, K. M. O'Sullivan, A. J. Kettle, N. Dickerhof, S. A. Summers, A. R. Kitching, and S. R. Holdsworth. 2015. Suppression of autoimmunity and renal disease in pristane-induced lupus by myeloperoxidase. *Arthritis Rheumatology* 67 (7):1868–1880.
101. Winterbourn, C. C., I. H. Buss, T. P. Chan, L. D. Plank, M. A. Clark, and J. A. Windsor. 2000. Protein carbonyl measurements show evidence of early oxidative stress in critically ill patients. *Critical Care Medicine* 28 (1):143–149.

102. Stamp, L. K., I. Khalilova, J. M. Tarr, R. Senthilmohan, R. Turner, R. C. Haigh, P. G. Winyard, and A. J. Kettle. 2012. Myeloperoxidase and oxidative stress in rheumatoid arthritis. *Rheumatology* 51:1796–1803.

103. Kettle, A. J., R. Turner, C. L. Gangell, D. T. Harwood, I. S. Khalilova, A. L. Chapman, C. C. Winterbourn, P. D. Sly, and C. F. Arest. 2014. Oxidation contributes to low glutathione in the airways of children with cystic fibrosis. *European Respiratory Journal* 44 (1):122–129.

104. Green, P. S., A. J. Mendez, J. S. Jacob, J. R. Crowley, W. Growdon, B. T. Hyman, and J. W. Heinecke. 2004. Neuronal expression of myeloperoxidase is increased in Alzheimer's disease. *Journal of Neurochemistry* 90 (3):724–733.

105. Nicholls, S. J. and S. L. Hazen. 2009. Myeloperoxidase, modified lipoproteins, and atherogenesis. *Journal of Lipid Research* 50 (Suppl):S346–S351.

106. Rymaszewski, A. L., E. Tate, J. P. Yimbesalu, A. E. Gelman, J. A. Jarzembowski, H. Zhang, K. A. Pritchard, Jr., and H. G. Vikis. 2014. The role of neutrophil myeloperoxidase in models of lung tumor development. *Cancers* 6 (2):1111–1127.

107. Nussbaum, C., A. Klinke, M. Adam, S. Baldus, and M. Sperandio. 2012. Myeloperoxidase: A leukocyte-derived protagonist of inflammation and cardiovascular disease. *Antioxidants & Redox Signaling* 18:692–713.

108. Marshall, C. J., M. Nallaratnam, T. Mocatta, D. W. Smyth, M. Richards, J. M. Elliot, J. Blake, C. C. Winterbourn, A. J. Kettle, and D. R. McClean. 2010. Factors influencing local and systemic levels of plasma myeloperoxidase in ST segment elevation acute myocardial infarction. *American Journal of Cardiology* 106 (3):316–322.

109. Forghani, R., H. J. Kim, G. R. Wojtkiewicz, L. Bure, Y. Wu, M. Hayase, Y. Wei, Y. Zheng, M. A. Moskowitz, and J. W. Chen. 2014. Myeloperoxidase propagates damage and is a potential therapeutic target for subacute stroke. *Journal of Cerebral Blood Flow and Metabolism* 35:485–493.

110. Wang, Q., Z. Xie, W. Zhang, J. Zhou, Y. Wu, M. Zhang, H. Zhu, and M. H. Zou. 2014. Myeloperoxidase deletion prevents high-fat diet-induced obesity and insulin resistance. *Diabetes* 63 (12):4172–4185.

111. Sirker, A., M. Zhang, and A. M. Shah. 2011. NADPH oxidases in cardiovascular disease: Insights from in vivo models and clinical studies. *Basic Research Cardiology* 106 (5):735–747.

112. Brandes, R. P., N. Weissmann, and K. Schroder. 2014. Nox family NADPH oxidases in mechano-transduction: Mechanisms and consequences. *Antioxidants & Redox Signaling* 20 (6):887–898.

113. Cantu-Medellin, N. and E. E. Kelley. 2013. Xanthine oxidoreductase-catalyzed reactive species generation: A process in critical need of reevaluation. *Redox Biology* 1 (1):353–358.

114. Segelmark, M., B. Persson, T. Hellmark, and J. Wieslander. 1997. Binding and inhibition of myeloperoxidase (MPO): A major function of ceruloplasmin? *Clinical Experimental Immunology* 108 (1):167–174.

115. Sokolov, A. V., M. O. Pulina, K. V. Ageeva, M. I. Ayrapetov, M. N. Berlov, G. N. Volgin, A. G. Markov et al. 2007. Interaction of ceruloplasmin, lactoferrin, and myeloperoxidase. *Biochemistry Biokhimiia* 72 (4):409–415.

116. Sokolov, A. V., K. V. Ageeva, O. S. Cherkalina, M. O. Pulina, E. T. Zakharova, V. N. Prozorovskii, D. V. Aksenov, V. B. Vasilyev, and O. M. Panasenko. 2010. Identification and properties of complexes formed by myeloperoxidase with lipoproteins and ceruloplasmin. *Chemistry and Physics of Lipids* 163 (4–5):347–355.

117. Chapman, A. L., T. J. Mocatta, S. Shiva, A. Seidel, B. Chen, I. Khalilova, M. E. Paumann-Page, G. N. Jameson, C. C. Winterbourn, and A. J. Kettle. 2013. Ceruloplasmin is an endogenous inhibitor of myeloperoxidase. *Journal of Biological Chemistry* 288 (9):6465–6477.

118. Abu-Soud, H. M. and S. L. Hazen. 2000. Nitric oxide is a physiological substrate for mammalian peroxidases. *Journal of Biological Chemistry* 275 (48):37524–37532.
119. Eiserich, J. P., S. Baldus, M. L. Brennan, W. Ma, C. Zhang, A. Tousson, L. Castro et al. 2002. Myeloperoxidase, a leukocyte-derived vascular NO oxidase. *Science* 296 (5577):2391–2394.
120. Ashby, M. T. 2008. Inorganic chemistry of defensive peroxidases in the human oral cavity. *Journal of Dental Research* 87 (10):900–914.
121. Stamp, L. K., R. Turner, I. S. Khalilova, M. Zhang, J. Drake, L. V. Forbes, and A. J. Kettle. 2014. Myeloperoxidase and oxidation of uric acid in gout: Implications for the clinical consequences of hyperuricaemia. *Rheumatology (Oxford)* 53 (11):1958–1965.

12 Myeloperoxidase-Derived Oxidants Hypochlorous Acid and Chloramines

Microbicidal Marvels and Inflammatory Mischief Makers

Margreet C.M. Vissers, Juliet M. Pullar, and Alexander V. Peskin

CONTENTS

INTRODUCTION

Neutrophils are the body's predominant phagocytes, with their primary function being the killing and disposal of invading microorganisms [1,2]. The early observation that the act of phagocytosis was accompanied by the consumption of vast amounts of oxygen—the oxidative burst [3]—was followed by the demonstration that the oxygen consumed was converted to H_2O_2 [4]. Seymour Klebanoff then reported that the antimicrobial capacity of H_2O_2 was enhanced by the neutrophil enzyme myeloperoxidase in the presence of a halide or pseudohalide [5,6]. Bacterial killing was mediated by the hypohalous acids, including hypochlorous acid (HOCl) [7]. That these white blood cells were endowed with the capacity to generate an oxidant such as HOCl (chlorine bleach), an "industrial" oxidant with broad reactivity against any number of microorganisms, could be considered an evolutionary masterstroke. Nothing would seem more fit-for-purpose: professional killer phagocytes that can generate a highly effective microbicidal agent, ensuring effective protection against invasion by a host of pathogens. This very attractive proposition, however, poses a problem in terms of protection from self-harm, as reactive oxidants cannot differentiate between foreign and host tissues. Other chapters in this book are devoted to the source of H_2O_2 generation in neutrophils, the NOX2 NADPH oxidase complex, and to the enzymology of myeloperoxidase. We will discuss the generation of HOCl, its reactivity, and the pathological consequences of its generation in inflammatory environments.

HYPOCHLOROUS ACID: GENERATION AND BIOLOGICAL REACTIVITY

The hypohalous acids are generated by the peroxidase-mediated oxidation of a halide X^- according to the following reaction:

$$H_2O_2 + X^- + H^+ \rightarrow HOX + H_2O \qquad (12.1)$$

The species of hypohalous acid formed will depend on both the substrate preference of the particular peroxidase and the relative availability of the halide substrate [8–15]. Myeloperoxidase is able to oxidize chloride (Cl^-), bromide (Br^-), iodide (I^-), and the pseudohalide thiocyanate (SCN^-) [16], and the variable availability of these substrates can result in the simultaneous generation of a mixture of products [8,9,17,18]. SCN^- is the preferred substrate for myeloperoxidase and although it is present at much lower levels than Cl^- in body fluids, a mixture of HOCl and hypothiocyanous acid (HOSCN) is likely in most biological settings [8]. When SCN^- levels are elevated, as in cigarette smokers, the generation of HOSCN may predominate [19–21]. The generation and reactivity of HOSCN and hypobromous acid (HOBr) is the topic of another chapter in this book.

HOCl Chemistry

HOCl exists in equilibrium with molecular chlorine and hypochlorite anion (OCl^-), as shown in Reactions 12.2 and 12.3. The amount of each species present in a physiological setting is dependent on pH and the availability of Cl^- [22]. At neutral pH, HOCl is in equilibrium with OCl^-, with the pK_a for Reaction 12.3 below 7.44 at 37°C, indicating that equimolar amounts of both species would be present in most biological fluids containing around 140 mM Cl^- [22,23]. At pH > 3 molecular chlorine (Reaction 12.3) is a minor species, and is therefore generally not considered to have a great influence as a reactive chlorine species in a physiological environment [22].

$$Cl_2 + H_2O \leftrightarrow HOCl + Cl^- + H^+ \qquad (12.2)$$

$$HOCl \leftrightarrow OCl^- + H^+ \qquad (12.3)$$

For ease of communication, and because HOCl is much more reactive than OCl^- [22,24], the $HOCl/OCl^-$ mixture present in body fluids is generally referred to simply as HOCl, and in this chapter we will also adopt this practice. Despite this, it should be remembered that the two species will exert a significantly different effect in a physiological environment due to differences in membrane permeability and reactivity [22,24]. The rapid reaction of HOCl with its targets, however, generally results in complete consumption of the oxidant and the depletion of OCl^- as the equilibrium of Reaction 12.3 shifts to the left.

HOCl Reactivity and Biological Targets

Thiols and Thioethers

HOCl is a highly reactive two-electron oxidant and is quite possibly the most prevalent oxidant in our bodies [2,22,25]. It has a broad reactivity range against its numerous targets [2,17,24], reacting most readily with thiols and thioethers to generate disulfides and higher oxidation products [2,24,26,27]. Whereas the rate constant for reaction of HOCl with cysteine and methionine residues is ~10^8 M^{-1} s^{-1} [17,28], this varies between thiols, with ionized thiols being found to be most reactive [20,23,29–31]. Glutathione (GSH) is a major target for HOCl in the cellular environment, generating both glutathione disulfide (GSSG) and glutathione sulfonamide, a unique and stable product [32–36]. The products of methionine oxidation depend on the location of this amino acid in the targeted protein. Reaction with an N-terminal residue results in ~80% dehydromethionine with the remainder being converted to methionine sulfoxide. Methionine residues at sites other than the N-terminal form methionine sulfoxide only [37].

Protein Targets

The rate constant for the reaction with protein targets other than thiols and thioethers, such as amino groups, lysine side chain amines, tryptophan and histidine residues, is orders of magnitude less than the thiol reaction ($k \sim 10^4$–10^5 M^{-1} s^{-1}) [17,28,30]. The reaction of HOCl with tyrosine and arginine is much slower again ($k < 100$ M^{-1} s^{-1}) [38], but the generation of a stable product such as chlorotyrosine means that this reaction can be used as a biomarker for HOCl formation *in vivo* [29,39–41]. The reactivity of other amino acids and the peptide bond itself is less well characterized and these groups are not considered to be preferred targets for HOCl. See Figure 12.1 for a graphic representation of HOCl reactivity.

Phospholipids and Alkenes

Alkenes, present in phospholipids and derived molecules, can undergo oxidative addition reactions with HOCl, generating chlorohydrins across the C=C double bond [42–46]. The physiological relevance of this reaction is being reconsidered, after early studies suggested that the reaction was slow and might therefore not be competitive in the thiol-rich cell environment [23,42,47,48]. However, there is significant potential for membrane disruption resulting from the saturation of phospholipid double bonds and the addition of a bulky chlorine, making the potential for this reaction of HOCl highly relevant in human pathology [49–53].

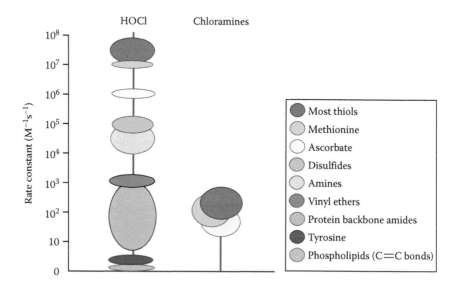

FIGURE 12.1 Schematic representation of the reactivity of HOCl and chloramines with varied biological targets. Rate constants are on a log scale. The rate constants given are from data provided in [20,23,30,46,149]. The rate constants for HOCl with protein–peptide bonds vary substantially depending on the local protein environment [17].

The accessibility of the reactive double bonds in phospholipids may influence their reactivity with HOCl. Plasmalogens are phospholipids containing a vinyl ether bond that lies close to the polar amine near the interface with the aqueous phase. They are present in cells of the vasculature and in heart muscle and their HOCl reaction products have been detected in tissues from atherosclerotic lesions [46,54]. This could be explained by the relative accessibility of the vinyl ether group compared with the double bond of phospholipid alkenes, which are buried deeper in the membrane hydrophobic environment [46,55], and the significantly faster reaction rate with vinyl ethers than with phospholipid alkenes (see Figure 12.1) [46,55].

Antioxidant and Other Cellular Targets

The rapid reaction of HOCl with thiols suggests that GSH will be the most effective antioxidant protection against this oxidant. In addition, ascorbate reacts readily with HOCl and, with intracellular ascorbate levels being in the low millimolar range [56–58], should significantly protect other cellular targets from HOCl-mediated damage. Vitamin E could also protect membrane phospholipids, as the reaction with Trolox occurs readily [59], although accessibility of vitamin E in a membrane environment is likely to affect this. Nucleic acids and free cellular nucleotides (e.g., NAD) also react with HOCl [59,60], but the biological effects of these reactions are poorly characterized.

Amines: Formation of Reactive Chloramines

HOCl reacts with amine groups (k ~ 10^4–10^5 M^{-1} s^{-1}) [2,17,22,29,59], generating chloramines that retain the two-electron oxidizing equivalents.

$$HOCl + R - NH_2 \rightarrow R - NHCl + H_2O \qquad (12.4)$$

Chloramines extend the reactivity of HOCl and will oxidize thiols and methionine residues [17,25]. Moreover, chloramines are much more selective than HOCl in reacting with thiol proteins and have a wide window for reactivity depending on thiol ionization state [61]. The biological activity of chloramines is determined by their site of generation, permeability, and stability and can vary greatly [28,30,62–64]. In addition to thiols and methionine residues, they also react with ascorbate, which may influence their biological activity [2]. Chloramines are variably permeable: some neutral small molecular weight compounds such as ammonia chloramine (NH_2Cl [monochloramine]) easily cross cell membranes and very low concentrations are able to affect cell function as readily as HOCl, with similarly high antimicrobial activity [67] and ability to oxidize intracellular thiols including GSH [25]. In contrast, other chloramines, such as taurine chloramine or those formed on proteins, are unable to oxidize intracellular thiols, even when present at high concentrations [25,65,66]. Chloramines formed on protein residues can break down to form carbonyls or NH_2Cl and ammonia dichloramine ($NHCl_2$) [67].

The variable permeability means that the reactivity of chloramines with cell targets is more complex than would be predicted by *in vitro* studies, being largely influenced by the site of generation and the compartmentalization within the cell. For example,

Midwinter et al. reported that the impermeable oxidant taurine chloramine was unable to oxidize intracellular thiols, even when present at high concentrations, but could still activate intracellular MAP kinases, likely through the oxidation of the EGF receptor on the plasma membrane [65,66]. The selectivity of chloramines for thiols and methionine residues could enhance the capacity for these oxidants to trigger signaling events. Cell-permeable chloramines were shown to influence the NFκB pathway through the modification of a single methionine residue in IκBα, thereby preventing degradation of IκBα by the proteasome [66]. That inactivation was due to the oxidation of a methionine to methionine sulfoxide was confirmed by reversal with methionine sulfoxide reductases. This reaction overrides the activation of NFκB and has the potential to exert a profound effect on the inflammatory response. Chloramines, like HOCl, can also oxidize N-terminal methionine residues to form dehydromethionine [37]. The formation of a cyclic product at protein N-termini by this reaction could have a profound effect on the interaction of the modified protein with other molecules. Proteins with N-terminal methionine residues include ubiquitin, which is involved in vital regulatory processes including protein degradation and N-terminal acetyltransferases that determine histone acetylation [68]. Chloramines have also been shown to induce p53-dependent growth arrest, stress responses, zinc mobilization, adhesion molecule expression, and apoptosis [69–71].

Which chloramines are formed by HOCl reaction *in vivo* will reflect the availability of the parent amine. Histamine is released from mast cells as an inflammatory mediator and therefore has the potential to react with HOCl from neutrophils. The resultant histamine chloramine is uncharged and is therefore relatively cell permeable [63]. It was also shown to be the most reactive of the biologically relevant chloramines [63]. Intracellular GSH was readily oxidized, but in contrast with HOCl that generates a number of reaction products from GSH, histamine chloramine produced only GSSG [63].

To add to the complexity of understanding the potential reactivity of HOCl and derived oxidants with cells, chloramine exchange was shown to occur and this was likely to affect their cellular interactions [72]. Transchlorination reactions between histamine, glycine, and taurine chloramines and protein chloramines occurred readily and influenced the reactivity with intracellular GAPDH, a highly oxidant-prone thiol enzyme [72].

$$R^1 - NH_2 + R^2 - NHCl \leftrightarrow R^1 - NHCl + R^2 - NH_2 \tag{12.5}$$

HOCl is a highly reactive but also selective oxidant and its diffusion distance from the site of generation will depend on the availability of target molecules. Generally in a cellular environment, with an abundance of available thiols such as GSH and proteins, HOCl could be considered to act over a very limited distance—estimated at around 1 μm in an intracellular space [73]. The chloramines are much less reactive and could persist in biological fluids for many minutes if available thiol substrates were restricted. This is possible in the extracellular milieu where thiol groups may be limiting. In such a setting, transchlorination reactions may contribute to the ability of the oxidant reaching intracellular targets [2,24,72,74].

HOCl GENERATION IN THE NEUTROPHIL PHAGOSOME
AND BACTERIAL KILLING

The phagocytosis of bacteria was one of the first observations of cells in action and was eloquently described by Metchnikoff in the late nineteenth century [75]. The demise of the engulfed bacterium was readily apparent but how this occurred remained an intriguing mystery for almost a century. The generation of superoxide and H_2O_2, together with the *in vitro* demonstrations of HOCl generation by myeloperoxidase in the 1960s and 1970s, led to the hypothesis that this was an essential component of neutrophil bacterial killing [7]. While this appeared to be a perfectly sensible proposal, demonstrating that this reaction actually does occur within the enclosed phagosome was an enormous challenge. When a bacterium is engulfed by a neutrophil, it becomes fully enclosed in a very confined space, and the chemical composition of this intraphagosomal space is determined (1) by the activation of the NOX2 complex in the membrane and generation of superoxide, (2) by fusion with cytoplasmic granules containing many and varied cytotoxic proteins, and (3) by active transport of electrolytes such as H^+ and Cl^- [76,77] (see Figure 12.2). Whether the phagosome provided the right conditions to support HOCl generation by myeloperoxidase was a matter of (sometimes fierce) debate. Is the optimal pH maintained? Is there sufficient Cl^-? Even if formed, is the HOCl able to contribute to bacterial killing? The essential nature of the oxidative burst for the microbicidal activity of neutrophils was clearly apparent from the clinical observations of individuals with chronic granulomatous disease who carry a mutation in the NOX2 enzyme complex and whose neutrophils lack an oxidative burst. These individuals are highly susceptible to bacterial infections that do not resolve, giving rise to granulomas and other pathologies [78–80]. However, in contrast with this phenotype in the absence of oxidant generation, myeloperoxidase deficiency is common but often goes undetected as the individuals do not exhibit clinical symptoms [81–84]. It was subsequently shown that H_2O_2 can build up to cytotoxic concentrations in myeloperoxidase-deficient neutrophils [85] and that myeloperoxidase deficiency does result in compromised killing capacity (reviewed in [83,86,87]). However, concrete evidence that HOCl was indeed formed in the phagosome in quantities sufficient to kill bacteria remains a challenging question that has kept us entertained and intrigued.

Evidence that superoxide and H_2O_2 are generated directly onto the phagocytosed target came from cytochemical studies [88,89]. Superoxide was calculated to be formed at 5–10 mM s^{-1} [90] and the myeloperoxidase concentration reached 1–2 mM, suggesting optimal conditions for substantial HOCl generation [91]. Demonstrations of the bleaching of fluorescein-coated beads following ingestion [92], the selective chlorination of fluorescein-coated particles [93], and chlorination of bacterial proteins [94] indicated that HOCl was indeed formed in the phagosome environment. Chlorinated bacterial and neutrophil proteins were detected following phagocytosis, with the vast majority (94%) being detected in the neutrophil itself [95]. This result reinforces the indiscriminate targeting of a reactive oxidant in a complex environment.

Direct visualization of oxidants generated in the phagosome has become possible with the development of HOCl-reactive fluorescent dyes. Although the absolute

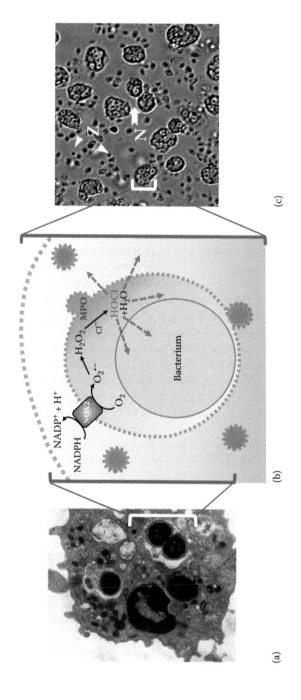

FIGURE 12.2   Generation of HOCl within the neutrophil phagosome. (a) Electron micrograph showing phagocytosed bacteria taken up into phagosomes. Note the limited space around the ingested organism. (b) Cartoon of the intracellular generation of HOCl, also indicating its potential to react both with the bacterial target and the neutrophil itself. (c) Live cell imaging shows the generation of HOCl within the phagosomes of neutrophils (N) following phagocytosis of zymosan particles (Z). HOCl was detected by reaction with the probe R19-S to produce its fluorescent product. Note also the heterogeneous response between phagosomes, with some showing positive red dye and others not. Image was taken using an Olympus IX81 live cell imaging microscope with an XM10 camera (20 times objective) and provided by Amelia Albrett, Tony Kettle, and Christine Winterbourn.

specificity of the probes for HOCl still needs to be chemically validated, real-time imaging indicates HOCl generation inside phagosomes following particle ingestion ([96–98] and Figure 12.2). This technology has also allowed visualization of individual phagosomes that suggests a heterogeneous response with respect to oxidant generation [96–98]. The implications of this variable reaction in phagosomes are unknown.

HOCl GENERATION IN THE EXTRACELLULAR SPACE

Myeloperoxidase can be released into the extracellular space after neutrophil degranulation [99] or from macrophages [100], and its high positive charge (pI ~ 11) encourages adherence to negatively charged surfaces and proteins. It has been detected in the subendothelial space of the vasculature [101,102] and administration of negatively charged heparin results in elevated plasma levels due to release of the myeloperoxidase enzyme from the endothelium [103]. It is also detected on neutrophil extracellular traps (cellular DNA material actively ejected from the cell) [104–106] and elevated myeloperoxidase is detected in plasma following myocardial infarction, presumably as a result of release from activated neutrophils [107]. Neutrophils also target host tissues in autoimmune diseases and are likely to release granule proteins directly into the extracellular space in these circumstances [108–110]. Taken together, these data indicate a high likelihood that HOCl could be formed in the tissues at inflammatory sites.

Biomarkers for HOCl Generation

Specific antibodies were used to detect the presence of HOCl-modified proteins colocalized with myeloperoxidase in atherosclerotic plaque tissue [111,112] and in human kidney [113]. Other biomarkers are stable products of HOCl reaction with proteins, GSH and phospholipids. Levels of 3-chlorotyrosine (Cl–Tyr) were elevated in patients with CVD and respiratory illness [41,114]. Glutathione sulfonamide, a unique oxidation product formed by reaction with GSH [33,115], is potentially useful as a biomarker because it is stable and is a specific signature molecule for HOCl. Glutathione sulfonamide is released from endothelial cells treated with nonlethal doses of HOCl [115] and has been detected in lung lavage fluid from children with cystic fibrosis [22,33,34].

Ready reaction with plasmalogens (see Figure 12.1), major phospholipids of endothelial and vascular smooth muscle cells and cardiac myocytes, generates excellent biomarkers: HOCl reacts with the plasmalogen vinyl ether group to produce 2-chlorofatty acids and 2-chlorofatty alcohols [116] that can be further metabolized to 2-chlorohexadonic acid and 2-chloradipic acid. These compounds retain the chlorine atom, are produced *in vivo* by myeloperoxidase [117,118], and can be detected using mass spectrometry. They have been detected in atheroma, in LDL from atherosclerotic lesions, and in the urine of humans and rats [54,119,120]. That 2-chloroadipic acid can be measured with assurance in urine makes this biomarker particularly attractive.

MECHANISMS OF HOCl-MEDIATED TISSUE INJURY

The detection of biomarkers of HOCl associated with a number of pathologies and its biological reactivity, as discussed above, provides a compelling case for its formation and potential *in vivo* impact. Consideration should therefore be given to the potential biological consequences of its formation. Much of the research with HOCl has been carried out *in vitro* and when interpreting these data, the reactivity of HOCl should be kept in mind. Many things are possible *in vitro*, but which reactions are physiologically relevant will depend on the *in vivo* sites of HOCl formation: because neutrophils are rarely activated in the circulation, targets are most likely to be host cells at inflammatory sites and the extracellular proteins in these same tissues. Plasma proteins, which are often studied, are unlikely to be affected, although they are of course potentially reactive.

REACTIVITY OF HOCl AND CHLORAMINES WITH CELLS

HOCl is more cytotoxic than most chloramines, reflecting its more indiscriminate reactivity and ability to react with numerous intracellular targets. Cellular antioxidant defense against HOCl and chloramines exists in the scavenging ability of thiol compounds, particularly GSH and the peroxiredoxins, and ascorbate, all abundant in cells and able to form a significant protective shield [25,71,115,121]. GSSG generated by HOCl or chloramines can be recycled to some extent [63] although recycling appears to be inefficient with higher concentrations of HOCl [25,115]. The peroxiredoxins (Prx) are ubiquitous and abundant cytoplasmic (Prx1 and Prx2) and mitochondrial (Prx3) proteins, and in endothelial cells and red blood cells, Prxs were oxidized by HOCl or cell-permeable chloramines to disulfide-linked dimers that could be reversed within 10–30 min [121,122]. That Prxs are able to limit the toxicity of chlorinated oxidants was suggested by the observation that pretreatment with H_2O_2 to oxidize Prxs increased the subsequent susceptibility to ammonia chloramine [121].

The thiol screen offered by GSH and the Prxs would be expected to limit the damage to other thiol-containing enzymes. However some enzymes have been shown to be highly susceptible and may have significant functional consequences. GAPDH is very readily oxidized in endothelial cells [32], with implications for cell metabolism, and a recent study has indicated alterations in lysosomal enzyme function, demonstrating selective inactivation of cys-dependent cathepsins B and L in J774A.1 macrophages [123]. In this same study, lysosomal acid lipase activity could also be decreased by exposure of the cells to HOCl-modified LDL, with implications for LDL processing by macrophages and the development of atherosclerosis [123].

The effects of cellular exposure to chlorinated oxidants is entirely dose dependent, with higher doses readily causing cell death and lower doses affecting numerous functional aspects of cell biology (reviewed by [21] and references therein). It should be noted that, when considering the cytotoxicity of a highly reactive oxidant such as HOCl, complete reaction with the oxidant will occur; hence, the *amount* of the oxidant that reacts with the target cell, rather than the *concentration*, will determine toxicity. For example, equivalent cytotoxicity could be achieved by adding

1 mL of a 100 μM solution to 10^6 cells as when adding 10 mL of a 10 μM solution to the same number of cells. In a similar manner, the medium in which the cells are contained will influence the reactivity of HOCl *in vitro*. Full culture medium containing free amino acids, serum proteins, and numerous other molecules will effectively act as a scavenger system for the added oxidant. In many *in vitro* experimental systems carried out in full medium, any observed effect on the cells is likely to be carried out by secondary chloramines formed from the proteins and free amino compounds present [25]. The cytotoxicity of chloramines reflects their permeability; impermeable agents such as taurine chloramine are almost completely nontoxic when added to intact cells [61,65], whereas only a very low dose of a highly cell-permeable agent such as monochloramine can cause extensive lysis [63].

HOCl- AND CHLORAMINE-INDUCED CELL DEATH

Necrotic cell death is initiated in vascular and aortic endothelial cells following exposure to >100 μM HOCl [124,125]. Oxidation of cell thiols does not appear to result in necrosis or cell lysis and is more likely to be associated with oxidation of alternative targets such as membrane proteins or lipids [124,126]. The toxic potential of HOCl is such that doses of HOCl that are able to cause rapid lysis (<1 h) may still lie within a physiologically relevant range [21]. The spectrum of dose-dependent toxicity continues such that lesser amounts are able to initiate the classical apoptotic pathway involving phosphatidylserine exposure, translocation of Bax to the mitochondrial membrane, caspase 3 activation and DNA fragmentation [20,124,125]. Sublethal oxidant exposure disrupts the cell cycle, inducing either growth arrest or proliferation [21,32,71,127]. These latter processes are initiated by the activation of intracellular signaling cascades.

EFFECT OF HOCl AND CHLORAMINES ON CELL SIGNALING

Cellular thiol proteins are readily oxidized upon exposure to HOCl or chloramines [28,32,65,128], with significant functional consequences. The EGF receptor on the cell surface and intracellular phosphatases have been identified as targets, with consequent activation of the MAP kinase pathways leading to apoptosis or growth arrest or altered endothelial cell stress responses [65,129]. More recently, increased ERK 1/2 phosphorylation was observed to initiate vasculogenesis in an *in vivo* mouse model of systemic sclerosis in which HOCl was administered daily for six weeks, generating a dermal and pulmonary sclerosis [130,131]. There was also increased activation of the membrane-bound GTPases Ras and Rho and increased VEGF expression. These changes could be prevented by the thiol reagent propylthiouracil, which reduced HOCl-mediated lipid peroxidation and prevented aortic thickening and myofibroblast differentiation [130,131].

   Perturbations in intracellular Ca^{2+} that were thiol-dependent were noted in rabbit ventricular myocytes within minutes of HOCl exposure [132], affecting cardiac contractility and sarcoplasmic reticulum Ca^{2+} ATPase activity [133]. Possibly associated with the changes in intracellular Ca^{2+} is a downstream effect on Ca^{2+} uptake into mitochondria that results in reduced mitochondrial membrane potential and mitochondrial disruption, this being observed in a number of cell types, including endothelial and epithelial cells, liver cells and neurons following low-level exposure to HOCl (reviewed in [21] and references therein).

A number of transcription factors are affected by HOCl or chloramines, with significant consequences for gene expression and cell function. Nuclear factor (NF) κB, localized in the cytoplasm as a dimer of p50–p65 subunits bound to the inhibitory protein IκB, regulates the gene expression of inflammatory cytokines and leukocyte adhesion proteins. Phosphorylation of IκB leads to its degradation by the proteasome and the release of active NFκB [134]. DNA binding by active NFκB requires an oxidant-sensitive cysteine residue in the p50 subunit to be maintained in its reduced state in the nucleus by thioredoxin [135]. Thus, loss of thioredoxin in an oxidizing environment can result in decreased transcriptional activity of NFκB [136]. Chloramines have been shown to oxidize IκBα by modification of a single methionine residue [66], preventing degradation of IκBα by the proteasome and overriding the activation of NFκB. These events have the potential to exert a profound effect on the inflammatory response.

The tumor suppressor p53 is a master regulatory transcription factor of cell life and death pathways [137,138]. The activity of p53 can be variously affected by oxidation of thiol residues that can affect DNA binding and transcription [139,140]. Low concentrations of reagent chloramines were able to activate p53 in fibroblasts and alter gene expression, upregulating p21 levels [69]. The nuclear factor erythroid 2-related factor (Nrf2) transcription factor upregulates the expression of antioxidant defense enzymes upon thiol oxidation of its inhibitor Keap-1. The demonstration of HOCl- and chloramine-mediated Nrf2 activation in macrophages [141] and epithelial cells [142], together with the upregulation of downstream Nrf2 gene targets such as heme oxygenase 1, is consistent with a thiol-specific oxidant and indicates that exposure to these oxidants can mediate a protective response in tissue cells.

EFFECT OF HOCl AND CHLORAMINES ON EXTRACELLULAR MATRIX

Myeloperoxidase released from neutrophils or macrophages has been detected in the subendothelial matrix of the blood vessels [101,102]. The proximity of the extracellular matrix (ECM) to the potential site of HOCl generation is strongly suggestive of this tissue being a biologically relevant target. The ECM not only provides a support for neighboring cells, but also influences the function of the adherent cells. Damage to the ECM can result in altered cell function and is integral to organ integrity, with alterations being reported in a number of inflammatory conditions including glomerulonephritis, rheumatoid arthritis, chronic obstructive pulmonary disease, and atherosclerosis (see the review by Chuang et al. [143] and references therein).

The ECM represents heterogeneous and site-specific structures with a backbone of collagen and elastin, together with numerous glycosylated proteins and proteoglycans and these components offer HOCl and chloramines a number of potential targets. The sugar moieties present can react with HOCl, but it seems that damage to the protein core of the glycoproteins and glycosaminoglycans predominates [143]. Many reactions are both possible and probable in the complex setting of the ECM, and demonstration of relative importance is challenging. However, despite this, there is strong evidence that modification of the ECM by HOCl alters cell adhesion, as shown with endothelial cells [144,145]. HOCl also promoted the detachment of endothelial cells from a Matrigel substratum [146], and the degradation of collagen by neutrophil proteinases has been shown to be enhanced following pretreatment of ECM with myeloperoxidase-derived oxidants [147].

SUMMARY AND CONCLUSIONS

From the information presented in this chapter, it is clear that neutrophils can, and do, generate HOCl when activated in an inflammatory setting. Whereas this property is extremely useful for the primary function of the neutrophil, that is, the killing of ingested microorganisms, the valiant effort by the cell to contain the oxidant into an enclosed space is not always successful, and biomarkers of chlorinated oxidants are readily detectable in acute inflammatory conditions such as cystic fibrosis, pneumonia and related respiratory infections, sepsis, and trauma such as burns [2,22]. In addition, the neutrophil is an ever-present cell in noninfectious inflammatory situations such as rheumatoid arthritis, chronic obstructive pulmonary disease, ischemia-reperfusion events (heart infarction and stroke), atherosclerosis, rheumatoid arthritis, glomerulonephritis, vasculitis, inflammatory bowel disease, and many more. Activation and release of myeloperoxidase into the tissues in these conditions will result in the generation of HOCl and the formation of downstream chloramines, the reactivity of which has been described here. The potential for these oxidants to react with highly susceptible thiols, proteins, and other vital cellular targets has prompted the search for effective inhibitors of myeloperoxidase, aimed at the prevention of this oxidant generation [22,148].

The fast reaction of the chlorinated oxidants with thiols has highlighted their potential for influencing cell signaling events as well as cell growth and survival that are often dependent on a thiol-mediated process. Transchlorination reactions may allow these oxidants to be relayed to targets far from where original oxidant was formed. In addition, the potential to generate dehydromethionine from N-terminal methionine residues, a reaction specific to HOCl and chloramines, may also impact on inflammation. These effects could be either positive or negative, and it is likely that the recyclable thiol-based antioxidant systems, the GSH/GSSG, and Prxs/thioredoxin systems are able to ameliorate many effects of HOCl exposure.

POSTSCRIPT

The authors of this chapter have shared an interest in neutrophil-mediated oxidant production and the biological consequences of HOCl generation with Professor Christine Winterbourn, to whom this book is dedicated, since the early 1980s. Interestingly, relatively few in the oxidative stress research area have ventured into the study of the halogenated oxidants. That Professor Winterbourn has embraced this topic reflects her willingness to engage with a complex and highly reactive oxidant with high biological relevance. Her knowledge of chemistry has combined well with our desire to understand the biological relevance of these oxidants and our shared interest in this topic persists. It has been a lot of fun!

REFERENCES

1. Klebanoff, S.J. and R.A. Clark. 1978. *The Neutrophil: Function and Clinical Disorders.* 1st edn. Amsterdam, the Netherlands: North-Holland Publishing Company.
2. Winterbourn, C.C., A.J. Kettle, and M.B. Hampton. 2016. Reactive oxygen species and neutrophil function. *Annu. Rev. Biochem.* 85:765–792. doi: 10.1146/annurev-biochem-060815-014442.
3. Sbarra, A.J. and M.L. Karnovsky. 1959. The biochemical basis of phagocytosis I. Metabolic changes during the ingestion of particles by polymorphonuclear leukocytes. *J. Biol. Chem.* 234:1355–1362.
4. Iyer, G.Y.N., M.F. Islam, and J.H. Quastel. 1961. Biochemical aspects of phagocytosis. *Nature* 192:535–541.
5. Klebanoff, S.J. 1967. A peroxidase-mediated antimicrobial system in leukocytes. *J. Clin. Invest.* 46:1078.
6. Klebanoff, S.J. 1968. Myeloperoxidase-halide-hydrogen peroxide antibacterial system. *J. Bacteriol.* 95:2131–2138.
7. Klebanoff, S.J. 1970. Myeloperoxidase: Contribution to the microbicidal activity of intact leukocytes. *Science* 169:1095–1097.
8. Van Dalen, C.J., M.W. Whitehouse, C.C. Winterbourn, and A.J. Kettle. 1997. Thiocyanate and chloride as competing substrates for myeloperoxidase. *Biochem. J.* 327 (Pt 2):487–492.
9. Van Dalen, C.J. and A.J. Kettle. 2001. Substrates and products of eosinophil peroxidase. *Biochem. J.* 358 (Pt 1):233–239.
10. Kettle, A.J. and C.C. Winterbourn. 1997. Myeloperoxidase: A key regulator of neutrophil oxidant production. *Redox Rep.* 3 (1):3–15.
11. Davies, M.J., C.L. Hawkins, D.I. Pattison, and M.D. Rees. 2008. Mammalian heme peroxidases: From molecular mechanisms to health implications. *Antioxid. Redox Signal* 10 (7):1199–1234. doi: 10.1089/ars.2007.1927.
12. Wijkstrom-Frei, C., S. El-Chemaly, R. Ali-Rachedi, C. Gerson, M.A. Cobas, R. Forteza, M. Salathe, and G.E. Conner. 2003. Lactoperoxidase and human airway host defense. *Am. J. Respir. Cell Mol. Biol.* 29 (2):206–212.
13. Nagy, P., S.S. Alguindigue, and M.T. Ashby. 2006. Lactoperoxidase-catalyzed oxidation of thiocyanate by hydrogen peroxide: A reinvestigation of hypothiocyanite by nuclear magnetic resonance and optical spectroscopy. *Biochemistry* 45 (41):12610–12616.
14. Ihalin, R., V. Loimaranta, M. Lenander-Lumikari, and J. Tenovuo. 1998. The effects of different (pseudo)halide substrates on peroxidase- mediated killing of *Actinobacillus actinomycetemcomitans. J Periodontal. Res.* 33 (7):421–427.
15. O'Brien, P.J. 2000. Peroxidases. *Chem. Biol. Interact.* 129 (1–2):113–139.
16. Winterbourn, C.C., M.C. Vissers, and A.J. Kettle. 2000. Myeloperoxidase. *Curr. Opin. Hematol.* 7 (1):53–58.
17. Pattison, D.I. and M.J. Davies. 2001. Absolute rate constants for the reaction of hypochlorous acid with protein side chains and peptide bonds. *Chem. Res. Toxicol.* 14 (10):1453–1464.
18. Pattison, D.I. and M.J. Davies. 2004. Kinetic analysis of the reactions of hypobromous acid with protein components: Implications for cellular damage and use of 3-bromotyrosine as a marker of oxidative stress. *Biochemistry* 43 (16):4799–4809.
19. Barrett, T.J. and C.L. Hawkins. 2012. Hypothiocyanous acid: Benign or deadly? *Chem. Res. Toxicol.* 25 (2):263–273. doi: 10.1021/tx200219s.
20. Pattison, D.I., M.J. Davies, and C.L. Hawkins. 2012. Reactions and reactivity of myeloperoxidase-derived oxidants: Differential biological effects of hypochlorous and hypothiocyanous acids. *Free Radic. Res.* 46 (8):975–995. doi: 10.3109/10715762.2012.667566.

21. Rayner, B.S., D.T. Love, and C.L. Hawkins. 2014. Comparative reactivity of myeloper-oxidase-derived oxidants with mammalian cells. *Free Radic. Biol. Med.* 71:240–255. doi: 10.1016/j.freeradbiomed.2014.03.004.

22. Kettle, A.J., A.M. Albrett, A.L. Chapman, N. Dickerhof, L.V. Forbes, I. Khalilova, and R. Turner. 2014. Measuring chlorine bleach in biology and medicine. *Biochim. Biophys. Acta* 1840 (2):781–793. doi: 10.1016/j.bbagen.2013.07.004.

23. Winterbourn, C.C. 2008. Reconciling the chemistry and biology of reactive oxygen spe-cies. *Nat. Chem. Biol.* 4 (5):278–286. doi: 10.1038/nchembio.85.

24. Pattison, D.I., C.L. Hawkins, and M.J. Davies. 2009. What are the plasma targets of the oxidant hypochlorous acid? A kinetic modeling approach. *Chem. Res. Toxicol.* 22 (5):807–817.

25. Pullar, J.M., M.C. Vissers, and C.C. Winterbourn. 2000. Living with a killer: The effects of hypochlorous acid on mammalian cells. *IUBMB Life* 50 (4–5):259–266.

26. den Hartog, G.J., G.R. Haenen, E. Vegt, W.J. van der Vijgh, and A. Bast. 2002. Efficacy of HOCl scavenging by sulfur-containing compounds: Antioxidant activity of glutathi-one disulfide? *Biol. Chem.* 383 (3–4):709–713.

27. Winterbourn, C.C. and S.O. Brennan. 1997. Characterisation of the oxidation prod-ucts of the reaction between reduced glutathione and hypochlorous acid. *Biochem. J.* 326:87–92.

28. Pattison, D.I., C.L. Hawkins, and M.J. Davies. 2007. Hypochlorous acid-mediated pro-tein oxidation: How important are chloramine transfer reactions and protein tertiary structure? *Biochemistry* 46 (34):9853–9864.

29. Green, J.N., A.J. Kettle, and C.C. Winterbourn. 2014. Protein chlorination in neutrophil phagosomes and correlation with bacterial killing. *Free Radic. Biol. Med.* 77:49–56. doi: 10.1016/j.freeradbiomed.2014.08.013.

30. Peskin, A.V. and C.C. Winterbourn. 2001. Kinetics of the reactions of hypochlorous acid and amino acid chloramines with thiols, methionine, and ascorbate. *Free Radic. Biol. Med.* 30:572–579.

31. Pattison, D.I. and M.J. Davies. 2006. Reactions of myeloperoxidase-derived oxidants with biological substrates: Gaining chemical insight into human inflammatory diseases. *Curr. Med. Chem.* 13 (27):3271–3290.

32. Pullar, J.M., C.C. Winterbourn, and M.C. Vissers. 1999. Loss of GSH and thiol enzymes in endothelial cells exposed to sublethal concentrations of hypochlorous acid. *Am. J. Physiol.* 277 (4 Pt 2):H1505–H1512.

33. Harwood, D.T., A.J. Kettle, and C.C. Winterbourn. 2006. Production of glutathione sul-fonamide and dehydroglutathione from GSH by myeloperoxidase-derived oxidants and detection using a novel LC-MS/MS method. *Biochem. J.* 399 (1):161–168.

34. Harwood, D.T., S.L. Nimmo, A.J. Kettle, C.C. Winterbourn, and M.T. Ashby. 2008. Molecular structure and dynamic properties of a sulfonamide derivative of glutathione that is produced under conditions of oxidative stress by hypochlorous acid. *Chem. Res. Toxicol.* 21 (5):1011–1016.

35. Carr, A.C. and C.C. Winterbourn. 1997. Oxidation of neutrophil glutathione and protein thiols by myeloperoxidase-derived hypochlorous acid. *Biochem. J.* 327 (Pt 1):275–281.

36. Winterbourn, C.C. 1997. Free radicals in biology: A personal perspective. *N. Z. Sci. Rev.* 54 (3–4):67–69.

37. Peskin, A.V., R. Turner, G.J. Maghzal, C.C. Winterbourn, and A.J. Kettle. 2009. Oxidation of methionine to dehydromethionine by reactive halogen species generated by neutrophils. *Biochemistry* 48 (42):10175–10182. doi: 10.1021/bi901266w.

38. Davies, M.J. 2016. Protein oxidation and peroxidation. *Biochem. J.* 473 (7):805–825. doi: 10.1042/bj20151227.

39. Kettle, A.J. 1996. Neutrophils convert tyrosyl residues in albumin to chlorotyrosine. *FEBS Lett.* 379:103–106.

40. Kettle, A.J. 1999. Detection of 3-chlorotyrosine in proteins exposed to neutrophil oxidants. *Methods Enzymol.* 300 (Part B):111–120.
41. Upston, J.M., X. Niu, A.J. Brown, R. Mashima, H. Wang, R. Senthilmohan, A.J. Kettle, R.T. Dean, and R. Stocker. 2002. Disease stage-dependent accumulation of lipid and protein oxidation products in human atherosclerosis. *Am. J. Pathol.* 160 (2):701–710.
42. Arnhold, J., A.N. Osipov, H. Spalteholz, O.M. Panasenko, and J. Schiller. 2001. Effects of hypochlorous acid on unsaturated phosphatidylcholines. *Free Radic. Biol. Med.* 31 (9):1111–1119.
43. Panasenko, O.M., H. Spalteholz, J. Schiller, and J. Arnhold. 2003. Myeloperoxidase-induced formation of chlorohydrins and lysophospholipids from unsaturated phosphatidylcholines. *Free Radic. Biol. Med.* 34 (5):553–562.
44. Schroter, J., H. Griesinger, E. Reuss, M. Schulz, T. Riemer, R. Suss, J. Schiller, and B. Fuchs. 2016. Unexpected products of the hypochlorous acid-induced oxidation of oleic acid: A study using high performance thin-layer chromatography-electrospray ionization mass spectrometry. *J. Chromatogr. A* 1439:89–96. doi: 10.1016/j.chroma.2015.11.059.
45. Schroter, J. and J. Schiller. 2016. Chlorinated phospholipids and fatty acids: (Patho) physiological relevance, potential toxicity, and analysis of lipid chlorohydrins. *Oxid. Med. Cell Longev.* 2016:8386362. doi: 10.1155/2016/8386362.
46. Skaff, O., D.I. Pattison, and M.J. Davies. 2008. The vinyl ether linkages of plasmalogens are favored targets for myeloperoxidase-derived oxidants: A kinetic study. *Biochemistry* 47 (31):8237–8245. doi: 10.1021/bi800786q.
47. Heinecke, J.W., W. Li, D.M. Mueller, A. Bohrer, and J. Turk. 1994. Cholesterol chlorohydrin synthesis by the myeloperoxidase-hydrogen peroxide-chloride system: Potential markers for lipoproteins oxidatively damaged by phagocytes. *Biochemistry* 33 (33):10127–10136.
48. Arnhold, J., O.M. Panasenko, J. Schiller, A. Vladimirov Yu, and K. Arnold. 1995. The action of hypochlorous acid on phosphatidylcholine liposomes in dependence on the content of double bonds. Stoichiometry and NMR analysis. *Chem. Phys. Lipids* 78 (1):55–64.
49. Carr, A.C., J.J. van den Berg, and C.C. Winterbourn. 1996. Chlorination of cholesterol in cell membranes by hypochlorous acid. *Arch. Biochem. Biophys.* 332 (1):63–69. doi: 10.1006/abbi.1996.0317.
50. Carr, A.C., J.J. van den Berg, and C.C. Winterbourn. 1998. Differential reactivities of hypochlorous and hypobromous acids with purified *Escherichia coli* phospholipid: Formation of haloamines and halohydrins. *Biochim. Biophys. Acta* 1392 (2–3):254–264.
51. Vissers, M.C., A.C. Carr, and C.C. Winterbourn. 2001. Fatty acid chlorohydrins and bromohydrins are cytotoxic to human endothelial cells. *Redox Rep.* 6 (1):49–55. doi: 10.1179/135100001101536030.
52. Vissers, M.C., A.C. Carr, and A.L. Chapman. 1998. Comparison of human red cell lysis by hypochlorous and hypobromous acids: Insights into the mechanism of lysis. *Biochem. J.* 330 (Pt 1):131–138.
53. Carr, A.C., M.C. Vissers, N.M. Domigan, and C.C. Winterbourn. 1997. Modification of red cell membrane lipids by hypochlorous acid and haemolysis by preformed lipid chlorohydrins. *Redox Rep.* 3 (5–6):263–271.
54. Ford, D.A. 2010. Lipid oxidation by hypochlorous acid: Chlorinated lipids in atherosclerosis and myocardial ischemia. *Clin. Lipidol.* 5 (6):835–852. doi: 10.2217/clp.10.68.
55. Skaff, O., D.I. Pattison, and M.J. Davies. 2007. Kinetics of hypobromous acid-mediated oxidation of lipid components and antioxidants. *Chem. Res. Toxicol.* 20 (12):1980–1988. doi: 10.1021/tx7003097.
56. Carr, A.C., S.M. Bozonet, J.M. Pullar, J.W. Simcock, and M.C. Vissers. 2013. Human skeletal muscle ascorbate is highly responsive to changes in vitamin C intake and plasma concentrations. *Am. J. Clin. Nutr.* 97 (4):800–807. doi: 10.3945/ajcn.112.053207.

57. Carr, A.C., S.M. Bozonet, J.M. Pullar, J.W. Simcock, and M.C. Vissers. 2013. A randomized steady-state bioavailability study of synthetic versus natural (kiwifruit-derived) vitamin C. *Nutrients* 5 (9):3684–3695. doi: 10.3390/nu5093684.
58. Kuiper, C., G.U. Dachs, M.J. Currie, and M.C. Vissers. 2014. Intracellular ascorbate enhances hypoxia-inducible factor (HIF)-hydroxylase activity and preferentially suppresses the HIF-1 transcriptional response. *Free Radic. Biol. Med.* 69:308–317. doi: 10.1016/j.freeradbiomed.2014.01.033.
59. Pattison, D.I., C.L. Hawkins, and M.J. Davies. 2003. Hypochlorous acid-mediated oxidation of lipid components and antioxidants present in low-density lipoproteins: Absolute rate constants, product analysis, and computational modeling. *Chem. Res. Toxicol.* 16 (4):439–449. doi: 10.1021/tx025670s.
60. Luxford, C., R.T. Dean, and M.J. Davies. 2002. Induction of DNA damage by oxidised amino acids and proteins. *Biogerontology* 3 (1–2):95–102.
61. Peskin, A.V. and C.C. Winterbourn. 2006. Taurine chloramine is more selective than hypochlorous acid at targeting critical cysteines and inactivating creatine kinase and glyceraldehyde-3-phosphate dehydrogenase. *Free Radic. Biol. Med.* 40 (1):45–53. doi: 10.1016/j.freeradbiomed.2005.08.019.
62. Thomas, E.L., M.B. Grisham, and M.M. Jefferson. 1986. Cytotoxicity of chloramines. *Methods Enzymol.* 132:585–593.
63. Peskin, A.V. and C.C. Winterbourn. 2003. Histamine chloramine reactivity with thiol compounds, ascorbate, and methionine and with intracellular glutathione. *Free Radic. Biol. Med.* 35 (10):1252–1260.
64. Peskin, A.V., R.G. Midwinter, D.T. Harwood, and C.C. Winterbourn. 2004. Chlorine transfer between glycine, taurine and histamine: Reaction rates and impact on cellular reactivity. *Free Radic. Biol. Med.* 15:1622–1630.
65. Midwinter, R.G., A.V. Peskin, M.C. Vissers, and C.C. Winterbourn. 2004. Extracellular oxidation by taurine chloramine activates ERK via the epidermal growth factor receptor. *J. Biol. Chem.* 279 (31):32205–32211.
66. Midwinter, R.G., F.C. Cheah, J. Moskovitz, M.C. Vissers, and C.C. Winterbourn. 2006. IkappaB is a sensitive target for oxidation by cell-permeable chloramines: Inhibition of NF-kappaB activity by glycine chloramine through methionine oxidation. *Biochem. J.* 396 (1):71–78.
67. Coker, M.S., W.P. Hu, S.T. Senthilmohan, and A.J. Kettle. 2008. Pathways for the decay of organic dichloramines and liberation of antimicrobial chloramine gases. *Chem. Res. Toxicol.* 21 (12):2334–2343. doi: 10.1021/tx800232v.
68. Magin, R.S., G.P. Liszczak, and R. Marmorstein. 2015. The molecular basis for histone H4- and H2A-specific amino-terminal acetylation by NatD. *Structure* 23 (2):332–341. doi: 10.1016/j.str.2014.10.025.
69. Vile, G.F., L.A. Rothwell, and A.J. Kettle. 2000. Initiation of rapid, p53-dependent growth arrest in cultured human skin fibroblasts by reactive chlorine species. *Arch. Biochem. Biophys.* 377 (1):122–128.
70. Tatsumi, T. and H. Fliss. 1994. Hypochlorous acid and chloramines increase endothelial permeability: Possible involvement of cellular zinc. *Am. J. Physiol.* 267 (36):H1597–H1607.
71. Vissers, M.C.M., W.-G. Lee, and M.B. Hampton. 2001. Regulation of apoptosis by vitamin C: Specific protection of the apoptotic machinery against exposure to chlorinated oxidants. *J. Biol. Chem.* 276 (50):46835–46840.
72. Peskin, A.V., R.G. Midwinter, D.T. Harwood, and C.C. Winterbourn. 2005. Chlorine transfer between glycine, taurine, and histamine: Reaction rates and impact on cellular reactivity. *Free Radic. Biol. Med.* 38 (3):397–405.
73. Winterbourn, C.C. and M.B. Hampton. 2008. Thiol chemistry and specificity in redox signaling. *Free Radic. Biol. Med.* 45 (5):549–561. doi: 10.1016/j.freeradbiomed.2008.05.004.

74. Pattison, D.I. and M.J. Davies. 2006. Evidence for rapid inter- and intramolecular chlorine transfer reactions of histamine and carnosine chloramines: Implications for the prevention of hypochlorous-acid-mediated damage. *Biochemistry* 45 (26):8152–8162. doi: 10.1021/bi060348s.

75. Pelner, L. 1969. Elie Metchnikoff, Ph.D. (1845–1916). Professor at Pasteur Institute. *N. Y. State J. Med.* 69 (17):2371–2376.

76. Menegazzi, R., S. Busetto, E. Decleva, R. Cramer, P. Dri, and P. Patriarca. 1999. Triggering of chloride ion efflux from human neutrophils as a novel function of leukocyte beta 2 integrins: Relationship with spreading and activation of the respiratory burst. *J. Immunol.* 162 (1):423–434.

77. Fittschen, C. and P.M. Henson. 1994. Linkage of azurophil granule secretion in neutrophils to chloride ion transport and endosomal transcytosis. *J. Clin. Invest.* 93:247–255.

78. Curnutte, J.T., D.M. Whitten, and B.M. Babior. 1974. Defective superoxide production by granulocytes from patients with chronic granulomatous disease. *N. Engl. J. Med.* 290 (11):593–597.

79. Smith, R.M. and J.T. Curnutte. 1991. Molecular basis of chronic granulomatous disease. *Blood* 77:673–686.

80. Winkelstein, J.A., M.C. Marino, R.B. Johnston, J. Boyle, J. Curnutte, J.I. Gallin, H.L. Malech et al. 2000. Chronic granulomatous disease. Report on a national registry of 368 patients. *Medicine* 79 (3):155–169.

81. Parry, M.F., R.K. Root, J.A. Metcalf, K.K. Delaney, L.S. Kaplow, and W.J. Richar. 1981. Myeloperoxidase deficiency: Prevalence and clinical significance. *Ann. Intern. Med.* 95 (3):293–301.

82. Nauseef, W.M., M. Cogley, S. Bock, and P.E. Petrides. 1998. Pattern of inheritance in hereditary myeloperoxidase deficiency associated with the R569W missense mutation. *J. Leukoc. Biol.* 63 (2):264–269.

83. Klebanoff, S.J. 1999. Myeloperoxidase. *Proc. Assoc. Am. Phys.* 111 (5):383–389.

84. Winterbourn, C.C. and A.J. Kettle. 2013. Redox reactions and microbial killing in the neutrophil phagosome. *Antioxid. Redox Signal* 18 (6):642–660. doi: 10.1089/ars.2012.4827.

85. Klebanoff, S.J. and S.H. Pincus. 1971. Hydrogen peroxide utilization in myeloperoxidase-deficient leukocytes: A possible microbicidal control mechanism. *J. Clin. Invest.* 50:2226–2229.

86. Klebanoff, S.J. 2005. Myeloperoxidase: Friend and foe. *J. Leukoc. Biol.* 77 (5):598–625.

87. Nauseef, W.M. 2007. How human neutrophils kill and degrade microbes: An integrated view. *Immunol. Rev.* 219:88–102.

88. Karnovsky, M.J. 1994. Cytochemistry and reactive oxygen species: A retrospective. *Histochemistry* 102:15–27.

89. Vissers, M.C., W.A. Day, and C.C. Winterbourn. 1985. Neutrophils adherent to a non-phagocytosable surface (glomerular basement membrane) produce oxidants only at the site of attachment. *Blood* 66 (1):161–166.

90. Hampton, M.B., A.J. Kettle, and C.C. Winterbourn. 1996. The involvement of super-oxide and myeloperoxidase in oxygen-dependent bacterial killing. *Infect. Immun.* 64:3512–3517.

91. Hampton, M.B., A.J. Kettle, and C.C. Winterbourn. 1998. Inside the neutrophil phagosome: Oxidants, myeloperoxidase and bacterial killing. *Blood* 92:3007–3017.

92. Hurst, J.K., J.M. Albrich, T.R. Green, H. Rosen, and S.J. Klebanoff. 1984. Myeloperoxidase-dependent fluorescein chlorination by stimulated neutrophils. *J. Biol. Chem.* 259 (8):4812–4821.

93. Jiang, Q. and J.K. Hurst. 1997. Relative chlorinating, nitrating, and oxidizing capabilities of neutrophils determined with phagocytosable probes. *J. Biol. Chem.* 272 (52):32767–32772.

94. Rosen, H., J.R. Crowley, and J.W. Heinecke. 2002. Human neutrophils use the myeloperoxidase-hydrogen peroxide-chloride system to chlorinate but not nitrate bacterial proteins during phagocytosis. *J. Biol. Chem.* 277 (34):30463–30468.

95. Chapman, A.L., M.B. Hampton, R. Senthilmohan, C.C. Winterbourn, and A.J. Kettle. 2002. Chlorination of bacterial and neutrophil proteins during phagocytosis and killing of *Staphylococcus aureus*. *J. Biol. Chem.* 277 (12):9757–9762.

96. Kenmoku, S., Y. Urano, H. Kojima, and T. Nagano. 2007. Development of a highly specific rhodamine-based fluorescence probe for hypochlorous acid and its application to real-time imaging of phagocytosis. *J. Am. Chem. Soc.* 129 (23):7313–7318. doi: 10.1021/ja068740g.

97. Chen, X., X. Wang, S. Wang, W. Shi, K. Wang, and H. Ma. 2008. A highly selective and sensitive fluorescence probe for the hypochlorite anion. *Chemistry* 14 (15):4719–4724.

98. Chen, X., K.A. Lee, E.M. Ha, K.M. Lee, Y.Y. Seo, H.K. Choi, H.N. Kim et al. 2011. A specific and sensitive method for detection of hypochlorous acid for the imaging of microbe-induced HOCl production. *Chem. Commun. (Camb.)* 47 (15):4373–4375. doi: 10.1039/c1cc10589b.

99. Rudolph, V., T.K. Rudolph, J.C. Hennings, S. Blankenberg, R. Schnabel, D. Steven, M. Haddad et al. 2007. Activation of polymorphonuclear neutrophils in patients with impaired left ventricular function. *Free Radic. Biol. Med.* 43 (8):1189–1196.

100. Daugherty, A., J.L. Dunn, D.L. Rateri, and J.W. Heinecke. 1994. Myeloperoxidase, a catalyst for lipoprotein oxidation, is expressed in human atherosclerotic lesions. *J. Clin. Invest.* 94:437–444.

101. Ballieux, B.E., K.T. Zondervan, P. Kievit, E.C. Hagen, L.A. van Es, F.J. van der Woude, and M.R. Daha. 1994. Binding of proteinase 3 and myeloperoxidase to endothelial cells: ANCA-mediated endothelial damage through ADCC? *Clin. Exp. Immunol.* 97 (1):52–60.

102. Eiserich, J.P., S. Baldus, M.L. Brennan, W. Ma, C. Zhang, A. Tousson, L. Castro et al. 2002. Myeloperoxidase, a leukocyte-derived vascular NO oxidase. *Science* 296 (5577):2391–2394.

103. Baldus, S., V. Rudolph, M. Roiss, W.D. Ito, T.K. Rudolph, J.P. Eiserich, K. Sydow et al. 2006. Heparins increase endothelial nitric oxide bioavailability by liberating vessel-immobilized myeloperoxidase. *Circulation* 113 (15):1871–1878. doi: 10.1161/circulationaha.105.590083.

104. Brinkmann, V., U. Reichard, C. Goosmann, B. Fauler, Y. Uhlemann, D.S. Weiss, Y. Weinrauch, and A. Zychlinsky. 2004. Neutrophil extracellular traps kill bacteria. *Science* 303 (5663):1532–1535.

105. Parker, H. and C.C. Winterbourn. 2012. Reactive oxidants and myeloperoxidase and their involvement in neutrophil extracellular traps. *Front. Immunol.* 3:424. doi: 10.3389/fimmu.2012.00424.

106. Stoiber, W., A. Obermayer, P. Steinbacher, and W.D. Krautgartner. 2015. The role of reactive oxygen species (ROS) in the formation of extracellular traps (ETs) in humans. *Biomolecules* 5 (2):702–723. doi: 10.3390/biom5020702.

107. Mocatta, T.J., A.P. Pilbrow, V.A. Cameron, R. Senthilmohan, C.M. Frampton, A.M. Richards, and C.C. Winterbourn. 2007. Plasma concentrations of myeloperoxidase predict mortality after myocardial infarction. *J. Am. Coll. Cardiol.* 49 (20):1993–2000.

108. Johnson, R.J., W.G. Couser, C.E. Alpers, M. Vissers, M. Schulze, and S.J. Klebanoff. 1988. The human neutrophil serine proteinases, elastase and cathepsin G, can mediate glomerular injury in vivo. *J. Exp. Med.* 168 (3):1169–1174.

109. Johnson, R.J., W.G. Couser, E.Y. Chi, S. Alder, and S.J. Klebanoff. 1987. New mechanism for glomerular injury. Myeloperoxidase-hydrogen peroxide-halide system. *J. Clin. Invest.* 79:1379–1387.

110. Johnson, R.J., S.J. Guggenheim, S.J. Klebanoff, R.F. Ochi, A. Wass, P. Baker, M. Schulze, and W.G. Couser. 1988. Morphologic correlates of glomerular oxidant injury induced by the myeloperoxidase-hydrogen peroxide-halide system of the neutrophil. *Lab. Invest.* 58:294–301.

111. Malle, E., L. Hazell, R. Stocker, W. Sattler, H. Esterbauer, and G. Waeg. 1995. Immunologic detection and measurement of hypochlorite-modified LDL with specific monoclonal antibodies. *Arterioscler. Thromb. Vasc. Biol.* 15 (7):982–989.

112. Malle, E., G. Waeg, R. Schreiber, E.F. Grone, W. Sattler, and H.J. Grone. 2000. Immunohistochemical evidence for the myeloperoxidase/H_2O_2/halide system in human atherosclerotic lesions. Colocalization of myeloperoxidase and hypochlorite-modified proteins. *Eur. J. Biochem.* 267 (14):4495–4503.

113. Malle, E., C. Woenckhaus, G. Waeg, H. Esterbauer, E.F. Grone, and H.-J. Grone. 1997. Immunological evidence for hypochlorite-modified proteins in human kidney. *Am. J. Pathol.* 150 (2):603–615.

114. Buss, I.H., R. Senthilmohan, B.A. Darlow, N. Mogridge, A.J. Kettle, and C.C. Winterbourn. 2003. 3-Chlorotyrosine as a marker of protein damage by myeloperoxidase in tracheal aspirates from preterm infants: Association with adverse respiratory outcome. *Pediatr. Res.* 53 (3):455–462.

115. Pullar, J.M., M.C. Vissers, and C.C. Winterbourn. 2001. Glutathione oxidation by hypochlorous acid in endothelial cells produces glutathione sulfonamide as a major product but not glutathione disulfide. *J. Biol. Chem.* 276 (25):22120–22125.

116. Wildsmith, K.R., C.J. Albert, D.S. Anbukumar, and D.A. Ford. 2006. Metabolism of myeloperoxidase-derived 2-chlorohexadecanal. *J. Biol. Chem.* 281 (25):16849–16860. doi: 10.1074/jbc.M602505200.

117. Anbukumar, D.S., L.P. Shornick, C.J. Albert, M.M. Steward, R.A. Zoeller, W.L. Neumann, and D.A. Ford. 2010. Chlorinated lipid species in activated human neutrophils: Lipid metabolites of 2-chlorohexadecanal. *J. Lipid Res.* 51 (5):1085–1092. doi: 10.1194/jlr.M003673.

118. Thukkani, A.K., J. McHowat, F.F. Hsu, M.L. Brennan, S.L. Hazen, and D.A. Ford. 2003. Identification of alpha-chloro fatty aldehydes and unsaturated lysophosphatidylcholine molecular species in human atherosclerotic lesions. *Circulation* 108 (25):3128–3133. doi: 10.1161/01.cir.0000104564.01539.6a.

119. Messner, M.C., C.J. Albert, J. McHowat, and D.A. Ford. 2008. Identification of lysophosphatidylcholine-chlorohydrin in human atherosclerotic lesions. *Lipids* 43 (3):243–249. doi: 10.1007/s11745-008-3151-z.

120. Brahmbhatt, V.V., C.J. Albert, D.S. Anbukumar, B.A. Cunningham, W.L. Neumann, and D.A. Ford. 2010. {Omega}-oxidation of {alpha}-chlorinated fatty acids: Identification of {alpha}-chlorinated dicarboxylic acids. *J. Biol. Chem.* 285 (53):41255–41269. doi: 10.1074/jbc.M110.147157.

121. Stacey, M.M., M.C. Vissers, and C.C. Winterbourn. 2012. Oxidation of 2-cys peroxiredoxins in human endothelial cells by hydrogen peroxide, hypochlorous acid, and chloramines. *Antioxid. Redox Signal* 17 (3):411–421. doi: 10.1089/ars.2011.4348.

122. Stacey, M.M., A.V. Peskin, M.C. Vissers, and C.C. Winterbourn. 2009. Chloramines and hypochlorous acid oxidize erythrocyte peroxiredoxin 2. *Free Radic. Biol. Med.* 47 (10):1468–1476. doi: 10.1016/j.freeradbiomed.2009.08.022.

123. Ismael, F.O., T.J. Barrett, D. Sheipouri, B.E. Brown, M.J. Davies, and C.L. Hawkins. 2016. Role of myeloperoxidase oxidants in the modulation of cellular lysosomal enzyme function: A contributing factor to macrophage dysfunction in atherosclerosis? *PLoS One* 11 (12):e0168844. doi: 10.1371/journal.pone.0168844.

124. Lloyd, M.M., M.A. Grima, B.S. Rayner, K.A. Hadfield, M.J. Davies, and C.L. Hawkins. 2013. Comparative reactivity of the myeloperoxidase-derived oxidants hypochlorous acid and hypothiocyanous acid with human coronary artery endothelial cells. *Free Radic. Biol. Med.* 65:1352–1362. doi: 10.1016/j.freeradbiomed.2013.10.007.

125. Vissers, M.C., J.M. Pullar, and M.B. Hampton. 1999. Hypochlorous acid causes caspase activation and apoptosis or growth arrest in human endothelial cells. *Biochem. J.* 344 (Pt 2):443–449.
126. Vissers, M.C., A. Stern, F. Kuypers, J. van den Berg, and C.C. Winterbourn. 1994. Membrane changes associated with lysis of red blood cells by hypochlorous acid. *Free Radic. Biol. Med.* 16 (6):703–712.
127. Whiteman, M., S.H. Chu, J.L. Siau, P. Rose, K. Sabapathy, J.T. Schantz, N.S. Cheung, J.P. Spencer, and J.S. Armstrong. 2007. The pro-inflammatory oxidant hypochlorous acid induces Bax-dependent mitochondrial permeabilisation and cell death through AIF-/EndoG-dependent pathways. *Cell. Signal.* 19 (4):705–714. doi: 10.1016/j.cellsig.2006.08.019.
128. Summers, F.A., A. Forsman Quigley, and C.L. Hawkins. 2012. Identification of proteins susceptible to thiol oxidation in endothelial cells exposed to hypochlorous acid and N-chloramines. *Biochem. Biophys. Res. Commun.* 425 (2):157–161. doi: 10.1016/j.bbrc.2012.07.057.
129. Midwinter, R.G., M.C. Vissers, and C.C. Winterbourn. 2001. Hypochlorous acid stimulation of the mitogen-activated protein kinase pathway enhances cell survival. *Arch. Biochem. Biophys.* 394 (1):13–20.
130. Bagnato, G., A. Bitto, N. Irrera, G. Pizzino, D. Sangari, M. Cinquegrani, W. Roberts et al. 2013. Propylthiouracil prevents cutaneous and pulmonary fibrosis in the reactive oxygen species murine model of systemic sclerosis. *Arthritis Res. Ther.* 15 (5):R120. doi: 10.1186/ar4300.
131. Bagnato, G., A. Bitto, G. Pizzino, W.N. Roberts, F. Squadrito, D. Altavilla, G. Bagnato, and A. Saitta. 2015. Propylthiouracil modulates aortic vasculopathy in the oxidative stress model of systemic sclerosis. *Vascul. Pharmacol.* 71:79–83. doi: 10.1016/j.vph.2014.12.006.
132. Eley, D.W., B. Korecky, H. Fliss, and M. Desilets. 1991. Calcium homeostasis in rabbit ventricular myocytes. *Circ. Res.* 69 (4):1132–1138.
133. Eley, D.W., J.M. Eley, B. Korecky, and H. Fliss. 1991. Impairment of cardiac contractility and sarcoplasmic reticulum Ca^{2+} ATPase activity by hypochlorous acid: Reversal by dithiothreitol. *Can. J. Physiol. Pharmacol.* 69:1677–1685.
134. Graham, B. and S.B. Gibson. 2005. The two faces of NFkappaB in cell survival responses. *Cell Cycle* 4 (10):1342–1345.
135. Li, X. and G.R. Stark. 2002. NFkappaB-dependent signaling pathways. *Exp. Hematol.* 30 (4):285–296.
136. Hirota, K., M. Murata, Y. Sachi, H. Nakamura, J. Takeuchi, K. Mori, and J. Yodoi. 1999. Distinct roles of thioredoxin in the cytoplasm and in the nucleus. A two-step mechanism of redox regulation of transcription factor NF-kappaB. *J. Biol. Chem.* 274 (39):27891–27897.
137. Bensaad, K. and K.H. Vousden. 2007. p53: New roles in metabolism. *Trends Cell Biol.* 17 (6):286–291.
138. Sablina, A.A., A.V. Budanov, G.V. Ilyinskaya, L.S. Agapova, J.E. Kravchenko, and P.M. Chumakov. 2005. The antioxidant function of the p53 tumor suppressor. *Nat. Med.* 11 (12):1306–1313.
139. Rainwater, R., D. Parks, M.E. Anderson, P. Tegtmeyer, and K. Mann. 1995. Role of cysteine residues in regulation of p53 function. *Mol. Cell Biol.* 15 (7):3892–3903.
140. Meplan, C., M.J. Richard, and P. Hainaut. 2000. Redox signalling and transition metals in the control of the p53 pathway. *Biochem. Pharmacol.* 59 (1):25–33.
141. Pi, J., Q. Zhang, C.G. Woods, V. Wong, S. Collins, and M.E. Andersen. 2008. Activation of Nrf2-mediated oxidative stress response in macrophages by hypochlorous acid. *Toxicol. Appl. Pharmacol.* 226 (3):236–243. doi: 10.1016/j.taap.2007.09.016.

142. Zhu, L., J. Pi, S. Wachi, M.E. Andersen, R. Wu, and Y. Chen. 2008. Identification of Nrf2-dependent airway epithelial adaptive response to proinflammatory oxidant-hypochlorous acid challenge by transcription profiling. *Am. J. Physiol. Lung Cell Mol. Physiol.* 294 (3):L469–L477. doi: 10.1152/ajplung.00310.2007.

143. Chuang, C.Y., G. Degendorfer, and M.J. Davies. 2014. Oxidation and modification of extracellular matrix and its role in disease. *Free Radic. Res.* 48 (9):970–989. doi: 10.3109/10715762.2014.920087.

144. Vissers, M.C. and C. Thomas. 1997. Hypochlorous acid disrupts the adhesive properties of subendothelial matrix. *Free Radic. Biol. Med.* 23 (3):401–411.

145. Woods, A.A. and M.J. Davies. 2003. Fragmentation of extracellular matrix by hypo-chlorous acid. *Biochem. J.* 376 (Pt 1):219–227. doi: 10.1042/bj20030715.

146. Sugiyama, S., K. Kugiyama, M. Aikawa, S. Nakamura, H. Ogawa, and P. Libby. 2004. Hypochlorous acid, a macrophage product, induces endothelial apoptosis and tissue fac-tor expression: Involvement of myeloperoxidase-mediated oxidant in plaque erosion and thrombogenesis. *Arterioscler. Thromb. Vasc. Biol.* 24 (7):1309–1314.

147. Vissers, M.C. and C.C. Winterbourn. 1986. The effect of oxidants on neutrophil-medi-ated degradation of glomerular basement membrane collagen. *Biochim. Biophys. Acta* 889 (3):277–286.

148. Hair, P.S., L.A. Sass, N.K. Krishna, and K.M. Cunnion. 2017. Inhibition of myeloper-oxidase activity in cystic fibrosis sputum by peptide inhibitor of complement C1 (PIC1). *PLoS One* 12 (1):e0170203. doi: 10.1371/journal.pone.0170203.

149. Storkey, C., M.J. Davies, and D.I. Pattison. 2014. Reevaluation of the rate constants for the reaction of hypochlorous acid (HOCl) with cysteine, methionine, and peptide deriva-tives using a new competition kinetic approach. *Free Radic. Biol. Med.* 73:60–66. doi: 10.1016/j.freeradbiomed.2014.04.024.

13 Peroxidase-Derived Oxidants Hypobromous Acid and Hypothiocyanous Acid

Clare L. Hawkins and Benjamin S. Rayner

CONTENTS

INTRODUCTION

Hypobromous acid (HOBr) and hypothiocyanous acid (HOSCN) are produced by a number of human peroxidase enzymes, including myeloperoxidase (MPO), eosinophil peroxidase (EPO), and lactoperoxidase (LPO), via the reaction of hydrogen peroxide with bromide (Br^-) and thiocyanate (SCN^-) ions, respectively [1–6]. HOBr and HOSCN are chemical oxidants that react with biological molecules and cells with quite different potency and selectivity [1,7]. In each case, these oxidants are critical in immune defense, as each has powerful antibacterial, antiviral, and antifungal properties (e.g., [8–13]). However, in some cases, the overproduction of these oxidants during chronic inflammation has been implicated in the development of disease [12,13]. For example, HOBr is believed to contribute to tissue damage in the lungs of patients with asthma, shown by the detection of elevated amounts of the biomarker 3-bromo-Tyr [12,14]. In contrast to HOBr, which is generally regarded as a damaging oxidant, the role of HOSCN in disease is unclear. Thus, some studies provide evidence to support cellular damage on exposure to this oxidant, which may exacerbate disease (reviewed [7,13]), while others support a potentially protective role of elevated SCN^- in inflammatory disease (reviewed [15,16]).

This chapter will review the formation and reactivity of HOBr and HOSCN in biological systems, and discuss the implications of these reactions in both immune defense and disease.

FORMATION OF HOBr AND HOSCN

The hypohalous acids HOBr and HOSCN are produced *in vivo* via the halogenation cycle of MPO, EPO, and LPO, where the halide or pseudohalide ions donate two electrons to generate the ferric form of the enzyme, which is termed Compound I. There are differences in both the specificity and second order rate constants for the reaction of Br^- and SCN^- with each peroxidase, which are shown in Table 13.1 (reviewed [1,17]). Under normal physiological concentrations of halide ions and at neutral pH, HOBr is thought to be produced *in vivo* primarily by EPO [2], though this species can also be produced by MPO, particularly at pH values > 7 [18]. Although at high concentrations, HOBr is known to be damaging, and can readily modify different biological targets (see below), the production of HOBr *in vivo* can also be beneficial. For example, formation of HOBr by EPO is believed to play a key role in the destruction of parasites, shown by isolated EPO in the presence of H_2O_2 efficiently killing the *Schistosoma mansoni* parasite [19], and a correlation between eosinophil levels and damaged parasites has been reported [20]. Other studies have shown that HOBr formed by peroxidasin is required for matrix synthesis, with bromine known to be essential for the correct assembly of collagen IV scaffolds, which are involved in tissue development and architecture [21,22].

HOSCN is produced *in vivo* by MPO, EPO, and LPO [2,3,6,23]. HOSCN may also be formed by the direct reaction of SCN^- with HOCl or HOBr, which occurs with a high rate constant in each case, and is likely to be relevant under physiological conditions [24,25]. It is well established that LPO is responsible for the production of HOSCN in the oral cavity and airway, where it plays a key antimicrobial role [13,15,26,27]. However, MPO is also believed to be an importance source of HOSCN, with reports that similar yields of HOCl and HOSCN are formed at physiological halide/pseudohalide levels, which is attributed largely to the difference in specificity constants of 1:730 for Cl^- compared to SCN^- [3].

TABLE 13.1

Apparent Second Order Rate Constants for the Halogenation Cycles of MPO, EPO, and LPO at pH 7.0 and 15°C

Peroxidase	Native Enzyme + H_2O_2 ($M^{-1} s^{-1}$)	Compound I + Cl^- ($M^{-1} s^{-1}$)	Compound I + Br^- ($M^{-1} s^{-1}$)	Compound I + SCN^- ($M^{-1} s^{-1}$)
MPO	1.4×10^7	2.5×10^4	1.1×10^6	9.6×10^6
EPO	4.3×10^7	3.1×10^3	1.9×10^7	1.0×10^8
LPO	1.1×10^7	—	4.1×10^4	2.0×10^8

Source: Taken from Furtmuller, P.G. et al., *Arch. Biochem. Biophys.*, 445(2), 199, 2006.

This has sparked increased interest in understanding the biological reactivity of HOSCN, given the key role of MPO in numerous inflammatory pathologies [28,29]. In addition, the concentration of SCN^- in the circulation dramatically influences the ratio of HOCl: HOSCN produced by MPO [16,30]. In smokers who have elevated SCN^- from the detoxification of cyanide in cigarette smoke, the shift to HOSCN production has been postulated to be detrimental, as greater damage is apparent in certain cell types exposed to HOSCN compared to HOCl (reviewed [7,31]). However, in other studies, supplementation with SCN^- is protective, with a reduction in oxidative damage, inflammation, and disease severity seen on diverting MPO to produce less HOCl [32–34].

REACTIVITY OF HOBr AND HOSCN WITH BIOLOGICAL MOLECULES

The reactivity and selectivity of HOBr and HOSCN with biological molecules are markedly different. HOBr, like HOCl, is a powerful oxidant that reacts rapidly with a wide range of biological molecules, including proteins, lipids, DNA, and carbohydrates [35], whereas HOSCN is a far less potent oxidant, which displays a marked selectivity for thiol and selenol moieties in biological environments [13]. The pK_a value of each oxidant is also markedly different, with the pK_a of HOBr reported as 8.7 [36], and that of HOSCN as 4.85–5.3 [37,38]. Thus, under normal physiological conditions HOBr will predominate over ^-OBr, whereas ^-OSCN will predominate over HOSCN, which may have implications for their cellular uptake *in vivo*. In each case, proteins are major targets for these oxidants, particularly HOBr, due to their abundance in biological systems and the presence of numerous reactive amino acid side chains [39]. Low-molecular-mass thiol and selenol species are also key targets, particularly for HOSCN, which has very limited reactivity with other biological targets [40,41]. The rate constants for the reaction of HOBr and HOSCN with various model protein components and the products formed in each case are collected in Tables 13.2 and 13.3. HOBr also reacts readily with DNA, lipids, and carbohydrates, particularly glycosaminoglycans, which are a key component of the extracellular matrix. With HOBr, the reaction with nitrogen-containing functional groups, particularly amine $(-NH_2)$ and amide $(-C(O)NH-)$ moieties, is also important in biological systems, as the resulting bromamines $(RR'NBr)$ and bromamides $(RC(O)N(R')Br)$ are able to brominate other substrates and propagate oxidative damage [1,42].

Reactivity with Proteins

The sulfur (and selenium)–containing functional groups of proteins are the most reactive sites for reaction with HOBr and HOSCN, with the second order rate constants for reaction of these oxidants with Cys reported as $1.2 \times 10^7 \, M^{-1} \, s^{-1}$ [1,35,39] and $7.8 \times 10^4 \, M^{-1} \, s^{-1}$ [31,38,40], respectively. The rate constants for these reactions are compared to other substrates and also the analogous reactions with HOCl in Table 13.2. With free Cys residues, these reactions result in the formation of sulfenyl halide (RS-X) species, which are unstable intermediates that can react further with other thiol residues to form disulfides, or be hydrolyzed to yield sulfenic acid

TABLE 13.2

Summary of Second Order Rate Constants for Reaction of HOCl, HOBr, and HOSCN with Selected Amino Acid, Peptide, and Protein Components

Substrate	k_2(HOCl)/M^{-1} s^{-1}	k_2(HOBr)/M^{-1} s^{-1}	k_2(HOSCN)/M^{-1} s^{-1}
Cysteine	—	—	7.8×10^{4d}
N-acetyl-Cys	ca. 3.2×10^{7a}	1.2×10^{7c}	7.3×10^{3d}
N-acetyl-Met-OMe	3.8×10^{7a}	3.6×10^{6c}	$\ll 10^{3d}$
(N-acetyl-Cys)$_2$	6.4×10^{3b}	3.4×10^{5c}	1.8×10^{3b}
3,3'-dithiodipropionic acid	$1.7 \times 10^{5a,b}$	$1.1 \times 10^{6b,c}$	1.9×10^{3b}
Nα-acetyl-Lys	ca. 7.9×10^{3a}	3.6×10^{5b}	—
N-acetyl-Tyr	47^a	2.6×10^{5b}	—
N-acetyl-Trp	7.8×10^{3a}	3.7×10^{6b}	—
Selenocysteine (Sec)	—	—	1.2×10^{6e}
Selenomethionine (SeMet)	—	—	2.8×10^{3e}

Apparent rate constants determined at pH 7.2–7.5 at 22°C–25°C.
[a] Pattison and Davies [124].
[b] Karimi et al. [125].
[c] Pattison and Davies [39].
[d] Skaff et al. [40].
[e] Skaff et al. [41].

(RSOH), sulfinic acid (RSO$_2$H), and sulfonic acid (RSO$_3$H) [1,13,43]. In general, the oxidation products formed on exposure of Cys residues to HOSCN are reversible, and can be repaired in the presence of a suitable reductant [44–46]. This has led to the hypothesis that SCN$^-$ may be able to modulate oxidative damage induced by MPO during chronic inflammatory conditions (e.g., [15,16]).

HOBr also reacts rapidly (k 4×10^6 M^{-1} s^{-1}) with the thioether Met, which forms methionine sulfoxide and methionine sulfone at high oxidant excesses [35,39,47,48], whereas HOSCN shows no apparent reactivity with Met under normal physiological conditions [3,40]. Reaction of HOBr with free Met or Met residues present at the protein N-terminus also results in the formation of dehydromethionine via a cyclization reaction [49]. This product can also be formed in reactions of Met with N-bromamines [49]. Selenomethionine (SeMet) is reactive with HOSCN, which results in the formation of methionine selenoxide (MetSeO) [41].

The nitrogen-containing amino acid side chains of Lys and His are also favorable targets for HOBr, which results in the formation of reactive N-brominated species, which are themselves oxidants [42]. N-bromamines (and bromamides) can react with other biological substrates to cause both oxidation (e.g., Met [50]) and bromination (e.g., Tyr [51]) reactions. HOBr can also convert Lys into its respective nitrile (2-amino-5-cyanopentanoic acid) [51]. There is evidence to show the decomposition of N-brominated species forming other reactive species, including nitrogen-centered radicals [48], which can rearrange and propagate damage to other sites and substrates [52,53]. The reaction of HOBr with Lys residues to form N-bromamines

TABLE 13.3
Summary of the Products Formed on Reaction of HOCl and HOSCN with Amino Acids, Peptides, and Related Species

Substrate	HOBr	HOSCN
Cysteine (Cys)	Sulfenyl chloride (RS-Br)	Sulfenyl thiocyanate (RS-SCN)
	Sulfenic acid (RS-OH)	Sulfenic acid (RS-OH)
	Sulfinic/sulfonic acids	Sulfinic/sulfonic acids
	Disulfides (RS-SR′)	Disulfides (RS-SR′)
Glutathione (GSH)	GSSG	GSSG
	Glutathione sulfonic acid	Mixed disulfides (GS-SR)
	Mixed disulfides (GS-SR)	
Methionine (Met)	Methionine sulfoxide (MetSO)	No reaction
	Methionine sulfone (MetSO$_2$)	
	Dehydromethionine	
Selenomethionine (SeMet)	Methionine selenoxide (MetSeO)	Methionine selenoxide (MetSeO)
Tryptophan (Trp)	Oxindolylalanine/2-	Oxindolylalanine/2-
	hydroxytryptophan	hydroxytryptophan
	Di-oxindolylalanine species	Di-oxindolylalanine species
Tyrosine (Tyr)	3-bromo-Tyr	Addition of SCN[a]
	3,5-dibromo-Tyr	
Histidine (His)	Ring bromamines (RR′N-Cl)	Thiocyanatimines (RR′N-SCN)
Amines (e.g., Lys, Arg,	N-brominated species	Amino thiocyanate species
α-amino group,	Bromamines (RR′NBr)	(RR′N-SCN)[b]
N-terminus)	Dibromamines (RNBr$_2$)	
Amides (e.g., Gln, Asn,	N-brominated species	No reaction
protein backbone)	Bromamides (RC(O)N(R′)Br)	

References are cited in the accompanying text.
[a] Believed to occur via the formation of (SCN)$_2$ [126], which may not be relevant at physiological pH [58].
[b] Unstable at physiological pH [37].

is also a pathway to the formation of sulfilimine cross-links between Lys and Met residues [50]. This may be a key mechanism to account for HOBr-induced protein cross-linking, and has particular significance in relation to the formation of sulfilimine cross-linked collagen IV scaffolds [54]. These structures are critical to matrix functionality and are formed by reactions involving HOBr, which is formed by peroxidasin, an enzyme utilizing Br⁻ as an essential cofactor [21].

In contrast, Lys and His residues are not believed to be favored targets for HOSCN, though there is some evidence for the formation of the analogous amino thiocyanate (RN–SCN) species on exposure of poly-Lys to HOSCN [45]. However, these products are not generally believed to be relevant *in vivo*, owing to their lack of stability at physiological pH [37]. Sulfonamides (R–SO$_2$–NH$_2$) and heterocyclic aromatic imines, including the imidazole side chain of His have been reported to react with HOSCN forming thiocyanatosulfonamide (R–SO$_2$–NH–SCN) and thiocyanatimine (ring RNH–SCN) species, respectively [37]. The backbone amide sites on proteins, together with the side chains of Gln and Asn, also show some reactivity with HOBr, which in the case of the protein backbone, can result in fragmentation reactions [39,55].

The aromatic side chains of Trp and Tyr are also highly reactive with HOBr (k 3.7×10^6 and 2.6×10^5 M^{-1} s^{-1}, respectively), with these reactions occurring rather more rapidly than the analogous reactions with HOCl [39]. For example, the HOBr-mediated bromination of Tyr residues occurs almost 5000 times faster than the chlorination of Tyr by HOCl, which has implications for the use of 3-bromo-Tyr and 3,5-dibromo-Tyr compared to the analogous chlorinated Tyr products as biomarkers to assess the formation of these oxidants *in vivo* [14,18,39,56]. With HOSCN, modification of Tyr residues has been observed either at low pH or in the presence of peroxidase/SCN$^-$ systems, which is attributed to the formation of thiocyanogen (SCN)$_2$, rather than reactions mediated by HOSCN [57].

Trp residues react with both HOBr [39,47] and HOSCN [45,58]. However, with HOSCN, this reaction again occurs predominately at low pH and likely involves (SCN)$_2$, and therefore may have limited relevance to biological systems [58]. With HOSCN, a number of Trp-derived products have been characterized, with evidence for the formation of oxindolyalanine (2-hydroxytryptophan) and related dioxygenated derivatives [45,58]. These products are also formed with HOCl, and are likely to be formed in the analogous reactions with HOBr [1,59]. There is also evidence for reactivity of various N-bromamines with Trp following exposure to proteins, which in some circumstances is reported to be more efficient than the reaction with the parent HOBr [60]. The targeting of Trp by HOBr, N-bromamines, and HOSCN has implications for both protein structure and function, with Trp oxidation linked to protein unfolding, aggregation, enzyme inactivation, and alteration of functionality [45,47,60,61].

REACTIVITY WITH NUCLEIC ACIDS

Exposure of nucleic acids to HOBr results in the formation of a series of both stable and unstable brominated products [62–65], as well as a number of other oxidized products [66,67], whereas there is no evidence to support the analogous reactions with HOSCN, either in isolated or cellular systems [68,69]. There is evidence for the HOBr-mediated bromination of both pyrimidine and purine nucleosides to form a range of products including 5-bromouracil, 5-bromouridine, 5-bromo-2'-deoxycytidine, 8-bromoadenine, and 8-bromo-2'-deoxyguanosine [62,63,65,66,70,71]. The mechanism involved is likely to be via electrophilic substitution reactions, as HOBr is a powerful electrophilic reagent as well as oxidant, with recent theoretical and experimental studies demonstrating that the reactivity of HOBr correlates well with its electrophilic strength [71]. There is also evidence for other nonbrominated products on exposure of nucleic acids to HOBr, which is particularly relevant for thymidine, where exposure to equimolar amounts of HOBr at physiological pH results in the formation of 5,6-dihydroxy-5,6-dihydrothymidine (thymine glycol) and 5-hydroxy-5,6-dihydrothymidine-6 phosphate [67]. The mechanism involved in this case is postulated to involve both bromohydrin and epoxide intermediates [67]. It has also been shown that the bromination of certain nucleosides by HOBr *in vitro*, particularly 2'-deoxyguanosine, can be enhanced in the presence of taurine by a mechanism involving the formation of taurine dibromamines [72].

The relevance of these brominated and oxidized nucleosides to health and disease is not well established, though 8-bromo-2'deoxyguanosine is present in inflammatory fluids early in the disease process, which has led to the suggestion that this compound may be a promising biomarker for early inflammation [70]. This modified nucleoside is also reported to have miscoding properties, suggesting that it may be able to play a role in driving mutagenesis [73]. Similarly, it has been postulated that the brominated pyrimidine nucleoside derivatives could have both cytotoxic and mutagenic effects, particularly at inflammatory sites that are rich in eosinophils [62,63], though further work is required to assess these reactions in detail.

REACTIVITY WITH LIPIDS

HOBr (and bromamines) react with the double bonds present in unsaturated fatty acid side chains and cholesterol to give bromohydrins (RCH=CHR' + HOBr → RCH(Br)–CH(OH)R'), which may be important *in vivo* as these species reportedly cause cell lysis and also have other cytotoxic properties [74–77]. The phospholipid head groups and plasmalogens, which contain a vinyl ether linkage, are also reactive with HOBr [78,79,80]. The rate constants for the reaction of HOBr with the model phospholipid head groups phosphoryl-serine and phosphoryl-ethanolamine and the model vinyl ether, ethylene glycol vinyl ether, are ca. 10^6 M^{-1} s^{-1}, which are higher than the corresponding reactions with aliphatic alkene models of phospholipids and most lipid-soluble antioxidants [78,79]. In the case of the plasmalogens, reaction with HOBr results in cleavage of the vinyl ether linkage, to give an α-bromo fatty aldehyde and lysophospholipids, which may have important implications *in vivo*, owing to the chemoattractant properties of the α-bromo fatty aldehyde [80]. Low-density lipoprotein (LDL) is another potential lipid-containing target for HOBr, with studies showing oxidation of both the apolipoprotein B moiety and the formation of fatty acid bromohydrins from the lipid moieties [81]. These modifications increased the relative electrophoretic mobility of LDL, which is consistent with the HOBr-induced modifications having potentially proatherogenic effects on cells [81]. These experimental observations are supported by kinetic modeling studies, where a marked difference in the selectivity between HOCl and HOBr on LDL is apparent [78].

In contrast to HOBr, it is rather less clear whether HOSCN plays a significant role in the modification of lipids *in vivo*. Thus, it has been shown that SCN^- can promote the MPO-dependent peroxidation of both LDL [82] and other plasma lipids [83], which in the latter case has been linked with dysfunctional lipid transport in patients with coronary heart disease [83,84]. The SCN^--dependent lipid peroxidation is suggested to involve the formation of SCN^--derived radicals, as it can be prevented by ascorbic acid [82], though there is some question as to whether this is thermodynamically favorable (reviewed [13]). The formation of lipid hydroperoxides, and other lipid-derived oxidation products, including 9-hydroxy-10,12-octadecadienoic acid (9-HODE) and F_2-isoprostanes, is also observed in studies with LDL and reagent

HOSCN, in addition to MPO/H$_2$O$_2$/SCN$^-$ systems [85]. Again, the mechanism involved in the HOSCN-mediated lipid oxidation is not certain. It may involve the formation of protein-derived intermediates such as radicals, as suggested previously by Zhang et al. [83], but further studies are warranted, given that no reactivity of HOSCN is seen with the double bonds of model isolated phospholipids [86].

CELLULAR REACTIVITY OF HOBr

By and large, the reactivity of HOBr with cellular systems has been far less extensively characterized compared to the other peroxidase-derived oxidants HOCl and HOSCN [7]. In early studies, liposomes were utilized as a model of oxidant membrane injury, where it was shown that HOBr, generated through the reaction of MPO and H$_2$O$_2$ with a bromide concentration of 1 mM, was more effective at causing membrane damage than HOCl at equivalent halide concentrations [87]. More recent studies with a variety of cell types, including red blood cells (RBCs) [75,88], endothelial [89,90], monocytes [88], macrophages [88,91], epithelial [92], osteoblasts [93], and tumor cells [94], support the membrane reactivity of HOBr, with evidence of rapid cell lysis upon exposure to HOBr at equivalent or lower concentrations than HOCl. With HOBr, RBC lysis occurred up to ten times more rapidly than with the equivalent concentration of HOCl. In addition, membrane modification was observed, characterized by extensive protein cross-linking and the formation of bromohydrin derivatives of phospholipids and cholesterol. The authors concluded that it was these HOBr-induced alterations to the membrane components of the RBC that were responsible for the evident cell lysis [75].

Subsequent experiments by these researchers demonstrated that exposure of human umbilical vein endothelial cells (HUVECs) to membrane-derived bromohydrins resulted in significant cell death, with the bromohydrins found to be approximately four times more cytotoxic than exposure to equivalent concentrations of chlorohydrins over the same time period. Furthermore, the evident necrosis demonstrated was preceded by a faster rate of uptake of the bromohydrins, compared to the chlorohydrins, by HUVECs indicative of heightened cellular reactivity of the HOBr-derived species [76]. The faster rates of RBC lysis seen with HOBr compared to HOCl has also been attributed to the formation of unstable membrane protein–derived bromamines that decompose to form membrane-bound radicals, which occurs to a much lesser extent with HOCl [88]. The data gathered on the susceptibility of RBCs to HOBr was subsequently compared to the extent of lysis evident within monocyte and macrophage cell cultures [88]. It was found that exposure of the THP-1 human monocyte cell line to 100 nmol HOBr per 10^6 cells caused up to 50% cell lysis within 2 h, while it took 400 nmol HOBr per 10^6 cells to achieve the same result within cultures of the J774A.1 murine macrophage cell line. The authors attributed this difference in susceptibility to HOBr between RBCs, monocytes, and macrophages to differences in cell membrane structure with no protein-derived, nitrogen-centered radicals detectable within either THP-1 or J774A.1 cultures [88].

Subsequent studies on J774A.1 macrophages exposed to concentrations of HOBr of up to 100 μM resulted in ~25% decrease in cellular GSH and a comparative decrease in total cellular thiols that was concurrent with a ~20% decrease in cell viability, in part attributable to apoptotic cell death pathways [91]. Likewise, exposure of the human leukemia cell line HL-60 for 15 min to physiological concentrations of between 100 and 200 µM HOBr resulted in the dose-dependent increase in the extent of apoptotic cell death as measured by DNA fragmentation at 24 h following exposure to the oxidant [94]. Consistent with the previous studies [88,91], the extent of apoptosis in this instance was attributed to the formation of bromamines through the reaction of HOBr with media constituents [94]. Taken together, these data suggest that, rather than having uniformly membrane-specific sequelae, the products generated upon HOBr exposure may also act intracellularly to exert detrimental effects.

It has been just over 30 years since Weiss and colleagues first demonstrated the preference of eosinophil peroxidase (EPO) to utilize bromide, over the more physiologically prevalent chloride ion, to generate HOBr [95]. It has further been shown that SCN^- is up to three times more preferred as a substrate for EPO over bromide, although this is dependent on the localized concentration of both of these substrates [14]. Although SCN^- is the preferred substrate at concentrations of the halides normally found in plasma and extracellular fluids, it has however been subsequently demonstrated that the extent of generation of brominated products, indicative of HOBr production, provides a useful diagnostic marker for the study of the activation of eosinophils within an *in vivo* setting [96,97].

ROLE OF HOBr IN DISEASE

One of the major defining steps within allergic inflammatory diseases such as chronic asthma is the recruitment and activation of eosinophils [14], with the degree of eosinophilia directly correlating with disease severity [98]. Two landmark publications by the Hazen laboratory [96,97] were the first to identify that the increased activity of EPO results in increased brominated tyrosine residues in bronchoalveolar lavage fluid from individuals with both mild [96] and severe [97] asthma, identifying 3-bromotyrosine as a molecular footprint for this inflammatory disease. Indeed, it has further been shown in the sputum from patients presenting with mild to moderate forms of the disease that the expression and activity of EPO is directly correlated with 3-bromotyrosine formation [14]. Furthermore, these studies also demonstrated no comparative increase in MPO expression or the HOCl marker 3-chlorotyrosine, indicating a role for eosinophils and not neutrophils within this disease setting [14].

Another disease with respiratory sequelae is cystic fibrosis (CF). Patients with CF have a mutation in the gene encoding the Cystic Fibrosis Transmembrane Conductance Regulator (CFTR) protein, a channel membrane protein responsible for transporting ions, including chloride, bromide, and thiocyanate, across epithelial cell membranes. As in the case with asthmatic patients, 3-bromotyrosine has been detected within the sputum from CF patients and the extent of 3-bromotyrosine formation correlates with both patient infection and respiratory status [99]. Unlike asthma, however, sputum from CF patients contained only MPO, with no other peroxidase (including EPO) detected, results that strongly implicate neutrophils and not eosinophils as the source of HOBr within this disease setting [99].

Utilizing an *ex vivo* model of isolated tracheas as an asthmatic airway model, Brottman and colleagues [100] demonstrated that EPO in the presence of both H_2O_2

and bromide resulted in the increased permeability of the epithelial layer, evidenced by the increased flux of molecules out of the tracheal lumen, without causing visible cell damage. The authors concluded that these changes in epithelial integrity demonstrated were the result of the effect of HOBr exposure on airway epithelium tight junctions, rather than the cells themselves, results that contradict the earlier *in vitro* studies on isolated epithelial cells that were lysed upon exposure to HOBr [92], as discussed here. The physiological relevance of these data implicates HOBr as playing a role in the modulation of airway hyper-reactivity within the setting of asthma by exposing underlying nerve endings.

The effect of HOBr on the extracellular matrix (ECM) has further been studied in reference to inflammatory disease conditions, including asthma [101], as well as within diabetic renal complications [102]. It has been shown that exposure of vascular smooth muscle cell–generated ECM to concentrations of HOBr of up to 200 μM resulted in the dose-dependent release of carbohydrate and protein components of the ECM over time, through the generation and subsequent decomposition of N-bromo intermediates [101]. Brown and coworkers [102] further demonstrated that, within the diabetic kidney, collagen IV is modified by HOBr, leading to the inhibition of $\alpha1\beta1$ integrin binding, directly affecting interaction of renal cells with the ECM in this setting [102]. Given this evidence of the degradation of ECM components by HOBr, it is conceivable that exposure to this oxidant, particularly during inflammatory diseases, may result in profound effects on the structural integrity of tissues and subsequent cellular and tissue function [101].

The presence of HOBr in the circulation, in addition to the direct effects detailed here, can mediate changes to other serum components, which may result in the further downstream cellular and tissue damage evident within inflammatory diseases. It has been established, within an *in vitro* setting at least, that HOBr is capable of modifying human serum albumin (HSA) by specifically targeting tryptophan residues on the protein [103]. This HOBr-modified HSA (HSA-Br) was subsequently shown to increase neutrophil superoxide production and lead to the dose-dependent increase in neutrophil degranulation concomitant with increased extracellular MPO activity, activating these mechanisms through a PI3K-dependent pathway [103]. While it is yet to be demonstrated upon direct exposure to HOBr, the activation of these cellular mechanisms following exposure to a physiologically achievable intermediate product, in this case modified HSA, illuminates the potential future identification of molecular processes for the targeted treatment of inflammatory diseases involving excessive HOBr production.

HOBr may also play a role in cardiomyopathies, particularly eosinophilic or Loeffler's endocarditis. This condition is characterized by the infiltration and adherence of eosinophils to the endocardium of the heart, resulting in progressive damage, first to the endocardial and then myocardial tissue, culminating in congestive heart failure associated with extensive fibrosis [104]. Slungaard and Mahoney [89], within an isolated rat heart model of the disease, demonstrated that EPO bound to heart muscle, in the presence of H_2O_2 and bromide, resulted in an abrupt decrease in aortic output consistent with the heart failure evident in this condition. Indeed, it has previously been shown that the presence of bromide is able to increase the TNFα-dependent toxicity of eosinophil interaction with endothelial cells, with the exogenous addition of 100 μM bromide to TNFα-activated eosinophils cocultured with HUVECs exacerbating endothelial cell death by ~30% [90]. The authors postulate that the generation of HOBr within this setting may contribute to the pathogenesis of Loeffler's endocarditis, although further studies confirming this hypothesis have yet to be undertaken.

CELLULAR REACTIVITY AND ROLE OF HOSCN IN DISEASE

Unlike the other peroxidase-derived oxidants HOCl and HOBr, which have been shown to be reactive with a range of biological targets, HOSCN is stable enough to penetrate intact mammalian cell membranes and induce intracellular oxidative stress [23,105], reacting with great specificity toward thiols, including glutathione (GSH), protein cysteine residues [40], and selenium-containing species [41]. It is well established that HOSCN acts as an antibacterial agent, able to prevent bacterial cell growth and perturb glucose transport through selective reactivity with thiols located within the active sites of key enzymes involved in bacterial glycolysis [106–109]. It has been reported that HOSCN demonstrates a relatively benign reactivity toward cells of mammalian origin, possibly due to the ability of HOSCN to be rapidly metabolized by thioredoxin reductase (TrxR), an enzyme absent in lower order organisms [110]. It has nonetheless been shown, particularly in cells of the vasculature, that HOSCN exposure elicits intracellular signaling cascades that can exert detrimental effects (reviewed [13,31]). An overview of the reactivity of HOSCN with various mammalian cells is provided in Table 13.4.

The endothelial lining of blood vessels is the primary point of exposure to oxidants derived from circulating leukocytes within the vasculature, with oxidative stress and damage to the endothelium the driving mechanism behind the development of atherosclerosis. Wang and colleagues were the first researchers to demonstrate that endothelial cell exposure to HOSCN results in the activation of cellular signaling mechanisms associated with thrombosis [23] and atherogenesis [105], key mitigating events in atherosclerotic lesion development. Tissue factor (TF) plays a pivotal role in the pathology of thrombosis *in vivo* by acting as the primary initiator of coagulation. Within *in vitro* cell culture models utilizing human umbilical vein endothelial cells (HUVEC), the researchers found that a 4 h exposure to concentrations of HOSCN between 10 and 100 μM resulted in up to a sevenfold increase in TF activity which was markedly increased in the presence of serum up to approximately thirtyfold over controls, consistent with the formation of HOSCN-derived stable sulfenyl thiocyanate (R-S-SCN) adducts or cysteinyl or other serum SH compounds such as GSH [23]. Furthermore, TF activity within the cell lysates was shown to be time dependent, increasing as early as 2 h following exposure to HOSCN, peaking at 6 h before a return to basal levels by 24 h. The researchers reasoned that this pattern of induction is suggestive of a transcriptional regulatory mechanism. Indeed, HOSCN-induced TF activity was shown to be dependent on NFκB activation, emanating from the phosphorylation of the extracellular signal–regulated kinases (ERK) 1/2 pathway. Subsequent studies using the ERK inhibitor U0126 or the irreversible NFκB antagonist andrographolide demonstrated the near total ablation of TF expression and activity in HUVEC exposed to HOSCN [23].

TABLE 13.4

Overview of the Reactivity of HOSCN with Various Mammalian Cell Types

Cell Type	Oxidant System	Observations	Reference
Gingival fibroblasts	LPO/H$_2$O$_2$/ SCN$^-$ and HOSCN	Concentrations up to 300 µM had no effect on proliferation as assessed by ^{3}H-thymidine incorporation.	[127]
RBC	LPO/H$_2$O$_2$/SCN$^-$	Hemolysis/GSH oxidation.	[128]
Porcine aortic endothelial cells	EPO/H$_2$O$_2$/SCN$^-$	SCN$^-$ reduces the extent of H$_2$O$_2$-induced cell lysis (^{51}Cr release) in the presence of Br$^-$.	[129]
RBC	EPO/H$_2$O$_2$/SCN$^-$	GSH oxidation/inactivation of ATPases.	[130]
HL-60 cells	MPO/H$_2$O$_2$/SCN$^-$	SCN$^-$ inhibits H$_2$O$_2$-induced apoptosis seen with Cl$^-$.	[94]
HUVECs	HOSCN	NFκB activation/increased expression of E-selectin, ICAM-1, and VCAM-1.	[105]
HUVECs	HOSCN	Increased tissue factor expression via NFκB and ERK1/2 activation.	[23]
J774A.1 cells	HOSCN	Protein thiol and GSH oxidation/cell lysis and apoptosis/necrosis.	[91]
Calu-3, Neuro-2A, Min-6, HCAECs	Glucose oxidase, MPO/H$_2$O$_2$/ Cl$^-$ with SCN$^-$	Protection against MPO/H$_2$O$_2$/Cl$^-$ induced toxicity, assessed by trypan blue exclusion assay.	[119]
HUVECs	HOSCN	Morphological changes and inhibition of apoptosis and caspase activation.	[116]
J774A.1 cells	HOSCN	Inhibition of PTPs and activation of MAPK signaling.	[113]
A549 cells	HOCl/SCN$^-$	Decreased LDH release with increasing SCN$^-$ concentrations.	[121]
HCAECs	HOSCN	Inhibition of SERCA/increased intracellular Ca^{2+}.	[111]
J774A.1 cells	HOSCN	GAPDH and creatine kinase inactivation/ protein sulfenic acid formation.	[43]
HCAECs	HOSCN	Protein thiol and GSH oxidation/GAPDH inactivation/mitochondrial dysfunction/ apoptosis.	[131]
HCAECs	HOSCN	Decreased eNOS activity and nitrite-nitrate formation/increased free Zn^{2+}.	[115]
J774A.1 cells	HOCl/SCN$^-$	Decreased HOCl-induced necrosis with increasing SCN$^-$ concentrations	[32]
J774A.1 cells	HOSCN	Targeting of thiol-containing metabolic proteins/inhibition of glycolysis.	[117]
HCAECs	HOSCN	Increased intracellular levels of chelatable iron/loss of aconitase activity/increased iron response protein-1 activity.	[114]
HMDMs	HOSCN	Increased PTP activity.	[112]

The expression of endothelial adhesion molecules during inflammation initiates the recruitment and extravasation of leukocytes at sites of tissue injury. As in the case of TF, the expression of a variety of adhesion molecules, including E-selectin, intercellular adhesion molecule 1 (ICAM-1), and vascular cell adhesion molecule 1 (VCAM-1), on the endothelial cell surface is regulated by NFκB activity. Parallel studies by Wang and colleagues [105] found that exposure to 150 μM HOSCN increased HUVEC cell surface expression of E-selectin, ICAM-1, and VCAM-1 up to approximately fivefold over controls. While all three molecules peaked expression following 4 h HOSCN exposure, the former rapidly returned to basal levels, while both ICAM-1 and VCAM-1 remained stably elevated over a prolonged 12 h period. These protein expression results were preceded by corresponding increases in mRNA expression, which peaked following 3 h HOSCN exposure. These results were compatible with a primarily transcriptional mechanism. As was the case for HOSCN-induced HUVEC TF expression and activity [23], the authors further showed in this study that andrographolide inhibition of NFκB activity negated adhesion molecule expression. Interestingly, the HOSCN-induced increased expression of E-selectin, ICAM-1, and VCAM-1 on HUVEC was further exacerbated through the inhibition of the phosphatidylinositol-3-kinase (PI3K)/Akt, suggesting a role for this signaling pathway in the suppression of these adhesion molecules during the immune response.

The chemotaxis of leukocytes to sites of stress or damage within the body is an essential mechanism in the inflammatory response. Upon an intraperitoneal injection of 150 μM HOSCN in mice, Wang and coworkers found that within 4 h there was an eightfold increase over controls in leukocyte extravasation within mouse peritoneal cavity lavage exudates, indicating a possible chemotaxic role for HOSCN [105]. In order to demonstrate the functional consequence of the convergence of both the ability of HOSCN to act as a chemoattractant and the upregulation of the adhesion molecules on HUVEC exposed to the oxidant, the authors of this study then preceded to coculture HOSCN-activated endothelial cells with isolated human neutrophils. They found that exposure of HUVEC to 150 μM HOSCN resulted in an eightfold increase in neutrophil adhesion to the endothelial cell surface [105]. Preincubation of HUVEC cultures with blocking antibodies against E-selectin or of neutrophils with an antibody against CD11b/CD18, the counterligand for ICAM-1, decreased the HOSCN-induced neutrophil adhesion by approximately 40%. Furthermore, andrographolide inhibition of NFκB activity severely attenuated this enhanced neutrophil adhesion induced by endothelial HOSCN exposure, data highlighting a potentially pivotal role for NFκB signaling in these HOSCN-mediated inflammatory processes [105].

The activation of NFκB-mediated pathways is not the only perturbation in intracellular molecular mechanisms induced by HOSCN that leads to vascular cell dysfunction that may result in further vessel patency complications. It has been demonstrated that HOSCN exposure is also able to invoke detrimental effects on crucial cellular processes such as calcium (Ca^{2+}) signaling [111], enzyme activity [112–116], and respiration [117] in a range of vascular cells. Within human coronary artery endothelial cell (HCAEC) cultures, Cook and coworkers demonstrated that exposure to up to 20 μM HOSCN resulted in the concentration-dependent increase in intracellular Ca^{2+} release from internal stores [111]. Experiments with isolated rat skeletal muscle sarcoplasmic reticulum vesicles exposed to HOSCN concentrations of between 10 and 100 µM for 2 h showed significant concentration-dependent inhibition of sarco/endoplasmic reticulum Ca^{2+}-ATPase (SERCA) activity, oxidation of SERCA thiol residues concurrent, and the loss of SERCA Cys residues [111]. Such HOSCN-mediated modification in intracellular Ca^{2+} storage may generate or exacerbate vascular dysfunction, which could have consequences for inflammatory disease settings such as those evident in atherosclerosis.

Modulation of intracellular Ca^{2+} is inextricably linked with permutations in the activity of endothelial nitric oxide synthetase (eNOS), the enzyme responsible for regulating vascular tone. Exposure of isolated recombinant human eNOS to increasing concentrations of HOSCN results in the dose-dependent deactivation of the enzyme, as evidenced by the conversion from the active dimer to the inactive monomer form of the molecule, concurrent with a decreased conversion of L-arginine to L-citrulline [115]. The results obtained from studies performed with isolated eNOS were confirmed within HCAEC and also intact rat aorta, showing that exposure to a HOSCN concentration as low as 50 µM results in the loss of protein thiols and subsequent decrease in L-citrulline formation and cGMP activity [115]. The mechanism of action of HOSCN on eNOS dysfunction was subsequently shown to involve the selective targeting by the oxidant of the Zn^{2+}-cysteine cluster of the eNOS enzyme, with the release of Zn^{2+} paralleling the loss of enzymatic activity [115].

The ability of HOSCN to directly affect the activity of intracellular enzymatic processes demonstrated in this eNOS study has been mirrored in research on the Fe^{2+}-sulfur protein aconitase, the inactivation of which by oxidants is implicated in mitochondrial dysfunction, intracellular iron accumulation and toxicity [114]. Activated aconitase released significant amounts of Fe^{2+} when exposed to ≥ 1.5 µM HOSCN over a 2 h time period, with concentrations of HOSCN as low as 3 µM per mg of protein releasing approximately 80% of Fe^{2+} concurrent with increased deactivation of the enzyme. These effects of HOSCN exposure were inhibited in the presence of citrate, suggesting that HOSCN inactivation of aconitase involves reactions at, or in close proximity to, the active site $[Fe\text{-}S]_4$ cluster. Indeed, peptide mass mapping revealed that HOSCN selectively targets cysteine residues of aconitase involved in binding the Fe^{2+}-sulfur cluster, located on the perimeter of the enzyme active site [114]. As in the case with studies on eNOS mentioned here, the authors here next sought to translate the results demonstrated with isolated enzyme into a functionally more relevant cellular system. Exposure of HCAEC to HOSCN concentrations of up to 200 µM for 2 h resulted in the dose-dependent decrease in HCAEC aconitase activity and the increase in intracellular free Fe^{2+} accumulation that occurred concurrent to an increase in expression of iron response protein 1 (IRP-1), upregulated in a bid to attenuate iron sequestration within the cells. From these studies, it is evident that HOSCN-induced changes to enzymatic activity may have far-reaching consequences.

One particular family of enzymes ubiquitously expressed in various tissue and cell types is protein tyrosine phosphatases (PTPs), responsible for phosphorylation events crucial in the regulation of signal transduction and therefore control of a multitude of cellular processes including cell cycle progression, activation, apoptosis, and metabolic homeostasis. Studies with murine J774A.1 macrophages [113] and with human monocyte-derived macrophages (HMDM) [112] have both demonstrated a HOSCN concentration–dependent loss of PTP activity. Assaying the effect of HOSCN on cell lysates, Lane and colleagues found that HOSCN concentrations as low as 10 μM resulted in the statistical significant inhibition of PTP activity that was only recoverable by DTT over short incubation periods [113]. Similar results were shown within intact macrophages [112,113], but only at the higher HOSCN concentrations ≥100 μM, differences that were attributed to the slow kinetics of the penetration of HOSCN into intact cells [112]. Inhibition of PTP activity at the higher concentrations of HOSCN was also associated with the activation of the p38 mitogen-activated kinase (MAPK) pathway [113] as well as the increased loss of cysteine residues and the formation of sulfenyl thiocyanates (RS-SCN) and/or sulfenic acids (RS-OH) on PTPs [112].

The formation of the reversible RS-SCN and RS-OH products by HOSCN has also been implicated in the disruption of macrophage cellular respiration following exposure to the oxidant [117]. Using the thiol-specific fluorescent probe IAF and MALDI-TOF MS to identify J774A.1 macrophage intracellular targets of HOSCN oxidation, Love and coworkers [117] found reversible thiol oxidation detectable on proteins involved in cell metabolism. These HOSCN-affected proteins included fructose-bisphosphate aldolase, triosephosphate isomerase (TPI), and GAPDH, which are essential for glycolysis, as well as the redox-active cysteine-dependent antioxidant enzyme peroxiredoxin-1. Further studies on macrophage metabolic activity in response to HOSCN exposure were undertaken using a Seahorse XF analyzer. Data obtained revealed that exposure to HOSCN resulted in a concentration-dependent decrease in the rate of glycolysis, subsequent lactate release from the cells, and the loss of cellular ATP stores [117]. The authors reasoned that these events were consistent with the inactivation of TPI and GAPDH, resulting in a rerouting of the glycolytic flux through the parallel pentose phosphate pathway. These perturbations in cellular respiration occurred in the absence of macrophage cell death and therefore may be involved in the promotion of inflammation, particularly in the development of atherosclerosis, in which macrophages are known to play a defined and continual role.

Macrophages are the key effector cell in atherosclerotic lesion development, infiltrating the vessel wall at sites of inflammation, consuming oxidized low density lipoprotein (oxLDL) present, and becoming the lipid-laden foam cells characteristic of the disease, eventually forming part of the necrotic core of advanced lesions. Exposure of LDL to HOSCN results in the formation of modified LDL, with physiological levels of the oxidant able to affect both the protein and lipid components of the LDL particle (see also above) [85]. Functional analysis of the effects of the HOSCN-mediated oxidation of LDL was subsequently undertaken using both J774A.1 macrophages and HMDM [85,118]. Incubation of macrophages with HOSCN-modified LDL resulted in the increased macrophage uptake of lipids and cellular accumulation of cholesterol and cholesterol esters compared to cells incubated with native LDL [85], and perturbed the activity of lysosomal enzymes involved in the detoxification of LDL and related materials [118]. These data implicate HOSCN in the oxidation of LDL within the setting of atherosclerosis, possibly contributing to macrophage lipid loading within this setting.

It is increasingly recognized that rather than being a benign oxidant [110], HOSCN is able to exert a range of effects on host cells and tissues that may have detrimental consequences and contribute toward disease development. Of particular significance is the ability of physiologically relevant concentrations of HOSCN to perturb signaling cascades, enzymatic activity, and cellular homeostasis across a range of cells of vascular origin. However, there is also evidence that the formation of HOSCN via addition of SCN^- to cell cultures exposed to the enzymatic $MPO/H_2O_2/Cl^-$ system or HOCl *in vitro* can reduce cellular damage and death [32,94,119,120]. Similarly, addition of bolus HOSCN to HUVECs protects them from undergoing apoptosis, though profound morphological changes to the cells was apparent following treatment [116].

Supplementation of animals with SCN^- protects lung tissue from damage in both wild type animals and murine models of cystic fibrosis (CF) [32,33,121]. The CF mice have low SCN^- levels in their epithelial lining fluid and suffer from enhanced tissue oxidation and chronic inflammation. This can be corrected by nebulized SCN^-, which decreases neutrophil infiltration and restores the tissue antioxidant levels. In addition, nebulized SCN^- decreased inflammation, cytokine release, bacterial load, and HOCl formation in the lungs of wild type and CF mice infected with *Pseudomonas aeruginosa* {Chandler, 2015 [32]; Chandler, 2013 [33]}, supporting the hypothesis that SCN^- acts as a protective molecule, by both boosting host defense and decreasing tissue injury and inflammation. Supplementation with SCN^- has also been shown to be protective toward the context of atherosclerosis, with human MPO transgenic atherosclerosis-prone mice supplemented with SCN^- showing a reduced extent of lesion formation [34]. Further studies are warranted to determine the mechanisms involved, and whether SCN^- can reduce the extent of MPO-induced tissue damage in these disease models, as in human studies, serum levels of SCN^- in smokers correlate with higher macrophage foam cell populations [122] and fatty streak formation [123], though these studies were performed in smokers, where other detrimental pathways may be a confounding factor.

REFERENCES

1. Davies, M. J., C. L. Hawkins, D. I. Pattison, and M. D. Rees. 2008. Mammalian heme peroxidases: From molecular mechanisms to health implications. *Antioxid Redox Signal* 10 (7):1199–1234. doi: 10.1089/ars.2007.1927.
2. van Dalen, C. J. and A. J. Kettle. 2001. Substrates and products of eosinophil peroxidase. *Biochem J* 358:233–239.
3. van Dalen, C. J., M. W. Whitehouse, C. C. Winterbourn, and A. J. Kettle. 1997. Thiocyanate and chloride as competing substrates for myeloperoxidase. *Biochem J* 327 (Pt 2):487–492.
4. Furtmuller, P. G., U. Burner, and C. Obinger. 1998. Reaction of myeloperoxidase compound I with chloride, bromide, iodide, and thiocyanate. *Biochemistry* 37:17923–17930.
5. Furtmuller, P. G., U. Burner, G. Regelsberger, and C. Obinger. 2000. Spectral and kinetic studies on the formation of eosinophil peroxidase compound I and its reaction with halides and thiocyanate. *Biochemistry* 39 (50):15578–15584.
6. Furtmuller, P. G., W. Jantschko, G. Regelsberger, C. Jakopitsch, J. Arnhold, and C. Obinger. 2002. Reaction of lactoperoxidase compound I with halides and thiocyanate. *Biochemistry* 41 (39):11895–11900.

7. Rayner, B. S., D. T. Love, and C. L. Hawkins. 2014. Comparative reactivity of myelo-peroxidase-derived oxidants with mammalian cells. *Free Radic Biol Med* 71:240–255. doi: 10.1016/j.freeradbiomed.2014.03.004.

8. Klebanoff, S. J. and R. W. Coombs. 1996. Virucidal effect of stimulated eosinophils on human immunodeficiency virus type 1. *AIDS Res Hum Retroviruses* 12 (1):25–29.

9. Gingerich, A., L. Pang, J. Hanson, D. Dlugolenski, R. Streich, E. R. Lafontaine, T. Nagy, R. A. Tripp, and B. Rada. 2016. Hypothiocyanite produced by human and rat respiratory epithelial cells inactivates extracellular H1N2 influenza A virus. *Inflamm Res* 65 (1):71–80. doi: 10.1007/s00011-015-0892-z.

10. Jong, E. C., W. R. Henderson, and S. J. Klebanoff. 1980. Bactericidal activity of eosinophil peroxidase. *J Immunol* 124 (3):1378–1382.

11. Klebanoff, S. J., W. H. Clem, and R. G. Luebke. 1966. Peroxidase-thiocyanate-hydrogen peroxide antimicrobial system. *Biochim Biophys Acta* 117 (1):63–72.

12. Wang, J. and A. Slungaard. 2006. Role of eosinophil peroxidase in host defence and disease pathology. *Arch Biochem Biophys* 445 (2):256–260.

13. Barrett, T. J. and C. L. Hawkins. 2012. Hypothiocyanous acid: Benign or deadly? *Chem Res Toxicol* 25 (2):263–273. doi: 10.1021/tx200219s.

14. Aldridge, R., T. Chan, C. van Dalen, R. Senthilmohan, M. Winn, P. Venge, G. Town, and A. Kettle. 2002. Eosinophil peroxidase produces hypobromous acid in the airways of stable asthmatics. *Free Radic Biol Med* 33 (6):847–856.

15. Chandler, J. D., and B. J. Day. 2012. Thiocyanate: A potentially useful therapeutic agent with host defense and antioxidant properties. *Biochem Pharmacol* 84 (11):1381–1387. doi: 10.1016/j.bcp.2012.07.029.

16. Chandler, J. D. and B. J. Day. 2015. Biochemical mechanisms and therapeutic potential of pseudohalide thiocyanate in human health. *Free Radic Res* 49 (6):695–710. doi: 10.3109/10715762.2014.1003372.

17. Furtmuller, P. G., M. Zederbauer, W. Jantschko, J. Helm, M. Bogner, C. Jakopitsch, and C. Obinger. 2006. Active site structure and catalytic mechanisms of human peroxidases. *Arch Biochem Biophys* 445 (2):199–213.

18. Senthilmohan, R. and A. J. Kettle. 2006. Bromination and chlorination reactions of myeloperoxidase at physiological concentrations of bromide and chloride. *Arch Biochem Biophys* 445:235–244.

19. Jong, E. C., A. A. F. Mahmoud, and S. J. Klebanoff. 1981. Peroxidase-mediated toxicity to schistosomula of *Schistosoma mansoni*. *J Immunol* 126 (2):468–471.

20. Meeusen, E. N. T. and A. Balic. 2000. Do eosinophils have a role in the killing of helminth parasites? *Parasitol Today* 16 (3):95–101.

21. McCall, A. S., C. F. Cummings, G. Bhave, R. Vanacore, A. Page-McCaw, and B. G. Hudson. 2014. Bromine is an essential trace element for assembly of collagen IV scaffolds in tissue development and architecture. *Cell* 157 (6):1380–1392. doi: 10.1016/j.cell.2014.05.009.

22. Bhave, G., C. F. Cummings, R. M. Vanacore, C. Kumagai-Cresse, I. A. Ero-Tolliver, M. Rafi, J. S. Kang et al. 2012. Peroxidasin forms sulfilimine chemical bonds using hypohalous acids in tissue genesis. *Nat Chem Biol* 8 (9):784–790. doi: 10.1038/nchembio.1038.

23. Wang, J. G., S. A. Mahmud, J. A. Thompson, J. G. Geng, N. S. Key, and A. Slungaard. 2006. The principal eosinophil peroxidase product, HOSCN, is a uniquely potent phagocyte oxidant inducer of endothelial cell tissue factor activity: A potential mechanism for thrombosis in eosinophilic inflammatory states. *Blood* 107 (2):558–565. doi: 10.1182/blood-2005-05-2152.

24. Ashby, M. T., A. C. Carlson, and M. J. Scott. 2004. Redox buffering of hypochlorous acid by thiocyanate in physiologic fluids. *J Am Chem Soc* 126:15976–15977.

25. Nagy, P., J. L. Beal, and M. T. Ashby. 2006. Thiocyanate is an efficient endogenous scavenger of the phagocytic killing agent hypobromous acid. *Chem Res Toxicol* 19 (4):587–593.

26. Wijkstrom-Frei, C., S. El-Chemaly, R. Ali-Rachedi, C. Gerson, M. A. Cobas, R. Forteza, M. Salathe, and G. E. Conner. 2003. Lactoperoxidase and human airway host defense. *Am J Respir Cell Mol Biol* 29 (2):206–212.

27. Tenovuo, J., B. Mansson-Rahemtulla, K. M. Pruitt, and R. Arnold. 1981. Inhibition of dental plaque acid production by the salivary lactoperoxidase antimicrobial system. *Infect Immun* 34 (1):208–214.

28. Davies, M. J. 2011. Myeloperoxidase-derived oxidation: Mechanisms of biological damage and its prevention. *J Clin Biochem Nutr* 48 (1):8–19. doi: 10.3164/jcbn.11-006FR.

29. Klebanoff, S. J. 2005. Myeloperoxidase: Friend and foe. *J Leukoc Biol* 77:598–625.

30. Morgan, P. E., D. I. Pattison, J. Talib, F. A. Summers, J. A. Harmer, D. S. Celermajer, C. L. Hawkins, and M. J. Davies. 2011. High plasma thiocyanate levels in smokers are a key determinant of thiol oxidation induced by myeloperoxidase. *Free Radic Biol Med* 51 (9):1815–1822. doi: 10.1016/j.freeradbiomed.2011.08.008.

31. Pattison, D. I., M. J. Davies, and C. L. Hawkins. 2012. Reactions and reactivity of myeloperoxidase-derived oxidants: Differential biological effects of hypochlorous and hypothiocyanous acids. *Free Radic Res* 46 (8):975–995. doi: 10.3109/10715762.2012.667566.

32. Chandler, J. D., E. Min, J. Huang, C. S. McElroy, N. Dickerhof, T. Mocatta, A. A. Fletcher et al. 2015. Antiinflammatory and antimicrobial effects of thiocyanate in a cystic fibrosis mouse model. *Am J Respir Cell Mol Biol* 53 (2):193–205. doi: 10.1165/rcmb.2014-0208OC.

33. Chandler, J. D., E. Min, J. Huang, D. P. Nichols, and B. J. Day. 2013. Nebulized thiocyanate improves lung infection outcomes in mice. *Br J Pharmacol* 169 (5):1166–1177. doi: 10.1111/bph.12206.

34. Morgan, P. E., R. P. Laura, R. A. Maki, W. F. Reynolds, and M. J. Davies. 2015. Thiocyanate supplementation decreases atherosclerotic plaque in mice expressing human myeloperoxidase. *Free Radic Res* 49 (6):743–749. doi: 10.3109/10715762.2015.1019347.

35. Pattison, D. I. and M. J. Davies. 2006. Reactions of myeloperoxidase-derived oxidants with biological substrates: Gaining chemical insight into human inflammatory diseases. *Curr Med Chem* 13:3271–3290.

36. Prutz, W. A., R. Kissner, W. H. Koppenol, and H. Ruegger. 2000. On the irreversible destruction of reduced nicotinamide nucleotides by hypohalous acids. *Arch Biochem Biophys* 380 (1):181–191.

37. Thomas, E. L. 1981. Lactoperoxidase-catalyzed oxidation of thiocyanate: Equilibria between oxidized forms of thiocyanate. *Biochemistry* 20 (11):3273–3280.

38. Nagy, P., G. N. Jameson, and C. C. Winterbourn. 2009. Kinetics and mechanisms of the reaction of hypothiocyanous acid with 5-thio-2-nitrobenzoic acid and reduced glutathione. *Chem Res Toxicol* 22 (11):1833–1840. doi: 10.1021/tx900249d.

39. Pattison, D. I. and M. J. Davies. 2004. A kinetic analysis of the reactions of hypobromous acid with protein components: Implications for cellular damage and the use of 3-bromotyrosine as a marker of oxidative stress. *Biochemistry* 43:4799–4809.

40. Skaff, O., D. I. Pattison, and M. J. Davies. 2009. Hypothiocyanous acid reactivity with low-molecular-mass and protein thiols: Absolute rate constants and assessment of biological relevance. *Biochem J* 422 (1):111–117. doi: 10.1042/BJ20090276.

41. Skaff, O., D. I. Pattison, P. E. Morgan, R. Bachana, V. K. Jain, K. I. Priyadarsini, and M. J. Davies. 2012. Selenium-containing amino acids are targets for myeloperoxidase-derived hypothiocyanous acid: Determination of absolute rate constants and implications for biological damage. *Biochem J* 441 (1):305–316. doi: 10.1042/BJ20101762.

42. Thomas, E. L., P. M. Bozeman, M. M. Jefferson, and C. C. King. 1995. Oxidation of bromide by the human leukocyte enzymes myeloperoxidase and eosinophil peroxidase. Formation of bromamines. *J Biol Chem* 270 (7):2906–2913.

43. Barrett, T. J., D. I. Pattison, S. E. Leonard, K. S. Carroll, M. J. Davies, and C. L. Hawkins. 2012. Inactivation of thiol-dependent enzymes by hypothiocyanous acid: Role of sulfenyl thiocyanate and sulfenic acid intermediates. *Free Radic Biol Med* 52 (6):1075–1085. doi: 10.1016/j.freeradbiomed.2011.12.024.

44. Aune, T. M. and E. L. Thomas. 1978. Oxidation of protein sulfhydryls by products of peroxidase-catalyzed oxidation of thiocyanate ion. *Biochemistry* 17 (6):1005–1010.

45. Hawkins, C. L., D. I. Pattison, N. R. Stanley, and M. J. Davies. 2008. Tryptophan residues are targets in hypothiocyanous acid-mediated protein oxidation. *Biochem J* 416 (3):441–452.

46. Thomas, E. L. 1985. Products of the lactoperoxidase-catalysed oxidation of thiocyanate and halides. In *The Lactoperoxidase System: Chemistry and Biological Significance*, edited by K. M. Pruitt and J. O. Tenovuo, pp. 31–53. New York: Marcel Dekker, Inc.

47. Hawkins, C. L. and M. J. Davies. 2005. The role of aromatic amino acid oxidation, protein unfolding, and aggregation in the hypobromous acid-induced inactivation of trypsin inhibitor and lysozyme. *Chem Res Toxicol* 18 (11):1669–1677.

48. Hawkins, C. L. and M. J. Davies. 2005. The role of N-bromo species and radical intermediates in hypobromous acid-induced protein oxidation. *Free Radic Biol Med* 39:900–912.

49. Peskin, A. V., R. Turner, G. J. Maghzal, C. C. Winterbourn, and A. J. Kettle. 2009. Oxidation of methionine to dehydromethionine by reactive halogen species generated by neutrophils. *Biochemistry* 48 (42):10175–10182. doi: 10.1021/bi901266w.

50. Ronsein, G. E., C. C. Winterbourn, P. Di Mascio, and A. J. Kettle. 2014. Cross-linking methionine and amine residues with reactive halogen species. *Free Radic Biol Med* 70:278–287. doi: 10.1016/j.freeradbiomed.2014.01.023.

51. Sivey, J. D., S. C. Howell, D. J. Bean, D. L. McCurry, W. A. Mitch, and C. J. Wilson. 2013. Role of lysine during protein modification by HOCl and HOBr: Halogen-transfer agent or sacrificial antioxidant? *Biochemistry* 52 (7):1260–1271. doi: 10.1021/bi301523s.

52. Hawkins, C. L., D. I. Pattison, and M. J. Davies. 2002. Reaction of protein chloramines with DNA and nucleosides: Evidence for the formation of radicals, protein-DNA cross-links and DNA fragmentation. *Biochem J* 365:605–615.

53. Hazell, L. J., M. J. Davies, and R. Stocker. 1999. Secondary radicals derived from chloramines of apolipoprotein B-100 contribute to HOCl-induced lipid peroxidation of low-density lipoproteins. *Biochem J* 339:489–495.

54. Vanacore, R., A. J. Ham, M. Voehler, C. R. Sanders, T. P. Conrads, T. D. Veenstra, K. B. Sharpless, P. E. Dawson, and B. G. Hudson. 2009. A sulfilimine bond identified in collagen IV. *Science* 325 (5945):1230–1234. doi: 10.1126/science.1176811.

55. Prutz, W. A. 1999. Consecutive halogen transfer between various functional groups induced by reaction of hypohalous acids: NADH oxidation by halogenated amide groups. *Arch Biochem Biophys* 371 (1):107–114.

56. Mitra, S. N., A. Slungaard, and S. L. Hazen. 2000. Role of eosinophil peroxidase in the origins of protein oxidation in asthma. *Redox Rep* 5 (4):215–224.

57. Aune, T. M. and E. L. Thomas. 1977. Accumulation of hypothiocyanite ion during peroxidase-catalyzed oxidation of thiocyanate ion. *Eur J Biochem* 80 (1):209–214.

58. Bonifay, V., T. J. Barrett, D. I. Pattison, M. J. Davies, C. L. Hawkins, and M. T. Ashby. 2014. Tryptophan oxidation in proteins exposed to thiocyanate-derived oxidants. *Arch Biochem Biophys* 564:1–11. doi: 10.1016/j.abb.2014.08.014.

59. Hawkins, C. L., D. I. Pattison, and M. J. Davies. 2003. Hypochlorite-induced oxidation of amino acids, peptides and proteins. *Amino Acids* 25 (3–4):259–274.

60. Petronio, M. S. and V. F. Ximenes. 2012. Effects of oxidation of lysozyme by hypoha-
    lous acids and haloamines on enzymatic activity and aggregation. *Biochim Biophys Acta*
    1824 (10):1090–1096. doi: 10.1016/j.bbapap.2012.06.013.
61. Hadfield, K. A., D. I. Pattison, B. E. Brown, L. Hou, K. A. Rye, M. J. Davies, and
    C. L. Hawkins. 2013. Myeloperoxidase-derived oxidants modify apolipoprotein A-I and
    generate dysfunctional high-density lipoproteins: Comparison of hypothiocyanous acid
    (HOSCN) with hypochlorous acid (HOCl). *Biochem J* 449 (2):531–542. doi: 10.1042/
    BJ20121210.
62. Henderson, J. P., J. Byun, M. V. Williams, M. L. McCormick, W. C. Parks, L. A. Ridnour,
    and J. W. Heinecke. 2001. Bromination of deoxycytidine by eosinophil peroxidase: A
    mechanism for mutagenesis by oxidative damage of nucleotide precursors. *Proc Natl
    Acad Sci USA* 98 (4):1631–1636.
63. Henderson, J. P., J. Byun, J. Takeshita, and J. W. Heinecke. 2003. Phagocytes produce
    5-chlorouracil and 5-bromouracil, two mutagenic products of myeloperoxidase, in
    human inflammatory tissue. *J Biol Chem* 278 (26):23522–23528.
64. Henderson, J. P., J. Byun, D. M. Mueller, and J. W. Heinecke. 2001. The eosinophil
    peroxidase -hydrogen peroxide-bromide system of human eosinophils generates
    5-bromouracil, a mutagenic thymine analogue. *Biochemistry* 40:2052–2059.
65. Shen, Z., S. N. Mitra, W. Wu, Y. Chen, Y. Yang, J. Qin, and S.L. Hazen. 2001. Eosinophil
    peroxidase catalyzes bromination of free nucleosides and double-stranded DNA.
    Biochemistry 40 (7):2041–2051.
66. Suzuki, T., A. Nakamura, and M. Inukai. 2013. Reaction of 3',5'-di-O-acetyl-2'-
    deoxyguansoine with hypobromous acid. *Bioorg Med Chem* 21 (13):3674–3679. doi:
    10.1016/j.bmc.2013.04.060.
67. Suzuki, T., A. Kitabatake, and Y. Koide. 2016. Reaction of thymidine with hypobromous
    acid in phosphate buffer. *Chem Pharm Bull (Tokyo)* 64 (8):1235–1238. doi: 10.1248/
    cpb.c16-00138.
68. White, W. E., Jr., K. M. Pruitt, and B. Mansson-Rahemtulla. 1983. Peroxidase-
    thiocyanate-peroxide antibacterial system does not damage DNA. *Antimicrob Agents
    Chemother* 23 (2):267–272.
69. Suzuki, T. and H. Ohshima. 2003. Modification by fluoride, bromide, iodide, thiocya-
    nate and nitrite anions of reaction of a myeloperoxidase-H_2O_2-Cl- system with nucleo-
    sides. *Chem Pharm Bull* 51 (3):301–304.
70. Asahi, T., H. Kondo, M. Masuda, H. Nishino, Y. Aratani, Y. Naito, T. Yoshikawa,
    S. Hisaka, Y. Kato, and T. Osawa. 2010. Chemical and immunochemical detection
    of 8-halogenated deoxyguanosines at early stage inflammation. *J Biol Chem* 285
    (12):9282–9291. M109.054213 [pii]. doi: 10.1074/jbc.M109.054213.
71. Ximenes, V. F., N. H. Morgon, and A. R. de Souza. 2015. Hypobromous acid, a pow-
    erful endogenous electrophile: Experimental and theoretical studies. *J Inorg Biochem*
    146:61–68. doi: 10.1016/j.jinorgbio.2015.02.014.
72. Asahi, T., Y. Nakamura, Y. Kato, and T. Osawa. 2015. Specific role of taurine in the
    8-brominated-2'-deoxyguanosine formation. *Arch Biochem Biophys* 586:45–50. doi:
    10.1016/j.abb.2015.10.002.
73. Sassa, A., T. Ohta, T. Nohmi, M. Honma, and M. Yasui. 2011. Mutational specificities
    of brominated DNA adducts catalyzed by human DNA polymerases. *J Mol Biol* 406
    (5):679–686. doi: 10.1016/j.jmb.2011.01.005.
74. Carr, A. C., C. C. Winterbourn, and J. J. M. van den Berg. 1996. Peroxidase-mediated
    bromination of unsaturated fatty acids to form bromohydrins. *Arch Biochem Biophys*
    327 (2):227–233.
75. Vissers, M. C., A. C. Carr, and A. L. Chapman. 1998. Comparison of human red cell
    lysis by hypochlorous and hypobromous acids: Insights into the mechanism of lysis.
    Biochem J 330 (Pt 1):131–138.

76. Vissers, M. C., A. C. Carr, and C. C. Winterbourn. 2001. Fatty acid chlorohydrins and bromohydrins are cytotoxic to human endothelial cells. *Redox Rep* 6 (1):49–55. doi: 10.1179/135100001101536030.

77. Carr, A. C., J. J. van den Berg, and C. C. Winterbourn. 1998. Differential reactivities of hypochlorous and hypobromous acids with purified *Escherichia coli* phospholipid: Formation of haloamines and halohydrins. *Biochim Biophys Acta* 1392 (2–3):254–264.

78. Skaff, O., D. I. Pattison, and M. J. Davies. 2007. Kinetics of hypobromous acid-mediated oxidation of lipid components and antioxidants. *Chem Res Toxicol* 20 (12):1980–1988. doi: 10.1021/tx7003097.

79. Skaff, O., D. I. Pattison, and M. J. Davies. 2008. The vinyl ether linkages of plasmalogens are favored targets for myeloperoxidase-derived oxidants: A kinetic study. *Biochemistry* 47 (31):8237–8245. doi: 10.1021/bi800786q.

80. Albert, C. J., A. K. Thukkani, R. M. Heuertz, A. Slungaard, S. L. Hazen, and D. A. Ford. 2003. Eosinophil peroxidase-derived reactive brominating species target the vinyl ether bond of plasmalogens generating a novel chemoattractant, alpha-bromo fatty aldehyde. *J Biol Chem* 278 (11):8942–8950.

81. Carr, A. C., E. A. Decker, Y. Park, and B. Frei. 2001. Comparison of low-density lipoprotein modification by myeloperoxidase-derived hypochlorous and hypobromous acids. *Free Radic Biol Med* 31:62–72.

82. Exner, M., M. Hermann, R. Hofbauer, B. Hartmann, S. Kapiotis, and B. Gmeiner. 2004. Thiocyanate catalyzes myeloperoxidase-initiated lipid oxidation in LDL. *Free Radic Biol Med* 37 (2):146–155.

83. Zhang, R., Z. Shen, W. M. Nauseef, and S. L. Hazen. 2002. Defects in leukocyte-mediated initiation of lipid peroxidation in plasma as studied in myeloperoxidase-deficient subjects: Systematic identification of multiple endogenous diffusible substrates for myeloperoxidase in plasma. *Blood* 99 (5):1802–1810.

84. Collins, P., G. M. Rosano, P. M. Sarrel, L. Ulrich, S. Adamopoulos, C. M. Beale, J. G. McNeill, and P. A. Poole-Wilson. 1995. 17 beta-Estradiol attenuates acetylcholine-induced coronary arterial constriction in women but not men with coronary heart disease. *Circulation* 92 (1):24–30.

85. Ismael, F. O., J. M. Proudfoot, B. E. Brown, D. M. van Reyk, K. D. Croft, M. J. Davies, and C. L. Hawkins. 2015. Comparative reactivity of the myeloperoxidase-derived oxidants HOCl and HOSCN with low-density lipoprotein (LDL): Implications for foam cell formation in atherosclerosis. *Arch Biochem Biophys* 573:40–51. doi: 10.1016/j.abb.2015.03.008.

86. Spalteholz, H., K. Wenske, and J. Arnhold. 2005. Interaction of hypohalous acids and heme peroxidases with unsaturated phosphatidylcholines. *Biofactors* 24:67–76.

87. Sepe, S. M. and R. A. Clark. 1985. Oxidant membrane injury by the neutrophil myeloperoxidase system. I. Characterization of a liposome model and injury by myeloperoxidase, hydrogen peroxide, and halides. *J Immunol* 134 (3):1888–1895.

88. Hawkins, C. L., B. E. Brown, and M. J. Davies. 2001. Hypochlorite- and hypobromite-mediated radical formation and its role in cell lysis. *Arch Biochem Biophys* 395 (2):137–145. doi: 10.1006/abbi.2001.2581.

89. Slungaard, A. and J. R. Mahoney, Jr. 1991. Bromide-dependent toxicity of eosinophil peroxidase for endothelium and isolated working rat hearts: A model for eosinophilic endocarditis. *J Exp Med* 173 (1):117–126.

90. Slungaard, A., G. M. Vercellotti, G. Walker, R. D. Nelson, and H. S. Jacob. 1990. Tumor necrosis factor alpha/cachectin stimulates eosinophil oxidant production and toxicity towards human endothelium. *J Exp Med* 171 (6):2025–2041.

91. Lloyd, M. M., D. M. van Reyk, M. J. Davies, and C. L. Hawkins. 2008. Hypothiocyanous acid is a more potent inducer of apoptosis and protein thiol depletion in murine macrophage cells than hypochlorous acid or hypobromous acid. *Biochem J* 414 (2):271–280. doi: 10.1042/BJ20080468.

92. Ayars, G. H., L. C. Altman, M. M. McManus, J. M. Agosti, C. Baker, D. L. Luchtel, D. A. Loegering, and G. J. Gleich. 1989. Injurious effect of the eosinophil peroxide-hydrogen peroxide-halide system and major basic protein on human nasal epithelium in vitro. *Am Rev Respir Dis* 140 (1):125–131. doi: 10.1164/ajrccm/140.1.125.

93. Maines, J., N. R. Khurana, K. Roman, D. Knaup, and M. Ahmad. 2006. Cytotoxic effects of activated bromine on human fetal osteoblasts in vitro. *J Endod* 32 (9):886–889. doi: 10.1016/j.joen.2006.03.006.

94. Wagner, B. A., K. J. Reszka, M. L. McCormick, B. E. Britigan, C. B. Evig, and C. P. Burns. 2004. Role of thiocyanate, bromide and hypobromous acid in hydrogen peroxide-induced apoptosis. *Free Radic Res* 38 (2):167–175.

95. Weiss, S. J., S. T. Test, C. M. Eckmann, D. Roos, and S. Regiani. 1986. Brominating oxidants generated by human eosinophils. *Science* 234 (4773):200–203.

96. Wu, W., M. K. Samoszuk, S. A. Comhair, M. J. Thomassen, C. F. Farver, R. A. Dweik, M. S. Kavuru, S. C. Erzurum, and S. L. Hazen. 2000. Eosinophils generate brominating oxidants in allergen-induced asthma. *J Clin Invest* 105 (10):1455–1463. doi: 10.1172/JCI9702.

97. MacPherson, J. C., S. A. Comhair, S. C. Erzurum, D. F. Klein, M. F. Lipscomb, M. S. Kavuru, M. K. Samoszuk, and S. L. Hazen. 2001. Eosinophils are a major source of nitric oxide-derived oxidants in severe asthma: Characterization of pathways available to eosinophils for generating reactive nitrogen species. *J Immunol* 166 (9):5763–5772.

98. Meijer, R. J., D. S. Postma, H. F. Kauffman, L. R. Arends, G. H. Koeter, and H. A. Kerstjens. 2002. Accuracy of eosinophils and eosinophil cationic protein to predict steroid improvement in asthma. *Clin Exp Allergy* 32 (7):1096–1103.

99. Thomson, E., S. Brennan, R. Senthilmohan, C. L. Gangell, A. L. Chapman, P. D. Sly, A. J. Kettle et al. 2010. Identifying peroxidases and their oxidants in the early pathology of cystic fibrosis. *Free Radic Biol Med* 49 (9):1354–1360. doi: 10.1016/j.freeradbiomed.2010.07.010.

100. Brottman, G. M., W. E. Regelmann, A. Slungaard, and O. D. Wangensteen. 1996. Effect of eosinophil peroxidase on airway epithelial permeability in the guinea pig. *Pediatr Pulmonol* 21 (3):159–166. doi: 10.1002/(SICI)1099-0496(199603)21:3<159::AID-PPUL2>3.0.CO;2-L.

101. Rees, M. D., T. N. McNiven, and M. J. Davies. 2007. Degradation of extracellular matrix and its components by hypobromous acid. *Biochem J* 401 (2):587–596. doi: 10.1042/BJ20061236.

102. Brown, K. L., C. Darris, K. L. Rose, O. A. Sanchez, H. Madu, J. Avance, N. Brooks et al. 2015. Hypohalous acids contribute to renal extracellular matrix damage in experimental diabetes. *Diabetes* 64 (6):2242–2253. doi: 10.2337/db14-1001.

103. Gorudko, I. V., D. V. Grigorieva, E. V. Shamova, V. A. Kostevich, A. V. Sokolov, E. V. Mikhalchik, S. N. Cherenkevich, J. Arnhold, and O. M. Panasenko. 2014. Hypohalous acid-modified human serum albumin induces neutrophil NADPH oxidase activation, degranulation, and shape change. *Free Radic Biol Med* 68:326–334. doi: 10.1016/j.freeradbiomed.2013.12.023.

104. Seguela, P. E., X. Iriart, P. Acar, M. Montaudon, R. Roudaut, and J. B. Thambo. 2015. Eosinophilic cardiac disease: Molecular, clinical and imaging aspects. *Arch Cardiovasc Dis* 108 (4):258–268. doi: 10.1016/j.acvd.2015.01.006.

105. Wang, J. G., S. A. Mahmud, J. Nguyen, and A. Slungaard. 2006. Thiocyanate-dependent induction of endothelial cell adhesion molecule expression by phagocyte peroxidases: A novel HOSCN-specific oxidant mechanism to amplify inflammation. *J Immunol* 177 (12):8714–8722.

106. Oram, J. D. and B. Reiter. 1966. The inhibition of streptococci by lactoperoxidase, thiocyanate and hydrogen peroxide. The oxidation of thiocyanate and the nature of the inhibitory compound. *Biochem J* 100 (2):382–388.

107. Carlsson, J., Y. Iwami, and T. Yamada. 1983. Hydrogen peroxide excretion by oral strep-tococci and effect of lactoperoxidase-thiocyanate-hydrogen peroxide. *Infect Immun* 40 (1):70–80.

108. Shin, K., H. Hayasawa, and B. Lonnerdal. 2001. Inhibition of *Escherichia coli* respira-tory enzymes by the lactoperoxidase-hydrogen peroxide-thiocyanate antimicrobial sys-tem. *J Appl Microbiol* 90 (4):489–493.

109. Shin, K., K. Yamauchi, S. Teraguchi, H. Hayasawa, and I. Imoto. 2002. Susceptibility of *Helicobacter pylori* and its urease activity to the peroxidase-hydrogen per-oxide-thiocyanate antimicrobial system. *J Med Microbiol* 51 (3):231–237. doi: 10.1099/0022-1317-51-3-231.

110. Chandler, J. D., D. P. Nichols, J. A. Nick, R. J. Hondal, and B. J. Day. 2013. Selective metabolism of hypothiocyanous acid by mammalian thioredoxin reductase promotes lung innate immunity and antioxidant defense. *J Biol Chem* 288 (25):18421–18428. doi: 10.1074/jbc.M113.468090.

111. Cook, N. L., H. M. Viola, V. S. Sharov, L. C. Hool, C. Schoneich, and M. J. Davies. 2012. Myeloperoxidase-derived oxidants inhibit sarco/endoplasmic reticulum Ca^{2+}-ATPase activity and perturb Ca^{2+} homeostasis in human coronary artery endothelial cells. *Free Radic Biol Med* 52 (5):951–961. doi: 10.1016/j.freeradbiomed.2011.12.001.

112. Cook, N. L., C. H. Moeke, L. I. Fantoni, D. I. Pattison, and M. J. Davies. 2016. The myeloperoxidase-derived oxidant hypothiocyanous acid inhibits protein tyrosine phos-phatases via oxidation of key cysteine residues. *Free Radic Biol Med* 90:195–205. doi: 10.1016/j.freeradbiomed.2015.11.025.

113. Lane, A. E., J. T. Tan, C. L. Hawkins, A. K. Heather, and M. J. Davies. 2010. The myeloperoxidase-derived oxidant HOSCN inhibits protein tyrosine phosphatases and modulates cell signalling via the mitogen-activated protein kinase (MAPK) pathway in macrophages. *Biochem J* 430 (1):161–169. doi: 10.1042/BJ20100082.

114. Talib, J. and M. J. Davies. 2016. Exposure of aconitase to smoking-related oxidants results in iron loss and increased iron response protein-1 activity: Potential mechanisms for iron accumulation in human arterial cells. *J Biol Inorg Chem* 21 (3):305–317. doi: 10.1007/s00775-016-1340-4.

115. Talib, J., J. Kwan, A. Suryo Rahmanto, P. K. Witting, and M. J. Davies. 2014. The smok-ing-associated oxidant hypothiocyanous acid induces endothelial nitric oxide synthase dysfunction. *Biochem J* 457 (1):89–97. doi: 10.1042/BJ20131135.

116. Bozonet, S. M., A. P. Scott-Thomas, P. Nagy, and M. C. Vissers. 2010. Hypothiocyanous acid is a potent inhibitor of apoptosis and caspase 3 activation in endothelial cells. *Free Radic Biol Med* 49 (6):1054–1063. S0891-5849(10)00393-X [pii]. doi: 10.1016/j. freeradbiomed.2010.06.028.

117. Love, D. T., T. J. Barrett, M. Y. White, S. J. Cordwell, M. J. Davies, and C. L. Hawkins. 2016. Cellular targets of the myeloperoxidase-derived oxidant hypothiocyanous acid (HOSCN) and its role in the inhibition of glycolysis in macrophages. *Free Radic Biol Med* 94:88–98. doi: 10.1016/j.freeradbiomed.2016.02.016.

118. Ismael, F. O., T. J. Barrett, D. Sheipouri, B. E. Brown, M. J. Davies, and C. L. Hawkins. 2016. Role of myeloperoxidase oxidants in the modulation of cellular lysosomal enzyme function: A contributing factor to macrophage dysfunction in atherosclerosis? *PLoS One* 11 (12):e0168844. doi: 10.1371/journal.pone.0168844.

119. Xu, Y., S. Szep, and Z. Lu. 2009. The antioxidant role of thiocyanate in the pathogenesis of cystic fibrosis and other inflammation-related diseases. *Proc Natl Acad Sci USA* 106 (48):20515–20519. doi: 10.1073/pnas.0911412106.

120. Rees, M. D., S. L. Maiocchi, A. J. Kettle, and S. R. Thomas. 2014. Mechanism and regu-lation of peroxidase-catalyzed nitric oxide consumption in physiological fluids: Critical protective actions of ascorbate and thiocyanate. *Free Radic Biol Med* 72:91–103. doi: 10.1016/j.freeradbiomed.2014.03.037.

121. Gould, N. S., S. Gauthier, C. T. Kariya, E. Min, J. Huang, and B. J. Day. 2010. Hypertonic saline increases lung epithelial lining fluid glutathione and thiocyanate: Two protective CFTR-dependent thiols against oxidative injury. *Respir Res* 11:119–128. doi: 10.1186/1465-9921-11-119. 1465-9921-11-119 [pii].
122. Botti, T. P., H. Amin, L. Hiltscher, and R. W. Wissler. 1996. A comparison of the quantitation of macrophage foam cell populations and the extent of apolipoprotein E deposition in developing atherosclerotic lesions in young people: High and low serum thiocyanate groups as an indication of smoking. *Atherosclerosis* 124 (2):191–202.
123. Scanlon, C. E. O., B. Berger, G. Malcom, and R. W. Wissler. 1996. Evidence for more extensive deposits of epitopes of oxidized low density lipoproteins in aortas of young people with elevated serum thiocyanate levels. *Atherosclerosis* 121 (1):23–33.
124. Pattison, D. I. and M. J. Davies. 2001. Absolute rate constants for the reaction of hypochlorous acid with protein side-chains and peptide bonds. *Chem Res Toxicol* 14:1453–1464.
125. Karimi, M., M. T. Ignasiak, B. Chan, A. K. Croft, L. Radom, C. H. Schiesser, D. I. Pattison, and M. J. Davies. 2016. Reactivity of disulfide bonds is markedly affected by structure and environment: Implications for protein modification and stability. *Sci Rep* 6:38572. doi: 10.1038/srep38572.
126. Aune, T. M., E. L. Thomas, and M. Morrison. 1977. Lactoperoxidase-catalyzed incorporation of thiocyanate ion into a protein substrate. *Biochemistry* 16 (21):4611–4615.
127. Tenovuo, J. and H. Larjava. 1984. The protective effect of peroxidase and thiocyanate against hydrogen peroxide toxicity assessed by the uptake of [3H]-thymidine by human gingival fibroblasts cultured in vitro. *Arch Oral Biol* 29 (6):445–451.
128. Grisham, M. B. and E. M. Ryan. 1990. Cytotoxic properties of salivary oxidants. *Am J Physiol* 258 (1 Pt 1):C115–C121.
129. Slungaard, A. and J. R. Mahoney, Jr. 1991. Thiocyanate is the major substrate for eosinophil peroxidase in physiologic fluids. Implications for cytotoxicity. *J Biol Chem* 266 (8):4903–4910.
130. Arlandson, M., T. Decker, V. A. Roongta, L. Bonilla, K. H. Mayo, J. C. MacPherson, S. L. Hazen, and A. Slungaard. 2001. Eosinophil peroxidase oxidation of thiocyanate— Characterization of major reaction products and a potential sulfhydryl-targeted cytotoxicity system. *J Biol Chem* 276 (1):215–224.
131. Lloyd, M. M., M. A. Grima, B. S. Rayner, K. A. Hadfield, M. J. Davies, and C. L. Hawkins. 2013. Comparative reactivity of the myeloperoxidase-derived oxidants hypochlorous acid and hypothiocyanous acid with human coronary artery endothelial cells. *Free Radic Biol Med* 65:1352–1362. doi: 10.1016/j.freeradbiomed.2013.10.007.

Section IV

H_2O_2 in Cellular Metabolism and Signaling

14 Hydrogen Peroxide-Dependent Redox Signaling

Basic Concepts and Unanswered Questions

Mark B. Hampton

CONTENTS

WHAT IS REDOX SIGNALING?

Cells use a considerable amount of energy to maintain an internal reducing environment. They synthesize proteins that can scavenge oxidants, chelate redox-active metals and repair oxidatively damaged lipids, proteins, and nucleic acids. Low molecular mass compounds are imported or synthesized to supply electrons to oxidized biomolecules. The complex antioxidant networks are highly coordinated, enabling removal of various oxidants and redox-active compounds, and any intermediates generated during the detoxification of an oxidizing species. Increased oxidative stress initiates a rapid response in cells, with alterations in antioxidant protein expression and diversion of more metabolic energy into reductive processes. The accumulation of oxidatively damaged biomolecules can be a signal to bolster antioxidant defenses, but more importantly, cells appear able to respond to subtle disturbances in homeostasis before damage occurs. This ability to sense and respond to increased oxidant levels is one form of redox signaling.

Inducible antioxidant defenses are likely to have evolved in the earliest life forms, protecting sensitive intracellular constituents as oxygen levels increased in the atmosphere [1]. Infact, it is hypothesized that these organisms had to cope with hydrogen peroxide, generated by the irradiation of water, prior to the evolution of photosynthesis and rise of atmospheric oxygen [2]. In addition to upregulating defenses, unicellular organisms respond to increased levels of hydrogen peroxide by migrating away from the source [3]. Mammalian cells are mostly sheltered from environmental extremes faced by unicellular organisms and plants, but they have maintained the ability to respond to hydrogen peroxide. Their major exposure comes from within, in particular, from metabolic processes that occur in peroxisomes, mitochondria, and endoplasmic reticulum. The constant production of hydrogen peroxide in cells is estimated to result in steady state levels in the low nanomolar range [4]. This will increase under various pathological conditions associated with oxidative stress, including ischemia-reperfusion, mitochondrial dysfunction, protein aggregate accumulation, and inflammation.

Phagocytic white blood cells are a major source of hydrogen peroxide during inflammation. The most abundant phagocyte is the neutrophil, which ingests pathogens into an intracellular compartment called a phagosome. The process of phagocytosis triggers an oxidative burst to assist in the destruction of the pathogen, but other soluble stimuli can also induce oxidant production [5]. Initiation of the oxidative burst involves the rapid assembly of an NADPH oxidase complex (NOX) that reduces oxygen in a one-electron reaction to generate superoxide. Neutrophil myeloperoxidase dismutates superoxide then uses the resulting hydrogen peroxide to generate the microbicidal oxidant hypochlorous acid. Although oxidant generation is predominantly contained within intracellular phagosomes, oxidants such as hydrogen peroxide can diffuse to neighboring cells [6]. While first discovered in phagocytes, a family of NOX enzymes was subsequently detected in nonphagocytic white blood cells and then in a variety of other cell types [7]. Expression is lower than in neutrophils, and NOX activity can either be constitutive or require activation by extracellular stimuli. This leads to an obvious question: Why do nonphagocytic cells intentionally generate superoxide and hydrogen peroxide?

The answer that has emerged over several years is that hydrogen peroxide acts as a signaling molecule—a second messenger generated in response to the activation of receptor-mediated signaling pathways that assists in transmitting or modulating signals through post-translational modification of redox-sensitive proteins (Figure 14.1). Redox signaling involving controlled or intentional oxidant generation is likely to have adapted the same processes that enable cells to respond to environmental oxidative stress. This chapter provides a general overview of these signaling processes, and identifies unresolved questions regarding the mechanisms and biological significance of redox signaling.

HOW DOES REDOX SIGNALING OCCUR?

Various growth factors, hormones, inflammatory mediators, and death receptor ligands are reported to increase NOX activity in cells, with the resultant hydrogen peroxide proposed to regulate many aspects of cell function ranging from proliferation and differentiation to cell death [8–10]. Hydrogen peroxide diffusing from mitochondria is also considered to play a signaling role, though the physiological triggers that lead to increased generation and/or release to the cytosol are not as well defined [11].

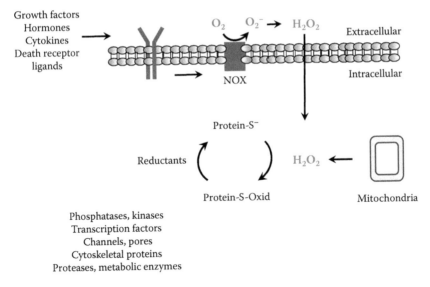

Growth factors
Hormones
Cytokines
Death receptor
ligands

O_2 O_2^- → H_2O_2 Extracellular

NOX

Intracellular

Protein-S$^-$

Reductants

Protein-S-Oxid

H_2O_2 ←

Mitochondria

Phosphatases, kinases
Transcription factors
Channels, pores
Cytoskeletal proteins
Proteases, metabolic enzymes

FIGURE 14.1   Redox signaling by hydrogen peroxide derived from NOX activation or mito-chondria. Protein-S$^-$ represents a reduced cysteine residue in a target protein, and Protein-S-Oxid represents the various reversible oxidation products that can be formed upon reaction with hydrogen peroxide, including sulfenic acid, intra- and intermolecular disulfides, and glutathionylated cysteine.

There are two important features of NOX activity that are sometimes overlooked. First, electrons from cytosolic NADPH are transferred to oxygen on the opposite side of the membrane. Second, the primary product of several NOX family members is superoxide, which then dismutates to hydrogen peroxide [7]. NOX assembly commonly occurs at the plasma membrane, with oxidant generation occurring outside the cells, but in some cases it occurs at the endoplasmic reticulum or endosomes [12]. Extracellular generation contrasts with typical intracellular second messenger systems such as cyclic nucleotides, phospholipid metabolites, and calcium. Hydrogen peroxide can, therefore, influence both the cell that generated it and neighboring cells. Extracellular hydrogen peroxide has to transport across the plasma membrane before being consumed in a cell, resulting in a concentration gradient. The controlled uptake of hydrogen peroxide via membrane-spanning aquaporins is proposed to regulate uptake by specific cells [13], and may influence their sensitivity to hydrogen peroxide. Aquaporins may also help to channel hydrogen peroxide to specific regions of the cells and even specific targets. Superoxide may play a direct role in signaling [14], though it does have limited membrane permeability. However, hydrogen peroxide levels could be controlled by the routing of superoxide away from dismutation, for example, reaction with other radicals such as nitric oxide.

Transmission of a redox signal requires reaction of hydrogen peroxide with specific target proteins. In the context of signaling, the sulfhydryl group of cysteine residues is the most studied target of hydrogen peroxide. Cysteine is a strong biological nucleophile, and in addition to its role in oxidoreductase enzymes, it is utilized at the active site of hydrolases, for example, cysteine proteases, and phosphate and methyl transferases, for example, tyrosine phosphatases. The reactive moiety is the thiolate anion. Its nucleophilicity at physiological pH is increased by positively charged residues in the active site that lower the pKa of cysteine from 8.5 to below 5 [15]. Cysteine is also used to chelate metals, such as iron and zinc in some respiratory chain complexes and transcription factors. Disulfide bonds play an important function in protein structure. With such a prominent role, post-translational oxidation and reduction of cysteine provides a means for the dynamic regulation of protein structure and function. Redox proteomics technology has revealed many proteins that undergo reversible oxidative modification upon exposure of cells to low levels of oxidants, or physiological and pathological increases in oxidant generation [16–18]. Lists include kinases, phosphatases, transcription factors, metabolic enzymes, and cytoskeletal and membrane proteins, thereby confirming the potential for redox signals to influence most major aspects of cell function.

It is important to mention, that while thiol proteomics methodology has dominated redox signaling cysteine-independent targets have typically been ignored. Methionine oxidation is known to influence protein function, and while selective oxidation has been reported in a physiological signaling pathway [19], sensitive labeling techniques have not been developed for global screening. The metal centers of the serine–threonine phosphatase PP1 are known to be sensitive to NOX-dependent inactivation [20]. Also, the reactivity of heme peroxidases with hydrogen peroxide means they have the potential to contribute to redox signaling, but there are no reported examples.

Why are the cysteines in some proteins oxidized during signaling, but not others? One possibility is a thermodynamic hierarchy based on the redox potential of the relevant cysteine residues. A problem with this scenario is that the thiol exchange reactions involved in establishing such equilibriums are slow, making them ineffective at responding to transient increases in hydrogen peroxide. These reactions will play an important role in the establishment and return to homeostasis following oxidant exposure. The signaling events themselves, however, appear to rely on a small subset of cysteines having the appropriate kinetic properties to react rapidly with hydrogen peroxide.

When the cellular proteins that become oxidized in cellular models are closely examined, some of these proteins are relatively unreactive with hydrogen peroxide. This issue was best illustrated in competition analyses performed by Christine Winterbourn. She demonstrated that the vast majority of hydrogen peroxide entering cells will react with an abundant family of thiol peroxidases called peroxiredoxins [15,21,22]. These proteins, which constitute up to 1% of cellular protein, have an active site peroxidatic cysteine that reacts with hydrogen peroxide at 10^6 to $10^7 \, M^{-1} \, s^{-1}$. This compares to lower abundance target proteins such as tyrosine phosphatases whose low pKa cysteines have rate constants of up to $10^2 \, M^{-1} \, s^{-1}$. It is important to note that this modeling assumes that the peroxiredoxins are in their reduced form. Oxidized peroxiredoxins are recycled by the thioredoxin system, and to some extent by glutaredoxin [23]. Since this recycling is slower than the initial oxidation step, at times of increased hydrogen peroxide oxidized peroxiredoxins will accumulate and won't be available to react with hydrogen peroxide. Even so, other reactive targets of the hydrogen peroxide, in particular the glutathione peroxidases and catalase, will account for much of the hydrogen peroxide. It seems that direct oxidation of phosphatases and other weakly reactive proteins is unlikely without some form of facilitation.

One simple explanation for the oxidation of low-reactivity thiols is that the rate constants derived from *in vitro* studies are significantly different from reactivity inside a cell. Interactions with other biomolecules could increase reactivity, enabling direct oxidation of sensitized target proteins. The peroxiredoxins have a specialized active site in which hydrogen bonding with conserved active site residues stabilizes the transition state intermediate, lowering the activation energy required to break the O–O bond [24,25]. Such a structure might be replicated through in situ protein-protein interactions. Another possibility is that only a very small fraction of the hydrogen peroxide generated as a signal actually reaches the appropriate target protein, and by extension, only a small fraction of the target protein would be modified. For this scenario to be biologically relevant there would have to be a "gain-in-function" for the target protein. If, for example, only a small proportion of cellular tyrosine phosphatases were inhibited, then it would be unlikely to have a significant impact on protein phosphorylation following addition of a growth factor.

Site-specific effects have been invoked as an alternate explanation for the selectivity observed during redox signaling. It is proposed that proteins closest to the site of hydrogen peroxide generation (or entry into the cell) are more likely to be directly oxidized. Indeed, NOXs have been reported to be colocalized with targets [26,27]. A variation on this theme hypothesizes that the peroxiredoxins, as major consumers of hydrogen peroxide, are inhibited within specific regions, enabling local hydrogen peroxide levels to increase and oxidize the less reactive targets. Two specific mechanisms of peroxiredoxin inactivation have been characterized: hyperoxidation of the active site cysteine by increased concentrations of hydrogen peroxide, which will completely inactivate peroxiredoxin activity [28], and phosphorylation of peroxiredoxins that decreases catalytic activity [29]. An alternate mechanism would be accumulation of the reversibly oxidized peroxiredoxin. These modifications have only been observed in a small fraction of cellular peroxiredoxins during signaling, which is consistent with it being restricted to specific regions, but there are still issues relating to the capacity of hydrogen peroxide to diffuse from these regions. Indeed, mathematical modeling indicates that local hydrogen peroxide concentrations would still not be high enough for direct oxidation of less reactive thiol proteins [30].

An alternate mechanism is emerging in which oxidation of target proteins occurs via thiol exchange reactions termed redox relays. In the most common scenario, the target protein interacts with a peroxiredoxin (or similarly reactive thiol peroxidase) that has been oxidized by hydrogen peroxide, forming a mixed disulfide with either the sulfenic acid formed at the peroxidatic cysteine of the peroxiredoxin (Figure 14.2), or via exchange with the disulfide formed between peroxidatic and resolving cysteines. These mechanisms were first discovered in yeast, and more recently examples have been reported in mammalian cells [31–33]. Relays are attractive because the sensing element is provided by proteins specialized in reacting with peroxide, and their physical interaction with target proteins provides an important layer of selectivity. These protein interactions could vary in different cell types or at different stages of the cell cycle, thereby altering the response of

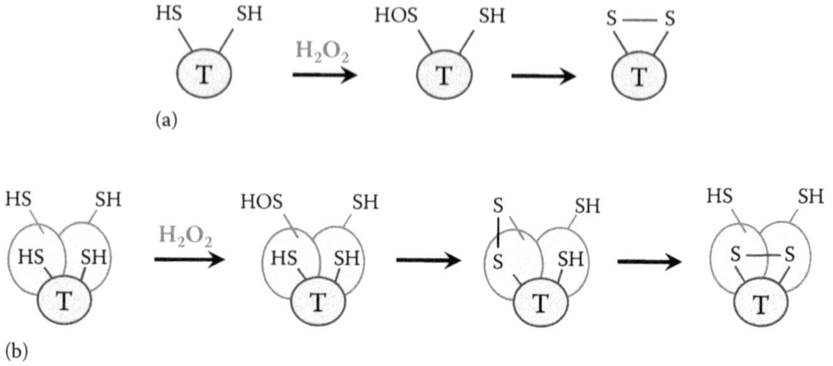

FIGURE 14.2 Direct and facilitated models for oxidation of a target protein by hydrogen peroxide. (a) Direct oxidation of a target protein (green) by hydrogen peroxide to form an intramolecular disulfide. (b) Facilitated oxidation between a peroxiredoxin (blue) and the target protein. In this example there is a physical interaction between the two proteins, and the sulfenic acid formed on the peroxiredoxin reacts with a cysteine on the target protein to form an intermolecular disulfide. This is resolved by a second cysteine on the target protein.

cells to hydrogen peroxide. The peroxiredoxins form complex oligomeric structures, including decameric or dodecameric rings that can stack upon each other, and are influenced by the redox state of the peroxiredoxin [34,35]. This is far more complex than required to consume hydrogen peroxide and supports a mechanism whereby peroxiredoxins channel the oxidizing equivalents from hydrogen peroxide into established signaling pathways. While an attractive hypothesis, it is too soon to know how widespread such a redox signaling mechanism is. The model makes a valuable prediction though; the oxidation of target proteins and subsequent signaling will be inhibited by removal of the peroxiredoxins, rather than increased as would be predicted by a competition model.

HOW IMPORTANT IS REDOX SIGNALING?

It is well established that cells respond to oxidative stress by increasing expression of their antioxidant defense systems. This can be mediated by the oxidation of transcription factors, some of which may be reactive enough to be directly oxidized, for example, bacterial OxyR [36], while others involve facilitated oxidation by peroxidases, for example, yeast Yap1 [37]. The activity of other transcription factors is influenced by the redox status of associated regulatory proteins, for example, mammalian Nrf2, which is regulated by the KEAP1 protein with reactive cysteines that respond to various oxidants and electrophiles [38]. Experiments in which the redox-sensitive residues are mutated clearly highlight these adaptive systems.

The biological relevance of controlled hydrogen peroxide production following receptor activation has proven more difficult to characterize. The first challenge has been detecting and quantifying the small changes in hydrogen peroxide production in these models. Chemical probes that fluoresce upon oxidation have been commonly used, but interpretation of results can be difficult [39]. In most cases these probes are not directly oxidized by hydrogen peroxide, and instead require a catalyst. Several intermediates can be generated during oxidation, and any changes in cells that impact on catalyst and intermediate levels will alter fluorescence independent of hydrogen peroxide levels. One of the probes most susceptible to artifacts is dichlorofluorescein [40], but this hasn't prevented it from also being the most widely used. Several new generation probes are emerging, including protein-based redox sensors, that will be much more valuable, particularly those that provide information on cell localization [41]. The peroxiredoxins themselves have been used as markers of oxidative stress, but in signaling models it can be difficult to observe global accumulation of oxidized forms without inhibition of reduction pathways [42].

Detection of redox changes provides evidence that a stimulus could be using oxidation as part of a signal transduction event, but does not reveal its importance. This requires the ability to specifically inhibit target oxidation and measure the consequences. It is extremely unlikely that hydrogen peroxide scavengers can be introduced in sufficient amounts to impact on redox signaling pathways. The one possible exception is catalase overexpression, which can effectively compete with the peroxiredoxins, and has been reported to have an inhibitory effect in some models [43]. One of the most used compounds in the redox signaling field is N-acetylcysteine (NAC). NAC is commonly described as a ROS scavenger, yet its reaction with hydrogen peroxide, the "ROS" usually invoked in the context of signaling, is slow and unlikely to be physiologically relevant. NAC increases intracellular glutathione concentrations, and is more likely to interfere with DCF oxidation through scavenging intermediate radicals rather than impacting on hydrogen peroxide levels. At the millimolar levels used in cell studies, NAC can reduce disulfides on extracellular receptors, which could explain inhibitory effects on signaling [44].

More direct effects have been observed by the use of NOX-knockout cells [20], and this approach should be employed in a broader range of signaling models. NOX inhibitors provide another useful experimental tool, but specificity is important. Earlier studies using the general flavoprotein inhibitor diphenyleneiodonium (DPI) need to be interpreted cautiously. To best unravel the biological relevance of redox signaling, it will be important to specifically disrupt the signaling event. This could be achieved by mutating relevant cysteines in the target protein to protect them from oxidation, as long as the cysteine is not critical for cell function. Alternatively, if signaling occurs through a relay mechanism, it may be feasible to uncouple the relay by introducing mutations that block protein–protein interactions. New approaches such as these are required to determine exactly which biological processes are influenced by redox signals.

A more detailed understanding of redox signaling mechanisms will bring with it the potential for practical applications. Oxidative stress is commonly linked to human disease through its ability to damage and kill cells. The effect of pathological oxidative stress on redox-sensitive signaling pathways is less well understood. These signaling pathways are fine-tuned to respond to small disturbances in redox homeostasis, and it is not clear what happens following acute stress such as ischemia/reperfusion injury or exposure to oxidants from neutrophils and their released peroxidases, or from more chronic oxidative stress such as accumulation of protein aggregates or dysfunctional mitochondria. Under increased oxidative stress are these pathways constantly stimulated, or is there a compensatory increase in thiol reductants making the pathways less responsive to stimulation? Are there significant changes to redox signaling networks in cancer cells, and can they be targeted for therapeutic purposes? How do these signaling pathways change during aging? These questions warrant detailed investigation.

To date, antioxidant therapies have been largely ineffective in treating disease. One of the key challenges is delivering sufficient amounts of scavengers to a site to protect other biomolecules. An alternate would be to limit excess generation in the first place, but this is challenging when the processes responsible have important physiological roles. If, as predicted here, redox signaling pathways are dependent on protein–protein interactions for selectivity and signal transduction, then this opens up a new approach for combating oxidative stress—the selective uncoupling of specific pathways with compounds that interfere with protein interactions. Such strategies should be explored now, not so much as therapeutic agents, but as experimental tools to provide crucial insight into the biochemical mechanisms and physiological significance of redox signaling.

SUMMARY

Redox signaling pathways enable cells to respond to oxidative stress, and to oxidants such as hydrogen peroxide that are generated in a controlled manner in cells to activate or modulate signal transduction pathways. In the best characterized systems, signal transmission occurs via transient oxidation of specific cysteine residues in regulatory proteins, including kinases, phosphatases, and transcription factors. A major issue in the field is to understand how selective oxidation occurs in cells, particularly when many of the target proteins are not reactive enough for direct oxidation to occur. Evidence is emerging that oxidizing equivalents can be channeled to specific targets through controlled protein–protein interactions termed redox relays. Disruption of these relays would provide a valuable tool to assess the importance of redox signaling in various biological processes. It is not clear exactly what impact pathological oxidative stress has on redox signaling pathways, but relay disruption may provide a novel approach for protecting cells from oxidative stress.

ACKNOWLEDGMENTS

I apologize to the many colleagues whose work was not cited in what is a brief overview of a very large field. As this chapter appears in a volume dedicated to Professor Christine Winterbourn, my selection of references is biased toward work that has come from her laboratory. I would like to gratefully acknowledge many enjoyable years of collaboration and mentoring from Christine. As a biologist with no significant undergraduate training in chemistry, I have benefited enormously from her rigor and application of fundamental chemical principles to her research. Even more important has been her infectious enthusiasm for science, and the establishment of an excellent culture within our Christchurch laboratory. There are still many unanswered questions as regards redox signaling, and I hope we can keep contributing to this intriguing field for some time yet. Our contributions to the field have primarily been funded by the Health Research Council of New Zealand and the Royal Society of New Zealand Marsden Fund.

REFERENCES

1. Fahey, R. C. 1977. Biologically important thiol-disulfide reactions and the role of cyst(e)ine in proteins: An evolutionary perspective. *Adv Exp Med Biol* 86A:1–30.
2. McKay, C. P. and H. Hartman. 1991. Hydrogen peroxide and the evolution of oxygenic photosynthesis. *Orig Life Evol Biosph* 21:157–163.
3. Benov, L. and I. Fridovich. 1996. *Escherichia coli* exhibits negative chemotaxis in gradients of hydrogen peroxide, hypochlorite, and N-chlorotaurine: Products of the respiratory burst of phagocytic cells. *Proc Natl Acad Sci USA* 93 (10):4999–5002.
4. Chance, B., H. Sies, and A. Boveris. 1979. Hydroperoxide metabolism in mammalian organs. *Physiol Rev* 59 (3):527–605.
5. Winterbourn, C. C., A. J. Kettle, and M. B. Hampton. 2016. Reactive oxygen species and neutrophil function. *Annu Rev Biochem* 85:765–792.
6. Bayer, S. B., G. Maghzal, R. Stocker, M. B. Hampton, and C. C. Winterbourn. 2013. Neutrophil-mediated oxidation of erythrocyte peroxiredoxin 2 as a potential marker of oxidative stress in inflammation. *FASEB J* 27 (8):3315–3322.
7. Lambeth, J. D. 2004. NOX enzymes and the biology of reactive oxygen. *Nat Rev Immunol* 4 (3):181–189.
8. Sundaresan, M., Z. Yu, V. J. Ferrans, K. Irani, and T. Finkel. 1995. Requirement for generation of H_2O_2 for platelet-derived growth factor signal transduction. *Science* 270:296–299.
9. Bae, Y. S., S. W. Kang, M. S. Seo, I. C. Baines, E. Tekle, P. B. Chock, and S. G. Rhee. 1997. Epidermal growth factor (EGF)-induced generation of hydrogen peroxide. Role in EGF receptor-mediated tyrosine phosphorylation. *J Biol Chem* 272:217–221.
10. Stone, J. R. and S. P. Yang. 2006. Hydrogen peroxide: A signaling messenger. *Antioxid Redox Signal* 8 (3–4):243–270.
11. Murphy, M. P. 2009. How mitochondria produce reactive oxygen species. *Biochem J* 417 (1):1–13.
12. Oakley, F. D., D. Abbott, Q. Li, and J. F. Engelhardt. 2009. Signaling components of redox active endosomes: The redoxosomes. *Antioxid Redox Signal* 11 (6):1313–1333.
13. Bienert, G. P. and F. Chaumont. 2014. Aquaporin-facilitated transmembrane diffusion of hydrogen peroxide. *Biochim Biophys Acta* 1840 (5):1596–1604.
14. Winterbourn, C. C. 2015. Are free radicals involved in thiol-based redox signaling? *Free Radic Biol Med* 80:164–170.
15. Winterbourn, C. C. and M. B. Hampton. 2008. Thiol chemistry and specificity in redox signaling. *Free Radic Biol Med* 45 (5):549–561.
16. Fratelli, M., H. Demol, M. Puype, S. Casagrande, I. Eberini, M. Salmona, V. Bonetto et al. 2002. Identification by redox proteomics of glutathionylated proteins in oxidatively stressed human T lymphocytes. *Proc Natl Acad Sci USA* 99 (6):3505–3510.
17. Baty, J. W., M. B. Hampton, and C. C. Winterbourn. 2005. Proteomic detection of hydrogen peroxide-sensitive thiol proteins in Jurkat cells. *Biochem J* 389:785–795.
18. Kumar, V., T. Kleffmann, M. B. Hampton, M. B. Cannell, and C. C. Winterbourn. 2013. Redox proteomics of thiol proteins in mouse heart during ischemia/reperfusion using ICAT reagents and mass spectrometry. *Free Radic Biol Med* 58:109–117.

19. Hung, R. J., C. W. Pak, and J. R. Terman. 2011. Direct redox regulation of F-actin assembly and disassembly by Mical. *Science* 334 (6063):1710–1713.

20. Jayavelu, A. K., J. P. Muller, R. Bauer, S. A. Bohmer, J. Lassig, S. Cerny-Reiterer, W. R. Sperr et al. 2016. NOX4-driven ROS formation mediates PTP inactivation and cell transformation in FLT3ITD-positive AML cells. *Leukemia* 30 (2):473–483.

21. Winterbourn, C. C. 2008. Reconciling the chemistry and biology of reactive oxygen species. *Nat Chem Biol* 4 (5):278–286.

22. Cox, A. G., C. C. Winterbourn, and M. B. Hampton. 2010. Mitochondrial peroxiredoxin involvement in antioxidant defence and redox signalling. *Biochem J* 425:313–325.

23. Peskin, A. V., A. G. Cox, P. Nagy, P. E. Morgan, M. B. Hampton, M. J. Davies, and C. C. Winterbourn. 2010. Removal of amino acid, peptide and protein hydroperoxides by reaction with peroxiredoxins 2 and 3. *Biochem J* 432 (2):313–321.

24. Hall, A., D. Parsonage, L. B. Poole, and P. A. Karplus. 2010. Structural evidence that peroxiredoxin catalytic power is based on transition-state stabilization. *J Mol Biol* 402 (1):194–209.

25. Nagy, P., A. Karton, A. Betz, A. V. Peskin, P. Pace, R. J. O'Reilly, M. B. Hampton, L. Radom, and C. C. Winterbourn. 2011. Model for the exceptional reactivity of peroxiredoxins 2 and 3 with hydrogen peroxide: A kinetic and computational study. *J Biol Chem* 286 (20):18048–18055.

26. Wu, R. F., Y. C. Xu, Z. Ma, F. E. Nwariaku, G. A. Sarosi, Jr., and L. S. Terada. 2005. Subcellular targeting of oxidants during endothelial cell migration. *J Cell Biol* 171 (5):893–904.

27. Chen, K., M. T. Kirber, H. Xiao, Y. Yang, and J. F. Keaney, Jr. 2008. Regulation of ROS signal transduction by NADPH oxidase 4 localization. *J Cell Biol* 181 (7):1129–1139.

28. Wood, Z. A., L. B. Poole, and P. A. Karplus. 2003. Peroxiredoxin evolution and the regulation of hydrogen peroxide signaling. *Science* 300 (5619):650–653.

29. Woo, H. A., S. H. Yim, D. H. Shin, D. Kang, D. Y. Yu, and S. G. Rhee. 2010. Inactivation of peroxiredoxin I by phosphorylation allows localized H(2)O(2) accumulation for cell signaling. *Cell* 140 (4):517–528.

30. Travasso, R. D., F. Sampaio Dos Aidos, A. Bayani, P. Abranches, and A. Salvador. 2017. Localized redox relays as a privileged mode of cytoplasmic hydrogen peroxide signaling. *Redox Biol* 12:233–245.

31. D'Autreaux, B. and M. B. Toledano. 2007. ROS as signalling molecules: Mechanisms that generate specificity in ROS homeostasis. *Nat Rev Mol Cell Biol* 8 (10):813–824.

32. Jarvis, R. M., S. M. Hughes, and E. C. Ledgerwood. 2012. Peroxiredoxin 1 functions as a signal peroxidase to receive, transduce, and transmit peroxide signals in mammalian cells. *Free Radic Biol Med* 53 (7):1522–1530.

33. Sobotta, M. C., W. Liou, S. Stocker, D. Talwar, M. Oehler, T. Ruppert, A. N. Scharf, and T. P. Dick. 2015. Peroxiredoxin-2 and STAT3 form a redox relay for H2O2 signaling. *Nat Chem Biol* 11 (1):64–70.

34. Wood, Z. A., L. B. Poole, R. R. Hantgan, and P. A. Karplus. 2002. Dimers to doughnuts: Redox-sensitive oligomerization of 2-cysteine peroxiredoxins. *Biochemistry* 41 (17):5493–5504.

35. Radjainia, M., H. Venugopal, A. Desfosses, A. J. Phillips, N. A. Yewdall, M. B. Hampton, J. A. Gerrard, and A. K. Mitra. 2015. Cryo-electron microscopy structure of human peroxiredoxin-3 filament reveals the assembly of a putative chaperone. *Structure* 23 (5):912–920.

36. Storz, G., L. A. Tartaglia, and B. N. Ames. 1990. Transcriptional regulator of oxidative stress-inducible genes: Direct activation by oxidation. *Science* 248 (4952):189–194.

37. Delaunay, A., D. Pflieger, M. B. Barrault, J. Vinh, and M. B. Toledano. 2002. A thiol peroxidase is an H2O2 receptor and redox-transducer in gene activation. *Cell* 111 (4):471–481.

38. Dinkova-Kostova, A. T., W. D. Holtzclaw, R. N. Cole, K. Itoh, N. Wakabayashi, Y. Katoh, M. Yamamoto, and P. Talalay. 2002. Direct evidence that sulfhydryl groups of Keap1 are the sensors regulating induction of phase 2 enzymes that protect against carcinogens and oxidants. *Proc Natl Acad Sci USA* 99:11908–11913.

39. Winterbourn, C. C. 2014. The challenges of using fluorescent probes to detect and quantify specific reactive oxygen species in living cells. *Biochim Biophys Acta* 1840 (2):730–738.

40. Wardman, P. 2007. Fluorescent and luminescent probes for measurement of oxidative and nitrosative species in cells and tissues: Progress, pitfalls, and prospects. *Free Radic Biol Med* 43 (7):995–1022.

41. Bilan, D. S. and V. V. Belousov. 2017. New tools for redox biology: From imaging to manipulation. *Free Radic Biol Med* 109:167–188.

42. Poynton, R. A. and M. B. Hampton. 2014. Peroxiredoxins as biomarkers of oxidative stress. *Biochim Biophys Acta* 1840 (2):906–912.

43. Brown, M. R., F. J. Miller, Jr., W. G. Li, A. N. Ellingson, J. D. Mozena, P. Chatterjee, J. F. Engelhardt et al. 1999. Overexpression of human catalase inhibits proliferation and promotes apoptosis in vascular smooth muscle cells. *Circ Res* 85 (6):524–533.

44. Hayakawa, M., H. Miyashita, I. Sakamoto, M. Kitagawa, H. Tanaka, H. Yasuda, M. Karin, and K. Kikugawa. 2003. Evidence that reactive oxygen species do not mediate NF-kappaB activation. *EMBO J* 22 (13):3356–3366.

15 Regulation of H₂O₂ Transport across Cell Membranes

Gerd Patrick Bienert, Iria Medraño-Fernandez, and Roberto Sitia

CONTENTS

INTRODUCTION

The senior author of this chapter remembers vividly the intense discussions with Christine Winterbourn in two *Gordon Research Conferences* on *Thiol signaling*. In the first one, Christine commented with skepticism on our data indicating that—at concentrations lower than 100 μM—exogenous H₂O₂ needs AQP8 to gain access into HeLa cells. Considering her knowledge and experience, my team and I were quite worried. Much happier were we two years later, when Christine accepted the role of

365

AQP8, maintaining however that membrane permeability to H_2O_2 was a controversial issue. We are now firmly convinced that peroxiporins regulate H_2O_2 transport and hence redox signaling, and hope that Christine's remaining doubts will disappear upon reading this chapter, which is warmly dedicated to her.

TRANSPORT OF H_2O_2 IS FACILITATED BY AQUAPORINS

ROS ARE ESSENTIAL FOR SIGNALING, BUT THEY CAN KILL

The term reactive oxygen species (ROS) groups a number of chemically reactive molecules and radicals (O_2^-, H_2O_2, OH^-) derived from the incomplete reduction of molecular oxygen. They most likely appeared on Earth together with the first oxygen molecules, about 2.5 billion years ago, matching the appearance of complex life forms [1]. In conditions of rising free oxygen concentration, organisms switched to a more efficient metabolism allowing the evolution of a huge diversity of forms of life. Not surprisingly, therefore, ROS were originally considered the toxic payback of sustaining an aerobic metabolism, to be swiftly removed for limiting the damage caused by hyperoxidation of macromolecular cell components [2]. However, recent studies revealed that ROS are used as key signal transduction molecules in most multicellular organisms [3,4], mainly via interacting with cysteine (cys) residues within target proteins and yielding a dynamic regulation based in either "on–off" or rheostatic sulfur switches [5]. These redox-mediated protein modifications modulate transcription, phosphorylation, and other important signaling events, and/or alter metabolic fluxes and reactions, playing an essential role in diverse physiological processes, including life–death decisions [3]. Moreover, a basal level of ROS seems to be required to support life (reviewed in [6]), highlighting the new view of ROS as being beneficial.

Owing to their Janus-like nature, ROS have traditionally been considered the double-edged sword of signaling (Figure 15.1). Cells maintain redox homeostasis by keeping a fine balance between their generation, diffusion, and scavenging via antioxidant systems, and removing oxidized cellular components via proteasomes and/or autophagy. However, if ROS pass a toxic threshold, they can overwhelm the antioxidant capacity of the cell, causing oxidative stress and cell death. The integration of all these different ROS-dependent reactions/signals determines the overall response of the cell to a particular stimulus.

H_2O_2 AS SIGNALING MOLECULE

A crucial feature of any signal is specificity. Thus, most free radical species of ROS do not fulfill the criteria necessary for signal transduction. Until recently, the reaction kinetics of ROS with their potential targets, in competition with the enzymes that remove them, has been given insufficient consideration. For instance, superoxide/ O_2^-, which can initiate a chain reaction, has a low oxidation rate constant—likely below 10^3 M^{-1} s^{-1} at pH 7.4 [7]—that would appear insignificant in biological systems in comparison to the rate constants for cytosolic and mitochondrial superoxide dismutases ($>10^9$ M^{-1} s^{-1}; [8]). On the other hand, hydroxyl radicals/HO have no

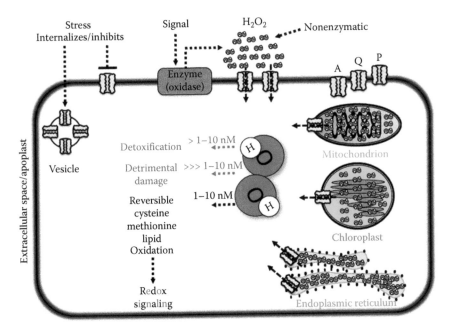

FIGURE 15.1 Impact of aquaporins on hydrogen peroxide–mediated redox signaling in cells. The main sources of hydrogen peroxide (H$_2$O$_2$) in cells are displayed: the extracellular space, mitochondria (blue), chloroplasts (green), and the endoplasmic reticulum (ER) (orange). Extracellularly generated (enzymatic or nonenzymatic) H$_2$O$_2$ enters cells via aquaporins (AQPs) (green) and reacts with cysteine thiols of select proteins to activate signal transduction cascades. H$_2$O$_2$ produced in chloroplasts, the ER, and mitochondria may also exit these organelles via AQPs. Depending on its concentration inside the cell, H$_2$O$_2$ can act as a signal molecule or as a stressor. Various stresses have been shown to inhibit AQP-mediated transport of H$_2$O$_2$, either by the internalization into vesicles or by a yet biochemically not identified posttranslational modification.

specificity as they react with almost any organic molecule with rate constants near the limit of diffusion. Remarkably, sequential univalent electron transfer occurs with many proteins that generate superoxide/O$_2^-$ within living cells, so that H$_2$O$_2$ is a common product of these reactions [9,10]. Moreover, reaction of oxygen radicals with scavenger molecules often results in their efficient conversion to H$_2$O$_2$ resulting in substantial generation of this molecule. But what really qualifies H$_2$O$_2$ as a versatile second messenger is (1) its rapid enzymatic generation in response to extracellular stimuli (2) its easy removal by numerous antioxidant cellular systems, and (3) its chemistry, that provides substrate specificity for thiol oxidation. Four oxidation states of cysteine can be generated upon H$_2$O$_2$ production. These are disulfides (-S-S-), sulfenic acid (–SOH), sulfinic acid (–SO$_2$H), and sulfonic acid (–SO$_3$H). While the latter is irreversible under physiological conditions [11], SO$_2$H species can be reduced by sulfiredoxin-1 (Srx1) in an ATP-dependent reaction [12]. Most importantly, reversible oxidation of thiols to disulfides or sulfenic acid residues can control biological functions in two nonmutually exclusive ways: by chemically altering active site cysteines, as in the case of tyrosine phosphatases, or by altering supramolecular interactions, as in the case of PKCs (reviewed [13,14]). Concerning the specificity of H_2O_2, a cysteine residue is a good target only if deprotonated into its thiolate anion form (–S⁻). Most cysteine residues in proteins have a pKa value of 8.5, which makes them unable to become anions at physiological pH values. However, if the cysteine residue is located in the vicinity of a positively charged amino acid, frequently an arginine, its lower pKa makes it a target for the oxidizing action of H_2O_2 [15]. Thus, the accessibility of deprotonated cysteine(s) dictates the specificity of H_2O_2 as second messenger.

COMPARTMENTALIZED H_2O_2 SIGNALING

An important feature of H_2O_2-driven signaling is its potential to site-localize oxidation in restricted zones of the cell, allowing spatially and temporally controlled signal transmission. Four main pathways generate H_2O_2 in living cells:

1. Oxidative phosphorylation in mitochondria
2. Electron transport chain in chloroplasts
3. Oxidative protein folding in the endoplasmic reticulum (ER)
4. NADPH oxidases (NOX), peroxidases, oxalate and amine oxidases residing in the plasma or endoplasmic reticulum (ER) membranes [16] (Figure 15.1)

In all cases, H_2O_2 must cross a membrane to reach its cytosolic targets. This characteristic can confer specificity to the redox signaling circuitry, ensuring physical insulation and kinetic control of ROS accumulation, and limit cell damage in case of dysregulation, assuming that biological membranes be impermeable to H_2O_2 [17].

Importantly, increasing evidence suggests that redox signaling and its control systems are specialized to organelle function; indeed, optimization of redox environments may have been an important component of evolution coupling chemistry and biological function in each compartment and at each stage of cellular life. In that sense, differences in H_2O_2 concentrations between compartments within a cellular system or between cells have been described for *Escherichia coli* [18], *Saccharomyces cerevisiae* [19,20], mammalian cell lines [21–23] (and references therein), and in plants [24]. Such chemical gradients could be built up due to differences in membrane lipid permeability and different H_2O_2 detoxification machineries within the cells or compartments. To make H_2O_2 a specific, rapid, and highly reactive signaling molecule and render its transport performance independent from slowly modifiable membrane compositions or enzyme activities, a facilitated diffusion via transport proteins would suit the physiological requirements for sufficiently rapid changes of transport processes and site-localized oxidation of protein targets. Cumulative evidence suggests that membrane diffusion of H_2O_2 is indeed facilitated by membrane proteins, namely by members of Major Intrinsic Proteins (MIPs) channel protein family, often referred to as aquaporins (AQPs) (Figure 15.1). AQPs have not only been shown to facilitate the transmembrane diffusion of H_2O_2, but are also physiologically important for the transmembrane transport regulation of both the essential H_2O_2 signaling molecule and the detrimental H_2O_2 reactive oxygen species [25] (Table 15.1).

TABLE 15.1
Functions of Aquaporins in H_2O_2-Mediated Transmembrane Signal Transduction Pathways

Signal Transduction Pathway— Physiological Mechanism in Which an Aquaporin Is Involved	Aquaporin	Site of Signal Action (Cell Type and/or Organism)	Reference
EGF-induced tyrosine kinase signaling		HeLa cells	[58]
NOX2 in B cell activation and differentiation		Normal and neoplastic B lymphocytes	[59]
NOX-, VEGF- AKT-dependent signaling pathways	AQP8	Human acute leukemia cells	[53]
ROS detoxification		Murine mitochondria	[66]
ROS detoxification		Sperm mitochondria (seabream)	[69]
Cell survival and stress resistance		HeLa cells	[25]
EGF- AKT kinase-mediated signal transduction		Human HEK293 cells	[50]
Actin polymerization T-lymphocyte migration		Murine and human T cells	[79]
CXCL12-induced AKT-mediated cell migration	AQP3	Human breast cancer cell lines	[52]
Pathogen defense signal transduction		Colonic epithelial cells	[55]
NF-kB-dependent psoriasis progression		Murine and human keratinocytes	[56]
Bleomycin-induced scleroderma		Murine fibroblasts	[57]
Pathogen defense signal transduction	AtPIP1;4	*Arabidopsis* leaf cells	[77]

AQUAPORINS

AQPs are membrane channel proteins consisting of six transmembrane helices connected by five loops and two termini extending into the cytosol. Two loops contain the conserved asparagine-proline-alanine (NPA) signature sequence or variants thereof. AQPs have a molecular weight between 20 and 40 kDa and assemble as tetramers within biological membranes [26]. Each channel within a tetramer is functional, and contains a narrow hydrophilic pathway with selectivity features generally excluding charged ions as well as large and bulky solutes, but permeable to water, glycerol, hydroxylated metalloid acids, urea, ammonia, carbon dioxide, lactic acid, acetic acid, nitric acid, and H_2O_2 [26]. In 1992, the laboratory of Peter Agre demonstrated that AQP1 from red blood cells, named CHIP28 at that time, is a functional water channel [27]. This study was a breakthrough in the description of single AQP isoforms. Due to the physiological significance of MIP protein family members in water transport processes, consistently described in the early MIP-characterizing studies, these proteins were designated as "water channels" or "aquaporins."

Peter Agre was awarded the Nobel Prize in 2003 for the discovery that AQPs represent physiologically important water channels. In the following years, the diversity and phylogeny of AQPs was elucidated [28], crystal structures for some members of the family were revealed [29,30] (and references therein), and questions such as subcellular localization of AQPs, their spatial and temporal expression patterns and regulation addressed [31]. In the last decade, knowledge about the importance of individual residues for pore selectivity, transport capacity, gating, and posttranslational modifications has blossomed [32], and consequences of posttranslational modifications and heteromerization (in plant AQPs) on the functional and trafficking regulation have been shown [31,33,34].

Long before the discovery of AQPs, the existence of proteins facilitating the transmembrane transport of water was speculated by various researchers [35,36] and reviewed by Finkelstein [37], based on experimental observations. Stein and Danielli postulated as early as in 1956 that facilitated diffusion through the cell plasma membrane takes place *"through a hydrogen-bonding stereochemically specific membrane component which extends through the thickness of the membrane."* The existence of "water channels" became obvious as the activation energy for the water passage through biological membranes was calculated much lower as for phospholipid-only bilayers. Moreover, permeability for water differed greatly between different cell types with similar membrane compositions. Furthermore, for a given cell type, membrane permeability varied in response to treatments with diuretic, antidiuretic, and mercury-containing agents. These chemicals, however, did not share the same inhibitory effect on water transport across synthetic membranes [38].

Interestingly, as it was just described for water, biophysical measurements and various experimental results suggested that (1) membranes are not freely permeable to H_2O_2 (though the permeability is quite high), (2) membrane permeability to H_2O_2 varies greatly between different cell types or in a time-dependent fashion, and (3) H_2O_2 membrane permeability is inhibited by AQP blockers (reviewed in [39,40]).

Despite these experimental indications that proteinaceous H_2O_2 transport facilitators exist, most likely AQPs, and in spite of the fact that several metabolic and signaling pathways require tight and fast regulation of the H_2O_2 membrane permeability to gain physiological effectiveness, the general assumption was that H_2O_2 crosses membranes via nonfacilitated diffusion, as it was assumed for water for a long time.

The presence of efficient H_2O_2 decomposing enzymes and molecules in living cells and the lack of robust assays to quantify and visualize H_2O_2 hampered many attempts to monitor its fluxes. In fact, it is a general problem to experimentally separate membrane permeation from reaction flows when the compound is highly reactive, takes part in chemical reactions, is rapidly decomposed, and has therewith a short half-life. Additionally, specific probes for H_2O_2 such as dyes or tracers were unavailable.

"Peroxiporin Hypothesis" of Henzler and Steudle

Nonetheless, like Danielli and Stein did for water, Henzler and Steudle hypothesized that *"some of the water channels in Chara (and, perhaps, in other species) serve as peroxiporins rather than as aquaporins"* [41]. Henzler and Steudle developed a mathematical model to describe H_2O_2 permeation through cell membranes taking the enzymatic decomposition into account. The model delivered a series of predictions, which were addressed experimentally using the algae *Chara corallina*. Comparing the permeability and reflection coefficients of water and H$_2$O$_2$ the authors suggested that H$_2$O$_2$ partly uses the same pathway through membranes as water does. Accordingly, the AQP inhibitor mercury chloride reduced the permeability coefficient for H$_2$O$_2$ permeation and the subsequent treatment with 2-mercaptoethanol reverted the inhibitory effects of the channel blocker [41]. This behavior is typical of AQP-mediated processes.

Then, the progress in the discovery of AQP-mediated H$_2$O$_2$ transport mechanisms resembled that of water. The analogy is also reasoned in the similar electro-chemical properties of water and H$_2$O$_2$ that allow permeation through membranes and AQPs [39], despite their different chemical reactivity.

SPECIFIC AQPS FACILITATE THE DIFFUSION OF H$_2$O$_2$ ACROSS MEMBRANES

In 2007, one of us and his coworkers demonstrated for the first time that specific AQP isoforms possess permeability to H$_2$O$_2$ [42]. In a comprehensive screen, AQP isoforms from different organisms and different subfamilies with different selectivity filters and substrate spectra were heterologously expressed in *Saccharomyces cerevisiae* mutants differing in sensitivity to exogenous H$_2$O$_2$. Dramatic differences in growth rates and cell survival were observed that nicely correlated with the amount of H$_2$O$_2$ added, the ability of the yeast strain to detoxify it, and, most importantly, the AQP isoform that is expressed. Most isoforms changed only slightly—if at all—the yeast sensitivity to H$_2$O$_2$ in a strain-specific manner [42]. In contrast, the expression of human AQP8 (hAQP8) and of *Arabidopsis* AtTIP1;1 and AtTIP1;2 significantly decreased cell growth and survival on a medium containing low amounts of H$_2$O$_2$. The addition of silver nitrate, an AQP inhibitor, did partly rescue the growth of yeast expressing hAQP8 as well as AtTIP1;1 and AtTIP1;2 on medium containing H$_2$O$_2$. More operative and direct methods were then used to confirm and compare the transport ability of the peroxiporins identified with toxicity assays. Yeast cells were loaded with a ROS-sensitive dye (5-(and-6)-chloromethyl-2,7-dichlorodihydrofluorescein diacetate acetyl ester [CM-H2DCFDA]) and incubated with or without H$_2$O$_2$ and the kinetics of fluorescence emission changes were monitored and quantified. In the absence of H$_2$O$_2$, dye-loaded yeast cells displayed similar fluorescent levels independent of whether they were transformed with an empty vector or with different AQPs. Likewise, addition of low H$_2$O$_2$ amounts to yeasts expressing AQPs that did not significantly impact growth rates increased the fluorescent signal only slightly, likely reflecting uptake of H$_2$O$_2$ via passive diffusion. In contrast, the signal increased significantly in yeast cells expressing hAQP8, AtTIP1;1, and AtTIP1;2, consistent with facilitated uptake of H$_2$O$_2$ mediated by these isoforms. Accordingly, fluorescence increase was prevented by prior treatment of yeast cells with AQP inhibitors. Single cell confocal microscopy assays confirmed that the H$_2$O$_2$-dependent fluorescence significantly increased only in cells transformed with hAQP8 and AtTIP1;1. Altogether these experiments provided compelling evidence demonstrating an AQP-mediated H$_2$O$_2$ transport [42].

Thereafter, yeast growth and survival assays in combination with H$_2$O$_2$ optical detection assays have been used to quantitatively assess whether and which AQPs are

TABLE 15.2

AQP Groups Demonstrated to Possess H₂O₂ Permeability Though No Physiological Signaling Function Has Been Identified Yet

Aquaporin-Group	Organism	Functional Assay System	References
Aquaporin-group	Mammal	Toxicity growth assay in yeast	[45,46,69]
Aquaglyceroporin-group	Mammal	Uptake assays in *Xenopus oocytes*	[70]
Aquaporin-group	Fungus	Toxicity growth assay in yeast	[80]
Aquaglyceroporin-group	Microbe	Toxicity growth assay in yeast	[46]
Aquaglyceroporin-group	Bacterium	Toxicity growth assay in yeast	[75]
Aquaporin-PIP2-group	Plant	Toxicity growth assay in yeast Uptake assay in yeast Molecular simulation study	[43–45,78]
Aquaporin-TIP-group	Plant	Toxicity growth assay in yeast Uptake assay in yeast	[42,44,81]
Aquaporin-NIP-group	Plant	Toxicity growth assay in yeast	[44,82]
Aquaporin-XIP-group	Plant	Toxicity growth assay in yeast	[26]

permeable to H_2O_2 (reviewed in [43]). Various AQP isoforms from bacteria, plants, and mammals belonging to different AQP subfamilies have been demonstrated in the following to be permeable to H_2O_2 [43] (see Tables 15.1 and 15.2).

Additionally, the results of molecular dynamic simulations on plant (SoPIP2;1) and mammalian (AQP1) crystal structures further corroborated that H_2O_2 represents a substrate for AQPs, indicating that H_2O_2 is permeating through these channels and exhibits flow behaviors similar to water [44,45]. These simulations revealed that the calculated energy barrier for H_2O_2 permeation was smaller in AQP-loaded bilayers than in pure bilayers, further suggesting a potential role of certain AQPs in physiological H_2O_2 transport processes. Moreover, experiments with AQP mutants implicated that water and H_2O_2 share similar transport routes through AQP pores [25,40,44,46], yielding similar results to molecular dynamic simulation studies. Detailed calculations showed that water and H_2O_2 have similar energy profiles and residence times throughout the AQP pore pathway. In contrast, H_2O_2 permeation via the "fifth pore" (the central cavity, which is formed in an AQP tetramers) was excluded, as calculated energy barriers for the fifth pore are either similar or even higher compared to those of protein-free bilayers [45].

PHYSIOLOGICAL FUNCTIONS OF MAMMALIAN AQPs IN H₂O₂ TRANSPORT

Once it was accepted that AQPs increase membrane permeability to H_2O_2 and the latter is a key signaling molecule, a number of studies in mammalian cell systems followed promptly to demonstrate that (1) AQPs facilitate transmembrane H_2O_2 diffusion in their native expression contexts, (2) AQP-mediated H_2O_2 transport is of physiological relevance, and (3) it can be regulated. Key to the success of these studies was the development of H_2O_2–specific probes of improved specificity, faster

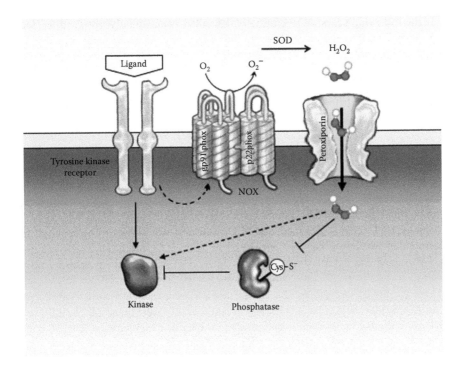

FIGURE 15.2 Signal amplification via AQP-mediated H$_2$O$_2$ transport. Upon receptor activation, H$_2$O$_2$ is produced by a NOX enzyme, frequently NOX2 [56,59], and reaches the cytosol by a peroxiporin, allowing site-localized modification of target proteins.

kinetics, and predetermined subcellular localization, of either chemical nature, as in the case of boronate-based fluorescent dyes [47] or genetically encoded fusion proteins between bacterial OxyR and a permutated fluorescent protein (*HyPer* and variants thereof) [48,49]. Using these technologies, it was possible to establish that AQP-mediated import of NOX-produced H$_2$O$_2$ has an important role amplifying diverse signal transduction cascades (Figure 15.2), as detailed here.

AQP3

The first AQP demonstrated to transport H$_2$O$_2$ in mammalian cells was AQP3 [50]. AQP3-expressing human embryonic kidney 293 (HEK293) cells, which were loaded with chemical or genetically encoded optical H$_2$O$_2$ sensors, responded with a significant higher fluorescent signal emission upon exposure to H$_2$O$_2$ than control cells [50]. Importantly, modifying AQP3 expression altered downstream intracellular H$_2$O$_2$ signaling. Thus, knockdown of hAQP3 expression using a specific shRNA decreased the influx of H$_2$O$_2$ upon extracellular application or after epidermal growth factor (EGF)–stimulated production. Simultaneously, downstream serine/threonine kinase AKT/protein kinase B phosphorylation circuits were reduced, implying that AQP3-mediated H$_2$O$_2$ import is important for intracellular redox signaling [50].

In the following years, AQP3, initially described as an aquaglyceroporin, became also the best-characterized peroxiporin and important functions were unveiled for its H_2O_2 transport activity, like the chemokine-dependent T lymphocyte migration during immune responses in mice [51]. CD4$^+$ T cells derived from AQP3$^{-/-}$ knockout mice showed impaired migration during cutaneous immune reactions when compared to cells from control animals. The authors showed that this phenotype was due to deficient H_2O_2—rather than water—transport capacity. In their setup, H_2O_2 was presumably produced extracellularly by a NOX enzyme after activation by chemokine ligands and then transported back to the cytoplasm via AQP3 [51]. Application of exogenous H_2O_2 in excess allowed passive uptake via nonfacilitated membrane diffusion and rescued CXCL12-stimulated migration ability in AQP3$^{-/-}$ cells. Accordingly, AQP3 silencing in human T cells impaired CXCL12-induced actin polymerization, used as a marker for cell motility [51].

Similar to T cells, AQP3-mediated H_2O_2 transport was shown to be required for migration of two breast cancer lines. In these cells, H_2O_2 produced by NOX2 uses AQP3 to enter the cytoplasm, where it oxidizes PTEN and protein tyrosine phosphatase 1B (PTP1B) allowing AKT activation and directional cell migration [52]. As discussed later in this chapter, the same pathway is activated in leukemia cells in an AQP8-dependent manner [53,54]. In addition, by overexpression of AQP3 and CXCL12 stimulation, the authors were able to show that the expression level of AQP3 positively correlated with cancer cell migration both *in vitro* and *in vivo*. On the other hand, AQP3 knockdown reduced spontaneous metastasis in orthotopic xenografts [52].

AQP3-mediated H_2O_2 transport is important also in colonic epithelium cells. These cells generate extracellular H_2O_2, which promotes signaling cascades initiating defense mechanisms or affecting barrier function in response to injury or microbial pathogens. Consequently, deregulation of intestinal H_2O_2 levels has been clinically connected to the early onset of inflammatory bowel disease and colon cancer. In this context, it was hypothesized that intestinal H_2O_2 transport across the cell membrane depends on AQP3 [55]. Accordingly, Caco-2 colonic epithelial cells expressing AQP3 accumulated detectable intracellular H_2O_2 significantly faster than AQP3-knockdown cells or cells treated with the AQP inhibitor silver nitrate. Upon wounding or exposure to the intestinal pathogen *Citrobacter rodentium*, colonic epithelial cells knocked down for AQP3 expression showed defective H_2O_2-associated responses. Analogously, AQP3$^{-/-}$ mice showed reduced colon wound healing and impaired immune responses against *C. rodentium* infection, compared to AQP3$^{+/+}$ mice [55].

Also, the pathogenesis of psoriasis is promoted by AQP3-dependent H_2O_2 channeling [56]. In AQP3$^{-/-}$ mice, H_2O_2 entry, nuclear factor-kB (NF-kB) activation, and IL-23-dependent disease progression are reduced. Concordantly, in TNFα-stimulated neonatal human and mouse keratinocytes, AQP3 allowed entry of NOX2-generated H_2O_2 molecules into the cytoplasm, where they oxidized and inactivated protein phosphatase 2A (PP2A) upstream of NF-kB activation [56]. Taken together, these data indicated that AQP3-facilitated H_2O_2 entry into keratinocytes is required for NF-kB activation and the consequent development of psoriasis.

AQP3 has been also shown to be involved in the pathogenesis of scleroderma, another autoimmune disease involving the skin. In a mouse model of inducible scleroderma, H$_2$O$_2$ levels increased upon bleomycin treatment of wild type but not AQP3-silenced fibroblasts [57]. These results leave little doubt that AQP3-dependent H$_2$O$_2$ transport impacts signaling.

AQP8

H$_2$O$_2$ Transport across the Plasma Membrane

In heterologous expression systems, AQP8 facilitates H$_2$O$_2$ diffusion across membranes [42]. The physiological importance of AQP8 in the transduction of H$_2$O$_2$ signals across the plasma membrane was demonstrated in human and murine cells [58,59]. Silencing or inhibiting hAQP8 in HeLa cells inhibited the activation of *HyPer* sensors after exposure to exogenous H$_2$O$_2$, suggesting that also AQP8 is an efficient peroxiporin. Moreover, hAQP8-silencing diminished EGF-induced transient cytosolic H$_2$O$_2$ peaks, dampening downstream signaling. The presence of catalase in the extracellular space phenocopied AQP8 silencing, implying that H$_2$O$_2$ generated in response to EGF was produced in the outer leaflet of the plasma membrane, hence accessible to the scavenger and needing to cross a lipid bilayer to promote signaling [58]. In an independent study, it was also shown that AQP8 is essential to import H$_2$O$_2$ generated by NOX2 upon B cell receptor or TLR4 cross-linking in B cells, and H$_2$O$_2$-mediated signal amplification is necessary for B cell activation and plasma cell differentiation [59].

Other examples of the involvement of AQP8 in signaling cascades can be found in studies performed using the human acute leukemia B1647 cell line [53]. In this model, inhibition of different AQP isoforms caused a decline in intracellular ROS, independently of whether H$_2$O$_2$ was externally applied or endogenously produced by NOX. The authors ranked the contribution to the transmembrane H$_2$O$_2$ transport as AQP8 > AQP3 ≫ AQP1. When *AQP*8 expression was positively or negatively modulated, downstream signaling was increased or decreased, respectively. Thus, AQP8-mediated H$_2$O$_2$ transport affects redox signaling and leukemia cell proliferation [53]. The biochemical consequences of this pathway were investigated also in B1647 cells that constitutively produce vascular endothelial growth factor (VEGF) [53]. Using cysteine sulfenylation as a readout for intracellular H$_2$O$_2$ reactivity, AQP8 expression levels were shown to influence the VEGF-induced oxidation of target proteins. Interestingly, increased AQP8 levels correlated with PTEN sulfenylation and prolonged AKT activity [53]. Taken together, these examples show that AQP8 can fine-tune cysteine modifications of proteins, modulating intracellular signaling pathways.

Transport across the Membranes of Organelles

H$_2$O$_2$ became a most intriguing matter for people interested in the biogenesis of secretory proteins when Jonathan Weismann and coworkers discovered that Ero1 flavoproteins use oxygen as an electron acceptor during oxidative folding in the ER, thus generating abundant H$_2$O$_2$ in the exocytic pathway [60]. It soon became clear that H$_2$O$_2$ is produced in equimolar amounts to the number of disulfide bonds inserted in secretory cargoes. Considering that a single plasma cell can produce several thousand immunoglobulins per second, each containing dozens of disulfides, it follows that up to 10^5 H$_2$O$_2$ molecules are generated each second in the ER, solely to sustain the antibody factory. Similar figures hallmark pancreatic beta cells and other professional secretors. As many other ER-resident enzymes have been also shown to be capable of producing H_2O_2, and/or transfer its oxidative power onto other proteins, whether and how transport of H_2O_2 from the ER lumen to the cytosol serves for physiological signaling purposes are open questions that remain to be solved [61].

Since AQP8 begins its folding schedule in the ER, it is not surprising that downregulation of the channel also had an impact on H_2O_2 transport across the membrane of this organelle, as shown using a *HyPer* probe specifically targeted to the ER lumen [58]. Thus, digitonin-treated HeLa cells showed a substantial reduction on ER H_2O_2 import when AQP8 expression was silenced, compared to control cells. In the same experimental setup, no differences were detected when the entry of H_2O_2 into mitochondria was analyzed. Considering that proteins that release H_2O_2 inside the ER, that is, NOX4, are able to oxidize and modulate the activity of cytoplasmic enzymes such as PTP1B [62], it is conceivable that AQP8-mediated ER H_2O_2 channeling be also used to provide specificity to interorganelle redox signaling [63].

Besides the secretory pathway and plasma membrane, AQP8 has been reported to reside also in mitochondria [64,65], a major source of ROS in animals. Mitochondria isolated from HepG2 hepatoma cells in which *AQP8* was silenced by a siRNA approach, displayed decreased H_2O_2 efflux ability and higher ROS levels compared to mitochondria from control cells. Partial permeabilization by digitonin abolished the differences between AQP8^{low} and wild type mitochondria. These data were interpreted to suggest that a fraction of AQP8 localizes in mitochondria mediating H_2O_2 release and that defective or deregulated AQP8 function may cause ROS-induced cell damage by opening the mitochondrial permeability transition pore [66]. However, independent studies in which water permeability was compared in mitochondrial membrane preparations from wild type *versus* AQP8-deficient mice argued against the presence of functional AQP8 in mitochondria [67]. The results were compatible with a diffusion rather than a facilitated mechanism. An intriguing possibility is that AQP8 be present in mitochondrial-associated membranes (MAMs), ER subregions that are tightly tethered to mitochondria. Facilitated H_2O_2 exchange between the two organelles could promote signaling, oxidative folding, and/or disposal by ER-resident scavengers, such as GPX8 [68].

Another study pointed at a role for mitochondrial AQP8b in ROS detoxification in spermatozoa of the marine teleost seabream [69]. As AQP8b facilitates H_2O_2 transport when heterologously expressed in *Xenopus laevis* oocytes, it could also have this function in sperm cells. Accordingly, when sperm are exposed to hypertonic conditions to mimic their release into sea water, AQP8b is rapidly phosphorylated and concentrates in the area occupied by the mitochondrion. The toxic effects observed under these hypertonic conditions were prevented by mitochondrial-targeted antioxidants. These results were interpreted to suggest that an AQP8-dependent facilitated efflux of H_2O_2 from mitochondria might prevent oxidative stress [69]. However, this report also showed that sperm mobility was arrested upon incubation with anti-AQP8 antibodies, a result suggesting a role for plasma membrane rather than mitochondrial AQP8.

Thus, the different subcellular localizations of AQPs and the underlying targeting mechanisms are relevant issues that deserve further studies. It remains to be established how AQP8, a protein destined to reach the plasma membrane through the secretory pathway, can be efficiently rerouted to mitochondria in certain cell types or—as interestingly if not more—within the same cell under different conditions. In our hands, a tagged version of human AQP8 did not localize to mitochondria upon overexpression (our unpublished results) and AQP8 silencing inhibited H$_2$O$_2$ entry into the ER, but not mitochondria, of semipermeabilized HeLa cells [58]. These results should be interpreted with caution, because appending a fluorescent tag could alter targeting of the chimeric AQP8, and digitonin treatment could increase mitochondrial permeability [66]. The development of targeted and faster H$_2$O$_2$ sensors of novel genetic tools and higher resolution imaging and cell fractionation assays will hopefully allow to establish under which conditions and how AQP8 (or other isoforms) are preferentially delivered to mitochondria or other organelles and the physiological consequences of interorganellar H$_2$O$_2$ transport.

AQP9

A reverse genetic approach using cell model systems (CHO-K1 and HepG2 cells) and AQP9 knockout mice have recently implicated AQP9 as an H$_2$O$_2$ transporter *in vivo*. The authors hypothesized that—like AQP3 and AQP8—this channel also might control physiologically important transmembrane H$_2$O$_2$ fluxes [70].

OTHER ISOFORMS

So far, the AQP family is more easily subdivided in two groups regarding their H$_2$O$_2$ channeling ability, those capable and those incapable of facilitating the transport of this molecule, rather than providing a ladder of decreasing efficiencies. A recent study however suggested that *"all AQPs are bona fide H$_2$O$_2$ channels with increasing H$_2$O$_2$ permeability from aquaglyceroporins via orthodox AQPs to aquaammoniaporins, with the latter displaying the optimal pore layout for H$_2$O$_2$ transport"* [46]. Transport capacities might be masked by limits in the detection assay or other experimental biases. Indeed, a variety of AQP isoforms seem to possess permeability to H$_2$O$_2$ (see Table 15.2). However, the court has still to decide whether these other AQPs transport H$_2$O$_2$ at rates of physiological interest in their native context.

MODULATION OF H$_2$O$_2$ TRANSPORT BY AQP GATING

Considering the detrimental roles of defective and excess of H$_2$O$_2$, it was to be expected that peroxiporin activities are tightly regulated. Recent lines of evidence fulfill this expectation [25,71]. Thus, we showed that heat, ER, hypoxia, and other types of cellular stress(es) block the transport of H$_2$O$_2$ and water through AQP8 (Figure 15.3). The blockade involves the generation of ROS, because transport was not impaired if inhibitors of ROS production (diphenyleneiodonium [DPI]) or scavenger molecules (N-acetyl cysteine, NAC) were present during stress induction. Moreover, a rapid exposure to reducing agents restored H$_2$O and H$_2$O$_2$ transport, suggesting redox-dependent inhibitory mechanisms [25]. Using HeLa transfectants expressing different AQP8 mutants, we identified cysteine 53—a residue located in the outer part of the pore—as the main target of the stress-induced redox regulation.

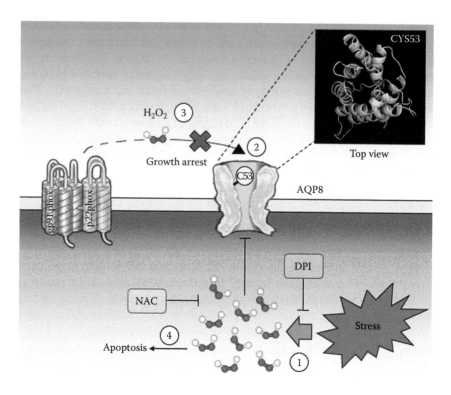

FIGURE 15.3 Regulation of signal intensity by AQP8 gating during cell stress. Cellular stresses exaggerate the production of intracellular ROS, ultimately H_2O_2, disrupting redox homeostasis (1). Oxidative modifications targeting cysteine 53 inhibit the AQP8-dependent bidirectional transport of water and H_2O_2 (2). Inhibitors of ROS production (DPI), or antioxidants (NAC), prevent AQP8 blockade. Stress-induced AQP8 gating prevents the NOX/H_2O_2–dependent amplification of growth factor–stimulated cascades (3), causing growth arrest. If stress persists, excess H_2O_2 accumulates, causing oxidative stress, macromolecular damage, and eventually cell death (4).

Interestingly, a cysteine occupies a similar position in the other AQPs known to exert peroxiporin activity. Excitingly, the expression of the noninhibitable mutant AQP8 C53S was advantageous for cell survival upon stress, treatment with chemotherapeutic drugs or radiation [25]. In normal cells, this mechanism might postpone proliferation and migration to resolution of the causes of stress. Considering that cancer cells frequently encounter hypoxia, starvation, or other stresses, bypassing AQP8 gating could favor tumor progression. As to the nature of the cysteine 53 chemical modification(s) that impede transport across AQP8, only negative results are so far available. Neither channel internalization, a phenomenon shown to inhibit AQP2 and AQP4 [72,73], nor formation of inter- or intramolecular disulfide bonds was observed [25].

Studies on plant cells clearly demonstrated that AtPIP2;1, an *Arabidopsis thaliana* peroxiporin, redistributed to late endosomes upon stress-induced oxidation. In this system, however, the internalization of many plasma membrane proteins could reflect bulk internalization rather than specific removal of AQP [74]. Plant Plasma Membrane Intrinsic Proteins (PIPs) possess highly conserved extracellular cysteine residues, which are theoretically prone to oxidative modification and reactions [45,75,76]. It would be of great interest to establish whether also these AQP isoforms are subjected to redox regulation.

PHYSIOLOGICAL FUNCTIONS OF PLANT AQPs IN H$_2$O$_2$ TRANSPORT

As in mammalian cells, plant cells contain numerous signaling pathways in which H$_2$O$_2$—extracellularly produced in the apoplast—subsequently fulfills its chemical function in the cytoplasm. One such example is the production of H$_2$O$_2$ by plant NADPH-dependent oxidases upon contact with pathogens. This H$_2$O$_2$ burst can activate signaling pathways resulting in systemic acquired resistance (SAR) or pathogen-associated molecular pattern (PAMP) triggered responses.

Expression of AtPIP1;4, a plant peroxiporin, is induced upon infection with *Pseudomonas syringae* pv *tomato* concomitantly with apoplastic H$_2$O$_2$ production. Arabidopsis *atpip1;4* knockout plants accumulate much less H$_2$O$_2$ than wild type and fail to promptly initiate SAR responses, thus failing to prevent bacterial pathogenicity [77]. Likewise, flagellin and chitin can induce efficient PAMP and SAR responses via apoplastic H$_2$O$_2$ production only in plant cells that express AtPIP1;4 [77]. Thus, AtPIP1;4-mediated H$_2$O$_2$ diffusion across the membrane seems to be essential for Arabidopsis to respond to pathogens.

PIPs are essential for plant water homeostasis, playing crucial roles in uptake from the soil, root-to-shoot translocation, transcellular transport, maintenance of cellular turgor pressure, and osmotic-driven processes [33]. Additional functions in CO$_2$ and H$_2$O$_2$ channeling have been hypothesized (reviewed in [26]).

Peroxiporin activity has been detected for several plant PIPs [44,75,78]. Dynowski and coworkers were the first to show that expression of plant PIP family members in yeast resulted in an enhanced H$_2$O$_2$ sensitivity and in an increased fluorescent signal of yeast loaded with a ROS-sensitive dye [44]. The authors further demonstrated that mutations that alter water transport–related gating and selectivity of PIPs also alter H$_2$O$_2$ permeability. Based on a PIP crystal structure, molecular dynamic simulations were calculated to assess the flow of H$_2$O$_2$ through PIP channel proteins. These analyses showed that H$_2$O$_2$ passes the pore but with a slightly less favorable conductance than water [44]. These calculations further pointed at certain channel regions and residues as more important for efficient H$_2$O$_2$ passage. Perhaps not surprisingly, some of these coincided with regions known to be critical for H$_2$O transport. Taken together, the literature suggests that PIPs regulate membrane resistance toward H$_2$O$_2$ diffusion. Thus, AtPIP1;4 will most likely not remain the only plant PIP that is functionally involved in transmembrane H$_2$O$_2$ signal transduction pathways.

CONCLUSIONS

The findings highlighted in this essay are the tip of an iceberg underlining the key roles that AQPs play in all kingdoms of life in regulating H_2O_2 transport across membranes. Considering the multilevel control of their expression, localization, and activity, AQPs emerge as potent signal rheostats besides being pure material flow facilitators. The interrelationships between AQP and ROS action are hence of central interest in both the plant and mammalian research fields, and deserve further intensive work.

ACKNOWLEDGMENTS

This work was supported by an Emmy Noether grant (1668/1-1) from the Deutsche Forschungsgemeinschaft to GPB and AIRC (IG14559), Fondazione Cariplo (2015-0591), Ministero della Salute (PE-2011-02352286), and Telethon (GGP15059) grants to RS. We thank the many scientists whose elegant work contributed to uncover the roles of aquaporins in H_2O_2 transport across cellular membranes. We apologize to the colleagues whose contributions to these research areas could not have been mentioned due to page restrictions.

ABBREVIATIONS

AQP	Aquaporin
cys	Cysteine
DPI	Diphenyleneiodonium
EGF	Epidermal growth factor
ER	Endoplasmic reticulum
IL-23	Interleukin-23
MAM	Mitochondrial-associated membrane
MIP	Major intrinsic protein
NAC	N-acetyl cysteine
NF-kB	Nuclear factor-kB
NIP	Nodulin26-like intrinsic protein
NOX	NADPH oxidase(s)
PAMP	Pathogen-associated molecular pattern
PIP	Plasma membrane intrinsic protein
PKC	Protein kinase C
PP2A	Protein phosphatase 2A
PTP1B	Protein tyrosine phosphatase 1B
ROS	Reactive oxygen species
SAR	Systemic acquired resistance
Srx1	Sulphiredoxin-1
TIP	Tonoplast intrinsic protein
TLR4	Toll-like receptor 4
TNFα	Tumor necrosis factor α
VEGF	Vascular endothelial growth factor
XIP	X-intrinsic protein

REFERENCES

1. Sessions, A. L., D. M. Doughty, P. V. Welander, R. E. Summons, and D. K. Newman. 2009. The continuing puzzle of the great oxidation event. *Curr Biol* 19 (14):R567–R574. doi: 10.1016/j.cub.2009.05.054.

2. McCord, J. M. 2000. The evolution of free radicals and oxidative stress. *Am J Med* 108 (8):652–659.

3. Holmstrom, K. M. and T. Finkel. 2014. Cellular mechanisms and physiological consequences of redox-dependent signalling. *Nat Rev Mol Cell Biol* 15 (6):411–421. doi:10.1038/nrm3801.

4. Sies, H. 2017. Hydrogen peroxide as a central redox signaling molecule in physiological oxidative stress: Oxidative eustress. *Redox Biol* 11:613–619. doi: 10.1016/j.redox.2016.12.035.

5. Paulsen, C. E. and K. S. Carroll. 2010. Orchestrating redox signaling networks through regulatory cysteine switches. *ACS Chem Biol* 5 (1):47–62. doi: 10.1021/cb900258z.

6. Mittler, R. 2017. ROS Are Good. *Trends Plant Sci* 22 (1):11–19. doi: 10.1016/j.tplants.2016.08.002.

7. Winterbourn, C. C. and D. Metodiewa. 1994. The reaction of superoxide with reduced glutathione. *Arch Biochem Biophys* 314 (2):284–290. doi: 10.1006/abbi.1994.1444.

8. Forman, H. J. and I. Fridovich. 1973. Superoxide dismutase: A comparison of rate constants. *Arch Biochem Biophys* 158 (1):396–400.

9. Fridovich, I. 1970. Quantitative aspects of the production of superoxide anion radical by milk xanthine oxidase. *J Biol Chem* 245 (16):4053–4057.

10. Landry, W. D. and T. G. Cotter. 2014. ROS signalling, NADPH oxidases and cancer. *Biochem Soc Trans* 42 (4):934–938. doi: 10.1042/bst20140060.

11. Huyer, G., S. Liu, J. Kelly, J. Moffat, P. Payette, B. Kennedy, G. Tsaprailis, M. J. Gresser, and C. Ramachandran. 1997. Mechanism of inhibition of protein-tyrosine phosphatases by vanadate and pervanadate. *J Biol Chem* 272 (2):843–851.

12. Biteau, B., J. Labarre, and M. B. Toledano. 2003. ATP-dependent reduction of cysteine-sulphinic acid by *S. cerevisiae* sulphiredoxin. *Nature* 425 (6961):980–984. doi: 10.1038/nature02075.

13. Rhee, S. G., S. W. Kang, W. Jeong, T. S. Chang, K. S. Yang, and H. A. Woo. 2005. Intracellular messenger function of hydrogen peroxide and its regulation by peroxiredoxins. *Curr Opin Cell Biol* 17 (2):183–189. doi: 10.1016/j.ceb.2005.02.004.

14. Knapp, L. T. and E. Klann. 2000. Superoxide-induced stimulation of protein kinase C via thiol modification and modulation of zinc content. *J Biol Chem* 275 (31):24136–24145. doi: 10.1074/jbc.M002043200.

15. Paulsen, C. E. and K. S. Carroll. 2013. Cysteine-mediated redox signaling: Chemistry, biology, and tools for discovery. *Chem Rev* 113 (7):4633–4679. doi: 10.1021/cr300163e.

16. Halliwell, B. and J. M. C. Gutteridge. 2015. *Free Radicals in Biology and Medicine*, 5th edn. New York: Oxford University Press.

17. Jones, D. P. 2008. Radical-free biology of oxidative stress. *Am J Physiol Cell Physiol* 295 (4):C849–C868. doi: 10.1152/ajpcell.00283.2008.

18. Seaver, L. C. and J. A. Imlay. 2001. Hydrogen peroxide fluxes and compartmentalization inside growing *Escherichia coli*. *J Bacteriol* 183 (24):7182–7189. doi: 10.1128/jb.183.24.7182-7189.2001.

19. Sousa-Lopes, A., F. Antunes, L. Cyrne, and H. S. Marinho. 2004. Decreased cellular permeability to H$_2$O$_2$ protects *Saccharomyces cerevisiae* cells in stationary phase against oxidative stress. *FEBS Lett* 578 (1–2):152–156. doi: 10.1016/j.febslet.2004.10.090.

20. Branco, M. R., H. S. Marinho, L. Cyrne, and F. Antunes. 2004. Decrease of H$_2$O$_2$ plasma membrane permeability during adaptation to H$_2$O$_2$ in *Saccharomyces cerevisiae*. *J Biol Chem* 279 (8):6501–6506. doi: 10.1074/jbc.M311818200.

21. Makino, N., K. Sasaki, K. Hashida, and Y. Sakakura. 2004. A metabolic model describing the H_2O_2 elimination by mammalian cells including H_2O_2 permeation through cytoplasmic and peroxisomal membranes: Comparison with experimental data. *Biochim Biophys Acta* 1673 (3):149–159. doi: 10.1016/j.bbagen.2004.04.011.
22. Antunes, F. and E. Cadenas. 2000. Estimation of H_2O_2 gradients across biomembranes. *FEBS Lett* 475 (2):121–126.
23. Marinho, H. S., L. Cyrne, E. Cadenas, and F. Antunes. 2013. The cellular steady-state of H_2O_2: Latency concepts and gradients. *Methods Enzymol* 527:3–19. doi: 10.1016/b978-0-12-405882-8.00001-5.
24. Noctor, G. and C. H. Foyer. 2016. Intracellular redox compartmentation and ROS-related communication in regulation and signaling. *Plant Physiol* 171 (3):1581–1592. doi: 10.1104/pp.16.00346.
25. Medrano-Fernandez, I., S. Bestetti, M. Bertolotti, G. P. Bienert, C. Bottino, U. Laforenza, A. Rubartelli, and R. Sitia. 2016. Stress regulates Aquaporin-8 permeability to impact cell growth and survival. *Antioxid Redox Signal* 24 (18):1031–1044. doi: 10.1089/ars.2016.6636.
26. Bienert, G. P., M. D. Bienert, T. P. Jahn, M. Boutry, and F. Chaumont. 2011. Solanaceae XIPs are plasma membrane aquaporins that facilitate the transport of many uncharged substrates. *Plant J* 66 (2):306–317. doi: 10.1111/j.1365-313X.2011.04496.x.
27. Preston, G. M., T. P. Carroll, W. B. Guggino, and P. Agre. 1992. Appearance of water channels in *Xenopus oocytes* expressing red cell CHIP28 protein. *Science* 256 (5055):385–387.
28. Abascal, F., I. Irisarri, and R. Zardoya. 2014. Diversity and evolution of membrane intrinsic proteins. *Biochim Biophys Acta* 1840 (5):1468–1481. doi: 10.1016/j.bbagen.2013.12.001.
29. Tani, K. and Y. Fujiyoshi. 2014. Water channel structures analysed by electron crystallography. *Biochim Biophys Acta* 1840 (5):1605–1613. doi: 10.1016/j.bbagen.2013.10.007.
30. Kirscht, A., S. S. Kaptan, G. P. Bienert, F. Chaumont, P. Nissen, B. L. de Groot, P. Kjellbom, P. Gourdon, and U. Johanson. 2016. Crystal structure of an ammonia-permeable aquaporin. *PLoS Biol* 14 (3):e1002411. doi: 10.1371/journal.pbio.1002411.
31. Hachez, C., A. Besserer, A. S. Chevalier, and F. Chaumont. 2013. Insights into plant plasma membrane aquaporin trafficking. *Trends Plant Sci* 18 (6):344–352. doi: 10.1016/j.tplants.2012.12.003.
32. Verkman, A. S., M. O. Anderson, and M. C. Papadopoulos. 2014. Aquaporins: Important but elusive drug targets. *Nat Rev Drug Discov* 13 (4):259–277. doi: 10.1038/nrd4226.
33. Maurel, C., Y. Boursiac, D. T. Luu, V. Santoni, Z. Shahzad, and L. Verdoucq. 2015. Aquaporins in plants. *Physiol Rev* 95 (4):1321–1358. doi: 10.1152/physrev.00008.2015.
34. Verdoucq, L., O. Rodrigues, A. Martiniere, D. T. Luu, and C. Maurel. 2014. Plant aquaporins on the move: Reversible phosphorylation, lateral motion and cycling. *Curr Opin Plant Biol* 22:101–107. doi: 10.1016/j.pbi.2014.09.011.
35. Stein, W. D. and J. F. Danielli . 1956. Structure and function in red cell permeability. *Discuss Faraday Soc* 21:238–251.
36. Benga, G., O. Popescu, V. I. Pop, and R. P. Holmes. 1986. p-(Chloromercuri)benzenesulfonate binding by membrane proteins and the inhibition of water transport in human erythrocytes. *Biochemistry* 25 (7):1535–1538.
37. Finkelstein, A. 1987. *Water Movement through Lipid Bilayers, Pores, and Plasma Membranes: Theory and Reality.* New York: Wiley.
38. Hachez, C. and F. Chaumont. 2010. Aquaporins: A family of highly regulated multifunctional channels. In *MIPs and Their Role in the Exchange of Metalloids*, G. P. Bienert and T. P. Jahn, eds., pp. 1–17. New York: Landes Bioscience and Springer.
39. Bienert, G. P., J. K. Schjoerring, and T. P. Jahn. 2006. Membrane transport of hydrogen peroxide. *Biochim Biophys Acta* 1758 (8):994–1003. doi: 10.1016/j.bbamem.2006.02.015.

40. Bienert, G. P., R. B. Heinen, M. C. Berny, and F. Chaumont. 2014. Maize plasma membrane aquaporin ZmPIP2;5, but not ZmPIP1;2, facilitates transmembrane diffusion of hydrogen peroxide. *Biochim Biophys Acta* 1838 (1 Pt B):216–222. doi: 10.1016/j.bbamem.2013.08.011.

41. Henzler, T. and E. Steudle. 2000. Transport and metabolic degradation of hydrogen peroxide in *Chara corallina*: Model calculations and measurements with the pressure probe suggest transport of H(2)O(2) across water channels. *J Exp Bot* 51 (353):2053–2066.

42. Bienert, G. P., A. L. Moller, K. A. Kristiansen, A. Schulz, I. M. Moller, J. K. Schjoerring, and T. P. Jahn. 2007. Specific aquaporins facilitate the diffusion of hydrogen peroxide across membranes. *J Biol Chem* 282 (2):1183–1192. doi: 10.1074/jbc.M603761200.

43. Bienert, G. P. and F. Chaumont. 2014. Aquaporin-facilitated transmembrane diffusion of hydrogen peroxide. *Biochim Biophys Acta* 1840 (5):1596–1604. doi: 10.1016/j.bbagen.2013.09.017.

44. Dynowski, M., G. Schaaf, D. Loque, O. Moran, and U. Ludewig. 2008. Plant plasma membrane water channels conduct the signalling molecule H$_2$O$_2$. *Biochem J* 414 (1):53–61. doi: 10.1042/bj20080287.

45. Cordeiro, R. M. 2015. Molecular dynamics simulations of the transport of reactive oxygen species by mammalian and plant aquaporins. *Biochim Biophys Acta* 1850 (9):1786–1794. doi: 10.1016/j.bbagen.2015.05.007.

46. Almasalmeh, A., D. Krenc, B. Wu, and E. Beitz. 2014. Structural determinants of the hydrogen peroxide permeability of aquaporins. *FEBS J* 281 (3):647–656. doi: 10.1111/febs.12653.

47. Lin, V. S., B. C. Dickinson, and C. J. Chang. 2013. Boronate-based fluorescent probes: Imaging hydrogen peroxide in living systems. *Methods Enzymol* 526:19–43. doi: 10.1016/b978-0-12-405883-5.00002-8.

48. Bilan, D. S. and V. V. Belousov. 2016. HyPer family probes: State of the art. *Antioxid Redox Signal* 24 (13):731–751. doi: 10.1089/ars.2015.6586.

49. Belousov, V. V., A. F. Fradkov, K. A. Lukyanov, D. B. Staroverov, K. S. Shakhbazov, A. V. Terskikh, and S. Lukyanov. 2006. Genetically encoded fluorescent indicator for intracellular hydrogen peroxide. *Nat Methods* 3 (4):281–286. doi: 10.1038/nmeth866.

50. Miller, E. W., B. C. Dickinson, and C. J. Chang. 2010. Aquaporin-3 mediates hydrogen peroxide uptake to regulate downstream intracellular signaling. *Proc Natl Acad Sci USA* 107 (36):15681–15686. doi: 10.1073/pnas.1005776107.

51. Hara-Chikuma, M., M. Chikuma, Y. Sugiyama, K. Kabashima, A. S. Verkman, S. Inoue, and Y. Miyachi. 2012. Chemokine-dependent T cell migration requires aquaporin-3-mediated hydrogen peroxide uptake. *J Exp Med* 209 (10):1743–1752. doi: 10.1084/jem.20112398.

52. Satooka, H. and M. Hara-Chikuma. 2016. Aquaporin-3 controls breast cancer cell migration by regulating hydrogen peroxide transport and its downstream cell signaling. *Mol Cell Biol* 36 (7):1206–1218. doi: 10.1128/mcb.00971-15.

53. Vieceli Dalla Sega, F., C. Prata, L. Zambonin, C. Angeloni, B. Rizzo, S. Hrelia, and D. Fiorentini. 2017. Intracellular cysteine oxidation is modulated by aquaporin-8-mediated hydrogen peroxide channeling in leukaemia cells. *Biofactors* 43 (2):232–242. doi: 10.1002/biof.1340.

54. Vieceli Dalla Sega, F., L. Zambonin, D. Fiorentini, B. Rizzo, C. Caliceti, L. Landi, S. Hrelia, and C. Prata. 2014. Specific aquaporins facilitate NOx-produced hydrogen peroxide transport through plasma membrane in leukaemia cells. *Biochim Biophys Acta* 1843 (4):806–814. doi: 10.1016/j.bbamcr.2014.01.011.

55. Thiagarajah, J. R., J. Chang, J. A. Goettel, A. S. Verkman, and W. I. Lencer. 2017. Aquaporin-3 mediates hydrogen peroxide-dependent responses to environmental stress in colonic epithelia. *Proc Natl Acad Sci USA* 114 (3):568–573. doi: 10.1073/pnas.1612921114.

56. Hara-Chikuma, M., H. Satooka, S. Watanabe, T. Honda, Y. Miyachi, T. Watanabe, and A. S. Verkman. 2015. Aquaporin-3-mediated hydrogen peroxide transport is required for NF-kappaB signalling in keratinocytes and development of psoriasis. *Nat Commun* 6:7454. doi: 10.1038/ncomms8454.

57. Luo, J., X. Liu, J. Liu, M. Jiang, M. Luo, and J. Zhao. 2016. Activation of TGF-beta1 by AQP3-mediated H_2O_2 transport into fibroblasts of a bleomycin-induced mouse model of scleroderma. *J Invest Dermatol* 136 (12):2372–2379. doi: 10.1016/j.jid.2016.07.014.

58. Bertolotti, M., S. Bestetti, J. M. Garcia-Manteiga, I. Medrano-Fernandez, A. Dal Mas, M. L. Malosio, and R. Sitia. 2013. Tyrosine kinase signal modulation: A matter of H_2O_2 membrane permeability? *Antioxid Redox Signal* 19 (13):1447–1451. doi: 10.1089/ars.2013.5330.

59. Bertolotti, M., G. Farinelli, M. Galli, A. Aiuti, and R. Sitia. 2016. AQP8 transports NOX2-generated H_2O_2 across the plasma membrane to promote signaling in B cells. *J Leukoc Biol* 100 (5):1071–1079. doi: 10.1189/jlb.2AB0116-045R.

60. Tu, B. P. and J. S. Weissman. 2002. The FAD- and O(2)-dependent reaction cycle of Ero1-mediated oxidative protein folding in the endoplasmic reticulum. *Mol Cell* 10 (5):983–994.

61. Ramming, T. and C. Appenzeller-Herzog. 2012. The physiological functions of mammalian endoplasmic oxidoreductin 1: On disulfides and more. *Antioxid Redox Signal* 16 (10):1109–1118. doi: 10.1089/ars.2011.4475.

62. Chen, K., M. T. Kirber, H. Xiao, Y. Yang, and J. F. Keaney, Jr. 2008. Regulation of ROS signal transduction by NADPH oxidase 4 localization. *J Cell Biol* 181 (7):1129–1139. doi: 10.1083/jcb.200709049.

63. Appenzeller-Herzog, C., G. Banhegyi, I. Bogeski, K. J. Davies, A. Delaunay-Moisan, H. J. Forman, A. Gorlach et al. 2016. Transit of H_2O_2 across the endoplasmic reticulum membrane is not sluggish. *Free Radic Biol Med* 94:157–160. doi: 10.1016/j.freeradbiomed.2016.02.030.

64. Calamita, G., A. Mazzone, A. Bizzoca, A. Cavalier, G. Cassano, D. Thomas, and M. Svelto. 2001. Expression and immunolocalization of the aquaporin-8 water channel in rat gastrointestinal tract. *Eur J Cell Biol* 80 (11):711–719. doi: 10.1078/0171-9335-00210.

65. Ferri, D., A. Mazzone, G. E. Liquori, G. Cassano, M. Svelto, and G. Calamita. 2003. Ontogeny, distribution, and possible functional implications of an unusual aquaporin, AQP8, in mouse liver. *Hepatology* 38 (4):947–957. doi: 10.1053/jhep.2003.50397.

66. Marchissio, M. J., D. E. Frances, C. E. Carnovale, and R. A. Marinelli. 2012. Mitochondrial aquaporin-8 knockdown in human hepatoma HepG2 cells causes ROS-induced mitochondrial depolarization and loss of viability. *Toxicol Appl Pharmacol* 264 (2):246–254. doi: 10.1016/j.taap.2012.08.005.

67. Yang, B., D. Zhao, and A. S. Verkman. 2006. Evidence against functionally significant aquaporin expression in mitochondria. *J Biol Chem* 281 (24):16202–16206. doi: 10.1074/jbc.M601864200.

68. Yoboue, E. D., A. Rimessi, T. Anelli, P. Pinton, and R. Sitia. 2017. Regulation of calcium fluxes by GPX8, a type-II transmembrane peroxidase enriched at the mitochondria-associated endoplasmic reticulum membrane. *Antioxid Redox Signal* 27 (9):583–595. doi: 10.1089/ars.2016.6866.

69. Chauvigné, F., M. Boj, R. N. Finn, and J. Cerdà. 2015. Mitochondrial aquaporin-8-mediated hydrogen peroxide transport is essential for teleost spermatozoon motility. *Sci Rep* 5:7789. doi: 10.1038/srep07789.

70. Watanabe, S., C. S. Moniaga, S. Nielsen, and M. Hara-Chikuma. 2016. Aquaporin-9 facilitates membrane transport of hydrogen peroxide in mammalian cells. *Biochem Biophys Res Commun* 471 (1):191–197. doi: 10.1016/j.bbrc.2016.01.153.

71. Laforenza, U., G. Pellavio, A. L. Marchetti, C. Omes, F. Todaro, and G. Gastaldi. 2016. Aquaporin-mediated water and hydrogen peroxide transport is involved in normal human spermatozoa functioning. *Int J Mol Sci* 18 (1). pii: E66. doi: 10.3390/ijms18010066.
72. Nielsen, S., C. L. Chou, D. Marples, E. I. Christensen, B. K. Kishore, and M. A. Knepper. 1995. Vasopressin increases water permeability of kidney collecting duct by inducing translocation of aquaporin-CD water channels to plasma membrane. *Proc Natl Acad Sci USA* 92 (4):1013–1017.
73. Potokar, M., M. Stenovec, J. Jorgacevski, T. Holen, M. Kreft, O. P. Ottersen, and R. Zorec. 2013. Regulation of AQP4 surface expression via vesicle mobility in astrocytes. *Glia* 61 (6):917–928. doi: 10.1002/glia.22485.
74. Wudick, M. M., X. Li, V. Valentini, N. Geldner, J. Chory, J. Lin, C. Maurel, and D. T. Luu. 2015. Subcellular redistribution of root aquaporins induced by hydrogen peroxide. *Mol Plant* 8 (7):1103–1114. doi: 10.1016/j.molp.2015.02.017.
75. Bienert, G. P., B. Desguin, F. Chaumont, and P. Hols. 2013. Channel-mediated lactic acid transport: A novel function for aquaglyceroporins in bacteria. *Biochem J* 454 (3):559–570. doi: 10.1042/bj20130388.
76. Kirscht, A., S. Survery, P. Kjellbom, and U. Johanson. 2016. Increased permeability of the aquaporin SoPIP2;1 by mercury and mutations in loop A. *Front Plant Sci* 7:1249. doi: 10.3389/fpls.2016.01249.
77. Tian, S., X. Wang, P. Li, H. Wang, H. Ji, J. Xie, Q. Qiu, D. Shen, and H. Dong. 2016. Plant aquaporin AtPIP1;4 links apoplastic H$_2$O$_2$ induction to disease immunity pathways. *Plant Physiol* 171 (3):1635–1650. doi: 10.1104/pp.15.01237.
78. Hooijmaijers, C., J. Y. Rhee, K. J. Kwak, G. C. Chung, T. Horie, M. Katsuhara, and H. Kang. 2012. Hydrogen peroxide permeability of plasma membrane aquaporins of *Arabidopsis thaliana*. *J Plant Res* 125 (1):147–153. doi: 10.1007/s10265-011-0413-2.
79. Hara-Chikuma, M., S. Watanabe, and H. Satooka. 2016. Involvement of aquaporin-3 in epidermal growth factor receptor signaling via hydrogen peroxide transport in cancer cells. *Biochem Biophys Res Commun* 471 (4):603–609. doi: 10.1016/j.bbrc.2016.02.010.
80. An, B., B. Li, H. Li, Z. Zhang, G. Qin, and S. Tian. 2016. Aquaporin8 regulates cellular development and reactive oxygen species production, a critical component of virulence in *Botrytis cinerea*. *New Phytol* 209 (4):1668–1680. doi: 10.1111/nph.13721.
81. Azad, A. K., N. Yoshikawa, T. Ishikawa, Y. Sawa, and H. Shibata. 2012. Substitution of a single amino acid residue in the aromatic/arginine selectivity filter alters the transport profiles of tonoplast aquaporin homologs. *Biochim Biophys Acta* 1818 (1):1–11. doi: 10.1016/j.bbamem.2011.09.014.
82. Katsuhara, M., S. Sasano, T. Horie, T. Matsumoto, J. Rhee, and M. Shibasaka. 2014. Functional and molecular characteristics of rice and barley NIP aquaporins transporting water, hydrogen peroxide and arsenite. *Plant Biotechnol* 31:213–219.

16 H$_2$O$_2$, Thioredoxin, and Signaling

Elias S.J. Arnér

CONTENTS

INTRODUCTION

It is generally accepted that one major step in the initiation of cellular signaling pathways involves an oxidative burst, that is, a transient surge in H$_2$O$_2$ derived from NADPH oxidases, which can be activated by extracellular stimuli constituted by binding of growth factors to their cognate receptors. Also mitochondria can lead to increases in H$_2$O$_2$ that may trigger signaling responses. Exactly how a burst of H$_2$O$_2$ can be mitigated to a distinct cellular response with altered transcriptional programs, ultimately resulting in a change of cellular phenotype is, however, less understood at the molecular level. In this chapter, the possible steps of redox control that can link an oxidative burst with modulations of the mammalian thioredoxin system will be discussed.

MAMMALIAN THIOREDOXIN SYSTEM IN MODULATION OF CELLULAR SIGNALING PATHWAYS

The two major reductive systems in mammalian cells are the thioredoxin and gluta-thione systems, which are generally complementary to each other in life supporting functions but may differ in the context of signaling [1–6]. Here we will focus on the thioredoxin system, which is constituted of isoenzymes of NADPH-dependent selenoprotein thioredoxin reductases with their primary substrates, which in turn modulate a large number of secondary target molecules by redox control. This system is well known to modulate a large number of signaling pathways in the cytosol, mitochondria, as well as nuclei, which has been extensively reviewed elsewhere [3,7–18] and shall thus not be repeated here. Hence we shall specifically discuss the possible molecular mechanisms of links between an oxidative burst of H_2O_2 and the modulation of cellular signaling pathways by the thioredoxin system. Before doing so, some of the key players of the mammalian thioredoxin system known to be linked to signaling pathways need to be briefly surveyed. See Figure 16.1 for a schematic summary and Table 16.1 for further information and key references.

KEY PLAYERS OF THE MAMMALIAN THIOREDOXIN SYSTEM

Proteins

The whole thioredoxin system becomes NADPH dependent due to the fact that it is propelled by NADPH-dependent oxidoreductases, mainly cytosolic TrxR1 and mitochondrial TrxR2. These enzymes are selenoproteins, that is, they are dependent upon a catalytic selenocysteine residue that in turn requires selenium for synthesis, and they are furthermore subject to extensive splicing that results in diverse variants of each of the two enzymes that mainly differ at their N-terminal domains [13,19–22]. Of note in relation to signaling events is the fact that the normally cytosolic TrxR1 can also be found in the nucleus in the form of TXNRD1_v2, where it can modulate the activities of several transcription factors such as the estrogen receptor isoforms [13,21], and in the form of TXNRD1_v3 at the cell membrane, where it colocalizes with membrane rafts and can promote formation of filopodia [13,23–25]. These splice variants of TrxR1 have not yet been examined at depth in relation to cellular signaling pathways, which should thus be a suitable topic for future studies.

The primary substrates of TrxR1 and TrxR2 are Trx1 and Trx2, respectively, and most actions of the mammalian thioredoxin system are likely to be carried out by the activities of these two proteins [11,13,26–30]. However, additional substrate proteins for TrxR1 such as TRP14 are also likely to be important in relation to signaling events [31–36]. It should furthermore be noted that although major signaling aberrations can be seen upon genetic deletion of TrxR1 or TrxR2 these are not necessarily related to effects on Trx1 or Trx2, respectively, suggesting crosstalk in redox pathways and indicating that specific effects on signaling depend upon more than solely an impairment of a general redox capacity [7,37–40].

Examples of secondary target proteins that are either requiring or being modulated by Trx1, Trx2, TRP14, or other members of the thioredoxin system, and which in turn regulate signaling pathways, are discussed further here.

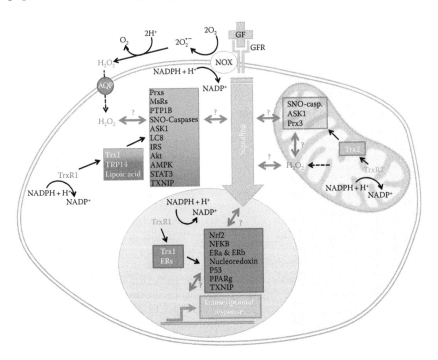

FIGURE 16.1 The mammalian thioredoxin system in relation to cellular signaling pathways. The figure schematically summarizes several levels of control whereby the mammalian thioredoxin system can modulate intracellular signaling pathways by redox control. Binding of a growth factor (GF) to its cognate receptor (GFR) typically activates NADPH-dependent NADPH oxidases (NOX, yellow), producing extracellular superoxide that dismutates to H$_2$O$_2$ and enters cells through aquaporins (AQP), as illustrated. This burst in H$_2$O$_2$ is part of a signaling cascade typically involving several levels of protein phosphorylation that converge on a nuclear transcriptional response (green). Several steps of this signaling can be modulated by the mammalian thioredoxin system, either in the cytosol as propelled through TrxR1 (light red, with primary and secondary targets boxed), in the nucleus through a specific nuclear splice variant of TrxR1 supporting nuclear Trx1 or estrogen receptors (ERs) (purple, with primary and secondary targets boxed), or in the mitochondria through TrxR2 and Trx2 (orange, with primary and secondary targets boxed). All of these factors are likely to modulate the outcome of signaling pathways. The exact molecular mechanisms for such interactions with H$_2$O$_2$ and other factors of cellular signaling are yet insufficiently known, with the overall outcome likely being dependent upon overall cellular context (red double-pointed arrows and question marks). See the text and Table 16.1 for further information and citations to relevant literature.

Low Molecular Weight Compounds

The mammalian thioredoxin reductases are, in addition to their primary protein substrates, also able to reduce a large number of low molecular weight substrates, as well as being inhibited by many compounds [9,13]. Several of these compounds are naturally occurring and are also known to modulate cellular signaling pathways, including lipoic acid [27,41,42], pyrroloquinoline quinone [43,44], other quinone

TABLE 16.1

Key Players of the Mammalian Thioredoxin System with Examples of Primary and Secondary Target Molecules Involved in Cellular Signaling Pathways

TrxR Isoform	Primary Substrate	Presumed Secondary Target Molecules Involved in Signaling	Effects of an Active Thioredoxin System on Signaling through the Primary Substrate and Secondary Target Molecules	References
TrxR1	Trx1	Prxs (mainly Prx1 and Prx2)	Supports antioxidant Prx functions, which counteracts H_2O_2 accumulation, but may also be directly involved in mediating signaling	[3,14,17,67,83,85,86]
		Msrs (MsrA and MsrB)	Reduces methionine sulfoxide residues that may be involved in signaling, e.g., through actin and Mical counteraction	[75,100,101]
		PTP1B	Activation of PTP1B, inhibition of receptor linked phosphorylation cascades	[33]
		Nitrosylated caspases	Either activation or inhibition of caspases through nitrosylation, depending upon facilitated denitrosylation or transnitrosylation reactions, respectively	[11,55,56,102,103]
		p53	Maturation of p53 (may be an effect that is independent of Trx1)	[37,76,77,104,105]
		Nrf2	Inhibition of Nrf2 (may be an effect that is independent of Trx1)	[9,106,107]
		Nucleoredoxin	Inhibition of Wnt signaling	[81,82]
		ASK1	Inhibition of ASK1 with prevention of JNK and p38 activation	[15,28,62,63]
		PPARγ	Suppression of PPARg activation (may be an effect that is independent of Trx1)	[37]

(Continued)

TABLE 16.1 (*Continued*)

Key Players of the Mammalian Thioredoxin System with Examples of Primary and Secondary Target Molecules Involved in Cellular Signaling Pathways

TrxR Isoform	Primary Substrate	Presumed Secondary Target Molecules Involved in Signaling	Effects of an Active Thioredoxin System on Signaling through the Primary Substrate and Secondary Target Molecules	References
		STAT3	Maintenance of STAT3 activity, since oxidized Prx2 inhibits STAT3	[85,86]
		TXNIP (VDUP1/TBP2)	Endogenous protein that acts as a dominant inhibitor of Trx1 activities, also acting independently of Trx1	[89–91]
	TRP14	PTP1B	Activation of PTP1B, inhibition of receptor linked phosphorylation cascades	[33]
		Dynein light chain LC8	Inhibition of NFκB signaling	[36]
	Lipoic acid	Insulin receptor, IRS-1, Akt, AMPK, p38, PTP1B	Stimulation of insulin-like signaling, with activation of Akt, AMPK and p38, and inhibition of PTP1B (molecular mechanisms unclear)	[41,42]
	ERα and ERβ	AP1, SP1, p53	Presumed activation of ERα and ERβ βψ διρεχτ interaction between the TrxR1 splice variant TXNRD_2 and the estrogen receptors in the nucleus	[21]
TrxR2	Trx2	Prx3	Supports antioxidant Prx functions, which counteracts H$_2$O$_2$ accumulation in the mitochondria, and may be more susceptible to inhibition than cytosolic Prxs	[97,98]

(Continued)

TABLE 16.1 (*Continued*)
Key Players of the Mammalian Thioredoxin System with Examples of Primary and Secondary Target Molecules Involved in Cellular Signaling Pathways

TrxR Isoform	Primary Substrate	Presumed Secondary Target Molecules Involved in Signaling	Effects of an Active Thioredoxin System on Signaling through the Primary Substrate and Secondary Target Molecules	References
		Nitrosylated mitochondrial caspase-3	Activation of mitochondrial nitrosylated caspase-3 resulting in either cellular apoptosis or promotion of glioma tumors	[54,56]
		ASK1 (mitochondrial)	Inhibition of mitochondria-associated ASK1 with prevention of JNK and Bcl-2 activation	[61]

This table summarizes primary substrates of TrxR1 and TrxR2, together with examples of secondary target molecules that are involved in modulation of cellular signaling pathways. The effects of an active thioredoxin system on these signaling pathways are also exemplified, together with citations to relevant literature for further information. See Figure 16.1 for a graphical summary of these pathways. Please note that the list is not extensive but should be seen as selected representative illustrations. Much further information on links between the mammalian thioredoxin system and cellular signaling pathways can be found in the literature. Please also note that several of these factors are also affected by the glutathione system, which is not covered in this chapter.

compounds [45–47], several electrophiles [9], and different selenium metabolites [13,18,48–53]. If or how effects on cellular signaling pathways linked to bursts of H_2O_2 are modulated by interactions of these low molecular compounds with the thioredoxin system needs to be better understood.

POSTTRANSLATIONAL MODIFICATIONS OF SIGNALING PROTEINS THAT MAY BE AFFECTED BY THE MAMMALIAN THIOREDOXIN SYSTEM

Thioredoxins are known as powerful and versatile protein disulfide reductases, and most of their actions are indeed likely to be mediated mainly by catalytic reduction of disulfides in target proteins. This holds true for Trx-mediated reactivation of classical Trx substrates such as oxidized ribonucleotide reductase, peroxiredoxins, methionine sulfoxide reductases (where MsrB1 however forms a selenenylsulfide that is reduced by Trx), and many other target proteins for the thioredoxin system [13,16,26,30]. Additional target motif entities of Trx1, Trx2, and TRP14 that are likely to be involved in modulation of signaling pathways include S-nitrosylated

[11,32,54–57] or persulfidated [31,58] Cys residues in target proteins, while in PTP1B it seems to be a sulfenylamide motif that can be reduced by Trx1 or TRP14 [33,59,60]. When it comes to regulation of the signaling kinase ASK1, reduced Trx1 binds to the protein in an inhibitory manner while oxidized Trx1 does not, thereby activating ASK1 under conditions where the active site motif of Trx1 becomes oxidized [15,28,61–63]. Thus, it is not merely through its potent disulfide reduction that the thioredoxin system asserts modulatory roles in signaling, but also through more elaborate molecular mechanisms. The final outcome in signaling will depend upon a combination of the nature of the posttranslational modification that is targeted, the change in function incurred by the modification, the exact signaling role(s) of the target protein(s), and the overall cellular context in which such modifications occur. Considering the complexity of the web of signaling networks in a cell, it becomes obvious that although it is clear that the thioredoxin system modulates the H₂O₂-linked signaling pathways in cells, it seems almost impossible to fully predict the final impact on cellular phenotypes that may be the result of a particular modification. In the next chapter we shall however briefly introduce some of the signaling pathways that are known to be modulated by the thioredoxin system.

SIGNALING PATHWAYS KNOWN TO BE MODULATED BY THE MAMMALIAN THIOREDOXIN SYSTEM

PROTEIN PHOSPHORYLATION CASCADES

Most of the growth factor receptor-linked signaling pathways are mediated by protein phosphorylation cascades, including those triggered by insulin, EGF, PDGF, FGF, Wnt, TGF, or TNF, such as exemplified by transmembrane receptors tyrosine kinases (RTKs) linked to mitogen-activated protein kinase (MAPK) pathways [64,65]. These pathways are also known to be redox modulated, that is, altered in their signaling capacities or profiles by redox-related events [66] and it is thereby not surprising that these pathways are also mediated by alterations in activities in the proteins of the thioredoxin system [36,67]. More specifically, it seems clear that oxidative inactivation of PTPs should be an important step to allow for phosphorylation cascades to occur [60,68,69] and in this context the opposing PTP-activating capacity of the thioredoxin system will be important in the inhibition of signaling [33,59]. The Trx-regulated kinase ASK1 is another important member of protein phosphorylation cascades as already mentioned, which is thus also inhibited by an active thioredoxin system.

ACTIN POLYMERIZATION

Rearrangements of actin polymerization and the cellular cytoskeleton are intimately linked with signaling pathways that result in an altered cellular phenotype [70]. Also here is redox regulation important, and the thioredoxin system is well known to modulate actin polymerization, either by reducing disulfides in actin or in actin-related proteins [71–74], or by supporting Msrs that counteract modulation of actin polymerization involving methionine residues that can be oxidized by the action of MICAL proteins [75]. Thus, not only the activities of cellular signaling pathways can be modulated by the thioredoxin system, but also cell shape and motility through direct regulation of actin polymerization.

TRANSCRIPTION FACTOR REGULATION

All signaling pathways that alter cellular phenotypes converge upon transcription factors that ultimately regulate cell fate as a result of altered transcriptional programs. Also in this context is the thioredoxin system important, as it directly regulates the activities of several transcription factors. Examples are many and include the observations that TrxR1 seems to be required for normal maturation of functional p53 [37,76–78] while the enzyme may also act as an important repressor of Nrf2 activation [9] and in the form of a specific splice variant affecting estrogen receptor signaling in the nucleus [21]. The TrxR1 substrate TRP14 is a repressor of NFκB signaling [35,36], and nuclear Trx1 modulates the activities of AP1 through Ref-1 [79], p21 through p53 [80], and Wnt/beta-catenin signaling through nucleoredoxin [81,82]. The thioredoxin system thereby carries out specific roles in the nuclei that directly affect transcriptional responses to cellular signaling pathways that may have been initiated at the cell membrane or in the cytosol, which in turn are also processes that may be modulated by the thioredoxin system, as discussed here. The complexity by which the thioredoxin system may affect signaling pathways is thus extensive. Next we shall discuss how a burst of H_2O_2 may be linked to these pathways.

HOW CAN A BURST IN H_2O_2 MODULATE SIGNALING PATHWAYS THROUGH INTERACTIONS WITH THE THIOREDOXIN SYSTEM?

Christine Winterbourn has convincingly shown that peroxidases, predominantly peroxiredoxins (Prxs), ought to be the first and major proteins to be oxidized upon a burst in H_2O_2, considering their high abundance in combination with their uniquely fast reactivity with H_2O_2 [14,83,84]. The question thereby becomes whether or how the oxidized Prxs in turn may transfer the signal to secondary target proteins, or, alternatively, whether yet unknown mechanisms increase the reactivity of specific signaling proteins with H_2O_2 to supersede that of Prxs. It was recently shown by Tobias Dick and colleagues that oxidized Prx2 can indeed transfer oxidative equivalents to the transcription factor STAT3 and thereby inhibit its activity [85]. As Winterbourn and Hampton conclude, this may either have constituted the first example of a general mammalian signaling mechanism, mediating the signal of a H_2O_2 burst to target proteins directly via Prxs, or, alternatively, a unique case, which should be a question that further studies need to address [86]. Moreover, it should be noted that because Prx2 is reactivated by Trx1, this reactivation will oxidize Trx1 and theoretically oxidized Trx1 may subsequently transfer oxidizing equivalents to other target proteins that take part in signaling [86]. Such a phenomenon can also be detected *in vitro*, whereby Trx1 can be shown to aggregate in inactive disulfide-linked dimers upon reactivity with oxidized Prx1, and it can be seen in cells treated with oxidants that the oxidized Trx1 dimers, which are not substrates of TrxR1, can subsequently be resolved by the glutaredoxin system [87]. With regard to STAT3 signaling, it is interesting to note that STAT3 activity upregulates TRP14 transcription, being encoded by the *TXNDC17* gene [88], and that thioredoxin-interacting protein (TXNIP), an endogenous Trx1 inhibitor [89–91], can inactivate STAT3 [92]. The example of STAT3 regulation by H$_2$O$_2$ and its modulation by the thioredoxin system, through several possible direct and indirect mechanisms *via* Prx2, Trx1, TRP14, and TXNIP, illustrates the intricate web of highly complex signaling pathways. Considering H$_2$O$_2$ as a signaling molecule, it is important to consider the mitochondria as alternative and powerful sources of H$_2$O$_2$, which may also affect signaling events occurring in the cytosol and nucleus [93–96]. Interestingly, some exogenous stimuli such as treatment of cells with auranofin or isothiocyanates, initially oxidize Prx3, the mitochondrial peroxiredoxin being regenerated by mitochondrial TrxR2 and Trx2, before oxidation of Prx1 or Prx2 occurs [97,98]. It is clear that H$_2$O$_2$ is a signaling molecule and that the thioredoxin system, either directly or indirectly, modulates the final outcome of H$_2$O$_2$-dependent signaling, but it is just as clear that the redox regulation of cellular signaling pathways is enormously complex.

CHALLENGES AHEAD: TO UNDERSTAND THE COMPLETE PICTURE

Based upon the many players of redox control and the major complexity of the molecular reactions that may lead from a burst of H$_2$O$_2$ to a final cellular response, as discussed and indicated in a highly simplified manner by Figure 16.1, it becomes challenging to ask and understand how a cell coordinates these signaling events. How can the seemingly simple molecular phenomena of ligand binding to a receptor that results in a burst of H$_2$O$_2$ lead to highly divergent responses dependent upon cell type, cell growth conditions, and specific ligand binding, as mediated by such complex cellular pathways? How can these pathways be redox regulated through the thioredoxin system in a specific manner, allowing transfer of signals from the cell surface to specific transcriptional responses and beyond? One answer might be that protein–protein interactions and compartmentalization effects separate specific signaling pathways from the cellular bulk of redox events. Another possibility is that different cell types react differently to stimuli because of divergent profiles in their redox pathways and signaling molecules, in terms of what exact variants that are expressed and at what levels. These two broad explanations are not exclusive of each other and are both possible as well as likely. Indeed, between individual cells of a single cell culture a high heterogeneity can be seen upon single stimuli with such diverse redox perturbing signaling compounds as auranofin, TNFα or hypoxia [99], suggesting highly dynamic events and inferring that the outcome of redox signaling in a complete organ or tissue will be dependent upon the combined outcome of rather heterogenic events on the individual cell level. To fully understand how a given stimulus resulting in a burst of H$_2$O$_2$ is mediated on the molecular level to a phenotypic response is still a highly challenging task. It seems clear that the thioredoxin system modulates these signaling pathways, but it requires many additional studies before the complete picture of these processes can be painted.

ACKNOWLEDGMENTS

Funding from the Swedish Research Council, the Swedish Cancer Society, the Knut and Alice Wallenberg Foundation, and Karolinska Institutet is thankfully acknowledged. The author also wishes to acknowledge the significant efforts by his own group members and by all talented and dedicated colleagues worldwide in the redox field that struggle to understand more of the detailed mechanisms of redox regulation. With this book being dedicated to Dr. Christine Winterbourn, the author finally wants to thank her specifically for being such a wonderful role model in the field, with her outstanding dedication, rigor, and stringency in science, combined with such a clear mind, bright ideas, and positive attitude toward science and life.

REFERENCES

1. Eriksson, S., J. R. Prigge, E. A. Talago, E. S. Arner, and E. E. Schmidt. 2015. Dietary methionine can sustain cytosolic redox homeostasis in the mouse liver. *Nat Commun* 6:6479.
2. Deponte, M. 2013. Glutathione catalysis and the reaction mechanisms of glutathione-dependent enzymes. *Biochim Biophys Acta* 1830 (5):3217–3266.
3. Brigelius-Flohe, R. and L. Flohe. 2011. Basic principles and emerging concepts in the redox control of transcription factors. *Antioxid Redox Signal* 15 (8):2335–2381.
4. Toppo, S., L. Flohe, F. Ursini, S. Vanin, and M. Maiorino. 2009. Catalytic mechanisms and specificities of glutathione peroxidases: Variations of a basic scheme. *Biochim Biophys Acta* 1790 (11):1486–1500.
5. Lillig, C. H., C. Berndt, and A. Holmgren. 2008. Glutaredoxin systems. *Biochim Biophys Acta* 1780 (11):1304–1317.
6. Fernandes, A. P. and A. Holmgren. 2004. Glutaredoxins: Glutathione-dependent redox enzymes with functions far beyond a simple thioredoxin backup system. *Antioxid Redox Signal* 6 (1):63–74.
7. Lei, X. G., J. H. Zhu, W. H. Cheng, Y. Bao, Y. S. Ho, A. R. Reddi, A. Holmgren, and E. S. Arner. 2016. Paradoxical roles of antioxidant enzymes: Basic mechanisms and health implications. *Physiol Rev* 96 (1):307–364.
8. Ye, Z. W., J. Zhang, D. M. Townsend, and K. D. Tew. 2015. Oxidative stress, redox regulation and diseases of cellular differentiation. *Biochim Biophys Acta* 1850 (8):1607–1621.
9. Cebula, M., E. E. Schmidt, and E. S. Arner. 2015. TrxR1 as a potent regulator of the Nrf2-Keap1 response system. *Antioxid Redox Signal* 23 (10):823–853.
10. Groitl, B. and U. Jakob. 2014. Thiol-based redox switches. *Biochim Biophys Acta* 1844 (8):1335–1343.
11. Sengupta, R. and A. Holmgren. 2013. Thioredoxin and thioredoxin reductase in relation to reversible S-nitrosylation. *Antioxid Redox Signal* 18 (3):259–269.
12. Kesarwani, P., A. K. Murali, A. A. Al-Khami, and S. Mehrotra. 2013. Redox regulation of T-cell function: From molecular mechanisms to significance in human health and disease. *Antioxid Redox Signal* 18 (12):1497–1534.
13. Arnér, E. S. J. 2009. Focus on mammalian thioredoxin reductases—Important selenoproteins with versatile functions. *Biochim Biophys Acta* 1790 (6):495–526.
14. Winterbourn, C. C. and M. B. Hampton. 2008. Thiol chemistry and specificity in redox signaling. *Free Radic Biol Med* 45 (5):549–561.
15. Matsuzawa, A. and H. Ichijo. 2008. Redox control of cell fate by MAP kinase: Physiological roles of ASK1-MAP kinase pathway in stress signaling. *Biochim Biophys Acta* 1780 (11):1325–1336.

16. Lillig, C. H. and A. Holmgren. 2007. Thioredoxin and related molecules-from biology to health and disease. *Antioxid Redox Signal* 9 (1):25–47.
17. Rhee, S. G., H. Z. Chae, and K. Kim. 2005. Peroxiredoxins: A historical overview and speculative preview of novel mechanisms and emerging concepts in cell signaling. *Free Radic Biol Med* 38 (12):1543–1552.
18. Rundlöf, A.-K. and E. S. J. Arnér. 2004. Regulation of the mammalian selenoprotein thioredoxin reductase 1 in relation to cellular phenotype, growth and signaling events. *Antiox Redox Signal* 6:41–52.
19. Su, D. and V. N. Gladyshev. 2004. Alternative splicing involving the thioredoxin reductase module in mammals: A glutaredoxin-containing thioredoxin reductase 1. *Biochemistry* 43 (38):12177–12188.
20. Rundlof, A. K., M. Janard, A. Miranda-Vizuete, and E. S. Arner. 2004. Evidence for intriguingly complex transcription of human thioredoxin reductase 1. *Free Radic Biol Med* 36 (5):641–656.
21. Damdimopoulos, A. E., A. Miranda-Vizuete, E. Treuter, J. Å. Gustafsson, and G. Spyrou. 2004. An alternative splicing variant of the selenoprotein thioredoxin reductase is a modulator of estrogen signaling. *J Biol Chem* 279 (37):38721–38729.
22. Sun, Q. A., F. Zappacosta, V. M. Factor, P. J. Wirth, D. L. Hatfield, and V. N. Gladyshev. 2001. Heterogeneity within animal thioredoxin reductases. Evidence for alternative first exon splicing. *J Biol Chem* 276:3106–3114.
23. Cebula, M., N. Moolla, A. Capovilla, and E. S. Arner. 2013. The rare TXNRD1_v3 ("v3") splice variant of human thioredoxin reductase 1 protein is targeted to membrane rafts by N-acylation and induces filopodia independently of its redox active site integrity. *J Biol Chem* 288 (14):10002–10011.
24. Damdimopoulou, P. E., A. Miranda-Vizuete, E. S. Arner, J. A. Gustafsson, and A. E. Damdimopoulos. 2009. The human thioredoxin reductase-1 splice variant TXNRD1_v3 is an atypical inducer of cytoplasmic filaments and cell membrane filopodia. *Biochim Biophys Acta* 1793 (10):1588–1596.
25. Dammeyer, P., A. E. Damdimopoulos, T. Nordman, A. Jimenez, A. Miranda-Vizuete, and E. S. Arner. 2008. Induction of cell membrane protrusions by the N-terminal glutaredoxin domain of a rare splice variant of human thioredoxin reductase 1. *J Biol Chem* 283 (5):2814–2821.
26. Gromer, S., S. Urig, and K. Becker. 2004. The thioredoxin system—From science to clinic. *Med Res Rev* 24 (1):40–89.
27. Nordberg, J. and E. S. Arner. 2001. Reactive oxygen species, antioxidants, and the mammalian thioredoxin system. *Free Radic Biol Med* 31 (11):1287–1312.
28. Nishiyama, A., H. Masutani, H. Nakamura, Y. Nishinaka, and J. Yodoi. 2001. Redox regulation by thioredoxin and thioredoxin-binding proteins. *IUBMB Life* 52 (1–2):29–33.
29. Nishinaka, Y., H. Masutani, H. Nakamura, and J. Yodoi. 2001. Regulatory roles of thioredoxin in oxidative stress-induced cellular responses. *Redox Rep* 6 (5):289–295.
30. Arner, E. S. and A. Holmgren. 2000. Physiological functions of thioredoxin and thioredoxin reductase. *Eur J Biochem* 267 (20):6102–6109.
31. Doka, E., I. Pader, A. Biro, K. Johansson, Q. Cheng, K. Ballago, J. R. Prigge et al. 2016. A novel persulfide detection method reveals protein persulfide- and polysulfide-reducing functions of thioredoxin and glutathione systems. *Sci Adv* 2 (1):e1500968.
32. Pader, I., R. Sengupta, M. Cebula, J. Xu, J. O. Lundberg, A. Holmgren, K. Johansson, and E. S. Arner. 2014. Thioredoxin-related protein of 14 kDa is an efficient L-cystine reductase and S-denitrosylase. *Proc Natl Acad Sci USA* 111 (19):6964–6969.
33. Dagnell, M., J. Frijhoff, I. Pader, M. Augsten, B. Boivin, J. Xu, P. K. Mandal et al. 2013. Selective activation of oxidized PTP1B by the thioredoxin system modulates PDGF-beta receptor tyrosine kinase signaling. *Proc Natl Acad Sci USA* 110 (33):13398–13403.

34. Woo, J. R., S. J. Kim, W. Jeong, Y. H. Cho, S. C. Lee, Y. J. Chung, S. G. Rhee, and S. E. Ryu. 2004. Structural basis of cellular redox regulation by human TRP14. *J Biol Chem* 279 (46):48120–48125.

35. Jeong, W., H. W. Yoon, S. R. Lee, and S. G. Rhee. 2004. Identification and characterization of TRP14, a thioredoxin-related protein of 14 kDa. New insights into the specificity of thioredoxin function. *J Biol Chem* 279 (5):3142–3150.

36. Jeong, W., T. S. Chang, E. S. Boja, H. M. Fales, and S. G. Rhee. 2004. Roles of TRP14, a thioredoxin-related protein in tumor necrosis factor-alpha signaling pathways. *J Biol Chem* 279 (5):3151–3159.

37. Peng, X., A. Gimenez-Cassina, P. Petrus, M. Conrad, M. Ryden, and E. S. Arner. 2016. Thioredoxin reductase 1 suppresses adipocyte differentiation and insulin responsiveness. *Sci Rep* 6:28080.

38. Prigge, J. R., L. Coppo, S. S. Martin, F. Ogata, C. G. Miller, M. D. Bruschwein, D. J. Orlicky et al. 2017. Hepatocyte hyperproliferation upon liver-specific co-disruption of thioredoxin-1, thioredoxin reductase-1, and glutathione reductase. *Cell Rep* 19 (13):2771–2781.

39. Prigge, J. R., S. Eriksson, S. V. Iverson, T. A. Meade, M. R. Capecchi, E. S. J. Arnér, and E. E. Schmidt. 2012. Hepatocyte DNA replication in growing liver requires either glutathione or a single allele of txnrd1. *Free Radic Biol Med* 52:803–810.

40. Rollins, M. F., D. M. van der Heide, C. M. Weisend, J. A. Kundert, K. M. Comstock, E. S. Suvorova, M. R. Capecchi, G. F. Merrill, and E. E. Schmidt. 2010. Hepatocytes lacking thioredoxin reductase 1 have normal replicative potential during development and regeneration. *J Cell Sci* 123 (Pt 14):2402–2412.

41. Packer, L. and E. Cadenas. 2011. Lipoic acid: Energy metabolism and redox regulation of transcription and cell signaling. *J Clin Biochem Nutr* 48 (1):26–32.

42. Arnér, E. S. J., J. Nordberg, and A. Holmgren. 1996. Efficient reduction of lipoamide and lipoic acid by mammalian thioredoxin reductase. *Biochem Biophys Res Commun* 225:268–274.

43. Xu, J. and E. S. Arner. 2012. Pyrroloquinoline quinone modulates the kinetic parameters of the mammalian selenoprotein thioredoxin reductase 1 and is an inhibitor of glutathione reductase. *Biochem Pharmacol* 83 (6):815–820.

44. Park, J. and J. E. Churchich. 1992. Pyrroloquinoline quinone (coenzyme PQQ) and the oxidation of SH residues in proteins. *Biofactors* 3 (4):257–260.

45. Xu, J., Q. Cheng, and E. S. Arner. 2016. Details in the catalytic mechanism of mammalian thioredoxin reductase 1 revealed using point mutations and juglone-coupled enzyme activities. *Free Radic Biol Med* 94:110–120.

46. Cenas, N., H. Nivinskas, Z. Anusevicius, J. Sarlauskas, F. Lederer, and E. S. J. Arnér. 2004. Interactions of quinones with thioredoxin reductase—A challenge to the antioxidant role of the mammalian selenoprotein. *J Biol Chem* 279:2583–2592.

47. Xia, L., T. Nordman, J. M. Olsson, A. Damdimopoulos, L. Björkhem-Bergman, I. Nalvarte, L. C. Eriksson, E. S. J. Arnér, G. Spyrou, and M. Björnstedt. 2003. The mammalian cytosolic selenoenzyme thioredoxin reductase reduces ubiquinone. A novel mechanism for defense against oxidative stress. *J Biol Chem* 278 (4):2141–2146.

48. Ueno, H., G. Hasegawa, R. Ido, T. Okuno, and K. Nakamuro. 2008. Effects of selenium status and supplementary seleno-chemical sources on mouse T-cell mitogenesis. *J Trace Elem Med Biol* 22 (1):9–16.

49. Mueller, A. S., S. D. Klomann, N. M. Wolf, S. Schneider, R. Schmidt, J. Spielmann, G. Stangl, K. Eder, and J. Pallauf. 2008. Redox regulation of protein tyrosine phosphatase 1B by manipulation of dietary selenium affects the triglyceride concentration in rat liver. *J Nutr* 138 (12):2328–2336.

50. Brigelius-Flohe, R. 2008. Selenium compounds and selenoproteins in cancer. *Chem Biodivers* 5 (3):389–395.

51. Papp, L. V., J. Lu, A. Holmgren, and K. K. Khanna. 2007. From selenium to seleno-proteins: Synthesis, identity, and their role in human health. *Antioxid Redox Signal* 9 (7):775–806.
52. Yoo, M. H., X. M. Xu, B. A. Carlson, V. N. Gladyshev, and D. L. Hatfield. 2006. Thioredoxin reductase 1 deficiency reverses tumor phenotype and tumorigenicity of lung carcinoma cells. *J Biol Chem* 281 (19):13005–13008.
53. Zheng, Y., L. Zhong, and X. Shen. 2005. Effect of selenium-supplement on the calcium signaling in human endothelial cells. *J Cell Physiol* 205 (1):97–106.
54. Shen, X., M. A. Burguillos, A. M. Osman, J. Frijhoff, A. Carrillo-Jimenez, S. Kanatani, M. Augsten et al. 2016. Glioma-induced inhibition of caspase-3 in microglia promotes a tumor-supportive phenotype. *Nat Immunol* 17 (11):1282–1290.
55. Hashemy, S. I. and A. Holmgren. 2008. Regulation of the catalytic activity and structure of human thioredoxin 1 via oxidation and S-nitrosylation of cysteine residues. *J Biol Chem* 283 (32):21890–21898.
56. Benhar, M., M. T. Forrester, D. T. Hess, and J. S. Stamler. 2008. Regulated protein deni-trosylation by cytosolic and mitochondrial thioredoxins. *Science* 320 (5879):1050–1054.
57. Tannenbaum, S. R. and F. M. White. 2006. Regulation and specificity of S-nitrosylation and denitrosylation. *ACS Chem Biol* 1 (10):615–618.
58. Wedmann, R., C. Onderka, S. Wei, I. A. Szijarto, J. L. Miljkovic, A. Mitrovic, M. Lange et al. 2016. Improved tag-switch method reveals that thioredoxin acts as depersulfidase and controls the intracellular levels of protein persulfidation. *Chem Sci* 7 (5):3414–3426.
59. Schwertassek, U., A. Haque, N. Krishnan, R. Greiner, L. Weingarten, T. P. Dick, and N. K. Tonks. 2014. Reactivation of oxidized PTP1B and PTEN by thioredoxin 1. *FEBS J* 281 (16):3545–3558.
60. Salmeen, A., J. N. Andersen, M. P. Myers, T. C. Meng, J. A. Hinks, N. K. Tonks, and D. Barford. 2003. Redox regulation of protein tyrosine phosphatase 1B involves a sul-phenyl-amide intermediate. *Nature* 423 (6941):769–773.
61. Lim, P. L., J. Liu, M. L. Go, and U. A. Boelsterli. 2008. The mitochondrial superoxide/thioredoxin-2/Ask1 signaling pathway is critically involved in troglitazone-induced cell injury to human hepatocytes. *Toxicol Sci* 101 (2):341–349.
62. Tobiume, K., A. Matsuzawa, T. Takahashi, H. Nishitoh, K. Morita, K. Takeda, O. Minowa, K. Miyazono, T. Noda, and H. Ichijo. 2001. ASK1 is required for sustained activations of JNK/p38 MAP kinases and apoptosis. *EMBO Rep* 2:222–228.
63. Saitoh, M., H. Nishitoh, M. Fujii, K. Takeda, K. Tobiume, Y. Sawada, M. Kawabata, K. Miyazono, and H. Ichijo. 1998. Mammalian thioredoxin is a direct inhibitor of apop-tosis signal-regulating kinase (ASK) 1. *EMBO J* 17:2596–2606.
64. Katz, M., I. Amit, and Y. Yarden. 2007. Regulation of MAPKs by growth factors and receptor tyrosine kinases. *Biochim Biophys Acta* 1773 (8):1161–1176.
65. Marshall, C. J. 1995. Specificity of receptor tyrosine kinase signaling: Transient versus sustained extracellular signal-regulated kinase activation. *Cell* 80 (2):179–185.
66. Liu, Y. and C. He. 2016. A review of redox signaling and the control of MAP kinase pathway in plants. *Redox Biol* 11:192–204.
67. Latimer, H. R. and E. A. Veal. 2016. Peroxiredoxins in regulation of MAPK signalling pathways; sensors and barriers to signal transduction. *Mol Cells* 39 (1):40–45.
68. Ostman, A., J. Frijhoff, A. Sandin, and F. D. Bohmer. 2011. Regulation of protein tyro-sine phosphatases by reversible oxidation. *J Biochem* 150 (4):345–356.
69. Cho, S. H., C. H. Lee, Y. Ahn, H. Kim, C. Y. Ahn, K. S. Yang, and S. R. Lee. 2004. Redox regulation of PTEN and protein tyrosine phosphatases in H(2)O(2) mediated cell signaling. *FEBS Lett* 560 (1–3):7–13.
70. Schmidt, A. and M. N. Hall. 1998. Signaling to the actin cytoskeleton. *Annu Rev Cell Dev Biol* 14:305–338. doi: 10.1146/annurev.cellbio.14.1.305.

71. Lassing, I., F. Schmitzberger, M. Bjornstedt, A. Holmgren, P. Nordlund, C. E. Schutt, and U. Lindberg. 2007. Molecular and structural basis for redox regulation of beta-actin. *J Mol Biol* 370 (2):331–348.

72. Landino, L. M., T. E. Skreslet, and J. A. Alston. 2004. Cysteine oxidation of tau and microtubule-associated protein-2 by peroxynitrite: Modulation of microtubule assembly kinetics by the thioredoxin reductase system. *J Biol Chem* 279 (33):35101–35105.

73. Yokomizo, A., M. Ono, H. Nanri, Y. Makino, T. Ohga, M. Wada, T. Okamoto, J. Yodoi, M. Kuwano, and K. Kohno. 1995. Cellular levels of thioredoxin associated with drug sensitivity to cisplatin, mitomycin C, doxorubicin, and etoposide. *Cancer Res* 55:4293–4296.

74. Khan, I. A. and R. F. Luduena. 1991. Possible regulation of the in vitro assembly of bovine brain tubulin by the bovine thioredoxin system. *Biochim Biophys Acta* 1076 (2):289–297.

75. Lee, B. C., Z. Peterfi, F. W. Hoffmann, R. E. Moore, A. Kaya, A. Avanesov, L. Tarrago et al. 2013. MsrB1 and MICALs regulate actin assembly and macrophage function via reversible stereoselective methionine oxidation. *Mol Cell* 51 (3):397–404.

76. Maillet, A. and S. Pervaiz. 2011. Redox regulation of p53, redox effectors regulated by p53: A subtle balance. *Antioxid Redox Signal* 16 (11):1285–1294.

77. Cassidy, P. B., K. Edes, C. C. Nelson, K. Parsawar, F. A. Fitzpatrick, and P. J. Moos. 2006. Thioredoxin reductase is required for the inactivation of tumor suppressor p53 and for apoptosis induced by endogenous electrophiles. *Carcinogenesis* 27 (12):2538–2549.

78. Moos, P. J., K. Edes, P. Cassidy, E. Massuda, and F. A. Fitzpatrick. 2003. Electrophilic prostaglandins and lipid aldehydes repress redox-sensitive transcription factors p53 and hypoxia-inducible factor by impairing the selenoprotein thioredoxin reductase. *J Biol Chem* 278 (2):745–750.

79. Hirota, K., M. Matsui, S. Iwata, A. Nishiyama, K. Mori, and J. Yodoi. 1997. AP-1 transcriptional activity is regulated by a direct association between thioredoxin and Ref-1. *Proc Natl Acad Sci USA* 94:3633–3638.

80. Ueno, M., H. Masutani, R. J. Arai, A. Yamauchi, K. Hirota, T. Sakai, T. Inamoto, Y. Yamaoka, J. Yodoi, and T. Nikaido. 1999. Thioredoxin-dependent redox regulation of p53-mediated p21 activation. *J Biol Chem* 274:35809–35815.

81. Funato, Y. and H. Miki. 2007. Nucleoredoxin, a novel thioredoxin family member involved in cell growth and differentiation. *Antioxid Redox Signal* 9 (8):1035–1057.

82. Kurooka, H., K. Kato, S. Minoguchi, Y. Takahashi, J. Ikeda, S. Habu, N. Osawa et al. 1997. Cloning and characterization of the nucleoredoxin gene that encodes a novel nuclear protein related to thioredoxin. *Genomics* 39:331–339.

83. Winterbourn, C. C. 2008. Reconciling the chemistry and biology of reactive oxygen species. *Nat Chem Biol* 4 (5):278–286.

84. Cox, A. G., C. C. Winterbourn, and M. B. Hampton. 2009. Mitochondrial peroxiredoxin involvement in antioxidant defence and redox signalling. *Biochem J* 425 (2):313–325. doi: 10.1042/BJ20091541.

85. Sobotta, M. C., W. Liou, S. Stocker, D. Talwar, M. Oehler, T. Ruppert, A. N. Scharf, and T. P. Dick. 2015. Peroxiredoxin-2 and STAT3 form a redox relay for H_2O_2 signaling. *Nat Chem Biol* 11 (1):64–70.

86. Winterbourn, C. C. and M. B. Hampton. 2015. Redox biology: Signaling via a peroxiredoxin sensor. *Nat Chem Biol* 11 (1):5–6.

87. Du, Y., H. Zhang, X. Zhang, J. Lu, and A. Holmgren. 2013. Thioredoxin 1 is inactivated due to oxidation induced by peroxiredoxin under oxidative stress and reactivated by the glutaredoxin system. *J Biol Chem* 288 (45):32241–32247.

88. Zhang, Z., A. Wang, H. Li, H. Zhi, and F. Lu. 2016. STAT3-dependent TXNDC17 expression mediates Taxol resistance through inducing autophagy in human colorectal cancer cells. *Gene* 584 (1):75–82.

89. Chong, C. R., W. P. Chan, T. H. Nguyen, S. Liu, N. E. Procter, D. T. Ngo, A. L. Sverdlov, Y. Y. Chirkov, and J. D. Horowitz. 2014. Thioredoxin-interacting protein: Pathophysiology and emerging pharmacotherapeutics in cardiovascular disease and diabetes. *Cardiovasc Drugs Ther* 28 (4):347–360.

90. Yoshihara, E., S. Masaki, Y. Matsuo, Z. Chen, H. Tian, and J. Yodoi. 2014. Thioredoxin/Txnip: Redoxisome, as a redox switch for the pathogenesis of diseases. *Front Immunol* 4:514.

91. Spindel, O. N., C. World, and B. C. Berk. 2012. Thioredoxin interacting protein: Redox dependent and independent regulatory mechanisms. *Antioxid Redox Signal* 16 (6):587–596.

92. Xu, G., J. Chen, G. Jing, and A. Shalev. 2013. Thioredoxin-interacting protein regulates insulin transcription through microRNA-204. *Nat Med* 19 (9):1141–1146.

93. Rhee, S. G. and I. S. Kil. 2016. Mitochondrial H₂O₂ signaling is controlled by the concerted action of peroxiredoxin III and sulfiredoxin: Linking mitochondrial function to circadian rhythm. *Free Radic Biol Med* 99:120–127.

94. Brand, M. D. 2016. Mitochondrial generation of superoxide and hydrogen peroxide as the source of mitochondrial redox signaling. *Free Radic Biol Med* 100:14–31.

95. Sies, H. 2014. Role of metabolic H₂O₂ generation: Redox signaling and oxidative stress. *J Biol Chem* 289 (13):8735–8741.

96. Murphy, M. P. 2012. Modulating mitochondrial intracellular location as a redox signal. *Sci Signal* 5 (242):pe39.

97. Cox, A. G., K. K. Brown, E. S. Arner, and M. B. Hampton. 2008. The thioredoxin reductase inhibitor auranofin triggers apoptosis through a Bax/Bak-dependent process that involves peroxiredoxin 3 oxidation. *Biochem Pharmacol* 76 (9):1097–1109.

98. Brown, K. K., S. E. Eriksson, E. S. Arner, and M. B. Hampton. 2008. Mitochondrial peroxiredoxin 3 is rapidly oxidized in cells treated with isothiocyanates. *Free Radic Biol Med* 45 (4):494–502.

99. Johansson, K., M. Cebula, O. Rengby, K. Dreij, K. E. Carlstrom, K. Sigmundsson, F. Piehl, and E. S. Arner. November 11, 2017. Cross talk in HEK293 cells between Nrf2, HIF, and NF-kappaB activities upon challenges with redox therapeutics characterized with single-cell resolution. *Antioxid Redox Signal* 26 (6):229–246.

100. Fomenko, D. E., S. V. Novoselov, S. K. Natarajan, B. C. Lee, A. Koc, B. A. Carlson, T. H. Lee, H. Y. Kim, D. L. Hatfield, and V. N. Gladyshev. 2009. MsrB1 (methionine-R-sulfoxide reductase 1) knock-out mice: Roles of MsrB1 in redox regulation and identification of a novel selenoprotein form. *J Biol Chem* 284 (9):5986–5993.

101. Oien, D. B. and J. Moskovitz. 2008. Substrates of the methionine sulfoxide reductase system and their physiological relevance. *Curr Top Dev Biol* 80:93–133.

102. Mitchell, D. A. and M. A. Marletta. 2005. Thioredoxin catalyzes the S-nitrosation of the caspase-3 active site cysteine. *Nat Chem Biol* 1 (3):154–158.

103. Stoyanovsky, D. A., Y. Y. Tyurina, V. A. Tyurin, D. Anand, D. N. Mandavia, D. Gius, J. Ivanova, B. Pitt, T. R. Billiar, and V. E. Kagan. 2005. Thioredoxin and lipoic acid catalyze the denitrosation of low molecular weight and protein S-nitrosothiols. *J Am Chem Soc* 127 (45):15815–15823.

104. Hedstrom, E., S. Eriksson, J. Zawacka-Pankau, E. S. Arner, and G. Selivanova. 2009. p53-dependent inhibition of TrxR1 contributes to the tumor-specific induction of apoptosis by RITA. *Cell Cycle* 8 (21):3576–3583.

105. Turunen, N., P. Karihtala, A. Mantyniemi, R. Sormunen, A. Holmgren, V. L. Kinnula, and Y. Soini. 2004. Thioredoxin is associated with proliferation, p53 expression and negative estrogen and progesterone receptor status in breast carcinoma. *APMIS* 112 (2):123–132.

106. Iverson, S. V., S. Eriksson, J. Xu, J. R. Prigge, E. A. Talago, T. A. Meade, E. S. Meade, M. R. Capecchi, E. S. Arner, and E. E. Schmidt. 2013. A Txnrd1-dependent metabolic switch alters hepatic lipogenesis, glycogen storage, and detoxification. *Free Radic Biol Med* 63:369–380.
107. Locy, M. L., L. K. Rogers, J. R. Prigge, E. E. Schmidt, E. S. Arner, and T. E. Tipple. 2012. Thioredoxin reductase inhibition elicits Nrf2-mediated responses in Clara cells: Implications for oxidant-induced lung injury. *Antioxid Redox Signal* 17 (10):1407–1416.

17 H$_2$O$_2$-Induced ERK 1/2 MAP Kinases Activity Mediating the Interaction between Thioredoxin-1 and the Thioredoxin Interacting Protein

Hugo P. Monteiro, Fernando T. Ogata, and Arnold Stern

CONTENTS

THIOREDOXIN-1

Thioredoxin-1 (Trx-1) plays a major role in maintaining cellular redox balance by functioning as a scavenger of reactive oxygen species [1]. Its molecular weight is 12 kDa and contains conserved cysteine residues at its active site (Cys 32 and Cys 35). Trx-1 functions within a thioredoxin system (Trx system), whose components in addition to Trx-1 are the mitochondrial isoform Trx-2, NADPH, and the thioredoxin reductases (TrxR) (see Arner, this book, [2]). The ubiquity of the Trx system implies that it can interact with a variety of targets. Peroxiredoxins are direct targets that function in the reduction of H$_2$O$_2$ and organic peroxides associated with redox signaling [3]. Ribonucleotide reductase is an indirect target of the Trx system. The Trx system supplies precursors for the synthesis of DNA [1].

Trx-1 is present in all cells, including tumor cells, where it serves as a positive regulator of cell survival [1–5]. It inhibits the apoptosis signal-regulated protein kinase ASK-1 [6] and regulates many redox-sensitive transcription factors [7–12].

Cellular compartmentation is an essential feature of Trx-1 mediated signaling events. Trx-1 is found in the cytoplasm and in the nucleus. When found in the nucleus, typically by transfer from the cytoplasm by redox-mediated signaling events [13–15], it functions as a regulator of transcription factors [7–10].

THIOREDOXIN INTERACTING PROTEIN

Thioredoxin interacting protein (Txnip) or thioredoxin binding protein 2 (TBP-2) is a vitamin D3 upregulated protein [16]. Txnip is a negative regulator of Trx-1 and may be a suppressor of Trx-1-mediated prosurvival signaling [17]. Txnip can interact with Trx-1 in vitro and in vivo, binding to the reduced, but not the oxidized or mutant form of Trx-1, where the two redox-active cysteine residues are substituted by serine residues [18]. Adenovirus-mediated overexpression of Txnip decreases Trx-1 activity in rat cardiomyocytes, leading to apoptosis [19]. Downregulation of Txnip in cells by H_2O_2 or exogenous NO increases Trx-1 activity [19,20].

Txnip and the Trx system in cells under oxidative stress are subjected to transcriptional regulation by microRNAs (miR). TrxRs, Trx-1, and Txnip each have almost 100 predicted miRs that can target their sequences, though only a few have been demonstrated to downregulate their mRNAs. TrxR1 and TrxR2 in non-small lung cancer cells are targets for miR-124 and for miR-17-3p, respectively [21,22]. Trx-1 is targeted by miR523-3p after exposure to ionizing radiation in several cell lines that are either from nontransformed or tumor-derived cells [23]. Txnip expression increases in senescent cells due to a decrease in miR-17-5p expression [24]. Melanoma-derived cell lines that induce expression of the miR 224/452 cluster associated with downregulation of Txnip drive the epithelial-mesenchymal transition [25]. MiR-20a targets Txnip in fibroblast-like synoviocytes after adjuvant-induced arthritis treatment in vitro [26].

TRX-1, TXNIP, ERK 1/2 MAP KINASES, AND H_2O_2

Trx1 is predominantly a cytosolic protein, but can be secreted from cells in two forms, a full-length and a truncated form [27]. In infection and inflammation both forms are released extracellularly where they behave as co-cytokines and chemokines [28,29].

Following challenge with H_2O_2 [10,15], UV irradiation [30], cis-diamine-dichloroplatinum II (cisplatin) [9], or NO [14,15]. Trx-1 migrates to the nucleus where it can regulate the DNA binding activity of NFkB, AP-1, p53, and the glucocorticoid receptor (GR) by reducing critical cysteine residues. Reduction of the Cys62 in the DNA binding loop of the p50 subunit of NFkB by Trx-1 enhances its transcriptional activity [11]. Conserved cysteine residues within the DNA binding domain of Fos and Jun undergo reduction by Trx-1 and reducing catalyst redox-factor-1 resulting in transcriptional activity of AP-1 [8]. Trx-1 regulates p53 transcriptional activity via redox regulation of the conserved DNA binding domain containing zinc and essential cysteine residues [9]. Ligand binding activity of the GR in the cytosol is maintained by Trx-1 mediated redox regulation, which also promotes GR translocation and its DNA binding activity [10].

H$_2$O$_2$, in a similar fashion as NO, causes the migration of cytoplasmic ERK 1/2 MAP kinases to the nucleus. Nuclear migration of ERK 1/2 MAP kinases results in downregulation of Txnip expression and activation of nuclear translocation of Trx-1. Inhibition of the ERK1/2 MAP kinases signaling pathway or transfection of cells with Protein Enriched in Astrocytes (PEA-15), a cytoplasmic anchor of the ERK 1/2 MAP kinases [31–33], results in inhibition of Trx-1 nuclear migration and down regulation of Txnip. Over expression of Txnip inhibits nuclear migration of Trx-1, while gene silencing of Txnip enhances nuclear migration of Trx-1, even in the absence of H$_2$O$_2$ [15].

CONCLUSION

Trx-1 and Txnip are critical to the redox regulation of cells. Regulation of the expression levels of Trx-1, Txnip, and of TrxRs by miRs is achieved under oxidative stress. H$_2$O$_2$, in a similar fashion as NO, affects nuclear migration of Trx-1. This is associated with intracellular compartmentalization and activation of ERK1/2 MAP kinases. The interaction of Trx-1 and Txnip coordinated by the ERK 1/2 MAP kinases is responsible for the nuclear migration of Trx-1 and cell survival. A general scheme illustrates how oxidative stress triggered by H$_2$O$_2$ or NO mediates Trx1/Txnip interactions associated with survival signaling (Figure 17.1).

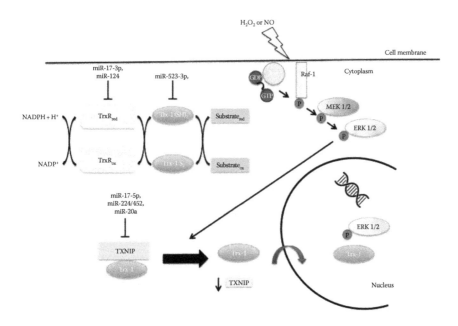

FIGURE 17.1 The Trx1/p21Ras-ERK1/2 MAP kinases/PEA-15/Txnip compartmentalized signaling pathway is stimulated by oxidative/nitrosative stress conditions. MiRs targeting Txnip and the Trx system determine signaling events occurring in the cytoplasm and the nucleus.

ACKNOWLEDGMENTS

Funding from the Brazilian Institutions FAPESP, CNPq, and CAPES is greatly acknowledged. The authors appreciate being a colleague and friend of Dr. Christine C. Winterbourn, to whom this book is dedicated. Dr. Winterbourn played a major role in helping to shape Hugo P. Monteiro's scientific career in its early phase.

REFERENCES

1. Holmgren, A. 1985. Thioredoxin. *Annu Rev Biochem* 54:237–271. doi: 10.1146/annurev. bi.54.070185.001321.
2. Arner, E. S. and A. Holmgren. 2000. Physiological functions of thioredoxin and thioredoxin reductase. *Eur J Biochem* 267 (20):6102–6109.
3. Lu, J. and A. Holmgren. 2014. The thioredoxin antioxidant system. *Free Radic Biol Med* 66:75–87. doi: 10.1016/j.freeradbiomed.2013.07.036.
4. Urig, S. and K. Becker. 2006. On the potential of thioredoxin reductase inhibitors for cancer therapy. *Semin Cancer Biol* 16 (6):452–465. doi: 10.1016/j.semcancer.2006.09.004.
5. Hanschmann, E. M., J. R. Godoy, C. Berndt, C. Hudemann, and C. H. Lillig. 2013. Thioredoxins, glutaredoxins, and peroxiredoxins—Molecular mechanisms and health significance: From cofactors to antioxidants to redox signaling. *Antioxid Redox Signal* 19 (13):1539–1605. doi: 10.1089/ars.2012.4599.
6. Saitoh, M., H. Nishitoh, M. Fujii, K. Takeda, K. Tobiume, Y. Sawada, M. Kawabata, K. Miyazono, and H. Ichijo. 1998. Mammalian thioredoxin is a direct inhibitor of apoptosis signal-regulating kinase (ASK) 1. *EMBO J* 17 (9):2596–2606. doi: 10.1093/emboj/17.9.2596.
7. Akamatsu, Y., T. Ohno, K. Hirota, H. Kagoshima, J. Yodoi, and K. Shigesada. 1997. Redox regulation of the DNA binding activity in transcription factor PEBP2. The roles of two conserved cysteine residues. *J Biol Chem* 272 (23):14497–14500.
8. Hirota, K., M. Matsui, S. Iwata, A. Nishiyama, K. Mori, and J. Yodoi. 1997. AP-1 transcriptional activity is regulated by a direct association between thioredoxin and Ref-1. *Proc Natl Acad Sci USA* 94 (8):3633–3638.
9. Ueno, M., H. Masutani, R. J. Arai, A. Yamauchi, K. Hirota, T. Sakai, T. Inamoto, Y. Yamaoka, J. Yodoi, and T. Nikaido. 1999. Thioredoxin-dependent redox regulation of p53-mediated p21 activation. *J Biol Chem* 274 (50):35809–35815.
10. Makino, Y., N. Yoshikawa, K. Okamoto, K. Hirota, J. Yodoi, I. Makino, and H. Tanaka. 1999. Direct association with thioredoxin allows redox regulation of glucocorticoid receptor function. *J Biol Chem* 274 (5):3182–3188.
11. Matthews, J. R., N. Wakasugi, J. L. Virelizier, J. Yodoi, and R. T. Hay. 1992. Thioredoxin regulates the DNA binding activity of NF-kappa B by reduction of a disulphide bond involving cysteine 62. *Nucleic Acids Res* 20 (15):3821–3830.
12. Grippo, J. F., A. Holmgren, and W. B. Pratt. 1985. Proof that the endogenous, heat-stable glucocorticoid receptor-activating factor is thioredoxin. *J Biol Chem* 260 (1):93–97.
13. Bai, J., H. Nakamura, Y. W. Kwon, I. Hattori, Y. Yamaguchi, Y. C. Kim, N. Kondo et al. 2003. Critical roles of thioredoxin in nerve growth factor-mediated signal transduction and neurite outgrowth in PC12 cells. *J Neurosci* 23 (2):503–509.
14. Arai, R. J., H. Masutani, J. Yodoi, V. Debbas, F. R. Laurindo, A. Stern, and H. P. Monteiro. 2006. Nitric oxide induces thioredoxin-1 nuclear translocation: Possible association with the p21Ras survival pathway. *Biochem Biophys Res Commun* 348 (4):1254–1260. doi: 10.1016/j.bbrc.2006.07.178.
15. Ogata, F. T., W. L. Batista, A. Sartori, T. F. Gesteira, H. Masutani, R. J. Arai, J. Yodoi, A. Stern, and H. P. Monteiro. 2013. Nitrosative/oxidative stress conditions regulate

thioredoxin-interacting protein (TXNIP) expression and thioredoxin-1 (TRX-1) nuclear localization. *PLoS One* 8 (12):e84588. doi: 10.1371/journal.pone.0084588.

16. Chen, K. S. and H. F. DeLuca. 1994. Isolation and characterization of a novel cDNA from HL-60 cells treated with 1,25-dihydroxyvitamin D-3. *Biochim Biophys Acta* 1219 (1):26–32.

17. Nishinaka, Y., H. Masutani, S. Oka, Y. Matsuo, Y. Yamaguchi, K. Nishio, Y. Ishii, and J. Yodoi. 2004. Importin alpha1 (Rch1) mediates nuclear translocation of thioredoxin-binding protein-2/vitamin D(3)-up-regulated protein 1. *J Biol Chem* 279 (36):37559–37565. doi: 10.1074/jbc.M405473200.

18. Nishiyama, A., M. Matsui, S. Iwata, K. Hirota, H. Masutani, H. Nakamura, Y. Takagi, H. Sono, Y. Gon, and J. Yodoi. 1999. Identification of thioredoxin-binding protein-2/vitamin D(3) up-regulated protein 1 as a negative regulator of thioredoxin function and expression. *J Biol Chem* 274 (31):21645–21650.

19. Wang, Y., G. W. De Keulenaer, and R. T. Lee. 2002. Vitamin D(3)-up-regulated protein-1 is a stress-responsive gene that regulates cardiomyocyte viability through interaction with thioredoxin. *J Biol Chem* 277 (29):26496–26500. doi: 10.1074/jbc.M202133200.

20. Schulze, P. C., H. Liu, E. Choe, J. Yoshioka, A. Shalev, K. D. Bloch, and R. T. Lee. 2006. Nitric oxide-dependent suppression of thioredoxin-interacting protein expression enhances thioredoxin activity. *Arterioscler Thromb Vasc Biol* 26 (12):2666–2672. doi: 10.1161/01.ATV.0000248914.21018.f1.

21. Hao, C., X. Xu, J. Ma, J. Xia, B. Dai, L. Liu, and Y. Ma. 2017. MicroRNA-124 regulates the radiosensitivity of non-small cell lung cancer cells by targeting TXNRD1. *Oncol Lett* 13 (4):8. doi: 10.3892/ol.2017.5701.

22. Tian, B., D. E. Maidana, B. Dib, J. B. Miller, P. Bouzika, J. W. Miller, D. G. Vavvas, and H. Lin. 2016. miR-17-3p exacerbates oxidative damage in human retinal pigment epithelial cells. *PLoS One* 11 (8):e0160887. doi: 10.1371/journal.pone.0160887.

23. Kraemer, A., Z. Barjaktarovic, H. Sarioglu, K. Winkler, F. Eckardt-Schupp, S. Tapio, M. J. Atkinson, and S. Moertl. 2013. Cell survival following radiation exposure requires miR-525-3p mediated suppression of ARRB1 and TXN1. *PLoS One* 8 (10):e77484. doi: 10.1371/journal.pone.0077484.

24. Zhuo de, X., X. H. Niu, Y. C. Chen, D. Q. Xin, Y. L. Guo, and Z. B. Mao. 2010. Vitamin D3 up-regulated protein 1(VDUP1) is regulated by FOXO3A and miR-17-5p at the transcriptional and post-transcriptional levels, respectively, in senescent fibroblasts. *J Biol Chem* 285 (41):31491–31501. doi: 10.1074/jbc.M109.068387.

25. Knoll, S., K. Furst, B. Kowtharapu, U. Schmitz, S. Marquardt, O. Wolkenhauer, H. Martin, and B. M. Putzer. 2014. E2F1 induces miR-224/452 expression to drive EMT through TXNIP downregulation. *EMBO Rep* 15 (12):1315–1329. doi: 10.15252/embr.201439392.

26. Li, X. F., W. W. Shen, Y. Y. Sun, W. X. Li, Z. H. Sun, Y. H. Liu, L. Zhang, C. Huang, X. M. Meng, and J. Li. 2016. MicroRNA-20a negatively regulates expression of NLRP3-inflammasome by targeting TXNIP in adjuvant-induced arthritis fibroblast-like synoviocytes. *Joint Bone Spine* 83 (6):695–700. doi: 10.1016/j.jbspin.2015.10.007.

27. Rubartelli, A., A. Bajetto, G. Allavena, E. Wollman, and R. Sitia. 1992. Secretion of thioredoxin by normal and neoplastic cells through a leaderless secretory pathway. *J Biol Chem* 267 (34):24161–24164.

28. Bertini, R., O. M. Howard, H. F. Dong, J. J. Oppenheim, C. Bizzarri, R. Sergi, G. Caselli et al. 1999. Thioredoxin, a redox enzyme released in infection and inflammation, is a unique chemoattractant for neutrophils, monocytes, and T cells. *J Exp Med* 189 (11):1783–1789.

29. Pekkari, K., J. Avila-Carino, A. Bengtsson, R. Gurunath, A. Scheynius, and A. Holmgren. 2001. Truncated thioredoxin (Trx80) induces production of interleukin-12 and enhances CD14 expression in human monocytes. *Blood* 97 (10):3184–3190.

30. Hirota, K., M. Murata, Y. Sachi, H. Nakamura, J. Takeuchi, K. Mori, and J. Yodoi. 1999. Distinct roles of thioredoxin in the cytoplasm and in the nucleus. A two-step mechanism of redox regulation of transcription factor NF-kappaB. *J Biol Chem* 274 (39):27891–27897.

31. Araujo, H., N. Danziger, J. Cordier, J. Glowinski, and H. Chneiweiss. 1993. Characterization of PEA-15, a major substrate for protein kinase C in astrocytes. *J Biol Chem* 268 (8):5911–5920.

32. Danziger, N., M. Yokoyama, T. Jay, J. Cordier, J. Glowinski, and H. Chneiweiss. 1995. Cellular expression, developmental regulation, and phylogenic conservation of PEA-15, the astrocytic major phosphoprotein and protein kinase C substrate. *J Neurochem* 64 (3):1016–1025.

33. Krueger, J., F. L. Chou, A. Glading, E. Schaefer, and M. H. Ginsberg. 2005. Phosphorylation of phosphoprotein enriched in astrocytes (PEA-15) regulates extracellular signal-regulated kinase-dependent transcription and cell proliferation. *Mol Biol Cell* 16 (8):3552–3561. doi: 10.1091/mbc.E04-11-1007.

18 Interactions between Nrf2 Activation and Glutathione in the Maintenance of Redox Homeostasis

Henry Jay Forman, Matilde Maiorino, and Fulvio Ursini

CONTENTS

CENTRAL ROLE OF GLUTATHIONE IN THE MAINTENANCE OF REDOX HOMEOSTASIS

The reduced form of glutathione (GSH) is best known for its roles in the prevention of cellular injury [1]. As a substrate for both the enzymes that reduce H_2O_2 and lipid hydroperoxides and the enzymes that attach GSH to xenobiotic compounds, GSH contributes to the maintenance of nontoxic concentrations of these potential toxicants [2,3]. But GSH is also important in the metabolism of endogenous molecules, including the leukotrienes [4], and in signal transduction, where it is used in the reversible glutathionylation of numerous kinases, protein phosphatases, and transcription factors [5–7]. While there has been a great amount of work on how elevating GSH synthesis contributes to the adaptation of oxidative or electrophilic stress, we will focus here on the role of GSH in maintaining redox homeostasis.

What Is Redox Homeostasis?

What is redox homeostasis? We recently published an article titled, "Redox homeostasis: The Golden Mean of healthy living [8]," in which we defined redox homeostasis operationally as "A steady-state redox status of the ensemble of redox couples [that] is maintained by metabolic fluxes and redox feedback where electrophiles produced by aerobic life stressors activate the mechanism reestablishing nucleophilic tone." While this appears to be similar to adaptation, it differs from adaptation because homeostasis involves a normal range in which the responses do not require a phenotypic switch. An example we used is the response to eating a meal and elevating blood glucose. Each time blood glucose is elevated it presents a metabolic challenge generating electrophiles and nucleophiles, but those are within a range that is not stressful to cells. These transient alterations in metabolism signal for transient changes in gene expression of the enzymes that maintain the steady state range of nucleophiles in the cell. This contrasts with the stress caused by prolonged elevation in glucose that can result in the establishment of offset steady states that are an adaptation to elevated glucose and allows viability, but are pathologically altered resulting in the metabolic syndrome.

Just as homeostasis, as defined by Bernard [9], is not defined by specific values for parameters, redox homeostasis is not defined by specific concentrations of nucleophiles and electrophiles. Rather it is a range of many nucleophiles that a biological system must maintain for normal function. To clarify what we mean by "normal" in this context we consider it as "the homeostatic range of multiple redox couples that define the steady state of a healthy organism." Stress disturbs the steady state and this activates a feedback nucleophilic response. If that stress is prolonged, the organism may adapt by moving to a new steady state, which allows survival, but is not normal. Obviously, this homeostatic condition will vary among cells. A more detailed explanation of redox homeostasis is in our recent article [8].

Several authors have offered definitions of "redox state," a term that would seem to be related to redox homeostasis, particularly focusing on the dominant intracellular nucleophile, GSH and using a ratio of the disulfide (GSSG) and reduced form ([GSSG]/2[GSH]) or the Nernst potential calculated from $[GSSG]/[GSH]^2$ [10,11]. Jones and Sies have also recently expanded upon this in a thought-provoking article concerning "the redox code" [12].

We have addressed in our reviews on redox signaling and redox homeostasis [8,13,14], our difference of opinion with the use of ratios to describe homeostasis. Simply, reversible reactions implied by redox potentials insufficiently account for the variation in the oxidized and reduced forms of glutathione and pyridine nucleotides in cells. This is because they do not factor in their *de novo* synthesis and other uses in metabolism. For example, NAD^+ is used by sirtuins and GSH is used by GSTs. Thus, the concentrations of GSH, GSSG, NADH, NAD^+, NADPH, and $NADP^+$ are a result of multiple enzymatic processes that maintain homeostasis rather their ratios being the driving force for it. Secondly, as Brigelius-Flohé [15] has pointed out, it is GSH concentration rather than the ratio of GSH/GSSG that is sensed by most of the enzymes that use GSH. Furthermore, the term "eustress" [16] seems problematic to us as it implies stress within the range of electrophilic challenge that we see as avoiding stress while maintaining the steady state range of nucleophiles, which we have also referred to as the "nucleophilic tone" [17]. We refer to the variations in nucleophiles and electrophiles that occur in the absence of stress as challenges within a homeostatic range. This discussion may to some extent seem like a semantic argument. Thus, we will hope the readers will decide for themselves whether eustress or redox homeostasis works for them.

While we have stated elsewhere that H_2O_2 is the most significant oxygen-centered species involved in signaling [13,14], other electrophiles are major contributors as well. Indeed, studies from numerous laboratories have indicated that many alkylating electrophiles are far more potent than H_2O_2 as activators of EpRE-regulated genes [18]. Electrophiles including 4-hydroxynonenal are well established as activators of signaling through EpRE as well as other pathways [19,20].

Similarly, GSH is not the only nucleophile that is important in the maintenance of redox homeostasis. Other major contributing nucleophiles are NADPH, NADH, and the reduced form of thioredoxin (Trx). Indeed, the Trx system may contribute as much to the reduction of hydroperoxides as the GSH system [21].

Both Prxs and GPxs appear to be capable of reducing H_2O_2 with variation in their kinetics among the enzymes and distribution among cells [22–25]. Lipid hydroperoxides can be reduced *in vitro* by several of the GPxs and Prx6 [23,24,26]. For lipid hydroperoxides in membranes, there is clear cut evidence only for GPx4, as deletion of Gpx4 has a phenotype from which there is no rescuing by other peroxidases [27]. For other GPxs and Prxs, the phenotype is rather weak [28].

Regardless, redox homeostasis is not about competition among the various systems for eliminating hydroperoxides, but an integration of the enzymatic systems for that purpose with all of metabolism and other cellular functions. While not the sole determinant of redox homeostasis, GSH does play several major roles in the cell, which will be briefly reviewed next.

ROLES FOR GSH

GSH in Metabolism, Cell Protection, and Protein Folding

Probably, the most commonly discussed role in cells for GSH is the reduction of hydroperoxides by GSH. There are eight mammalian GPxs, which, along with Prdx6, use GSH to reduce hydroperoxides in Reaction 18.1:

$$ROOH + 2GSH \rightarrow ROH + H_2O + GSSG \qquad (18.1)$$

where
  ROOH is a hydroperoxide or H_2O_2
  ROH is the corresponding alcohol or another molecule of water
  GSSG is glutathione disulfide, often imprecisely referred to as oxidized glutathione

Five of the eight known mammalian GPxs (1–4 and 6) use a selenocysteine in their catalysis of hydroperoxide reduction while GPxs 5, 7, and 8 use a catalytic cysteine [24]. Prdx6 is an unusual protein having two independently functional active sites, one being operatively a GPx, while the other has phospholipase A2 activity [29].

In the physiological state, almost all glutathione is maintained as GSH. GSSG is reduced by glutathione reductase using NADPH:

$$GSSG + NADPH + H^+ \rightarrow 2GSH + NADP^+ \tag{18.2}$$

The other enzymes that eliminate hydroperoxides are catalase, which dismutates H_2O_2, and the other five mammalian Prdxs, which use reduced Trx, which is also maintained by reduction using NADPH. NADPH is maintained, primarily by the pentose shunt enzymes, glucose-6-phosphate dehydrogenase and 6-phosphogluconate dehydrogenase, but also by the action of cytosolic isocitrate dehydrogenase and malic enzyme 1.

Another important role for GSH is in the conjugation and elimination of electrophiles in addition (Reaction 18.3) or substitution (Reaction 18.4) reactions catalyzed by glutathione S-transferases (GSTs):

$$RC = CR' + GSH \rightarrow RC(SG)CHR' \tag{18.3}$$

$$RX + GSH \rightarrow RSG + HX \tag{18.4}$$

While these reactions are important in the elimination of a vast assortment of xenobiotic compounds or their electrophilic metabolites [30], GSTs are also involved in the conjugation of endogenously produced lipid peroxidation products including 4-hydroxy-2-nonenal [31] and the production of leukotriene C4, which involves the adduction to an epoxide, a third type of GST catalyzed reaction [32]. Interestingly, the GSTs are generally not impressive in their catalytic activity. Indeed, the nonenzymatic conjugation of GSH to some electrophiles can be faster at a high GSH physiologically relevant concentration than is catalyzed by the GST, which is saturated by GSH and at its maximum rate [33].

GSH also participates in protein folding. Protein disulfide isomerases (PDIs) and glutaredoxins (GRxs) are enzymes having a thioredoxin-like structure (thioredoxin fold). They catalyze exchange of protein thiols and disulfides allowing the tertiary structures of proteins to change. GRxs can reduce protein–glutathione mixed disulfides, forming a GRx–glutathione mixed disulfide that reacts with GSH to form GSSG and the reduced GRx:

$$GRx - SH + Protein - SSG \leftrightarrow GRx - SSG + Protein - SH \tag{18.5}$$

$$GSH + GRx - SSG \leftrightarrow GRx - SH + GSSG \tag{18.6}$$

The reactions are reversible and thereby provide a mechanism for protein glutathionylation (see next section on signaling), but in mammalian cells, which maintain high GSH, the reaction tends to go to the right.

Based upon *in vitro* studies, GSH may also be used to form a mixed disulfide with PDI that releases GSH to become an intramolecular disulfide, but the precise mechanisms whereby GSH is used by PDI in protein folding is uncertain [34]. Recently, Maiorino and her colleagues demonstrated that the glutathione peroxidase activity of GPx7 leads to PDI oxidation in competition with GSH [35].

GSH in Signaling

GSH may play several roles in signal transduction. GPxs and Prdx6 activities use GSH to maintain a low concentration of the second messengers, H_2O_2 and lipid hydroperoxides. This is analogous to the actions of the phosphodiesterases that hydrolyze phosphate bonds in cyclic AMP, cyclic GMP, and the inositol phosphates. In this way, the peroxidases act as limiters of signaling by the hydroperoxide second messengers. There is a hypothesis that overwhelming the peroxidases would allow signaling by nonenzymatic oxidation of protein thiolates [36], but this ignores both the abundance of enzymes, which includes the other Prdxs as well as catalase, and the reality that the intracellular GSH is alone sufficient to outcompete any protein thiolate that does not have a high rate of peroxidase activity or is bound to zinc or another metal [37].

The peroxidases may also act in signaling in a more direct manner. As a post-translational modification, glutathionylation can result in the increased or decreased activity of signaling proteins. The involvement of protein glutathionylation has been previously proposed in multiple cellular processes including inhibition of the activities of sodium-potassium ATPase, glyceraldehyde phosphate dehydrogenase, caspase 3, decreased actin polymerization, increased Hsp70 chaperone function, and inhibition of DNA binding of the transcription factors NF-κB and c-Jun [7,38–40]. In addition, the activities of the signaling proteins protein tyrosine phosphatase 1B (PTP1B) and STAT3 by glutathionylation have also been previously proposed as targets [41,42]. A decade ago, Gallogly and Mieyal proposed that glutathionylation was a common reversible mechanism in redox signaling and suggested several enzymes that might catalyze glutathionylation, which included GPx-like enzymes and GRx5 [5]. Recently, we proposed several mechanisms for the glutathionylation of signaling proteins by GPxs, Prdxs, GSTs, and GRxs [37]. For the latter two, which might facilitate disulfide exchange of GSSG with protein thiols, we pointed out a GPx or Prdx6 would need to be physically very close to provide sufficient GSSG to push the reaction in the direction of protein glutathionylation (see Reactions 18.5 and 18.6 and discussion above).

This section concerned the multiple roles of glutathione, a major metabolite that is present at relatively high (1–10 mM) concentration in cells. Usually, thought of as a protector against oxidative stress and xenobiotic toxicity, GSH clearly is involved in many other important functions. It is therefore not surprising that the regulation of its synthesis is well controlled. The next section concerns both how GSH synthesis and many of the enzymes that use it as a substrate are regulated by one of the cell's most important transcription factors, Nrf2, but also how GSH regulates Nrf2 activation.

Nrf2 ACTIVATION AND ITS CENTRAL ROLE IN GLUTATHIONE METABOLISM

MECHANISMS OF Nrf2 ACTIVATION

The transcription factor, nuclear factor (erythroid-derived 2)-like 2, better known as Nrf2, acts as a master regulator in the maintenance of redox homeostasis. Nrf2 is referred to as a cap"n"collar basic-region leucine zipper transcription factor. It is often stated that Nrf2 regulates cellular defenses against oxidative, electrophilic and environmental stress, but it also has essential roles in modulating basal expression of enzymes in intermediary metabolism, mitochondrial biogenesis, and the proteasome [43]. Recently, a special issue of *Free Radical Biology and Medicine* for which reference [43] served as the introduction described in great detail the regulation of Nrf2 and its roles in multiple aspects of cell function. Therefore, here we will only briefly describe how it is activated and then focus on how GSH affects Nrf2 activation.

In the cytosol, the protein called Keap1, which is an adaptor subunit of Cullin 3-based E3 ubiquitin ligase, is a direct sensor of electrophilic molecules including H_2O_2 [44,45]. Keap1 contains several cysteines that have been implicated as the sites modified by either direct alkylation or oxidation [46], with cys 151 being the cys that forms a disulfide bond between two Keap1 molecules when oxidized by H_2O_2 [47]. It is likely that the reactivity of these cysteines is facilitated by binding of zinc [37,46]. When Keap1 is active, which it is constitutively, it facilitates the rapid and constant degradation of Nrf2 in the cytosol. That prevents most Nrf2 from being activated and moving to the nucleus. But while nuclear Nrf2 is constitutively low, it is not absent. When the concentration of an electrophile is elevated, it will alkylate specific cysteine residues of Keap1 [45]. This modification of Keap1 inhibits its facilitation of Nrf2 degradation [45]. Although it was thought that Nrf2 would then dissociate and be free to enter the nucleus, it now appears that it is newly synthesized Nrf2 that escapes binding to Keap1 and travels to the nucleus where it can signal for the transcription of genes [48]. Part of the facilitation of Nrf2 degradation appears to require its phosphorylation by GSK3, which in turn can be inhibited by its phosphorylation by Akt. Akt is activated by PI3K, and this action is prevented by PTEN. Electrophiles inhibit PTEN, adding another point at which electrophiles signal for increased Nrf2 activity [49]. But there are several other signaling pathways activated by electrophiles that contribute to Nrf2 activation, including the activation of PKCδ [50], which may phosphorylate Nrf2, and TAK1, which phosphorylates p62/SQSTM1 [51]. Then, p62, which sequesters Keap1 and facilitates its degradation, enables Nrf2 to escape degradation mediated by Keap1 [52]. Figure 18.1 summarizes aspects of Nrf2 activation described here.

Nrf2 translocates to the nucleus where it binds to antioxidant/electrophile response element (ARE/EpRE) sequences in the promoter regions of many genes. Hayes and coworkers provided a recent list of a number of the Nrf2 regulated genes and their functions [53]. In its DNA binding, Nrf2 partners with other proteins including small Mafs or c-Jun that may facilitate or inhibit transcription depending upon the gene [54–56] probably as a consequence of the non-consensus nature of functional EpRE sequences that varies among genes [57]. Further regulation of Nrf2

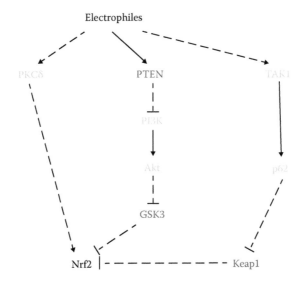

FIGURE 18.1 Activation of Nrf2 by electrophiles including H_2O_2. See text. Proteins that promote Nrf2 activation are in green while those that inhibit Nrf2 activation are shown in red. Solid arrows designate a direct stimulatory effect. Dashed arrows designate an indirect stimulatory effect. Solid perpendicular lines designate direct inhibitory effect. Dashed perpendicular lines designate direct inhibitory effect.

function occurs in the nucleus, where it competes for binding to EpRE with Bach1 [58], which is a repressor, different forms of Nrf1 that act as either repressors or alternative activators [59], and c-Myc, which inhibits transcription and facilitates Nrf2 degradation [60].

REGULATION OF Nrf2 ACTIVATION BY GSH

As described here, modification of Keap1 by addition of electrophiles to its cysteines causes Keap1 to be unable to facilitate Nrf2 degradation. This is considered a major mechanism in Nrf2 activation. There are reports demonstrating that glutathionylation of Keap1 can be also lead to Nrf2 activation [61,62]. The latter report suggests that Keap1 glutathionylation is facilitated by GST-Pi [47]. As noted here, glutathionylation of Keap1 by GST-Pi would require that both a source of H_2O_2 and a GPx or Prdx6 be in the immediate vicinity to provide GSSG. But the oxidation of Keap1 by H_2O_2 does not require glutathionylation because a disulfide can be formed between two Keap1 molecules [47], facilitated by zinc binding to reactive cysteines in Keap1 [46]. Although glutathionylation or disulfide formation in Keap1 provides a direct mechanism for activating Nrf2 using H_2O_2, alkylation by electrophiles rather than oxidation by H_2O_2 appears to be the predominant mechanism of Nrf2 activation [63,64]. Thus, the connection between GSH and Nrf2 activation is more likely a competition between the removal of electrophiles by GSTs using GSH to prevent Nrf2 activation and alkylation of Keap1 to activate Nrf2.

A major connection between Nrf2 activation and GSH is that the transcription of the four genes involved in GSH synthesis is dependent upon Nrf2/EpRE binding [65]. The gene products are the two glutamate cysteine ligase subunits, GCLC and GCLM [66,67], glutathione reductase, and the cystine/glutamate transporter XCT [68]. Many studies have been made of Nrf2 activation using the GCLC as a model. But it is interesting to note that while GCLC has 13 EpRE consensus sequences, binding of Nrf2 to only one of them appears to increase transcription while three others appear to be involved in basal but not electrophile-inducible transcription [57]. This evidence provides a clear warning about inferring regulation when a consensus sequence is found in the promoter of a gene. Thus, for some of the genes cited in this section, the evidence based on binding of Nrf2 implies only the potential for electrophilic inducibility through Nrf2.

Nrf2 has also been suggested to regulate several of the genes encoding enzymes that use GSH to reduce hydroperoxides. Both Prdx6 [69] and GPx2 [70] have been shown to be regulated by Nrf2. It is then not surprising that the enzyme the catalyzes reduction of GSSG to GSH, glutathione reductase [71], and enzymes that maintain NADPH, its reducing substrate, glucose-6-phosphate dehydrogenase [72], malic enzyme 1 [73] and isocitrate dehydrogenase 1 [74] are also regulated by Nrf2. Gorrini et al. [65] has discussed these relationships and their association with cellular anticancer defenses. It is interesting that Prdx1 [75], which uses Trx to reduce H_2O_2, Trx [76], and Trx reductases 1 and 2 [77], are also regulated by Nrf2, complementing the effect of Nrf2 on GSH.

As noted here, GSTs catalyze the conjugation of endogenous and exogenous electrophiles, but also appear to play a role in signal transduction. Nrf2 regulates the transcription of at least eight GSTs [65]: GSTA1 [78], GSTA2 [79], GSTA3 [72], GSTA5 [79], GSTM1 [72], GSTM2 [72], GSTM3 [80], GSTM4 [80], and GSTP1 [81]. Finally, the export of molecules conjugated to glutathione is facilitated by members of the multidrug resistance protein (MRP3) family of which at least four, MRP1 [82], MRP2 [83], MRP3 [84], and MRP4 [85] are regulated by Nrf2.

CONCLUSION

In this brief chapter, we have addressed the relationship of GSH to Nrf2. It seems clear that there is both feedback and feed-forward signaling between Nrf2 activation and the use of GSH in protection of cells. Nrf2 activation leads to the transcription of enzymes that synthesize GSH, maintain it and NADPH in their reduced forms, and the use of GSH in protecting cells from hydroperoxides and other electrophiles. While glutathionylation of Keap1, which indicates an increase in hydroperoxide production, can signal for Nrf2 activation, GSH elimination of electrophiles and hydroperoxides decreases Nrf2 activation. These interactions contribute to redox homeostasis and, thereby, maintenance of the golden mean that defines a healthy life.

REFERENCES

1. Meister, A. 1991. Glutathione deficiency produced by inhibition of its synthesis, and its reversal; applications in research and therapy. *Pharm. Ther.* 51:155–194.
2. Sies, H. 1997. Oxidative stress: Oxidants and antioxidants. *Exp. Physiol.* 82:291–295. doi:10.1113/expphysiol.1997.sp004024.

3. Jakoby, W.B. 1978. The glutathione S-transferases: A group of multifunctional detoxification proteins. *Adv. Enzymol. Relat. Areas Mol. Biol.* 46:383–414.
4. Rouzer, C.A., W.A. Scott, O.W. Griffith, A.L. Hamill, and Z.A. Cohn. 1981. Depletion of glutathione selectively inhibits synthesis of leukotriene C by macrophages. *Proc. Natl. Acad. Sci. USA* 78:2532–2536.
5. Gallogly, M.M. and J.J. Mieyal. 2007. Mechanisms of reversible protein glutathionylation in redox signaling and oxidative stress. *Curr. Opin. Pharmacol.* 7:381–391. doi:10.1016/j.coph.2007.06.003.
6. Ghezzi, P. 2005. Regulation of protein function by glutathionylation. *Free Radic. Res.* 39:573–580. doi:10.1080/10715760500072172.
7. Dalle-Donne, I., R. Rossi, D. Giustarini, R. Colombo, and A. Milzani. 2007. S-glutathionylation in protein redox regulation. *Free Radic. Biol. Med.* 43:883–898. doi:10.1016/j.freeradbiomed.2007.06.014.
8. Ursini, F., M. Maiorino, and H.J. Forman. 2016. Redox homeostasis: The Golden Mean of healthy living. *Redox Biol.* 8:205–215. doi:10.1016/j.redox.2016.01.010.
9. Bernard, C. n.d. *Introduction à l'étude de la médecine expérimentale*, 1865; English translation, Macmillan & Co. Ltd, Paris, France.
10. Schafer, F.Q. and G.R. Buettner. 2001. Redox environment of the cell as viewed through the redox state of the glutathione disulfide/glutathione couple. *Free Radic. Biol. Med.* 30:1191–1212.
11. Go, Y.-M. and D.P. Jones. 2010. Redox clamp model for study of extracellular thiols and disulfides in redox signaling. *Methods Enzymol.* 474:165–179. doi:10.1016/S0076-6879(10)74010-6.
12. Jones, D.P. and H. Sies. 2015.The redox code. *Antioxid. Redox Signal.* 23:734–746. doi:10.1089/ars.2015.6247.
13. Forman, H.J., F. Ursini, and M. Maiorino. 2014. An overview of mechanisms of redox signaling. *J. Mol. Cell. Cardiol.* 73:2–9. doi:10.1016/j.yjmcc.2014.01.018.
14. Forman, H.J., M. Maiorino, and F. Ursini. 2010. Signaling functions of reactive oxygen species. *Biochemistry* 49:835–842. doi:10.1021/bi9020378.
15. Brigelius-Flohe, R. and L. Flohe. 2011. Basic principles and emerging concepts in the redox control of transcription factors. *Antioxid. Redox Signal.* 15:2335–2381. doi:10.1089/ars.2010.3534.
16. Sies, H. 2017. Hydrogen peroxide as a central redox signaling molecule in physiological oxidative stress: Oxidative eustress. *Redox Biol.* 11:613–619. doi:10.1016/j.redox.2016.12.035.
17. Forman, H.J., K.J.A. Davies, and F. Ursini. 2014. How do nutritional antioxidants really work: Nucleophilic tone and para-hormesis versus free radical scavenging in vivo. *Free Radic. Biol. Med.* 66:24–35. doi:10.1016/j.freeradbiomed.2013.05.045.
18. Talalay, P., J.W. Fahey, W.D. Holtzclaw, T. Prestera, and Y. Zhang. 1995. Chemoprotection against cancer by phase 2 enzyme induction. *Toxicol Lett.* 82–83:173–179.
19. Poli, G., R.J. Schaur, W.G. Siems, and G. Leonarduzzi. 2008. 4-hydroxynonenal: A membrane lipid oxidation product of medicinal interest. *Med. Res. Rev.* 28:569–631. doi:10.1002/med.20117.
20. Zhang, H. and H.J. Forman. 2009. Signaling pathways involved in phase II gene induction by alpha, beta-unsaturated aldehydes. *Toxicol. Ind. Health.* 25:269–278. doi:10.1177/0748233709102209.
21. Lu, J. and A. Holmgren. 2014. The thioredoxin antioxidant system. *Free Radic. Biol. Med.* 66:75–87. doi:10.1016/j.freeradbiomed.2013.07.036.
22. Rhee, S.G., H.Z. Chae, and K. Kim. 2005. Peroxiredoxins: A historical overview and speculative preview of novel mechanisms and emerging concepts in cell signaling. *Free Radic. Biol. Med.* 38:1543–1552.
23. Arthur, J.R. 2000. The glutathione peroxidases. *Cell. Mol. Life Sci.* 57:1825–1835. doi:10.1007/PL00000664.

24. Brigelius-Flohé, R. and M. Maiorino. 2013. Glutathione peroxidases. *Biochim. Biophys. Acta—Gen. Subj.* 1830:3289–3303. doi:10.1016/j.bbagen.2012.11.020.

25. Flohe, L., S. Toppo, G. Cozza, and F. Ursini. 2011. A comparison of thiol peroxidase mechanisms. *Antioxid. Redox Signal.* 15:763–780. doi:10.1089/ars.2010.3397.

26. Fisher, A.B., C. Dodia, Y. Manevich, J.W. Chen, and S.I. Feinstein. 1999. Phospholipid hydroperoxides are substrates for non-selenium glutathione peroxidase. *J. Biol. Chem.* 274:21326–21334.

27. Imai, H. 2004. Biological significance of lipid hydroperoxide and its reducing enzyme phospholipid hydroperoxide glutathione peroxidase in mammalian cells. *Yakugaku Zasshi.* 124:937–957.

28. Brigelius-Flohé, R. and A. Kipp. 2009. Glutathione peroxidases in different stages of carcinogenesis. *Biochim. Biophys. Acta—Gen. Subj.* 1790:1555–1568. doi:10.1016/j.bbagen.2009.03.006.

29. Chen, J.W., C. Dodia, S.I. Feinstein, M.K. Jain, and A.B. Fisher. 2000. 1-Cys peroxiredoxin, a bifunctional enzyme with glutathione peroxidase and phospholipase A2 activities. *J. Biol. Chem.* 275:28421–28427.

30. Sheehan, D., G. Meade, V.M. Foley, and C.A. Dowd. 2001. Structure, function and evolution of glutathione transferases: Implications for classification of non-mammalian members of an ancient enzyme superfamily. *Biochem. J.* 360:1–16. doi:10.1042/0264-6021:3600001.

31. Cheng, J.Z., Y. Yang, S.P. Singh, S.S. Singhal, S. Awasthi, S.S. Pan, S.V. Singh, P. Zimniak, and Y.C. Awasthi. 2001. Two distinct 4-hydroxynonenal metabolizing glutathione S-transferase isozymes are differentially expressed in human tissues. *Biochem. Biophys. Res. Commun.* 282:1268–1274. doi:10.1006/bbrc.2001.4707. S0006-291X(01)94707-2 (pii).

32. Murphy, R.C., S. Hammarström, and B. Samuelsson. 1979. Leukotriene C: A slow-reacting substance from murine mastocytoma cells. *Proc. Natl. Acad. Sci. USA* 76:4275–4279.

33. Coles, B., I. Wilson, P. Wardman, J.A. Hinson, S.D. Nelson, and B. Ketterer. 1988. The spontaneous and enzymatic reaction of N-acetyl-p-benzoquinonimine with glutathione: A stopped-flow kinetic study. *Arch. Biochem. Biophys.* 264:253–260.

34. Hatahet, F. and L.W. Ruddock. 2009. Protein disulfide isomerase: A critical evaluation of its function in disulfide bond formation. *Antioxid. Redox Signal.* 11:2807–2850. doi:10.1089/ars.2009.2466.

35. Bosello-Travain, V., M. Conrad, G. Cozza, A. Negro, S. Quartesan, M. Rossetto, A. Roveri, S. Toppo, F. Ursini, M. Zaccarin, and M. Maiorino. 2013. Protein disulfide isomerase and glutathione are alternative substrates in the one Cys catalytic cycle of glutathione peroxidase 7. *Biochim. Biophys. Acta.* 1830:3846–3857. doi:10.1016/j.bbagen.2013.02.017.

36. Wood, Z.A., L.B. Poole, and P.A. Karplus. 2003. Peroxiredoxin evolution and the regulation of hydrogen peroxide signaling. *Science* 300:650–653.

37. Forman, H.J., M.J. Davies, A.C. Krämer, G. Miotto, M. Zaccarin, H. Zhang, and F. Ursini. 2017. Protein cysteine oxidation in redox signaling: Caveats on sulfenic acid detection and quantification. *Arch. Biochem. Biophys.* 617:26–37 doi:10.1016/j.abb.2016.09.013.

38. Ghezzi, P. 2013. Protein glutathionylation in health and disease. *Biochim. Biophys. Acta—Gen. Subj.* 1830:3165–3172. doi:10.1016/j.bbagen.2013.02.009.

39. Mieyal, J.J., M.M. Gallogly, S. Qanungo, E.A. Sabens, and M.D. Shelton. 2008. Molecular mechanisms and clinical implications of reversible protein S-glutathionylation. *Antioxid. Redox Signal.* 10:1941–1988. doi:10.1089/ars.2008.2089.

40. Lillig, C.H., C. Berndt, and A. Holmgren. 2008. Glutaredoxin systems. *Biochim. Biophys. Acta—Gen. Subj.* 1780:1304–1317. doi:10.1016/j.bbagen.2008.06.003.

41. Barrett, W.C., J.P. DeGnore, S. Konig, H.M. Fales, Y.F. Keng, Z.Y. Zhang, M.B. Yim, and P.B. Chock. 1999. Regulation of PTP1B via glutathionylation of the active site cysteine 215. *Biochemistry* 38:6699–6705.
42. Sobotta, M.C., W. Liou, S. Stöcker, D. Talwar, M. Oehler, T. Ruppert, A.N.D. Scharf, and T.P. Dick. 2015. Peroxiredoxin-2 and STAT3 form a redox relay for H_2O_2 signaling. *Nat. Chem. Biol.* 11:64–70. doi:10.1038/nchembio.1695.
43. Mann, G.E. and H.J. Forman. 2015. Introduction to special issue on Nrf2 regulated redox signaling and metabolism in physiology and medicine. *Free Radic. Biol. Med.* 88:91–92. doi:10.1016/j.freeradbiomed.2015.08.002.
44. Itoh, K., N. Wakabayashi, Y. Katoh, T. Ishii, K. Igarashi, J.D. Engel, and M. Yamamoto. 1999. Keap1 represses nuclear activation of antioxidant responsive elements by Nrf2 through binding to the amino-terminal Neh2 domain. *Genes Dev.* 13:76–86.
45. Canning, P., F.J. Sorrell, and A.N. Bullock. 2015. Structural basis of Keap1 interactions with Nrf2. *Free Radic. Biol. Med.* 88:101–107. doi:10.1016/j.freeradbiomed.2015.05.034.
46. Dinkova-Kostova, A.T., W.D. Holtzclaw, and N. Wakabayashi. 2005. Keap1, the sensor for electrophiles and oxidants that regulates the phase 2 response, is a zinc metalloprotein. *Biochemistry* 44:6889–6899. doi:10.1021/bi047434h.
47. Fourquet, S., R. Guerois, D. Biard, and M.B. Toledano. 2010. Activation of NRF2 by nitrosative agents and H_2O_2 involves KEAP1 disulfide formation. *J. Biol. Chem.* 285:8463–8471. doi:10.1074/jbc.M109.051714.
48. McMahon, M., N. Thomas, K. Itoh, M. Yamamoto, and J.D. Hayes. 2006. Dimerization of substrate adaptors can facilitate Cullin-mediated ubiquitylation of proteins by a "tethering" mechanism: A two-site interaction model for the Nrf2-Keap1 complex. *J. Biol. Chem.* 281:24756–24768. doi:10.1074/jbc.M601119200.
49. Rada, P., A.I. Rojo, N. Evrard-Todeschi, N.G. Innamorato, A. Cotte, T. Jaworski, J.C. Tobón-Velasco et al. 2012. Structural and functional characterization of Nrf2 degradation by the glycogen synthase kinase 3/β-TrCP axis. *Mol. Cell. Biol.* 32:3486–3499. doi:10.1128/MCB.00180-12.
50. Zhang, H. and H.J. Forman. 2008. Acrolein induces heme oxygenase-1 through PKC-delta and PI3K in human bronchial epithelial cells. *Am. J. Respir. Cell Mol. Biol.* 38:483–490. doi:10.1165/rcmb.2007-0260OC.
51. Hashimoto, K., A.N. Simmons, R. Kajino-sakamoto, Y. Tsuji, and J. Ninomiya-tsuji. 2016. TAK1 regulates the Nrf2 antioxidant system through modulating p62/SQSTM1. *Antioxid. Redox Signal.* 25:1–52. doi:10.1089/ars.2016.6663.
52. Jiang, T., B. Harder, M. Rojo de la Vega, P.K. Wong, E. Chapman, and D.D. Zhang. 2015. p62 links autophagy and Nrf2 signaling. *Free Radic. Biol. Med.* 88:199–204. doi:10.1016/j.freeradbiomed.2015.06.014.
53. Tebay, L.E., H. Robertson, S.T. Durant, S.R. Vitale, T.M. Penning, A.T. Dinkova-Kostova, and J.D. Hayes. 2015. Mechanisms of activation of the transcription factor Nrf2 by redox stressors, nutrient cues, and energy status and the pathways through which it attenuates degenerative disease. *Free Radic. Biol. Med.* 88:108–146. doi:10.1016/j.freeradbiomed.2015.06.021.
54. Itoh, K., T. Chiba, S. Takahashi, T. Ishii, K. Igarashi, Y. Katoh, T. Oyake et al. 1997. An Nrf2/small Maf heterodimer mediates the induction of phase II detoxifying enzyme genes through antioxidant response elements. *Biochem. Biophys. Res. Commun.* 236:313–322. S0006291X97969436 (pii).
55. Yang, H., T. Liu, J. Wang, T.W.H. Li, W. Fan, H. Peng, A. Krishnan, G.J. Gores, J.M. Mato, and S.C. Lu. 2016. Deregulated methionine adenosyltransferase α1, c-Myc, and Maf proteins together promote cholangiocarcinoma growth in mice and humans. *Hepatology* 64:439–455. doi:10.1002/hep.28541.
56. Levy, S., A.K. Jaiswal, and H.J. Forman. 2009. The role of c-Jun phosphorylation in EpRE activation of phase II genes. *Free Radic. Biol. Med.* 47:1172–1179. doi:10.1016/j.freeradbiomed.2009.07.036.

57. Zhang, H. and H.J. Forman. 2010. Reexamination of the electrophile response element sequences and context reveals a lack of consensus in gene function. *Biochim. Biophys. Acta* 1799:496–501. doi:10.1016/j.bbagrm.2010.05.003.
58. Dhakshinamoorthy, S., A.K. Jain, D.A. Bloom, and A.K. Jaiswal. 2005. Bach1 competes with Nrf2 leading to negative regulation of the antioxidant response element (ARE)-mediated NAD(P)H:quinone oxidoreductase 1 gene expression and induction in response to antioxidants. *J. Biol. Chem.* 280:16891–16900. doi:10.1074/jbc.M500166200. M500166200 (pii).
59. Bugno, M., M. Daniel, N.L. Chepelev, and W.G. Willmore. 2015. Changing gears in Nrf1 research, from mechanisms of regulation to its role in disease and prevention. *Biochim. Biophys. Acta—Gene Regul. Mech.* 1849:1260–1276. doi:10.1016/j.bbagrm.2015.08.001.
60. Levy, S. and H.J. Forman. 2010. C-Myc is a Nrf2-interacting protein that negatively regulates phase II genes through their electrophile responsive elements. *IUBMB Life* 62:237–246. doi:10.1002/iub.314.
61. Gambhir, L., R. Checker, M. Thoh, R.S. Patwardhan, D. Sharma, M. Kumar, and S.K. Sandur. 2014. 1,4-Naphthoquinone, a pro-oxidant, suppresses immune responses via KEAP-1 glutathionylation. *Biochem. Pharmacol.* 88:95–105. doi:10.1016/j.bcp.2013.12.022.
62. Carvalho, A.N., C. Marques, R.C. Guedes, M. Castro-Caldas, E. Rodrigues, J. Van Horssen, and M.J. Gama. 2016. S-Glutathionylation of Keap1: A new role for glutathione S-transferase pi in neuronal protection. *FEBS Lett.* 590:1455–1466. doi:10.1002/1873-3468.12177.
63. Prochaska, H.J. and P. Talalay. 1988. Regulatory mechanisms of monofunctional and bifunctional anticarcinogenic enzyme inducers in murine liver. *Cancer Res.* 48:4776–4782.
64. Friling, R.S., A. Bensimon, Y. Tichauer, and V. Daniel. 1990. Xenobiotic-induced expression of murine glutathione S-transferase Ya subunit gene is controlled by an electrophile-responsive element. *Proc. Natl. Acad. Sci. USA* 87:6258–6262.
65. Gorrini, C., I.S. Harris, and T.W. Mak. 2013. Modulation of oxidative stress as an anticancer strategy. *Nat. Rev. Drug Discov.* 12:931–947. doi:10.1038/nrd4002.
66. Mulcahy, R.T. and J.J. Gipp. 1995. Identification of a putative antioxidant response element in the 5′-flanking region of the human g-glutamylcycteine synthetase heavy subunit gene. *Biochem. Biophys. Res. Commun.* 209:227–233.
67. Moinova, H.R. and R.T. Mulcahy. 1998. An electrophile responsive element (EpRE) regulates b-naphthoflavone induction of the human g-glutamylcysteine synthetase regulatory subunit gene. Constitutive expression is mediated by an adjacent AP-1 site. *J. Biol. Chem.* 273:14683–14689.
68. Sasaki, H., H. Sato, K. Kuriyama-Matsumura, K. Sato, K. Maebara, H. Wang, M. Tamba, K. Itoh, M. Yamamoto, and S. Bannai. 2002. Electrophile response element-mediated induction of the cystine/glutamate exchange transporter gene expression. *J. Biol. Chem.* 277:44765–44771. doi:10.1074/jbc.M208704200.
69. Chowdhury, I., Y. Mo, L. Gao, A. Kazi, A.B. Fisher, and S.I. Feinstein. 2009. Oxidant stress stimulates expression of the human peroxiredoxin 6 gene by a transcriptional mechanism involving an antioxidant response element. *Free Radic. Biol. Med.* 46:146–153. doi:10.1016/j.freeradbiomed.2008.09.027.
70. Banning, A., S. Deubel, D. Kluth, Z. Zhou, and R. Brigelius-Flohé. 2005. The GI-GPx gene is a target for Nrf2. *Mol. Cell. Biol.* 25:4914–4923. doi:10.1128/MCB.25.12.4914-4923.2005.
71. Hübner, R.-H., J.D. Schwartz, P. De Bishnu, B. Ferris, L. Omberg, J.G. Mezey, N.R. Hackett, and R.G. Crystal. 2009. Coordinate control of expression of Nrf2-modulated genes in the human small airway epithelium is highly responsive to cigarette smoking. *Mol. Med.* 15:203–219. doi:10.2119/molmed.2008.00130.

72. Thimmulappa, R.K., K.H. Mai, S. Srisuma, T.W. Kensler, M. Yamamoto, and S. Biswal. 2002. Identification of Nrf2-regulated genes induced by the chemopreventive agent sulforaphane by oligonucleotide microarray. *Cancer Res.* 62:5196–5203. doi:10.1080/01635589209514201.

73. Qaisiya, M., C.D. Coda Zabetta, C. Bellarosa, and C. Tiribelli. 2014. Bilirubin mediated oxidative stress involves antioxidant response activation via Nrf2 pathway. *Cell. Signal.* 26:512–520. doi:10.1016/j.cellsig.2013.11.029.

74. Kanamori, M., T. Higa, Y. Sonoda, S. Murakami, M. Dodo, H. Kitamura, K. Taguchi et al. 2015. Activation of the NRF2 pathway and its impact on the prognosis of anaplastic glioma patients. *Neuro. Oncol.* 17:555–565. doi:10.1093/neuonc/nou282.

75. Kim, Y.J., J.Y. Ahn, P. Liang, C. Ip, Y. Zhang, and Y.M. Park. 2007. Human prx1 gene is a target of Nrf2 and is up-regulated by hypoxia/reoxygenation: Implication to tumor biology. *Cancer Res.* 67:546–554. doi:10.1158/0008-5472.CAN-06-2401.

76. Kim, Y.-C., Y. Yamaguchi, N. Kondo, H. Masutani, and J. Yodoi. 2003. Thioredoxin-dependent redox regulation of the antioxidant responsive element (ARE) in electrophile response. *Oncogene* 22:1860–1865. doi:10.1038/sj.onc.1206369.

77. Hintze, K.J. and E.C. Theil. 2005. DNA and mRNA elements with complementary responses to hemin, antioxidant inducers, and iron control ferritin-L expression. *Proc. Natl. Acad. Sci. USA* 102:15048–15052. doi:10.1073/pnas.0505148102.

78. Chorley, B.N., M.R. Campbell, X. Wang, M. Karaca, D. Sambandan, F. Bangura, P. Xue, J. Pi, S.R. Kleeberger, and D.A. Bell. 2012. Identification of novel NRF2-regulated genes by ChiP-Seq: Influence on retinoid X receptor alpha. *Nucleic Acids Res.* 40:7416–7429. doi:10.1093/nar/gks409.

79. Yates, M.S., M.K. Kwak, P.A. Egner, J.D. Groopman, S. Bodreddigari, T.R. Sutter, K.J. Baumgartner et al. 2006. Potent protection against aflatoxin-induced tumorigenesis through induction of Nrf2-regulated pathways by the triterpenoid 1-(2-cyano-3-,12-dioxooleana-1, 9(11)-dien-28-oyl)imidazole. *Cancer Res.* 66:2488–2494. doi:10.1158/0008-5472.CAN-05-3823.

80. Chanas, S.A., Q. Jiang, M. McMahon, G.K. McWalter, L.I. McLellan, C.R. Elcombe, C.J. Henderson et al. 2002. Loss of the Nrf2 transcription factor causes a marked reduction in constitutive and inducible expression of the glutathione S-transferase Gsta1, Gsta2, Gstm1, Gstm2, Gstm3 and Gstm4 genes in the livers of male and female mice. *Biochem. J.* 365:405–416. doi:10.1042/BJ20020320.

81. Montano, M.M., H. Deng, M. Liu, X. Sun, and R. Singal. 2004. Transcriptional regulation by the estrogen receptor of antioxidative stress enzymes and its functional implications. *Oncogene* 23:2442–2453. doi:10.1038/sj.onc.1207358.

82. Hayashi, A., H. Suzuki, K. Itoh, M. Yamamoto, and Y. Sugiyama. 2003. Transcription factor Nrf2 is required for the constitutive and inducible expression of multidrug resistance-associated protein 1 in mouse embryo fibroblasts. *Biochem. Biophys. Res. Commun.* 310:824–829. S0006291X03018886 (pii).

83. Vollrath, V., A.M. Wielandt, M. Iruretagoyena, and J. Chianale. 2006. Role of Nrf2 in the regulation of the *Mrp2* (*ABCC2*) gene. *Biochem. J.* 395:599–609. doi:10.1042/BJ20051518.

84. Mahaffey, C.M., H. Zhang, A. Rinna, W. Holland, P.C. Mack, and H.J. Forman. 2009. Multidrug-resistant protein-3 gene regulation by the transcription factor Nrf2 in human bronchial epithelial and non-small-cell lung carcinoma. *Free Radic. Biol. Med.* 46:1650–1657. doi:10.1016/j.freeradbiomed.2009.03.023.

85. Aleksunes, L.M., A.L. Slitt, J.M. Maher, L.M. Augustine, M.J. Goedken, J.Y. Chan, N.J. Cherrington, C.D. Klaassen, and J.E. Manautou. 2008. Induction of Mrp3 and Mrp4 transporters during acetaminophen hepatotoxicity is dependent on Nrf2. *Toxicol. Appl. Pharmacol.* 226:74–83. doi:10.1016/j.taap.2007.08.022. S0041-008X(07)00394-8 (pii).

19 Roles of Hydrogen Peroxide in the Regulation of Vascular Tone

Christopher P. Stanley, Ghassan J. Maghzal, and Roland Stocker

CONTENTS

REGULATION OF VASCULAR TONE

Arterial tone describes the balance between the constriction and relaxation imposed upon an artery, a dynamic state that is governed by endocrine, paracrine, and autocrine stimuli. Such control of arterial tone is important in the maintenance of total peripheral resistance and thus the regulation of blood pressure under physiological and pathological conditions. While arterial constriction and control of arterial tone by the endocrine system are important topics [1], the focus of this chapter is the role of hydrogen peroxide (H_2O_2) as a paracrine and autocrine regulator of arterial relaxation/dilation. However, it is fitting to first briefly outline the key effectors in arterial relaxation.

Three major mechanisms have been well characterized as key endothelial cell-derived inducers of arterial relaxation. These are: prostacyclin (or PGI_2) proposed initially by Moncada et al. [2]; nitric oxide ($^\bullet NO$) discovered by Furchgott and Zewadzki [3], Palmer et al. [4], Ignarro et al. [5]; and the endothelium-dependent hyperpolarization that is independent of $^\bullet NO$ and PGI_2 described initially by Feletou and Vanhoutte [6], Chen et al. [7]. Since the initial discovery of arterial relaxation induced by endothelium-derived hyperpolarization, several mediators have been shown to produce this response. While direct electrical coupling between junctional proteins of endothelial and smooth muscle cells (termed endothelial derived hyperpolarization [EDH]) has been suggested as a regulator of smooth muscle tone, there is also evidence for the presence of a diffusible, endothelium-derived factor capable of hyperpolarizing smooth muscle cells (termed endothelial derived hyperpolarizing factor [EDHF]). Key mechanisms of EDH and EDHF are described in Table 19.1. While there is evidence for the existence of both these phenomena, the relative contributions of each in the control of vascular tone are still unclear and, therefore, this review will group EDH and EDHF (EDH/EDHF) as one entity, except where references have specifically defined EDH or EDHF.

The relative contribution of $^\bullet NO$, PGI_2, and EDH/EDHF to the regulation of arterial tone varies between species, vascular beds, and experimental setting. However, it is often observed that EDH/EDHF responses are more prominent in resistance vessels [8]. It has long been known that H_2O_2 causes arterial relaxation, with more recent reports proposing H_2O_2 as an EDH/EDHF. Therefore, the aim of this chapter is to describe the potential role of H_2O_2 in the regulation of arterial tone and to evaluate its arterial effects, origins, mechanisms of action, and its regulation within the arterial tissue and cells.

H_2O_2-INDUCED ARTERIAL RELAXATION

This section firstly describes the key features of relaxant responses to exogenous H_2O_2, and secondly summarizes the evidence in support of stimuli-induced formation of H_2O_2 as an effector molecule in arterial relaxation. Details on the potential sources of arterial H_2O_2 and the mechanisms of H_2O_2-induced relaxation are described in sections "Sources of H_2O_2 in the Vasculature" and "Targets of H_2O_2 in the Modulation of Arterial Tone," respectively.

TABLE 19.1

Key Modulators of Arterial Relaxation

Pathway	Early Pathway Events	Downstream Signaling	References
Arachidonic acid (COX metabolism)	↑Prostacyclin and stimulation of IP receptor	↑AC, cAMP, PKA, VSMC hyperpolarization, and ↓VSMC $[Ca^{2+}]_I$	[2,9–11]
	↑PGE_2 and stimulation of EP_2 and EP_4 receptors	↑cAMP, PKA, VSMC hyperpolarization, and ↓VSMC $[Ca^{2+}]_I$	[12,13]
Nitric oxide	Conversion of L-arginine to citrulline and ˙NO and activation of sGC	↑cGMP, PKG, VSMC hyperpolarization, ↓VSMC $[Ca^{2+}]_I$, and MLCK Ca^{2+} sensitivity	[3,14–17]
Carbon monoxide	Activation of sGC	↑cGMP, PKG, activation of BK_{Ca} channels VSMC hyperpolarization, ↓VSMC $[Ca^{2+}]_I$	[18–21]
EDH/EDHF	K^+ efflux through endothelial calcium-activated K^+ channels	Activation of VSMC K_{ir} and Na/K^+ ATPase channels, VSMC hyperpolarization ↓VSMC $[Ca^{2+}]_I$	[6,7,22–24]
	Electrical conductance through junctional proteins	Connexins mediate electrical conduction of hyperpolarization along EC and VSMC leading to ↓VSMC $[Ca^{2+}]_I$	[25–28]
	Small molecule movement through myoendothelial gap junctions	IP_3 or Ca^{2+} movement through myoendothelial gap junctions, increase in EC $[Ca^{2+}]_I$, activation of eNOS and/or IK_{Ca} channels, subsequent activation of pathways described here	[29]
	Arachidonic acid (CYP450 metabolism)	Production of EETs: ↑EC $[Ca^{2+}]_i$ through TRP receptors, activation of endothelial K_{ca} channels, and/or EET diffusion to VSMC activating unidentified receptors. ↑VSMC hyperpolarization and ↓VSMC $[Ca^{2+}]_I$	[30–34]
	Arachidonic acid (lipoxygenase metabolism)	Formation of THETAs, HEETAs, and hydroxyeicosatetraenoic acids, ↑VSMC hyperpolarization through K^+ efflux	[35–38]
	H_2S	Formation of sulfide anion, nitrogen hybrid groups, enhanced ˙NO activity, oxidation of PKG1α, ↑VSMC hyperpolarization, and ↓VSMC $[Ca^{2+}]_I$	[39–42]
	Bradykinin	Activation of B_1 and B_2 receptors, ↑EC $[Ca^{2+}]_i$, EC hyperpolarization, formation of EETs, activation of BK_{Ca} channels	[43,44]

Abbreviations: AC, adenylate cyclase; BK_{ca}, big conductance calcium-activated potassium channel; cAMP, cyclic adenosine monophosphate; cGMP, cyclic guanosine monophosphate; EC, endothelial cell; eNOS, endothelium derived nitric oxide synthase; EETs, epoxyeicosatrienoic acids; HEETAs, hydroxyepoxyeicosatrienoic acids; H_2S, hydrogen sulfide; IP_3, inositol 1,4,5-triphosphate; IK_{ca}, intermediate conductance calcium-activated potassium channel; K_{ir}, inward rectifying potassium channel; NO, nitric oxide; PKA, protein kinase A; PKG, protein kinase G; Na^+/K^+ ATPase, sodium potassium exchanger; sGC, soluble guanylyl cyclase; TRP, transient receptor potential channel; THETAs, trihydroxyeicosatrienoic acids; VSMC, vascular smooth muscle cell.

ARTERIAL RELAXATION IN RESPONSE TO EXOGENOUS H₂O₂

Needleman et al. [45] were the first to report an arterial relaxant response to exogenously added H_2O_2. They observed H_2O_2 to cause relaxation of rabbit aortic strips with an EC_{50} of 600 μM. Since then, H_2O_2 added exogenously at micromolar to millimolar concentrations has been shown to cause relaxation of conduit and resistance canine coronary arteries [46], bovine pulmonary [47], rat cerebral [48], rat thoracic aorta [49], mouse mesenteric [50], porcine coronary [51], human submucosal intestinal [52], and human coronary arteries [53,54]. In general, H_2O_2 is more potent in relaxing resistance than conduit arteries [55]. The initial report of Needleman et al. [9] predates the discovery of the importance of the endothelium in the regulation of arterial tone and therefore the role of the endothelium was not tested in that work. Many subsequent reports have demonstrated however that the responses to H_2O_2 are largely endothelium independent. Nevertheless, removal of the endothelium reduces the potency of H_2O_2 responses in canine coronary [46] and basilar artery [56], canine and human submucosal arteries [52], and rabbit aorta [57].

STIMULI THAT INDUCE H₂O₂ FORMATION AND ARTERIAL RELAXATION

A range of stimuli is thought to induce the formation of H_2O_2 that then acts as an EDH/EDHF. Importantly, however, for H_2O_2 to be considered an EDH/EDHF, relaxation must fulfill two criteria: H_2O_2-induced relaxation must be shown to be independent of ·NO and prostaglandin synthesis, and H_2O_2 must be shown to hyperpolarize and relax the arterial smooth muscle. This situation is distinct from H_2O_2 causing arterial relaxation via the release of ·NO and/or prostaglandins (discussed in section "Targets of H_2O_2 in the Modulation of Arterial Tone").

Early studies invoking a role for endogenous H_2O_2 in arterial relaxation showed that bradykinin and sodium arachidonate caused dilation of cat pial arterioles in a catalase-sensitive manner [58]. As catalase does not enter cells [59], these findings suggested that bradykinin and sodium arachidonate elicited a release of H_2O_2 to the extra-cellular space, and that such released H_2O_2 then diffused and crossed cell membranes to reach the smooth muscle cells. Subsequent studies, using patch clamp techniques, showed that H_2O_2 activates vascular smooth muscle ion channels, causing increased open probability of key channels involved in EDH/EDHF responses [60,61]. In the first comprehensive study posing H_2O_2 as an EDH/EDHF, Matoba et al. [50] reported acetylcholine-induced relaxation and smooth muscle cell hyperpolarization in mouse mesenteric arteries to be almost completely inhibited by catalase. This inhibitory effect of catalase was abrogated completely by aminotriazole, implying that catalytic removal of H_2O_2 was required for catalase to prevent arterial relaxation and smooth muscle cell hyperpolarization. Matoba et al. [50] also showed similar membrane potential profiles for acetylcholine (ACh) and H_2O_2. Specifically, ACh induced changes in membrane potential of around -7 and -2 mV in the absence and presence of catalase, respectively. Similarly, exogenous H_2O_2 caused hyperpolarization of -5 mV. Using similar techniques, H_2O_2 was proposed as an EDH/EDHF in bradykinin-stimulated human mesenteric arteries [62] and porcine coronary arterioles [51].

Using isolated porcine coronary arteries stimulated with bradykinin, Matoba and colleagues [51] provided indirect evidence for the release of H_2O_2 from the endothelium. The H_2O_2 in the medium was determined by electron paramagnetic resonance (EPR) spectroscopy. The assay is based on the formation of para-acetamidophenoxyl radical by compound I derived from a reaction of H_2O_2 with horseradish peroxidase (HRP), and the resulting phenoxyl radical then being trapped by a nitroxide spin trap (1-hydroxy-2,2,5,5,-tetramethyl-3-imidazoline-3-oxide) resulting in an EPR-active nitroxide radical species [63]. These experiments showed bradykinin to evoke an increase in EPR signal that was abolished almost completely when catalase was present at the time of bradykinin addition, or when the endothelium was removed from the arteries prior to bradykinin addition [51]. Catalase also inhibited flow-induced dilation of cannulated human coronary resistance arteries isolated from patients with coronary artery disease [64]. Under these experimental conditions, H_2O_2 release from the arteries was suggested, based on EPR studies using 5-*tert*-butoxycarbonyl 5-methyl-1-pyrroline *N*-oxide (BMPO) as a spin trap [65]. Specifically, flow induced the appearance of a BMPO-OH signal in the collected arterial perfusate, whereas there was no detectable EPR signal under static conditions. In addition to BMPO-OH, flow was also associated with the appearance of ubisemiquinone radical, and both EPR signals were attenuated partially in the presence of superoxide dismutase (SOD), catalase, or rotenone [65]. The authors concluded that the flow-mediated BMPO-OH signal was in part due to H_2O_2 derived from superoxide radical anion ($O_2^{\cdot-}$) formed from the autoxidation of mitochondrial ubisemiquinone radical [65]. This interpretation is problematic, as it is unclear why the appearance of the ubisemiquinone radical in the perfusate required the addition of BMPO, and how mitochondria containing the ubisemiquinone radical can arise in the perfusate of the cannulated arteries in the reported absence of flow-mediated endothelial denudation [65]. The role of flow-induced production of H_2O_2 and subsequent dilation has been investigated further using a bioassay in which an upstream artery (endothelium intact arteriole) acting as effluent donor was connected to an effluent detector (endothelium denuded arteriole) via luminal perfusion [53]. In this assay, dilation of the "detector artery" via flow from the "donor artery" was inhibited by catalase or insertion of a catalase-containing column between the donor and effector artery.

While there is a strong argument to be made for the role of H_2O_2 as an arterial relaxant and for stimulus-induced H_2O_2 as an EDH/EDHF, there are some concerns that deserve consideration in the interpretation of present results. First, much of the evidence supporting the role of H_2O_2 in stimulus-induced relaxation relies on the use of catalase or PEG-catalase. A potential pitfall with this approach is that at least some commercial catalase preparations contain impurities that have been reported to also inhibit arterial relaxation mediated by epoxyeicosatrienoic acids (EETs) [66]. Therefore, adequate controls for relaxant pathways in addition to that induced by H_2O_2 are required to ensure specificity of the catalase effect. Second, by removing extracellular H_2O_2, exogenously added catalase is assumed to create a H_2O_2 concentration gradient, thereby decreasing intracellular H_2O_2 [67]. In this case, however, catalase might also be expected to affect constrictor responses to contractile agents if H_2O_2 plays a prominent role in the regulation of basal tone. To date, however, such an effect of catalase has not been reported. Third, much of the evidence implicating

H_2O_2 as an EDHF relies on the use of indirect, EPR-based methods (see Reference [29]) or fluorescence readouts using probes such as 2′,7′-dichlorodihydrofluorescein diacetate or hydroethidine that are not specific and may lead to erroneous conclusions (for review see [68,69]). Therefore, while current evidence suggests that, upon stimulation of arterial segments, H_2O_2 is formed/released to induce relaxation, more specific and accurate quantification of H_2O_2 remains elusive. However, such specific and accurate methods to quantify H_2O_2 are required to conclusively show, or refute, H_2O_2 as an endogenous arterial signaling molecule.

SOURCES OF H_2O_2 IN THE VASCULATURE

The main source of H_2O_2 in the vasculature is thought to be $O_2^{\cdot-}$ that dismutates spontaneously or is catalyzed by SOD to H_2O_2. Superoxide can be formed by various enzymes in endothelial and smooth muscle cells, as well as adventitial fibroblasts. These enzymatic sources of $O_2^{\cdot-}$ include different isoforms of NAPDH oxidases (NOX), uncoupled endothelial nitric oxide synthase (eNOS), xanthine oxidoreductase, and mitochondrial respiration complexes. In addition, enzymes involved in the metabolism of arachidonic acid, that is, lipoxygenases, cyclooxygenases, and cytochrome P450, have also been described to generate $O_2^{\cdot-}$. Most studies addressing the source of $O_2^{\cdot-}$ and H_2O_2 in the vasculature are based on *in vitro* studies using isolated cells, with very few studies using intact arterial segments to determine the precise location and nature of these enzymatic systems involved.

NADPH OXIDASES

NADPH oxidases are the only enzymes whose primary function is to produce $O_2^{\cdot-}$/ H_2O_2. They constitute a family of seven transmembrane proteins that catalyze the transfer of electrons from NADPH across membranes to molecular oxygen resulting in formation of $O_2^{\cdot-}$ [70]. They have varied tissue distribution, with only NOX 1, 2, 4, and 5 found in the vasculature. The subcellular localization of the NOX isoforms is distinctive, suggesting varied functions dependent on the cell type. For example, NOX1 localizes to the caveolae on the cell surface of rat and human smooth muscle cells, while NOX4 is located primarily in the focal adhesions of such cells [71]. In endothelial cells, NOX4 predominately localizes to the endoplasmic reticulum, while NOX2 associates with membrane ruffles and the cytoskeleton [72], perinuclear/nuclear membranes, and the endoplasmic reticulum [73,74]. Based on mRNA expression, NOX4 is the most abundant isoform in endothelial cells, with markedly higher levels of its mRNA compared with NOX2 (20 to 5000-fold) and NOX1 (>300-fold) [75–77].

While most NOX enzymes form $O_2^{\cdot-}$, NOX4 has been reported to primarily form H_2O_2, based on peroxidase-dependent assays such as Amplex® Red and luminol/ HRP to detect H_2O_2 [78]. However, a recent study using HPLC with fluorescence detection of 2-hydroxyethidium as a specific assay for $O_2^{\cdot-}$, reported NOX4 in endothelial cells to generate $O_2^{\cdot-}$, especially under shear stress conditions [79], where it may be involved in the activation of eNOS and hence control of vascular tone. In bovine pulmonary arteries, where H_2O_2 is proposed to regulate basal tone, siRNA silencing of NOX4, but not NOX2, has been shown to enhance basal and constrictor tone [80]. In human coronary arteries, apocynin and the NOX peptide inhibitor gp91ds-tat have been reported to inhibit bradykinin-induced stimulation of H_2O_2 and subsequent relaxation [81]. These studies suggest a role for NOX-derived $O_2^{\cdot-}/H_2O_2$ in arterial relaxation, although it is important to point out that apocynin interferes with peroxidase-dependent assays for $O_2^{\cdot-}/H_2O_2$ rather than being a NOX inhibitor.

All components of NOX2, also known as the phagocytic NADPH oxidase, are found in endothelial cells [82]. However, unlike the phagocytic form that is stimulated to assemble on the plasma membrane and produce bursts of large quantities of $O_2^{\cdot-}$, the components of vascular NOX2 are constitutively expressed intracellularly [74], resulting in constant generation of $O_2^{\cdot-}$ [82]. This characteristic of endothelial NOX2 has been proposed to be crucial for compartmentalized redox signaling, by physically separating formation of $O_2^{\cdot-}$ from that of $^\cdot$NO near the plasma and Golgi membranes, thereby maintaining $^\cdot$NO bioavailability [83].

Compared with the ubiquitous expression of NOX2 and 4 in the vasculature, reports of NOX1 and 5 in vascular cells suggest that they are present at relatively lower amounts. While present in smooth muscle cells [71,84], the occurrence of low levels of NOX1 has been reported in cultured human [76] and rat aortic endothelial cells. NOX5, a Ca^{2+}-dependent enzyme, has been detected in human endothelial [85] and smooth muscle cells [86], predominantly on the plasma membrane and in the ER.

Uncoupled eNOS

Endothelial nitric oxide synthase (eNOS) is a heme- and flavin-containing enzyme that reduces oxygen and incorporates it into L-arginine to form L-citrulline and $^\cdot$NO. In its active form, the enzyme is present as a homodimer and utilizes tetrahydrobiopterin (BH_4) as a cofactor. However, under certain conditions, including lack of BH_4 [87] or the substrate L-arginine, as well during pathological conditions such as vascular disease and associated oxidative stress [88], the eNOS dimer dissociates to monomers. Such "uncoupled" eNOS forms $O_2^{\cdot-}$ instead of $^\cdot$NO. Formation of $O_2^{\cdot-}$ by eNOS has also been suggested to occur under physiological conditions in the context of host defense as treatment of allografts with the BH_4 precursor sepiapterin decreased $O_2^{\cdot-}$ (assessed by lucigenin-enhanced chemiluminescence) and increased $^\cdot$NO in allografts [89]. Interestingly, pharmacological inhibition of eNOS using L-N^G-nitroarginine methyl ester (L-NAME) has also been shown to uncouple the enzyme leading to the formation of H_2O_2 as measured by the H_2O_2-specific molecular probe HyPer [90].

Xanthine Oxidoreductase

Another potential source of $O_2^{\cdot-}$ and H_2O_2 in the vasculature is xanthine oxidoreductase that contains flavins, iron-sulfur clusters, and molybdenum. The enzyme catalyzes the last steps of purine metabolism, converting xanthine or hypoxanthine to uric acid and forming $O_2^{\cdot-}$ and H_2O_2 as by-products [91]. Xanthine oxidoreductase is derived from xanthine dehydrogenase by thiol oxidation and/or proteolytic cleavage, usually under pathological situations such as vascular disease including atherosclerosis [92]. In fact, xanthine oxidase activity is increased in diseased human coronary arteries [93]. Xanthine oxidase avidly binds to the endothelium via interaction with sulfated glycosaminoglycans, suggesting that vascular wall xanthine oxidase is derived from circulating enzyme [93,94].

MITOCHONDRIA

Mitochondria are a well-known source of cellular $O_2^{\bullet-}$ that is formed at multiple sites in the mitochondrial electron transport chain, including the ubiquinone-binding sites in complexes I and III, glycerol 3-phosphate dehydrogenase, flavin in complex I, the flavoprotein Q oxidoreductase involved in fatty acid β-oxidation, and pyruvate and 2-oxoglutarate dehydrogenases [95]. These sites utilize electrons from NADH or reduced flavin to generate predominantly $O_2^{\bullet-}$ into the mitochondrial matrix, where MnSOD converts it to H_2O_2. Some sites, such as complex III, are also able to release $O_2^{\bullet-}$ into both the mitochondrial matrix and intermembrane space [95]. While mitochondria generate $O_2^{\bullet-}$ at a steady rate under normal physiological conditions, there is also evidence for enhanced generation of $O_2^{\bullet-}$ following exposure to certain stimuli or altered cellular states. For example, in coronary arteries from patients with coronary artery disease, inhibitors of mitochondrial complex I (rotenone) and III (myxothiazol) have been implicated in H_2O_2 inducing arterial dilation [65]. It has also been shown that angiotensin II increases mitochondrial H_2O_2 in aortic endothelial cells [96]. Similarly, CD40 treatment of human endothelial cells has been reported to increase cellular $O_2^{\bullet-}$, suggested to be of mitochondrial origin [97].

TARGETS OF H_2O_2 IN THE MODULATION OF ARTERIAL TONE

ENDOTHELIAL CELL $\bullet$NO

Relaxation in response to exogenously applied H_2O_2 has been reported to be partly dependent on the endothelium and $\bullet$NO in the rabbit aorta [57,98,99], canine basilar artery [56], rat aorta [56,100], and rat gracilis [101]. These studies suggest $\bullet$NO to be a potential mediator of H_2O_2-induced arterial relaxation. Indeed, several *in vitro* studies using endothelial cells derived from conduit arteries support an interaction between H_2O_2 and the proteins involved in $\bullet$NO formation either directly (NOS) or indirectly (via signaling). For example, acute exposure to ~100 μM H_2O_2 increases eNOS activity via phosphorylation/dephosphorylation mediated by Akt and phosphoinositide 3-kinase (PI3K) [102] and MEK/ERK1/2 [103]. Hydrogen peroxide also increases eNOS expression through calmodulin kinase II and JAK2 signaling [103,104], and increases the concentrations of the eNOS cofactor BH_4 through JAK/STAT signaling [105,106]. The initiation of such H_2O_2 signaling remains unclear. For instance, kinases such as Akt, PI3K, MEK and JAK2 have been proposed to be under direct control of H_2O_2 (reviewed in [107]), whereas [108] observed H_2O_2-mediated activation of the JNK pathway involves both Src family kinases and transactivation of the epidermal growth factor receptor.

Another potential unifying feature is that several of these findings are dependent on Ca^{2+} [102–104]. Indeed, endothelial $^{\bullet}NO$ production and $^{\bullet}NO$-mediated arterial relaxation in response to H_2O_2 are decreased by removal of extracellular and intracellular Ca^{2+} [56]. Additionally, H_2O_2 (0.1–10 mM) was shown to increase endothelial Ca^{2+} in a manner that was sensitive to intracellular store depletion [109], inhibition of transient receptor potential (TRP) channel [110], and inhibition of phospholipase C and inositol triphosphate receptor (IP_3R) [111]. However, the mechanisms underlying these effects remain unclear. Interestingly, H_2O_2 (100 μM) also increases IP_3R sensitivity to IP_3-induced Ca^{2+} release and Ca^{2+}-induced Ca^{2+} release via glutathionylation of IP_3R_1 [112]. Indeed, oxidation of certain cysteine residues of all IP_3R subtypes has been shown to increase receptor sensitivity [113]. Taken together, these studies suggest that H_2O_2 is able to stimulate and facilitate $^{\bullet}NO$ signaling in isolated endothelial cells, possibly via redox modulation of protein kinases and alteration in intracellular Ca^{2+}. Importantly, however, the relevance of these cellular studies for arterial relaxation remains questionable. Thus, the relaxation of mouse aorta and mesenteric arteries by H_2O_2 was reported to be endothelium independent and hence unlikely to involve eNOS/$^{\bullet}NO$ [114]. Similarly, pre-exposure of conduit arteries to H_2O_2 limits rather than enhances ACh-induced $^{\bullet}NO$ release and relaxation [99].

ENDOTHELIAL EDH/EDHF

In nearly all resistance vessels, arterial relaxation in response to H_2O_2 appears to be independent of the endothelium, with the possible exception of human submucosal intestinal arterioles [52]. Despite this, however, endogenously formed H_2O_2 could conceivably interact with endothelial cell proteins in a manner that potentially enhances EDH/EDHF type responses (see Figure 19.1).

In rabbit aortic valve endothelial cells, H_2O_2 (10 μM) enhanced Ca^{2+} release from the endoplasmic reticulum induced by cyclopiazonic acid, an inhibitor of the sarcoplasmic/endoplasmic reticulum Ca^{2+}-ATPase [115]. This enhanced Ca^{2+} release was suppressed by cell-permeable glutathione monoethylester, suggesting redox modification of IP_3R via glutathionylation (see Endothelial Cell $^{\bullet}NO$ section). Irrespective of the precise mechanisms, an enhancement of Ca^{2+} release by H_2O_2 could conceivably enhance EDH/EDHF responses. Work from the same group also showed that pre-exposure to H_2O_2 (10 μM) enhanced endothelium-dependent arterial relaxation in response to a range of agents, including cyclopiazonic acid. In this situation, activation of SK_{ca} and IK_{ca} channels, and connexin 43-mediated gap junction communication partially explained arterial relaxation [115,116]. Similarly, in isolated endothelial tubes, H_2O_2 (200 μM) induced hyperpolarization in a manner that was inhibited by pharmacological inhibitors of SK_{ca} and IK_{ca} channels [117]. The same group later proposed SK_{ca} and IK_{ca} activation to be mediated by an increase in cellular Ca^{2+} via redox regulation of TRP channels [118]. Despite both studies showing H_2O_2 to propagate endothelial hyperpolarization, the underlying redox reactions remain to be firmly established, just as it remains to be determined whether the hyperpolarization observed extends to vascular smooth muscle. Hydrogen peroxide has been shown to increase IP_3R-mediated Ca^{2+} release [112] and TRP channels have been shown to be redox active [119–121]. A further possibility is that the "open probability" of gap junctions increases upon

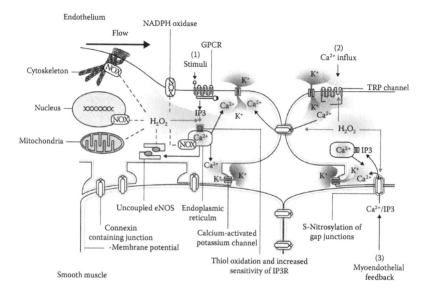

FIGURE 19.1 Mechanisms by which H_2O_2 has been proposed to potentiate EDH/EDHF responses in arterial endothelium. Initial components of the EDH/EDHF response involve increases in endothelial Ca^{2+}. This occurs either through receptor stimulation and subsequent release of Ca^{2+} from internal stores (1), influx of external Ca^{2+} (2), or myoendothelial feedback (3). Increased endothelial Ca^{2+} can then stimulate Ca^{2+}-activated K^+ channels causing an efflux of K^+, which results in membrane hyperpolarization. This hyperpolarizing response travels between endothelial cells via connexin-containing junctions or via activation of further ion channels and pumps. Once the vascular smooth muscle becomes hyperpolarized, voltage-gated Ca^{2+}-channels close and this ultimately decreases arterial tone. Upon release from any of its potential sources (denoted by dotted lines), H_2O_2 may increase IP_3-mediated release of Ca^{2+} from the endoplasmic reticulum and calcium entry through TRP channels (denoted in red). H_2O_2 may also facilitate myoendothelial feedback by aiding passage of IP_3 and/or Ca^{2+} through connexin-containing junctions. Such influx of Ca^{2+} would then counteract arterial constriction through stimulating arterial relaxant pathways.

redox modification, such as *S*-nitrosylation [122]. The H_2O_2-induced opening of gap junctions may then aid in potentiating hyperpolarization by allowing passage of IP_3 or Ca^{2+} from the smooth muscle to the endothelium to stimulate endoplasmic reticulum or Ca^{2+}-activated potassium channels, ultimately potentiating EDH/EDHF type responses [123]. However, such spread of hyperpolarization would be expected to be slower than the rates of conduction via "electrotonic spread."

Additional roles for H_2O_2 in the endothelium may involve epoxyeicosatrienoic acids (EETs). Hydrogen peroxide and EETs (also proposed as EDHFs, see Table 19.1) share a complex relationship. Two separate studies from the Gutterman laboratory showed flow-mediated dilation to be attenuated by catalase and inhibitors of cytochrome P450 [53,124]. This observation is unlikely explained by H_2O_2 stimulating EET production, as H_2O_2 decreases cytochrome P450-mediated metabolism of arachidonic acid to EETs [125]. Rather it was speculated that flow/shear stress-mediated EET production precedes formation of H_2O_2 [126]. Accordingly, shear stress increases phospholipase A_2 activity, with the arachidonic acid released being metabolized to EETs by cytochrome P450. EETs then stimulate Ca^{2+} influx through TRP channels that in turn enhances mitochondrial production of reactive species, such as $O_2^{\bullet-}$, which is readily converted to H_2O_2 (reviewed [127]). Indeed, flow-induced relaxation of cannulated coronary resistance arteries isolated from patients with coronary artery disease is attenuated by pharmacological inhibition of mitochondrial complex I and III [65]. Similarly, using endothelium intact "donor" and endothelium-denuded "detector" human coronary arteries from coronary artery disease patients, bradykinin induced relaxation that is attenuated by inhibitors of CYP450 and the EET receptor only if catalase was present in the donor artery chamber [125]. Follow-up studies from the same group confirmed a role for H_2O_2 suppression of arachidonic acid metabolism to EETs [81].

SOLUBLE GUANYLYL CYCLASE IN SMOOTH MUSCLE CELLS

As indicated, arterial relaxation in response to exogenous H_2O_2 has been shown to be independent of the endothelium in the range of vascular beds including bovine pulmonary and coronary arteries [47], human coronary arteries [54], mouse mesenteric arteries [50], and human mesenteric arteries [62]. Early mechanistic studies focused on the role of soluble guanylyl cyclase (sGC), as H_2O_2 was a candidate endogenous activator of sGC prior to the discovery of $^\bullet$NO [128]. A substantive body of works by Wolin and coworkers showed that H_2O_2-mediated relaxation of pulmonary arteries was inhibited by methylene blue, a nonspecific inhibitor of sGC used at that time [47]. Using sGC isolated from bovine lung homogenates, bolus H_2O_2 or H_2O_2 formed continuously by glucose/glucose oxidase, increased cGMP formation [47,129]. Surprisingly, catalase enhanced this increase in cGMP and this effect was inhibited by aminotriazole, suggesting that catalase compound I was responsible for sGC activation [47,129]. Furthermore, short-term pretreatment of pulmonary arteries with the sGC inhibitor ODQ (1H-[1,2,4]oxadiazolo[4,3-a]quinoxalin-1-one) inhibited $^\bullet$NO donor-mediated relaxation, but had no effect on relaxation induced by H_2O_2 [130,131]. By contrast, decreasing sGC protein (via prolonged exposure to ODQ or siRNA directed against sGC β1 subunit) attenuated H_2O_2-mediated relaxation. These observations have been interpreted as H_2O_2 continuously regulating arterial tone so that conditions associated with decreased H_2O_2 lead to arterial constriction [129–131] (see Figure 19.2).

PROTEIN KINASE G1α IN SMOOTH MUSCLE CELLS

H_2O_2-induced arterial relaxation has also been proposed to occur via oxidative activation of protein kinase G1α (PKG1α), independent of the endothelium, sGC and cGMP [132]. Oxidative activation of PKG1α results in the formation of an interprotein disulfide bond and homodimer, involving cysteine residue 42 in the leucine zipper domain [132]. In isolated murine hearts, 100 μM H_2O_2 was shown to dimerize PKG1α with a half maximal time ($t_{1/2}$) of 1 min, without reversal at 4 min [132]. Interestingly, kinetic studies with purified protein revealed that H_2O_2-oxidized PKG1α had lower K_m and V_{max}-values for substrate compared with cGMP-activated PKG1α [132].

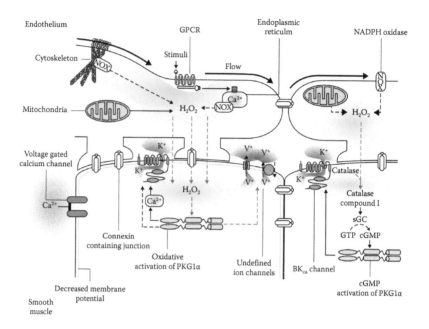

FIGURE 19.2 Mechanisms by which H_2O_2 is proposed to induce relaxation of vascular smooth muscle. *H_2O_2-induced arterial relaxation via oxidative activation of PKG1α* (left section). H_2O_2 formed in the endothelium diffuses to the smooth muscle via an unidentified pathway. Once in the smooth muscle, H_2O_2 interacts with cysteine 42 located in the leucine zipper region of PKG1α. The subsequent signaling of PKG1α is as yet unidentified; however, initial reports suggest interaction with the large conductance Ca^{2+}-activated K^+ channel (BK_{ca}) either directly or through Ca^{2+} sparks. Oxidized PKG1α may also activate a range of as yet undetermined K^+ channels. *H_2O_2-induced arterial relaxation via compound I formation and activation of sGC* (right section). H_2O_2 is proposed to form compound I upon reaction with catalase. Once formed, compound I stimulates sGC leading to activation of PKG1α via increased cGMP, which can then signal through BK_{ca} channels. Activation of BK_{ca} channels in both cases would then induce hyperpolarization and subsequent closure of voltage-gated Ca^{2+}-channel.

Support for a physiological role of H_2O_2-induced oxidative activation of PKG1α was obtained using a genetically modified, "redox dead" mouse in which cysteine 42 within the leucine zipper domain of PKG1α was mutated to serine [55]. Experiments with arteries isolated from redox dead versus wild type mice showed PKG1α cysteine 42 to be required for H_2O_2-induced relaxation of resistance and conduit vessels, ACh-induced relaxation of mesenteric arteries, and hyperpolarizing responses to H_2O_2 [55]. Using this mouse model, oxidative activation of PKG1α was shown to contribute to the regulation of physiological blood pressure [55], decreased myogenic tone in mouse mesenteric arteries [133], and contributed to hypotension in mouse models of sepsis [134]. A role for H_2O_2-induced oxidative PKG1α activation has also been implicated in mediating flow-induced dilation of human coronary arteries from patients with coronary artery disease [54], the regulation of tonic contractile tone in bovine pulmonary arteries, and hypoxia-induced arterial relaxation of bovine coronary arteries [131]. At present, the best indicator of oxidative activation PKG1α is formation of PKG1α homodimer [135], although known pharmacological inhibitors of cGMP-activated PKG (e.g., Rp-8-Br-PET-cGMP, DT-2) also inhibit H_2O_2-oxidized PKG1α [54,55,132,133,136], and are therefore potentially useful tools to assess the roles of "oxidatively" activated PKG1α.

Oxidative activation of PKG1α may also help explain the activation of several channels by H_2O_2. Thus, numerous studies have reported pharmacological inhibition of channels to block H_2O_2-mediated arterial relaxation and/or hyperpolarization that is otherwise attenuated in the "redox-dead PKG1α knock-in mouse." For example, there is convincing pharmacological and electrophysiological evidence for oxidized PKG1α causing hyperpolarization and arterial relaxation via activation of big conductance calcium-activated K^+ channels (BK_{ca}) [54,55,131,133,136]. Other K^+ channels have been implicated in H_2O_2-induced hyperpolarization of vascular smooth muscle [50,51,60,62,64,137]. These additional K^+ channels include K_v, K_{ATP}, and the sodium potassium exchanger, as assessed by pharmacological studies employing specific inhibitors [137–140]. While an involvement of oxidized PKG1α in the H_2O_2-induced activation of these channels is yet to be shown, patch clamp experiments employing either intact cells or cell membrane excisions indicate that H_2O_2 mediates channel activation indirectly via intracellular signaling [54,60,61,141–147] (see Figure 19.2).

REGULATION OF VASCULAR H_2O_2 CONCENTRATIONS AND H_2O_2-MEDIATED ARTERIAL RELAXATION

In principal, there are three ways to regulate arterial wall concentrations of H_2O_2 and thus its ability to regulate vascular tone. The first involves inhibition of H_2O_2 formation, by suppressing the sources of $O_2^{\cdot-}/H_2O_2$ (discussed in section "Sources of H_2O_2 in the Vasculature") including SOD. Such inhibition may occur at the level of "activators" of these enzymes. For example, an abundance of substrate (arginine) and cofactor (BH_4) will help maintain eNOS in the coupled, dimeric state, and thereby facilitate formation of $\cdot NO$ instead of $O_2^{\cdot-}$. Inhibiting the dismutation of $O_2^{\cdot-}$ to H_2O_2 can decrease endothelium-dependent hyperpolarization, as EDHF-mediated relaxation of mesenteric arteries is decreased but not eliminated in Cu,Zn-SOD$^{-/-}$ compared with wild type mice, and catalase has no further effect on the attenuated relaxation or hyperpolarization in these Cu,Zn-SOD$^{-/-}$ mice [148]. These results suggest that cytosolic $O_2^{\cdot-}$ in arterial cells is partially responsible for catalase-accessible and hence extracellular H_2O_2, which then causes relaxation and hyperpolarization.

The second mechanism decreasing H_2O_2-mediated relaxation relates to the scavenging/detoxification of arterial H_2O_2. The major H_2O_2-detoxification enzymes are peroxiredoxins, glutathione peroxidases, and catalase, all of which react rapidly with H_2O_2.

PEROXIREDOXINS

The major cellular detoxifiers of H_2O_2 are the cysteine peroxidases, peroxiredoxins (Prx). These enzymes show high reactivity for H_2O_2 with a rate constant in the 10^7–10^8 M^{-1} s^{-1} range, similar to catalase and glutathione peroxidases [149]. However, Prx are more abundant than glutathione peroxidases and catalase, and they are present in different cell compartments [150]. There are six different Prx isoforms in mammals with many found in vascular cells in mice [151], rats [152], and humans [153]. Peroxiredoxins obtain their reducing equivalents from the NADPH/ thioredoxin reductase/thioredoxin system that regenerates the catalytically active cysteine residue in Prx [154]. The importance of this redox system in H_2O_2-mediated relaxation has been highlighted in a recent study, which showed mesenteric arteries from thioredoxin transgenic mice to have increased Prx activity and higher EDH relaxation compared with arteries from control mice [155].

The fact that increased Prx activity resulting from the thioredoxin transgene increases rather than decreases EDH relaxation [155] suggests that Prx may not always act as an "H_2O_2 detoxifier." Indeed, it is increasingly recognized that Prx are critical for H_2O_2 to act as a cellular signaling molecule, including in the vasculature where Prx2 has been shown to be involved in platelet-derived growth factor signaling [156] and VEGF receptor oxidation [157]. This is because Prx can interact with a number of different proteins [158], thereby providing the basis for potential "redox relays" that transduce the H_2O_2 "message" through sequential thiol oxidation [159]. This is highlighted by a recent report demonstrating Prx2 with oxidized catalytic cysteine residue to transfer oxidative equivalents to the redox-controlled transcription factor STAT3 [160], generating disulfide-linked STAT3 oligomers with attenuated transcriptional activity. The finding of [155], described here, suggests that Prx also participates in redox relays in the arterial wall in the context of H_2O_2 acting as an EDH, an area of research worth future considerations.

Glutathione Peroxidases (GPx)

GPxs are selenocysteine-containing proteins that act as antioxidants by using glutathione as a cofactor to reduce H_2O_2 and a wide range of organic hydroperoxides to water and alcohols, respectively. There are eight isozymes in humans, with GPx1 being the most abundant. The importance of these enzymes in detoxifying H_2O_2 *in vivo* has been shown in animals, as GPx1 gene knockout mice exhibit increased susceptibility to oxidative stress-inducing agents including H_2O_2 [161]. Conversely, induction of GPx1 in human endothelial cells affords protection against H_2O_2-induced damage [162]. Evidence supporting a role for GPx in regulating H_2O_2-mediated signaling comes from data showing that abolition of GPx1 in mice enhances oxidative inactivation of phosphatase and tensin homolog, and thereby increases PI3K/Akt signaling [163], while overexpression of GPx1 in mice diminishes growth factor or Akt signaling [164]. Therefore, one would expect that in the vasculature too, a decrease in GPx would enhance H_2O_2-mediated relaxation while overexpression of GPx would be expected to have the opposite effect. Contrary to this expectation, however, mesenteric arterioles from GPx1$^{-/-}$ mice contracted in response to bradykinin [165], while transgenic GPx1 expression improved vascular reactivity as assessed by *in vivo* videomicroscopy [166]. The reasons for this apparent discrepancy are currently not known.

CATALASE

Unlike Prx or Gpx that rely on a reactive cysteine or seleno-cysteine, respectively, catalase is a heme-containing enzyme that converts H_2O_2 to water and oxygen. Catalase is predominantly found in peroxisomes where it detoxifies H_2O_2 formed as a by-product of the catabolism of fatty acids and D-amino acids [167].

Global overexpression of catalase in mice does not have a profound effect on arterial H_2O_2 concentration or blood pressure [168]. However, targeting catalase expression to vascular tissues in healthy mice strongly reduced steady-state concentration of vascular H_2O_2 and surprisingly caused hypotension, independent of the endothelial 'NO bioavailability and completely reversible by treatment with the catalase inhibitor aminotriazole [169]. Overall, these findings appear to argue against *endogenous* H_2O_2 acting as an EDH/EDHF in healthy mice and suggest that it might act as a vasoconstrictor rather than relaxant, since scavenging of H_2O_2 by catalase would be expected to increase blood pressure.

DISRUPTING H_2O_2 SIGNALING DOWNSTREAM OF H_2O_2

The final mechanism of modulation of H_2O_2-mediated arterial relaxation entails the disruption of its signaling and downstream pathways that lead to hyperpolarization of the smooth muscle and thus relaxation. One such inhibitor is 'NO, which is known to negatively affect EDH/EDHF. Nearly 20 years ago, 'NO donors were shown to attenuate EDH-mediated responses in isolated rabbit carotid and porcine coronary arteries [170] and in the canine coronary microvasculature [171]. While the mechanism of this is not fully understood, one possible pathway is cGMP-dependent activation of PKG1α outcompeting H_2O_2-mediated oxidative activation of PKG1α [49,132]. Indeed, cGMP-dependent activation of PKG1α has been shown to desensitize vascular smooth muscle cells to H_2O_2-mediated relaxation [172]. Conversely, pharmacological inhibition of sGC sensitizes conduit vessels to H_2O_2-mediated relaxation [172]. While these studies are contrary to research suggesting that H_2O_2 increases endothelial 'NO (see section "Targets of H_2O_2 in the Modulation of Arterial Tone"), they highlight the issue of species and vascular bed differences. Nonetheless, the findings from the Eaton group (Burgoyne et al. [132], Prysyazhna et al. [55], Burgoyne et al. [172]) suggest that at least in the mouse excessive endothelium-derived 'NO inhibits the actions of H_2O_2 as an EDHF.

CONCLUSIONS AND PERSPECTIVES

Overall, existing literature strongly supports a role for exogenously added H_2O_2 to act as an arterial relaxant through hyperpolarization of arterial smooth muscle. There is also substantial evidence for endogenous H_2O_2 to act in a similar way, although this evidence relied on indirect methods for the detection of the H_2O_2 putatively involved. The major underlying problem with this is the inherent difficulty in the unambiguous determination of H_2O_2 in complex biological systems. Additional indirect evidence for a role of H_2O_2 as EDH/EDHF comes from studies reporting oxidative modifications of downstream targets of H_2O_2, such as PKG1α. The potential problem with this interpretation is, however, that oxidative modification of these targets is not specific to H_2O_2, and that H_2O_2 itself has only a limited capacity to directly oxidize protein thiols [150]. Future studies should focus on the role of *endogenous* H_2O_2 in arterial relaxation, with the challenge being the direct detection of this oxidant that *in vivo* is rapidly metabolized by numerous systems and hence likely present at only low concentrations. Consideration may also be given to the possibility that the role of H_2O_2 as an arterial relaxant might differ between healthy and diseased states.

REFERENCES

1. Levick, J. R. 2010. *An Introduction to Cardiovascular Physiology*. CRC Press, Boca Raton, FL.
2. Moncada, S., R. Gryglewski, S. Bunting, and J. R. Vane. 1976. An enzyme isolated from arteries transforms prostaglandin endoperoxides to an unstable substance that inhibits platelet aggregation. *Nature* 263 (5579):663–665.
3. Furchgott, R. F. and J. V. Zawadzki. 1980. The obligatory role of endothelial cells in the relaxation of arterial smooth muscle by acetylcholine. *Nature* 288 (5789):373–376.
4. Palmer, R. M., A. G. Ferrige, and S. Moncada. 1987. Nitric oxide release accounts for the biological activity of endothelium-derived relaxing factor. *Nature* 327 (6122):524–526.
5. Ignarro, L. J., R. E. Byrns, G. M. Buga, and K. S. Wood. 1987. Endothelium-derived relaxing factor from pulmonary artery and vein possesses pharmacologic and chemical properties identical to those of nitric oxide radical. *Circ Res* 61 (6):866–879.
6. Feletou, M. and P. M. Vanhoutte. 1988. Endothelium-dependent hyperpolarization of canine coronary smooth muscle. *Br J Pharmacol* 93 (3):515–524.
7. Chen, G., H. Suzuki, and A. H. Weston. 1988. Acetylcholine releases endothelium-derived hyperpolarizing factor and EDRF from rat blood vessels. *Br J Pharmacol* 95 (4):1165–1174.
8. Shimokawa, H., H. Yasutake, K. Fujii, M. K. Owada, R. Nakaike, Y. Fukumoto, T. Takayanagi et al. 1996. The importance of the hyperpolarizing mechanism increases as the vessel size decreases in endothelium-dependent relaxations in rat mesenteric circulation. *J Cardiovasc Pharmacol* 28 (5):703–711.
9. Ignarro, L. J., R. G. Harbison, K. S. Wood, M. S. Wolin, D. B. McNamara, A. L. Hyman, and P. J. Kadowitz. 1985. Differences in responsiveness of intrapulmonary artery and vein to arachidonic acid: Mechanism of arterial relaxation involves cyclic guanosine $3':5'$-monophosphate and cyclic adenosine $3':5'$-monophosphate. *J Pharmacol Exp Ther* 233 (3):560–569.
10. Corriu, C., M. Feletou, E. Canet, and P. M. Vanhoutte. 1996. Endothelium-derived factors and hyperpolarization of the carotid artery of the guinea-pig. *Br J Pharmacol* 119 (5):959–964.
11. Corriu, C., M. Feletou, G. Edwards, A. H. Weston, and P. M. Vanhoutte. 2001. Differential effects of prostacyclin and iloprost in the isolated carotid artery of the guinea-pig. *Eur J Pharmacol* 426 (1–2):89–94.
12. Purdy, K. E. and W. J. Arendshorst. 2000. EP_1 and EP_4 receptors mediate prostaglandin E_2 actions in the microcirculation of rat kidney. *Am J Physiol Renal Physiol* 279 (4):F755–F764.
13. Abran, D., I. Dumont, P. Hardy, K. Peri, D. Y. Li, S. Molotchnikoff, D. R. Varma, and S. Chemtob. 1997. Characterization and regulation of prostaglandin E_2 receptor and receptor-coupled functions in the choroidal vasculature of the pig during development. *Circ Res* 80 (4):463–472.
14. Sauzeau, V., H. Le Jeune, C. Cario-Toumaniantz, A. Smolenski, S. M. Lohmann, J. Bertoglio, P. Chardin, P. Pacaud, and G. Loirand. 2000. Cyclic GMP-dependent protein kinase signaling pathway inhibits RhoA-induced Ca^{2+} sensitization of contraction in vascular smooth muscle. *J Biol Chem* 275 (28):21722–21729.

15. Cohen, R. A., R. M. Weisbrod, M. Gericke, M. Yaghoubi, C. Bierl, and V. M. Bolotina. 1999. Mechanism of nitric oxide-induced vasodilatation: Refilling of intracellular stores by sarcoplasmic reticulum Ca^{2+} ATPase and inhibition of store-operated Ca^{2+} influx. *Circ Res* 84 (2):210–219.

16. Robertson, B. E., R. Schubert, J. Hescheler, and M. T. Nelson. 1993. cGMP-dependent protein kinase activates Ca-activated K channels in cerebral artery smooth muscle cells. *Am J Physiol* 265 (1 Pt 1):C299–C303.

17. Bolotina, V. M., S. Najibi, J. J. Palacino, P. J. Pagano, and R. A. Cohen. 1994. Nitric oxide directly activates calcium-dependent potassium channels in vascular smooth muscle. *Nature* 368 (6474):850–853.

18. Wang, R., Z. Wang, and L. Wu. 1997. Carbon monoxide-induced vasorelaxation and the underlying mechanisms. *Br J Pharmacol* 121 (5):927–934.

19. Naik, J. S. and B. R. Walker. 2003. Heme oxygenase-mediated vasodilation involves vascular smooth muscle cell hyperpolarization. *Am J Physiol Heart Circ Physiol* 285 (1):H220–H228.

20. Furchgott, R. F. and D. Jothianandan. 1991. Endothelium-dependent and -independent vasodilation involving cyclic GMP: Relaxation induced by nitric oxide, carbon monoxide and light. *Blood Vessels* 28 (1–3):52–61.

21. Jones, A. W., W. Durante, and R. J. Korthuis. 2010. Heme oxygenase-1 deficiency leads to alteration of soluble guanylate cyclase redox regulation. *J Pharmacol Exp Ther* 335 (1):85–91.

22. Knot, H. J. and M. T. Nelson. 1998. Regulation of arterial diameter and wall $[Ca^{2+}]$ in cerebral arteries of rat by membrane potential and intravascular pressure. *J Physiol* 508:199–209.

23. Knot, H. J., P. A. Zimmermann, and M. T. Nelson. 1996. Extracellular K^+-induced hyperpolarizations and dilatations of rat coronary and cerebral arteries involve inward rectifier K^+ channels. *J Physiol* 492:419–430.

24. Crane, G. J., N. Gallagher, K. A. Dora, and C. J. Garland. 2003. Small- and intermediate-conductance calcium-activated K^+ channels provide different facets of endothelium-dependent hyperpolarization in rat mesenteric artery. *J Physiol* 553 (Pt 1):183–189.

25. Yamamoto, Y., M. F. Klemm, F. R. Edwards, and H. Suzuki. 2001. Intercellular electrical communication among smooth muscle and endothelial cells in guinea-pig mesenteric arterioles. *J Physiol* 535 (Pt 1):181–195.

26. Emerson, G. G. and S. S. Segal. 2001. Electrical activation of endothelium evokes vasodilation and hyperpolarization along hamster feed arteries. *Am J Physiol Heart Circ Physiol* 280 (1):H160–H167.

27. Chaytor, A. T., L. M. Bakker, D. H. Edwards, and T. M. Griffith. 2005. Connexin-mimetic peptides dissociate electrotonic EDHF-type signalling via myoendothelial and smooth muscle gap junctions in the rabbit iliac artery. *Br J Pharmacol* 144 (1):108–114.

28. Mather, S., K. A. Dora, S. L. Sandow, P. Winter, and C. J. Garland. 2005. Rapid endothelial cell-selective loading of connexin 40 antibody blocks endothelium-derived hyperpolarizing factor dilation in rat small mesenteric arteries. *Circ Res* 97 (4):399–407.

29. Tran, C. H., M. S. Taylor, F. Plane, S. Nagaraja, N. M. Tsoukias, V. Solodushko, E. J. Vigmond, T. Furstenhaupt, M. Brigdan, and D. G. Welsh. 2012. Endothelial Ca^{2+} wavelets and the induction of myoendothelial feedback. *Am J Physiol Cell Physiol* 302 (8):C1226–C1242.

30. Campbell, W. B., D. Gebremedhin, P. F. Pratt, and D. R. Harder. 1996. Identification of epoxyeicosatrienoic acids as endothelium-derived hyperpolarizing factors. *Circ Res* 78 (3):415–423.

31. Huang, A., D. Sun, A. Jacobson, M. A. Carroll, J. R. Falck, and G. Kaley. 2005. Epoxyeicosatrienoic acids are released to mediate shear stress-dependent hyperpolarization of arteriolar smooth muscle. *Circ Res* 96 (3):376–383.

32. Fisslthaler, B., R. Popp, L. Kiss, M. Potente, D. R. Harder, I. Fleming, and R. Busse. 1999. Cytochrome P450 2C is an EDHF synthase in coronary arteries. *Nature* 401 (6752):493–497.

33. Fleming, I., A. Rueben, R. Popp, B. Fisslthaler, S. Schrodt, A. Sander, J. Haendeler et al. 2007. Epoxyeicosatrienoic acids regulate Trp channel dependent Ca^{2+} signaling and hyperpolarization in endothelial cells. *Arterioscler Thromb Vasc Biol* 27 (12):2612–2618.

34. Vriens, J., G. Owsianik, B. Fisslthaler, M. Suzuki, A. Janssens, T. Voets, C. Morisseau et al. 2005. Modulation of the Ca^{2+} permeable cation channel TRPV4 by cytochrome P450 epoxygenases in vascular endothelium. *Circ Res* 97 (9):908–915.

35. Gauthier, K. M., Y. Chawengsub, D. H. Goldman, R. E. Conrow, S. Anjaiah, J. R. Falck, and W. B. Campbell. 2008. 11(R),12(S),15(S)-trihydroxyeicosa-5(Z),8(Z),13(E)-trienoic acid: An endothelium-derived 15-lipoxygenase metabolite that relaxes rabbit aorta. *Am J Physiol Heart Circ Physiol* 294 (3):H1467–H1472.

36. Gauthier, K. M., D. H. Goldman, N. T. Aggarwal, Y. Chawengsub, J. R. Falck, and W. B. Campbell. 2011. Role of arachidonic acid lipoxygenase metabolites in acetylcholine-induced relaxations of mouse arteries. *Am J Physiol Heart Circ Physiol* 300 (3):H725–H735.

37. Gauthier, K. M., N. Spitzbarth, E. M. Edwards, and W. B. Campbell. 2004. Apamin-sensitive K^+ currents mediate arachidonic acid-induced relaxations of rabbit aorta. *Hypertension* 43 (2):413–419.

38. Zhang, D. X., K. M. Gauthier, Y. Chawengsub, and W. B. Campbell. 2007. ACh-induced relaxations of rabbit small mesenteric arteries: Role of arachidonic acid metabolites and K^+. *Am J Physiol Heart Circ Physiol* 293 (1):H152–H159.

39. Stubbert, D., O. Prysyazhna, O. Rudyk, J. Scotcher, J. R. Burgoyne, and P. Eaton. 2014. Protein kinase G Iα oxidation paradoxically underlies blood pressure lowering by the reductant hydrogen sulfide. *Hypertension* 64 (6):1344–1351.

40. Cortese-Krott, M. M., B. O. Fernandez, J. L. Santos, E. Mergia, M. Grman, P. Nagy, M. Kelm, A. Butler, and M. Feelisch. 2014. Nitrosopersulfide ($SSNO^-$) accounts for sustained NO bioactivity of S-nitrosothiols following reaction with sulfide. *Redox Biol* 2:234–244.

41. Cortese-Krott, M. M., G. G. Kuhnle, A. Dyson, B. O. Fernandez, M. Grman, J. F. DuMond, M. P. Barrow et al. 2015. Key bioactive reaction products of the NO/H_2S interaction are S/N-hybrid species, polysulfides, and nitroxyl. *Proc Natl Acad Sci USA* 112 (34):E4651–E4660.

42. Liang, G. H., Q. Xi, C. W. Leffler, and J. H. Jaggar. 2012. Hydrogen sulfide activates Ca^{2+} sparks to induce cerebral arteriole dilatation. *J Physiol* 590 (11):2709–2720.

43. Edwards, G., M. Feletou, M. J. Gardener, C. D. Glen, G. R. Richards, P. M. Vanhoutte, and A. H. Weston. 2001. Further investigations into the endothelium-dependent hyperpolarizing effects of bradykinin and substance P in porcine coronary artery. *Br J Pharmacol* 133 (7):1145–1153.

44. Weston, A. H., M. Feletou, P. M. Vanhoutte, J. R. Falck, W. B. Campbell, and G. Edwards. 2005. Bradykinin-induced, endothelium-dependent responses in porcine coronary arteries: Involvement of potassium channel activation and epoxyeicosatrienoic acids. *Br J Pharmacol* 145 (6):775–784.

45. Needleman, P., B. Jakschik, and E. M. Johnson, Jr. 1973. Sulfhydryl requirement for relaxation of vascular smooth muscle. *J Pharmacol Exp Ther* 187 (2):324–331.

46. Rubanyi, G. M. and P. M. Vanhoutte. 1986. Oxygen-derived free radicals, endothelium, and responsiveness of vascular smooth muscle. *Am J Physiol* 250 (5 Pt 2):H815–H821.

47. Burke, T. M. and M. S. Wolin. 1987. Hydrogen peroxide elicits pulmonary arterial relaxation and guanylate cyclase activation. *Am J Physiol* 252 (4 Pt 2):H721–H732.

48. Sobey, C. G., D. D. Heistad, and F. M. Faraci. 1997. Mechanisms of bradykinin-induced cerebral vasodilatation in rats. Evidence that reactive oxygen species activate K$^+$ channels. *Stroke* 28 (11):2290–2295.

49. Mian, K. B. and W. Martin. 1997. Hydrogen peroxide-induced impairment of reactivity in rat isolated aorta: Potentiation by 3-amino-1,2,4-triazole. *Br J Pharmacol* 121 (4):813–819.

50. Matoba, T., H. Shimokawa, M. Nakashima, Y. Hirakawa, Y. Mukai, K. Hirano, H. Kanaide, and A. Takeshita. 2000. Hydrogen peroxide is an endothelium-derived hyperpolarizing factor in mice. *J Clin Invest* 106 (12):1521–1530.

51. Matoba, T., H. Shimokawa, K. Morikawa, H. Kubota, I. Kunihiro, L. Urakami-Harasawa, Y. Mukai, Y. Hirakawa, T. Akaike, and A. Takeshita. 2003. Electron spin resonance detection of hydrogen peroxide as an endothelium-derived hyperpolarizing factor in porcine coronary microvessels. *Arterioscler Thromb Vasc Biol* 23 (7):1224–1230.

52. Hatoum, O. A., D. G. Binion, H. Miura, G. Telford, M. F. Otterson, and D. D. Gutterman. 2005. Role of hydrogen peroxide in ACh-induced dilation of human submucosal intestinal microvessels. *Am J Physiol Heart Circ Physiol* 288 (1):H48–H54.

53. Liu, Y., A. H. Bubolz, S. Mendoza, D. X. Zhang, and D. D. Gutterman. 2011. H$_2$O$_2$ is the transferrable factor mediating flow-induced dilation in human coronary arterioles. *Circ Res* 108 (5):566–573.

54. Zhang, D. X., L. Borbouse, D. Gebremedhin, S. A. Mendoza, N. S. Zinkevich, R. Li, and D. D. Gutterman. 2012. H$_2$O$_2$-induced dilation in human coronary arterioles: Role of protein kinase G dimerization and large-conductance Ca^{2+}-activated K$^+$ channel activation. *Circ Res* 110 (3):471–480.

55. Prysyazhna, O., O. Rudyk, and P. Eaton. 2012. Single atom substitution in mouse protein kinase G eliminates oxidant sensing to cause hypertension. *Nat Med* 18 (2):286–290.

56. Yang, Z. W., A. Zhang, B. T. Altura, and B. M. Altura. 1998. Endothelium-dependent relaxation to hydrogen peroxide in canine basilar artery: A potential new cerebral dilator mechanism. *Brain Res Bull* 47 (3):257–263.

57. Zembowicz, A., R. J. Hatchett, A. M. Jakubowski, and R. J. Gryglewski. 1993. Involvement of nitric oxide in the endothelium-dependent relaxation induced by hydrogen peroxide in the rabbit aorta. *Br J Pharmacol* 110 (1):151–158.

58. Kontos, H. A. 1985. George E. Brown memorial lecture. Oxygen radicals in cerebral vascular injury. *Circ Res* 57 (4):508–516.

59. Beckman, J. S., R. L. Minor, Jr., C. W. White, J. E. Repine, G. M. Rosen, and B. A. Freeman. 1988. Superoxide dismutase and catalase conjugated to polyethylene glycol increases endothelial enzyme activity and oxidant resistance. *J Biol Chem* 263 (14):6884–6892.

60. Barlow, R. S. and R. E. White. 1998. Hydrogen peroxide relaxes porcine coronary arteries by stimulating BK$_{ca}$ channel activity. *Am J Physiol* 275 (4 Pt 2):H1283–H1289.

61. Barlow, R. S., A. M. El-Mowafy, and R. E. White. 2000. H$_2$O$_2$ opens BK$_{ca}$ channels via the PLA$_2$-arachidonic acid signaling cascade in coronary artery smooth muscle. *Am J Physiol Heart Circ Physiol* 279 (2):H475–H483.

62. Matoba, T., H. Shimokawa, H. Kubota, K. Morikawa, T. Fujiki, I. Kunihiro, Y. Mukai, Y. Hirakawa, and A. Takeshita. 2002. Hydrogen peroxide is an endothelium-derived hyperpolarizing factor in human mesenteric arteries. *Biochem Biophys Res Commun* 290 (3):909–913.

63. Matsuo, T., H. Shinzawa, H. Togashi, M. Aoki, K. Sugahara, K. Saito, T. Saito et al. 1998. Highly sensitive hepatitis B surface antigen detection by measuring stable nitroxide radical formation with ESR spectroscopy. *Free Radic Biol Med* 25 (8):929–935.

64. Miura, H., J. J. Bosnjak, G. Ning, T. Saito, M. Miura, and D. D. Gutterman. 2003. Role for hydrogen peroxide in flow-induced dilation of human coronary arterioles. *Circ Res* 92 (2):e31–e40.
65. Liu, Y., H. Zhao, H. Li, B. Kalyanaraman, A. C. Nicolosi, and D. D. Gutterman. 2003. Mitochondrial sources of H_2O_2 generation play a key role in flow-mediated dilation in human coronary resistance arteries. *Circ Res* 93 (6):573–580.
66. Gauthier, K. M., L. Olson, A. Harder, M. Isbell, J. D. Imig, D. D. Gutterman, J. R. Falck, and W. B. Campbell. 2011. Soluble epoxide hydrolase contamination of specific catalase preparations inhibits epoxyeicosatrienoic acid vasodilation of rat renal arterioles. *Am J Physiol Renal Physiol* 301 (4):F765–F772.
67. Halliwell, B. and J. M. C. Gutteridge. 2007. *Free Radicals in Biology and Medicine*. Oxford University Press, Oxford, U.K.
68. Wardman, P. 2007. Fluorescent and luminescent probes for measurement of oxidative and nitrosative species in cells and tissues: Progress, pitfalls, and prospects. *Free Radic Biol Med* 43 (7):995–1022.
69. Kalyanaraman, B., M. Hardy, R. Podsiadly, G. Cheng, and J. Zielonka. 2017. Recent developments in detection of superoxide radical anion and hydrogen peroxide: Opportunities, challenges, and implications in redox signaling. *Arch Biochem Biophys* 617:38–47.
70. Bedard, K. and K. H. Krause. 2007. The NOX family of ROS-generating NADPH oxidases: Physiology and pathophysiology. *Physiol Rev* 87 (1):245–313.
71. Hilenski, L. L., R. E. Clempus, M. T. Quinn, J. D. Lambeth, and K. K. Griendling. 2004. Distinct subcellular localizations of Nox1 and Nox4 in vascular smooth muscle cells. *Arterioscler Thromb Vasc Biol* 24 (4):677–683.
72. Van Buul, J. D., M. Fernandez-Borja, E. C. Anthony, and P. L. Hordijk. 2005. Expression and localization of NOX2 and NOX4 in primary human endothelial cells. *Antioxid Redox Signal* 7 (3–4):308–317.
73. Li, J. M. and A. M. Shah. 2002. Intracellular localization and preassembly of the NADPH oxidase complex in cultured endothelial cells. *J Biol Chem* 277 (22):19952–19960.
74. Bayraktutan, U., L. Blayney, and A. M. Shah. 2000. Molecular characterization and localization of the NAD(P)H oxidase components gp91-phox and p22-phox in endothelial cells. *Arterioscler Thromb Vasc Biol* 20 (8):1903–1911.
75. Ago, T., T. Kitazono, H. Ooboshi, T. Iyama, Y. H. Han, J. Takada, M. Wakisaka, S. Ibayashi, H. Utsumi, and M. Iida. 2004. Nox4 as the major catalytic component of an endothelial NAD(P)H oxidase. *Circulation* 109 (2):227–233.
76. Sorescu, D., D. Weiss, B. Lassegue, R. E. Clempus, K. Szocs, G. P. Sorescu, L. Valppu et al. 2002. Superoxide production and expression of nox family proteins in human atherosclerosis. *Circulation* 105 (12):1429–1435.
77. Datla, S. R., H. Peshavariya, G. J. Dusting, K. Mahadev, B. J. Goldstein, and F. Jiang. 2007. Important role of Nox4 type NADPH oxidase in angiogenic responses in human microvascular endothelial cells in vitro. *Arterioscler Thromb Vasc Biol* 27 (11):2319–2324.
78. Takac, I., K. Schroder, L. Zhang, B. Lardy, N. Anilkumar, J. D. Lambeth, A. M. Shah, F. Morel, and R. P. Brandes. 2011. The E-loop is involved in hydrogen peroxide formation by the NADPH oxidase Nox4. *J Biol Chem* 286 (15):13304–13313.
79. Sanchez-Gomez, F. J., E. Calvo, R. Breton-Romero, M. Fierro-Fernandez, N. Anilkumar, A. M. Shah, K. Schroder, R. P. Brandes, J. Vazquez, and S. Lamas. 2015. NOX4-dependent Hydrogen peroxide promotes shear stress-induced SHP2 sulfenylation and eNOS activation. *Free Radic Biol Med* 89:419–430.
80. Ahmad, M., M. R. Kelly, X. Zhao, S. Kandhi, and M. S. Wolin. 2010. Roles for Nox4 in the contractile response of bovine pulmonary arteries to hypoxia. *Am J Physiol Heart Circ Physiol* 298 (6):H1879–H1888.

81. Larsen, B. T., A. H. Bubolz, S. A. Mendoza, K. A. Pritchard, Jr., and D. D. Gutterman. 2009. Bradykinin-induced dilation of human coronary arterioles requires NADPH oxidase-derived reactive oxygen species. *Arterioscler Thromb Vasc Biol* 29 (5):739–745.

82. Jones, S. A., V. B. O'Donnell, J. D. Wood, J. P. Broughton, E. J. Hughes, and O. T. Jones. 1996. Expression of phagocyte NADPH oxidase components in human endothelial cells. *Am J Physiol* 271 (4 Pt 2):H1626–H1634.

83. Thomas, S. R., P. K. Witting, and G. R. Drummond. 2008. Redox control of endothelial function and dysfunction: Molecular mechanisms and therapeutic opportunities. *Antioxid Redox Signal* 10 (10):1713–1765.

84. Lassegue, B., D. Sorescu, K. Szocs, Q. Yin, M. Akers, Y. Zhang, S. L. Grant, J. D. Lambeth, and K. K. Griendling. 2001. Novel gp91phox homologues in vascular smooth muscle cells: Nox1 mediates angiotensin II-induced superoxide formation and redox-sensitive signaling pathways. *Circ Res* 88 (9):888–894.

85. BelAiba, R. S., T. Djordjevic, A. Petry, K. Diemer, S. Bonello, B. Banfi, J. Hess, A. Pogrebniak, C. Bickel, and A. Gorlach. 2007. NOX5 variants are functionally active in endothelial cells. *Free Radic Biol Med* 42 (4):446–459.

86. Guzik, T. J., W. Chen, M. C. Gongora, B. Guzik, H. E. Lob, D. Mangalat, N. Hoch et al. 2008. Calcium-dependent NOX5 nicotinamide adenine dinucleotide phosphate oxidase contributes to vascular oxidative stress in human coronary artery disease. *J Am Coll Cardiol* 52 (22):1803–1809.

87. Vasquez-Vivar, J., B. Kalyanaraman, P. Martasek, N. Hogg, B. S. Masters, H. Karoui, P. Tordo, and K. A. Pritchard, Jr. 1998. Superoxide generation by endothelial nitric oxide synthase: The influence of cofactors. *Proc Natl Acad Sci USA* 95 (16):9220–9225.

88. Stocker, R. and J. F. Keaney, Jr. 2004. Role of oxidative modifications in atherosclerosis. *Physiol Rev* 84 (4):1381–1478.

89. Huisman, A., I. Vos, E. E. van Faassen, J. A. Joles, H. J. Grone, P. Martasek, A. J. van Zonneveld, A. F. Vanin, and T. J. Rabelink. 2002. Anti-inflammatory effects of tetrahydrobiopterin on early rejection in renal allografts: Modulation of inducible nitric oxide synthase. *FASEB J* 16 (9):1135–1137.

90. Jin, B. Y., J. L. Sartoretto, V. N. Gladyshev, and T. Michel. 2009. Endothelial nitric oxide synthase negatively regulates hydrogen peroxide-stimulated AMP-activated protein kinase in endothelial cells. *Proc Natl Acad Sci USA* 106 (41):17343–17348.

91. Jarasch, E. D., C. Grund, G. Bruder, H. W. Heid, T. W. Keenan, and W. W. Franke. 1981. Localization of xanthine oxidase in mammary-gland epithelium and capillary endothelium. *Cell* 25 (1):67–82.

92. Guzik, T. J., J. Sadowski, B. Guzik, A. Jopek, B. Kapelak, P. Przybylowski, K. Wierzbicki, R. Korbut, D. G. Harrison, and K. M. Channon. 2006. Coronary artery superoxide production and Nox isoform expression in human coronary artery disease. *Arterioscler Thromb Vasc Biol* 26 (2):333–339.

93. Spiekermann, S., U. Landmesser, S. Dikalov, M. Bredt, G. Gamez, H. Tatge, N. Reepschlager, B. Hornig, H. Drexler, and D. G. Harrison. 2003. Electron spin resonance characterization of vascular xanthine and NAD(P)H oxidase activity in patients with coronary artery disease: Relation to endothelium-dependent vasodilation. *Circulation* 107 (10):1383–1389.

94. White, C. R., V. Darley-Usmar, W. R. Berrington, M. McAdams, J. Z. Gore, J. A. Thompson, D. A. Parks, M. M. Tarpey, and B. A. Freeman. 1996. Circulating plasma xanthine oxidase contributes to vascular dysfunction in hypercholesterolemic rabbits. *Proc Natl Acad Sci USA* 93 (16):8745–8749.

95. Brand, M. D. 2010. The sites and topology of mitochondrial superoxide production. *Exp Gerontol* 45 (7–8):466–472.

96. Doughan, A. K., D. G. Harrison, and S. I. Dikalov. 2008. Molecular mechanisms of angiotensin II-mediated mitochondrial dysfunction: Linking mitochondrial oxidative damage and vascular endothelial dysfunction. *Circ Res* 102 (4):488–496.

97. Davis, B. and M. H. Zou. 2005. CD40 ligand-dependent tyrosine nitration of prostacyclin synthase in vivo. *Circulation* 112 (14):2184–2192.
98. Bharadwaj, L. and K. Prasad. 1995. Mediation of H_2O_2-induced vascular relaxation by endothelium-derived relaxing factor. *Mol Cell Biochem* 149–150:267–270.
99. Thomas, S. R., E. Schulz, and J. F. Keaney, Jr. 2006. Hydrogen peroxide restrains endothelium-derived nitric oxide bioactivity—Role for iron-dependent oxidative stress. *Free Radic Biol Med* 41 (4):681–688.
100. Mian, K. B. and W. Martin. 1995. The inhibitory effect of 3-amino-1,2,4-triazole on relaxation induced by hydroxylamine and sodium azide but not hydrogen peroxide or glyceryl trinitrate in rat aorta. *Br J Pharmacol* 116 (8):3302–3308.
101. Cseko, C., Z. Bagi, and A. Koller. 2004. Biphasic effect of hydrogen peroxide on skeletal muscle arteriolar tone via activation of endothelial and smooth muscle signaling pathways. *J Appl Physiol* 97 (3):1130–1137.
102. Thomas, S. R., K. Chen, and J. F. Keaney, Jr. 2002. Hydrogen peroxide activates endothelial nitric-oxide synthase through coordinated phosphorylation and dephosphorylation via a phosphoinositide 3-kinase-dependent signaling pathway. *J Biol Chem* 277 (8):6017–6024.
103. Cai, H., M. E. Davis, G. R. Drummond, and D. G. Harrison. 2001. Induction of endothelial NO synthase by hydrogen peroxide via a Ca^{2+}/calmodulin-dependent protein kinase II/janus kinase 2-dependent pathway. *Arterioscler Thromb Vasc Biol* 21 (10):1571–1576.
104. Drummond, G. R., H. Cai, M. E. Davis, S. Ramasamy, and D. G. Harrison. 2000. Transcriptional and posttranscriptional regulation of endothelial nitric oxide synthase expression by hydrogen peroxide. *Circ Res* 86 (3):347–354.
105. Shimizu, S., M. Ishii, Y. Miyasaka, Y. Wajima, T. Negoro, T. Hagiwara et al. 2005. Possible involvement of hydroxyl radical on the stimulation of tetrahydrobiopterin synthesis by hydrogen peroxide and peroxynitrite in vascular endothelial cells. *Int J Biochem Cell Biol* 37 (4): 864–875.
106. Shimizu, S., T. Hiroi, M. Ishii, T. Hagiwara, T. Wajima, A. Miyazaki et al. 2008. Hydrogen peroxide stimulates tetrahydrobiopterin synthesis through activation of the Jak2 tyrosine kinase pathway in vascular endothelial cells. *Int J Biochem Cell Biol* 40 (4): 755–765.
107. Knock, G. A. and J. P. Ward. 2011. Redox regulation of protein kinases as a modulator of vascular function. *Antioxid Redox Signal* 15 (6):1531–1547.
108. Chen, K., J. A. Vita, B. C. Berk, and J. F. Keaney, Jr. 2001. c-Jun N-terminal kinase activation by hydrogen peroxide in endothelial cells involves SRC-dependent epidermal growth factor receptor transactivation. *J Biol Chem* 276 (19):16045–16050.
109. Hu, Q., S. Corda, J. L. Zweier, M. C. Capogrossi, and R. C. Ziegelstein. 1998. Hydrogen peroxide induces intracellular calcium oscillations in human aortic endothelial cells. *Circulation* 97 (3):268–275.
110. Hecquet, C. M., G. U. Ahmmed, S. M. Vogel, and A. B. Malik. 2008. Role of TRPM2 channel in mediating H_2O_2-induced Ca^{2+} entry and endothelial hyperpermeability. *Circ Res* 102 (3):347–355.
111. Sun, L., H. Y. Yau, O. C. Lau, Y. Huang, and X. Yao. 2011. Effect of hydrogen peroxide and superoxide anions on cytosolic Ca^{2+}: Comparison of endothelial cells from large-sized and small-sized arteries. *PLoS One* 6 (9):e25432.
112. Lock, J. T., W. G. Sinkins, and W. P. Schilling. 2012. Protein S-glutathionylation enhances Ca^{2+}-induced Ca^{2+} release via the IP_3 receptor in cultured aortic endothelial cells. *J Physiol* 590 (15):3431–3447.
113. Csordas, G. and G. Hajnoczky. 2009. SR/ER-mitochondrial local communication: Calcium and ROS. *Biochim Biophys Acta* 1787 (11):1352–1362.

114. Ellis, A., M. Pannirselvam, T. J. Anderson, and C. R. Triggle. 2003. Catalase has negligible inhibitory effects on endothelium-dependent relaxations in mouse isolated aorta and small mesenteric artery. *Br J Pharmacol* 140 (7):1193–1200.

115. Edwards, D. H., Y. Li, and T. M. Griffith. 2008. Hydrogen peroxide potentiates the EDHF phenomenon by promoting endothelial Ca^{2+} mobilization. *Arterioscler Thromb Vasc Biol* 28 (10):1774–1781.

116. Garry, A., D. H. Edwards, I. F. Fallis, R. L. Jenkins, and T. M. Griffith. 2009. Ascorbic acid and tetrahydrobiopterin potentiate the EDHF phenomenon by generating hydrogen peroxide. *Cardiovasc Res* 84 (2):218–226.

117. Behringer, E. J., R. L. Shaw, E. B. Westcott, M. J. Socha, and S. S. Segal. 2013. Aging impairs electrical conduction along endothelium of resistance arteries through enhanced Ca^{2+}-activated K^+ channel activation. *Arterioscler Thromb Vasc Biol* 33 (8):1892–1901.

118. Socha, M. J., E. M. Boerman, E. J. Behringer, R. L. Shaw, T. L. Domeier, and S. S. Segal. 2015. Advanced age protects microvascular endothelium from aberrant Ca^{2+} influx and cell death induced by hydrogen peroxide. *J Physiol* 593 (9):2155–2169.

119. Macpherson, L. J., A. E. Dubin, M. J. Evans, F. Marr, P. G. Schultz, B. F. Cravatt, and A. Patapoutian. 2007. Noxious compounds activate TRPA1 ion channels through covalent modification of cysteines. *Nature* 445 (7127):541–545.

120. Kozai, D., N. Ogawa, and Y. Mori. 2014. Redox regulation of transient receptor potential channels. *Antioxid Redox Signal* 21 (6):971–986.

121. Pires, P. W. and S. Earley. 2016. Redox regulation of transient receptor potential channels in the endothelium. *Microcirculation* 24:e12329.

122. Retamal, M. A., C. J. Cortes, L. Reuss, M. V. Bennett, and J. C. Saez. 2006. S-nitrosylation and permeation through connexin 43 hemichannels in astrocytes: Induction by oxidant stress and reversal by reducing agents. *Proc Natl Acad Sci USA* 103 (12):4475–4480.

123. Behringer, E. J. and S. S. Segal. 2012. Spreading the signal for vasodilatation: Implications for skeletal muscle blood flow control and the effects of ageing. *J Physiol* 590 (24):6277–6284.

124. Miura, H., R. E. Wachtel, Y. Liu, F. R. Loberiza, Jr., T. Saito, M. Miura, and D. D. Gutterman. 2001. Flow-induced dilation of human coronary arterioles: Important role of Ca^{2+}-activated K^+ channels. *Circulation* 103 (15):1992–1998.

125. Larsen, B. T., D. D. Gutterman, A. Sato, K. Toyama, W. B. Campbell, D. C. Zeldin, V. L. Manthati, J. R. Falck, and H. Miura. 2008. Hydrogen peroxide inhibits cytochrome p450 epoxygenases: Interaction between two endothelium-derived hyperpolarizing factors. *Circ Res* 102 (1):59–67.

126. Ellinsworth, D. C., S. L. Sandow, N. Shukla, Y. Liu, J. Y. Jeremy, and D. D. Gutterman. 2016. Endothelium-derived hyperpolarization and coronary vasodilation: Diverse and integrated roles of epoxyeicosatrienoic acids, hydrogen peroxide, and gap junctions. *Microcirculation* 23 (1):15–32.

127. Brookes, P. S., Y. Yoon, J. L. Robotham, M. W. Anders, and S. S. Sheu. 2004. Calcium, ATP, and ROS: A mitochondrial love-hate triangle. *Am J Physiol Cell Physiol* 287 (4):C817–C833.

128. White, A. A., K. M. Crawford, C. S. Patt, and P. J. Lad. 1976. Activation of soluble guanylate cyclase from rat lung by incubation or by hydrogen peroxide. *J Biol Chem* 251 (23):7304–7312.

129. Wolin, M. S. and T. M. Burke. 1987. Hydrogen peroxide elicits activation of bovine pulmonary arterial soluble guanylate cyclase by a mechanism associated with its metabolism by catalase. *Biochem Biophys Res Commun* 143 (1):20–25.

130. Neo, B. H., S. Kandhi, and M. S. Wolin. 2010. Roles for soluble guanylate cyclase and a thiol oxidation-elicited subunit dimerization of protein kinase G in pulmonary artery relaxation to hydrogen peroxide. *Am J Physiol Heart Circ Physiol* 299 (4):H1235–H1241.

131. Neo, B. H., S. Kandhi, and M. S. Wolin. 2011. Roles for redox mechanisms controlling protein kinase G in pulmonary and coronary artery responses to hypoxia. *Am J Physiol Heart Circ Physiol* 301 (6):H2295–H2304.

132. Burgoyne, J. R., M. Madhani, F. Cuello, R. L. Charles, J. P. Brennan, E. Schroder, D. D. Browning, and P. Eaton. 2007. Cysteine redox sensor in PKGIα enables oxidant-induced activation. *Science* 317 (5843):1393–1397.

133. Khavandi, K., R. L. Baylie, S. A. Sugden, M. Ahmed, V. Csato, P. Eaton, D. C. Hill-Eubanks, A. D. Bonev, M. T. Nelson, and A. S. Greenstein. 2016. Pressure-induced oxidative activation of PKG enables vasoregulation by Ca²⁺ sparks and BK channels. *Sci Signal* 9 (449):ra100.

134. Rudyk, O., A. Phinikaridou, O. Prysyazhna, J. R. Burgoyne, R. M. Botnar, and P. Eaton. 2013. Protein kinase G oxidation is a major cause of injury during sepsis. *Proc Natl Acad Sci USA* 110 (24):9909–9913.

135. Burgoyne, J. R. and P. Eaton. 2013. Detecting disulfide-bound complexes and the oxidative regulation of cyclic nucleotide-dependent protein kinases by H_2O_2. *Methods Enzymol* 528:111–128.

136. Ohashi, J., A. Sawada, S. Nakajima, K. Noda, A. Takaki, and H. Shimokawa. 2012. Mechanisms for enhanced endothelium-derived hyperpolarizing factor-mediated responses in microvessels in mice. *Circ J* 76 (7):1768–1779.

137. Wong, P. S., M. J. Garle, S. P. Alexander, M. D. Randall, and R. E. Roberts. 2014. A role for the sodium pump in H_2O_2-induced vasorelaxation in porcine isolated coronary arteries. *Pharmacol Res* 90:25–35.

138. Nishijima, Y., S. Cao, D. S. Chabowski, A. Korishettar, A. Ge, X. Zheng, R. Sparapani, D. D. Gutterman, and D. X. Zhang. 2017. Contribution of $K_V1.5$ channel to H_2O_2-induced human arteriolar dilation and its modulation by coronary artery disease. *Circ Res* 120 (4):658–669.

139. Nacitarhan, C., Z. Bayram, B. Eksert, C. Usta, I. Golbasi, and S. S. Ozdem. 2007. The effect of hydrogen peroxide in human internal thoracic arteries: Role of potassium channels, nitric oxide and cyclooxygenase products. *Cardiovasc Drugs Ther* 21 (4):257–262.

140. Gao, Y. J., S. Hirota, D. W. Zhang, L. J. Janssen, and R. M. Lee. 2003. Mechanisms of hydrogen-peroxide-induced biphasic response in rat mesenteric artery. *Br J Pharmacol* 138 (6):1085–1092.

141. Ohashi, M., F. Faraci, and D. Heistad. 2005. Peroxynitrite hyperpolarizes smooth muscle and relaxes internal carotid artery in rabbit via ATP-sensitive K⁺ channels. *Am J Physiol Heart Circ Physiol* 289 (5):H2244–H2250.

142. Chai, Y., D. M. Zhang, and Y. F. Lin. 2011. Activation of cGMP-dependent protein kinase stimulates cardiac ATP-sensitive potassium channels via a ROS/calmodulin/CaMKII signaling cascade. *PLoS One* 6 (3):e18191.

143. Tang, X. D., M. L. Garcia, S. H. Heinemann, and T. Hoshi. 2004. Reactive oxygen species impair Slo1 BK channel function by altering cysteine-mediated calcium sensing. *Nat Struct Mol Biol* 11 (2):171–178.

144. Liu, B., X. Sun, Y. Zhu, L. Gan, H. Xu, and X. Yang. 2010. Biphasic effects of H_2O_2 on BK_{Ca} channels. *Free Radic Res* 44 (9):1004–1012.

145. Yang, Y., W. Shi, N. Cui, Z. Wu, and C. Jiang. 2010. Oxidative stress inhibits vascular K_{ATP} channels by S-glutathionylation. *J Biol Chem* 285 (49):38641–38648.

146. Bannister, J. P., B. A. Young, M. J. Main, A. Sivaprasadarao, and D. Wray. 1999. The effects of oxidizing and cysteine-reactive reagents on the inward rectifier potassium channels $K_{ir}2.3$ and $K_{ir}1.1$. *Pflugers Arch* 438 (6):868–878.

147. Papreck, J. R., E. A. Martin, P. Lazzarini, D. Kang, and D. Kim. 2012. Modulation of $K_{2P}3.1$ (TASK-1), $K_{2P}9.1$ (TASK-3), and TASK-1/3 heteromer by reactive oxygen species. *Pflugers Arch* 464 (5):471–480.

148. Morikawa, K., H. Shimokawa, T. Matoba, H. Kubota, T. Akaike, M. A. Talukder, M. Hatanaka et al. 2003. Pivotal role of Cu,Zn-superoxide dismutase in endothelium-dependent hyperpolarization. *J Clin Invest* 112 (12):1871–1879.
149. Randall, L. M., G. Ferrer-Sueta, and A. Denicola. 2013. Peroxiredoxins as preferential targets in H_2O_2-induced signaling. *Methods Enzymol* 527:41–63.
150. Winterbourn, C. C. 2013. The biological chemistry of hydrogen peroxide. *Methods Enzymol* 528:3–25.
151. Godoy, J. R., M. Funke, W. Ackermann, P. Haunhorst, S. Oesteritz, F. Capani, H. P. Elsasser, and C. H. Lillig. 2011. Redox atlas of the mouse. Immunohistochemical detection of glutaredoxin-, peroxiredoxin-, and thioredoxin-family proteins in various tissues of the laboratory mouse. *Biochim Biophys Acta* 1810 (1):2–92.
152. Lee, C. K., H. J. Kim, Y. R. Lee, H. H. So, H. J. Park, K. J. Won, T. Park, K. Y. Lee, H. M. Lee, and B. Kim. 2007. Analysis of peroxiredoxin decreasing oxidative stress in hypertensive aortic smooth muscle. *Biochim Biophys Acta* 1774 (7):848–855.
153. Stacey, M. M., M. C. Vissers, and C. C. Winterbourn. 2012. Oxidation of 2-cys peroxiredoxins in human endothelial cells by hydrogen peroxide, hypochlorous acid, and chloramines. *Antioxid Redox Signal* 17 (3):411–421.
154. Yamawaki, H., J. Haendeler, and B. C. Berk. 2003. Thioredoxin: A key regulator of cardiovascular homeostasis. *Circ Res* 93 (11):1029–1033.
155. Hilgers, R. H. and K. C. Das. 2015. Role of in vivo vascular redox in resistance arteries. *Hypertension* 65 (1):130–139.
156. Choi, M. H., I. K. Lee, G. W. Kim, B. U. Kim, Y. H. Han, D. Y. Yu, H. S. Park et al. 2005. Regulation of PDGF signalling and vascular remodelling by peroxiredoxin II. *Nature* 435 (7040):347–353.
157. Kang, D. H., D. J. Lee, K. W. Lee, Y. S. Park, J. Y. Lee, S. H. Lee, Y. J. Koh et al. 2011. Peroxiredoxin II is an essential antioxidant enzyme that prevents the oxidative inactivation of VEGF receptor-2 in vascular endothelial cells. *Mol Cell* 44 (4):545–558.
158. Bertoldi, M. 2016. Human Peroxiredoxins 1 and 2 and their interacting protein partners; through structure toward functions of biological complexes. *Protein Pept Lett* 23 (1):69–77.
159. Jarvis, R. M., S. M. Hughes, and E. C. Ledgerwood. 2012. Peroxiredoxin 1 functions as a signal peroxidase to receive, transduce, and transmit peroxide signals in mammalian cells. *Free Radic Biol Med* 53 (7):1522–1530.
160. Sobotta, M. C., W. Liou, S. Stocker, D. Talwar, M. Oehler, T. Ruppert, A. N. Scharf, and T. P. Dick. 2015. Peroxiredoxin-2 and STAT3 form a redox relay for H_2O_2 signaling. *Nat Chem Biol* 11 (1):64–70.
161. de Haan, J. B., C. Bladier, P. Griffiths, M. Kelner, R. D. O'Shea, N. S. Cheung, R. T. Bronson et al. 1998. Mice with a homozygous null mutation for the most abundant glutathione peroxidase, Gpx1, show increased susceptibility to the oxidative stress-inducing agents paraquat and hydrogen peroxide. *J Biol Chem* 273 (35):22528–22536.
162. Zhang, Y., D. E. Handy, and J. Loscalzo. 2005. Adenosine-dependent induction of glutathione peroxidase 1 in human primary endothelial cells and protection against oxidative stress. *Circ Res* 96 (8):831–837.
163. Loh, K., H. Deng, A. Fukushima, X. Cai, B. Boivin, S. Galic, C. Bruce et al. 2009. Reactive oxygen species enhance insulin sensitivity. *Cell Metab* 10 (4):260–272.
164. McClung, J. P., C. A. Roneker, W. Mu, D. J. Lisk, P. Langlais, F. Liu, and X. G. Lei. 2004. Development of insulin resistance and obesity in mice overexpressing cellular glutathione peroxidase. *Proc Natl Acad Sci USA* 101 (24):8852–8857.
165. Forgione, M. A., N. Weiss, S. Heydrick, A. Cap, E. S. Klings, C. Bierl, R. T. Eberhardt, H. W. Farber, and J. Loscalzo. 2002. Cellular glutathione peroxidase deficiency and endothelial dysfunction. *Am J Physiol Heart Circ Physiol* 282 (4):H1255–H1261.

166. Weiss, N., Y. Y. Zhang, S. Heydrick, C. Bierl, and J. Loscalzo. 2001. Overexpression of cellular glutathione peroxidase rescues homocyst(e)ine-induced endothelial dysfunction. *Proc Natl Acad Sci USA* 98 (22):12503–12508.

167. Bonekamp, N. A., A. Volkl, H. D. Fahimi, and M. Schrader. 2009. Reactive oxygen species and peroxisomes: Struggling for balance. *Biofactors* 35 (4):346–355.

168. Yang, H., M. Shi, H. VanRemmen, X. Chen, J. Vijg, A. Richardson, and Z. Guo. 2003. Reduction of pressor response to vasoconstrictor agents by overexpression of catalase in mice. *Am J Hypertens* 16 (1):1–5.

169. Suvorava, T. and G. Kojda. 2009. Reactive oxygen species as cardiovascular mediators: Lessons from endothelial-specific protein overexpression mouse models. *Biochim Biophys Acta* 1787 (7):802–810.

170. Bauersachs, J., R. Popp, M. Hecker, E. Sauer, I. Fleming, and R. Busse. 1996. Nitric oxide attenuates the release of endothelium-derived hyperpolarizing factor. *Circulation* 94 (12):3341–3347.

171. Nishikawa, Y., D. W. Stepp, and W. M. Chilian. 2000. Nitric oxide exerts feedback inhibition on EDHF-induced coronary arteriolar dilation in vivo. *Am J Physiol Heart Circ Physiol* 279 (2):H459–H465.

172. Burgoyne, J. R., O. Prysyazhna, O. Rudyk, and P. Eaton. 2012. cGMP-dependent activation of protein kinase G precludes disulfide activation: Implications for blood pressure control. *Hypertension* 60 (5):1301–1308.

Index

Milton Keynes UK
Ingram Content Group UK Ltd.
UKHW051013071024
449327UK00012B/228

9 780367 657581